Problems and Solutions in INTRODUCTORY MECHANICS

David Morin

Harvard University

© David Morin 2014
All rights reserved

ISBN-10: 1482086921
ISBN-13: 978-1482086928

Printed by CreateSpace

Cover image: iStock photograph by Vernon Wiley

Additional resources located at:
www.people.fas.harvard.edu/~djmorin/book.html

Contents

Preface		**vii**
1	**Problem-solving strategies**	**1**
1.1	Basic strategies	1
	1.1.1 Solving problems symbolically	1
	1.1.2 Checking units/dimensions	2
	1.1.3 Checking limiting/special cases	3
	1.1.4 Taylor series	4
1.2	List of strategies	6
	1.2.1 Getting started	6
	1.2.2 Solving the problem	8
	1.2.3 Troubleshooting	9
	1.2.4 Finishing up	12
	1.2.5 Looking ahead	12
1.3	How to use this book	13
1.4	Multiple-choice questions	14
1.5	Problems	16
1.6	Multiple-choice answers	17
1.7	Problem solutions	20
2	**Kinematics in 1-D**	**27**
2.1	Introduction	27
2.2	Multiple-choice questions	29
2.3	Problems	31
2.4	Multiple-choice answers	32
2.5	Problem solutions	36
3	**Kinematics in 2-D (and 3-D)**	**42**
3.1	Introduction	42
3.2	Multiple-choice questions	44
3.3	Problems	46
3.4	Multiple-choice answers	50
3.5	Problem solutions	52
4	**F=ma**	**68**
4.1	Introduction	68
4.2	Multiple-choice questions	72
4.3	Problems	77
4.4	Multiple-choice answers	82
4.5	Problem solutions	87

5 Energy — 104
- 5.1 Introduction — 104
- 5.2 Multiple-choice questions — 108
- 5.3 Problems — 110
- 5.4 Multiple-choice answers — 116
- 5.5 Problem solutions — 118

6 Momentum — 142
- 6.1 Introduction — 142
- 6.2 Multiple-choice questions — 145
- 6.3 Problems — 149
- 6.4 Multiple-choice answers — 154
- 6.5 Problem solutions — 158

7 Torque — 177
- 7.1 Introduction — 177
- 7.2 Multiple-choice questions — 180
- 7.3 Problems — 183
- 7.4 Multiple-choice answers — 189
- 7.5 Problem solutions — 193

8 Angular momentum — 220
- 8.1 Introduction — 220
- 8.2 Multiple-choice questions — 222
- 8.3 Problems — 224
- 8.4 Multiple-choice answers — 228
- 8.5 Problem solutions — 231

9 Statics — 251
- 9.1 Introduction — 251
- 9.2 Multiple-choice questions — 252
- 9.3 Problems — 253
- 9.4 Multiple-choice answers — 255
- 9.5 Problem solutions — 256

10 Oscillations — 265
- 10.1 Introduction — 265
- 10.2 Multiple-choice questions — 268
- 10.3 Problems — 268
- 10.4 Multiple-choice answers — 272
- 10.5 Problem solutions — 272

11 Gravity — 287
- 11.1 Introduction — 287
- 11.2 Multiple-choice questions — 289
- 11.3 Problems — 291
- 11.4 Multiple-choice answers — 295
- 11.5 Problem solutions — 297

12 Fictitious forces — 314
- 12.1 Introduction — 314
- 12.2 Multiple-choice questions — 316
- 12.3 Problems — 319
- 12.4 Multiple-choice answers — 320
- 12.5 Problem solutions — 325

13 Appendices   331

- 13.1 Appendix A: Vectors 331
  - 13.1.1 Basics 331
  - 13.1.2 Cartesian coordinates 331
  - 13.1.3 Polar coordinates 333
  - 13.1.4 Adding and subtracting vectors 334
  - 13.1.5 Using components 336
  - 13.1.6 Dot product 337
  - 13.1.7 Cross product 338
- 13.2 Appendix B: Taylor series 339
  - 13.2.1 Basics 339
  - 13.2.2 How many terms to keep? 340
  - 13.2.3 Dimensionless quantities 341
- 13.3 Appendix C: Limiting cases, scientific method 342
- 13.4 Appendix D: Problems requiring calculus 344

Preface

This problem book grew out of a "freshman physics" mechanics course taught at Harvard University during the past decade and a half. Most of the problems are from exams or problem sets, although I have added others to round out the distribution of topics. Some of the problems are standard ones, but many are off the beaten path. In the end, there is a finite number of principles in introductory mechanics, so the problems inevitably start looking familiar after a while. Two topics from the course that aren't included in this book are relativity and damped/driven oscillatory motion. Perhaps these will appear in a future edition, along with other topics such as fluids and precessional angular momentum.

This book will be helpful to both high-school students and college students taking courses in introductory physics (just mechanics, not electricity and magnetism). Calculus is used throughout the book, although it turns out that only a sixth of the problems actually require it. This subset of problems is listed in Appendix D. If you haven't studied calculus yet, just steer clear of those problems, and you can view this book as an algebra-based one. The problems are generally on the level of the one-star or two-star problems in my *Introduction to Classical Mechanics* textbook,[1] which covers a number of more advanced topics such as Lagrangians, normal modes, gyroscopic motion, etc. I will occasionally refer you to that book if you are interested in delving further into various topics.

It is important to note that this book *should not be thought of as a textbook*. Although there is an introduction to each chapter where the basics are presented, this introduction is brief. It is no substitute for the text in a chapter in a standard introductory textbook. This book is therefore designed to be used in tandem with a normal textbook. You can think of this book as supplementing a textbook by providing a stockpile of additional problems. Or you can think of a textbook as supplementing this book by providing additional background.

In most chapters the first few problems are foundational ones. These problems cover basic results and theorems that you can use when solving other problems. When a basic result is stated in the introduction to each chapter, you will generally be referred to a foundational problem for the proof. The book is self contained, in that we derive everything we need. It's just that many of the derivations are shifted to the problems.

A set of multiple-choice questions precedes the problems in each chapter. These questions are usually conceptual ones that you can do in your head. In the rare case where they require a calculation, it is a very minor one. The book contains about 150 multiple-choice questions, in addition to nearly 250 free-response problems.

Depending on how you use this book, it can be an invaluable resource — or a complete waste of time. So here is some critical advice on using the solutions to the problems: If you are having trouble solving a problem, it is imperative that you don't look at the solution too soon. Brood over it for a while. If you do finally look at the solution, *don't* just read it through. Instead, cover it up with a piece of paper and read one line at a time until you get a hint to get started. Then set the book aside and work out the problem for real. Repeat this process as necessary. Actively solving the problem is the only way it will sink in. This piece of advice on how to use this book is so important that I'm going to repeat it and display it prominently in a box:

---

[1] *Introduction to Classical Mechanics, With Problems and Solutions*, David Morin, Cambridge University Press, 2008. This will be referred to as "Morin (2008)."

> If you need to look at the solution to a problem to get a hint (after having thought about it for a while), cover it up with a piece of paper and read one line at a time until you can get started. Then set the book aside and work things out for real. You will learn a great deal this way. If you instead read a solution straight through without having first solved the problem, you will learn very little.

The only scenario in which you should ever read a solution straight through is where you've already solved the problem. However, even in this case you should be careful. If I've given an alternative solution, then you should again just read one line at a time until you can get started and solve it that way too.

To belabor the point, it is quite astonishing how unhelpful it is to simply read a solution instead of solving a problem. You'd *think* it would do some good, but in fact it is completely ineffective in raising your understanding to the next level. Of course, a careful reading of the introductions is necessary to get the basics down. But once that is accomplished, it's time to start solving problems. If Level 1 is understanding the basic concepts, and Level 2 is being able to *apply* those concepts, then you can read and read until the cows come home, and you'll never get past Level 1.

A few informational odds and ends: We'll use the standard mks (meter-kilogram-second) system of units in this book. Concerning notation, a dot above a letter, such as $\dot{x}$, denotes a time derivative. A boldface letter, such as **v**, denotes a vector. Chapter 13 consists of appendices: Appendix A gives a review of vectors, Appendix B covers Taylor series, Appendix C is an aside on the scientific method, and Appendix D lists the problems that require calculus. There are 364 figures in the book, which coincidentally is the total number of gifts given during the 12 days of Christmas, and which ironically is one gift for every day of the year except Christmas!

It was the fall semester of 2000 when I first taught the course on which this book is based, so it would be an understatement to say that I have benefitted over the years from the input of many people, including roughly 1,000 students. I would particularly like to thank Carey Witkov for carefully reading through the entire book and offering many valuable suggestions. Other friends and colleagues whose input I am grateful for are (with my memory being skewed toward more recent years): Jacob Barandes, Allen Crockett, Howard Georgi, Doug Goodale, Theresa Morin Hall, Rob Hart, Paul Horowitz, Randy Kelley, Andrew Milewski, Prahar Mitra, Joon Pahk, Dave Patterson, Joe Peidle, Courtney Peterson, Daniel Rosenberg, Wolfgang Rueckner, Alexia Schulz, Nils Sorensen, Joe Swingle, Corri Taylor, and Rebecca Taylor.

Despite careful editing, there is zero probability that this book is error free. If anything looks amiss, please check the webpage *www.people.fas.harvard.edu/~djmorin/book.html* for a list of typos, updates, additional material, etc. And please let me know if you discover something that isn't already posted. Suggestions are always welcome. Happy problem solving!

David Morin
Cambridge, MA

Chapter 1

Problem-solving strategies

TO THE READER: This book is available as both a paperback and an eBook. I have made a few chapters available on the web, but it is possible (based on past experience) that a pirated version of the complete book will eventually appear on file-sharing sites. In the event that you are reading such a version, I have a request:

If you don't find this book useful (in which case you probably would have returned it, if you had bought it), or if you do find it useful but aren't able to afford it, then no worries; carry on. However, if you do find it useful and are able to afford the Kindle eBook[1] (priced somewhere between $7 and $10), then please consider purchasing it (available on Amazon). I chose to self-publish this book so that I could keep the cost low. The resulting price of around $10, which is very inexpensive for a 350-page physics book, is less than a movie and a bag of popcorn, with the added bonus that the book lasts for more than two hours and has zero calories (if used properly!).

– David Morin

1.1 Basic strategies

In view of the fact that this is a problem book, it makes sense to start off by arming you with some strategies for solving problems. This is the subject of the present chapter. We'll begin with a few strategies that are discussed somewhat in depth, and then we'll provide a long list of 30-ish strategies. You obviously shouldn't try to memorize all of them. Just remember that the list is there, and refer back to it every now and then.

1.1.1 Solving problems symbolically

If you are solving a problem where the given quantities are specified numerically, it is highly advantageous to immediately change the numbers to letters and then solve the problem in terms of the letters. After you obtain a symbolic answer in terms of these letters, you can plug in the actual numerical values to obtain a numerical answer. There are many advantages to using letters:

- IT IS QUICKER. It's much easier to multiply a g by an ℓ by writing them down on a piece of paper next to each other, than it is to multiply their numerical values on a calculator. If solving a problem involves five or ten such operations, the time would add up if you performed all the operations on a calculator.

- YOU ARE LESS LIKELY TO MAKE A MISTAKE. It's very easy to mistype an 8 for a 9 in a calculator, but you're probably not going to miswrite a q for an a on a piece of paper. But even if you

---

[1] If you don't already have the Kindle reading app for your computer, you can download it free from Amazon.

do, you'll quickly realize that it should be an *a*. You certainly won't just give up on the problem and deem it unsolvable because no one gave you the value of *q*!

- YOU CAN DO THE PROBLEM ONCE AND FOR ALL. If someone comes along and says, oops, the value of ℓ is actually 2.4 m instead of 2.3 m, then you won't have to do the whole problem again. You can simply plug the new value of ℓ into your symbolic answer.

- YOU CAN SEE THE GENERAL DEPENDENCE OF YOUR ANSWER ON THE VARIOUS GIVEN QUANTITIES. For example, you can see that it grows with quantities *a* and *b*, decreases with *c*, and doesn't depend on *d*. There is *much* more information contained in a symbolic answer than in a numerical one. And besides, symbolic answers nearly always look nice and pretty.

- YOU CAN CHECK UNITS AND SPECIAL CASES. These checks go hand-in-hand with the previous "general dependence" advantage. We'll discuss these very important checks below.

Two caveats to all this: First, occasionally there are times when things get messy when working with letters. For example, solving a system of three equations in three unknowns might be rather cumbersome unless you plug in the actual numbers. But in the vast majority of problems, it is highly advantageous to work entirely with letters. Second, if you solve a problem that was posed with letters instead of numbers, it's always a good idea to pick some values for the various parameters to see what kinds of numbers pop out, just to get a general sense of the size of things.

1.1.2 Checking units/dimensions

The words *dimensions* and *units* are often used interchangeably, but there is technically a difference: dimensions refer to the general qualities of mass, length, time, etc., whereas units refer to the specific way we quantify these qualities. For example, in the standard meters-kilogram-second (mks) system of units we use in this book, the meter is the unit associated with the dimension of length, the joule is the unit associated with the dimension of energy, and so on. However, we'll often be sloppy and ignore the difference between units and dimensions.

The consideration of units offers two main benefits:

- Considering the units of the relevant quantities before you start solving a problem can tell you roughly what the answer has to look like, up to numerical factors. This practice is called *dimensional analysis*.

- Checking units at the end of a calculation (which is something you should *always* do) can tell you if your answer has a chance at being correct. It won't tell you that your answer is definitely correct, but it might tell you that your answer is definitely *incorrect*. For example, if your goal in a problem is to find a length, and if you end up with a mass, then you know that it's time to look back over your work.

In the mks system of units, the three fundamental mechanical units are the meter (m), kilogram (kg), and second (s). All other units in mechanics, for example the joule (J) or the newton (N), can be built up from these fundamental three. If you want to work with dimensions instead of units, then you can write everything in terms of length (L), mass (M), and time (T). The difference is only cosmetic.

As an example of the above two benefits of considering units, consider a pendulum consisting of a mass m hanging from a massless string with length ℓ; see Fig. 1.1. Assume that the pendulum swings back and forth with an angular amplitude θ_0 that is small; that is, the string doesn't deviate far from vertical. What is the period, call it T_0, of this oscillatory motion? (The period is the time of a full back-and-forth cycle.)

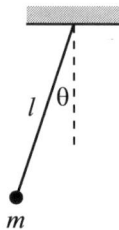

Figure 1.1

With regard to the first of the above benefits, what can we say about the period T_0, by looking only at units and not doing any calculations? Well, we must first make a list of all the quantities the period can possibly depend on. The mass m (with units of kg), the length ℓ (with units of m), and the angular amplitude θ_0 (which is unitless) are given, but additionally there might be dependence on g (the acceleration due to gravity, with units of m/s^2). If you think for a little

1.1. BASIC STRATEGIES

while, you'll come to the conclusion that there really isn't anything else the period can depend on (assuming that we ignore air resistance).

So the question becomes: How does T_0 depend on m, ℓ, θ_0, and g? Or equivalently: How can we produce a quantity with units of seconds from four quantities with units of kg, m, 1, and m/s^2? (The 1 signifies no units.) We quickly see that the answer can't involve the mass m, because there would be no way to get rid of the units of kg. We then see that if we want to end up with units of seconds, the answer must be proportional to $\sqrt{\ell/g}$, because this gets rid of the meters and leaves one power of seconds in the numerator. Therefore, by looking only at the units involved, we have shown that $T_0 \propto \sqrt{\ell/g}$.[2]

This is all we can say by considering units. For all we know, there is a numerical factor out front, and also an arbitrary function of θ_0 (which won't mess up the units, because θ_0 is unitless). The correct answer happens to be $T_0 = 2\pi\sqrt{\ell/g}$, but there is no way to know this without solving the problem for real.[3] However, even though we haven't produced an exact result, there is still a great deal of information contained in our $T_0 \propto \sqrt{\ell/g}$ statement. For example, we see that the period is independent of m; a small mass and a large mass swing back and forth at the same rate. We also see that if we quadruple the length of the string, then the period gets doubled. And if we place the same pendulum on the moon, where the g factor is 1/6 of that on the earth, the period increases by a factor of $\sqrt{6} \approx 2.4$; the pendulum swings back and forth more slowly. Not bad for doing nothing other than considering units!

While this is all quite interesting, the second of the above two benefits (checking the units of an answer) is actually the one you will get the most mileage out of when solving problems, mainly because you should make use of it *every* time you solve a problem. It only takes a second. In the present example with the pendulum, let's say that you solved the problem correctly and ended up with $T_0 = 2\pi\sqrt{\ell/g}$. You should then immediately check the units, which do indeed correctly come out to be seconds. If you had made a mistake in your solution, such as flipping the square root upside down (so that you instead had $\sqrt{g/\ell}$), then your units check would yield the incorrect units of s^{-1}. You would then know to go back and check over your work.

Throughout this book, we often won't bother to explicitly write down the units check if the check is a simple one (as with the above pendulum). But you should of course always do the check in your head. In more complicated cases where it actually takes a little algebra to show that the units work out, we'll write things out explicitly.

1.1.3 Checking limiting/special cases

As with units, the consideration of limiting cases (or perhaps we should more generally say special cases) offers two main benefits. First, it can help you get started on a problem. If you are having trouble figuring out how a given system behaves, then you can imagine making, for example, a certain length become very large or very small, and then you can see what happens to the behavior. Having convinced yourself that the length actually affects the system in extreme cases (or perhaps you will discover that the length doesn't affect things at all), it will then be easier to understand how it affects the system in general. This will then make it easier to write down the relevant quantitative equations (conservation laws, $F = ma$ equations, etc.), which will allow you to fully solve the problem. In short, modifying the various parameters and seeing the effects on the system can lead to an enormous amount of information.

Second, as with checking units, checking limiting cases (or special cases) is something you should *always* do at the end of a calculation. As with units, checking limiting cases won't tell you that your answer is definitely correct, but it might tell you that your answer is definitely

---

[2] In this setup it was easy to determine the correct combination of the given parameters. But in more complicated setups, you might find it simpler to write down a general product of the given dimensionful quantities raised to arbitrary powers, and then solve a system of equations to determine these powers. An example of this method is given in the solution to Problem 1.4.

[3] This $T_0 = 2\pi\sqrt{\ell/g}$ result holds in the approximation where the amplitude θ_0 is small. For a general value of θ_0, the period actually *does* involve a function of θ_0. This function can't be written in closed form, but it starts off as $1 + \theta_0^2/16 + \cdots$. It takes a lot of work to show this, though. See Exercise 4.23 in *Introduction to Classical Mechanics, With Problems and Solutions*, David Morin, Cambridge University Press, 2008; henceforth referred to as "Morin (2008)."

incorrect. Your intuition about limiting cases is generally *much* better than your intuition about generic values of the parameters. You should use this to your advantage.

As an example, consider the trigonometric formula for $\tan(\theta/2)$. The formula can be written in many different ways. Let's say that you're trying to derive it, but you keep making mistakes and getting different answers. However, let's assume that you're pretty sure it takes the form of $\tan(\theta/2) = A(1 \pm \cos\theta)/\sin\theta$, where A is a numerical coefficient. Can you determine the correct form of the answer by checking special cases? Indeed you can, because you know what $\tan(\theta/2)$ equals for a few special values of θ:

- $\theta = 0$: We know that $\tan(0/2) = 0$, so this immediately rules out the $(1 + \cos\theta)/\sin\theta$ form, because this isn't zero when $\theta = 0$; it actually goes to infinity at $\theta = 0$. The answer must therefore take the form of $A(1 - \cos\theta)/\sin\theta$. (This appears to be $0/0$ when $\theta = 0$, but it does indeed go to zero, as you can check by using the Taylor series for $\sin\theta$ and $\cos\theta$; see the subsection on Taylor series below.)

- $\theta = 90°$: We know that $\tan(90°/2) = 1$, which quickly gives $A = 1$. So the correct answer must be $\tan(\theta/2) = (1 - \cos\theta)/\sin\theta$.

- $\theta = 180°$: If you want to feel better about this result, you can note that it gives the correct answer for another special value of θ; it correctly goes to infinity when $\theta = 180°$.

Of course, none of what we've done here demonstrates that $(1 - \cos\theta)/\sin\theta$ is actually the correct answer. But checking the above special cases does two things: it rules out some incorrect answers, and it makes us feel better about the correct answer.

A type of approximation that often comes up involves expressions of the form $ab/(a+b)$, that is, a product over a sum. For example, the equivalent mass in Problem 4.5 turns out to be

$$M = \frac{4m_1 m_2}{m_1 + m_2}. \tag{1.1}$$

What does M look like in the limit where m_1 is much smaller than m_2? In this limit we can ignore the m_1 in the denominator, but we *can't* ignore it in the numerator. So we obtain $M \approx 4m_1 m_2/(0 + m_2) = 4m_1$. Why can we can ignore one of the m_1's but not the other? We can ignore the m_1 in the denominator because it appears there as an *additive* term. If m_1 is small, then erasing it essentially doesn't change the value of the denominator. However, in the numerator m_1 appears as a *multiplicative* term. Even if m_1 is small, its value certainly affects the value of the numerator. Decreasing m_1 by a factor of 10 would decrease the numerator by the same factor of 10. So we certainly can't just erase it. (That would change the units of M anyway.)

Alternatively, you can obtain the $M \approx 4m_1$ result in the limit of small m_1 by applying a Taylor series (discussed below) to M. But this would be overkill. It's much easier to just erase the m_1 in the denominator. In any case, if you're ever unsure about which terms you should keep and which terms you can ignore, just plug some very small numbers (or very large numbers, depending on what limit you're dealing with) into a calculator to see how the expression depends on the various parameters.

It should be noted that there is no need to wait until the end of a solution to check limiting cases (or units, too). Whenever you arrive at an intermediate result that lends itself to checking limiting cases, you should check them. If you find that something is amiss, this will prevent you from wasting time carrying onward with incorrect results.

1.1.4 Taylor series

A tool that often comes up when checking limiting cases is the Taylor series. A Taylor series expresses a function $f(x)$ as a series expansion in x (that is, a sum of terms involving different powers of x). Perhaps the most well-known Taylor series is the one for the function $f(x) = e^x$:

$$e^x = 1 + x + \frac{x^2}{2!} + \frac{x^3}{3!} + \frac{x^4}{4!} + \cdots. \tag{1.2}$$

1.1. BASIC STRATEGIES

A number of other Taylor series are listed near the beginning of Appendix B (Section 13.2). The rest of Appendix B contains a discussion of Taylor series and various issues that arise when using them. If you've never seen Taylor series before, you should take a moment and read the appendix. For the present purposes, we'll just take the above expression for e^x as given and see where it leads us.[4]

As an example of the utility of Taylor series, consider a beach ball that is dropped from rest. It can be shown that if air drag is taken into account, and if the drag force is proportional to the velocity (so that it takes the form $F_d = -bv$, where b is the drag coefficient), then the ball's velocity (with upward taken as positive) as a function of time equals

$$v(t) = -\frac{mg}{b}\left(1 - e^{-bt/m}\right). \tag{1.3}$$

This is a somewhat complicated expression, so you might be a little doubtful of its validity. Let's therefore look at some limiting cases. If these limiting cases yield expected results, then we can feel more confident that the expression is actually correct.

If t is very small (more precisely, if $bt/m \ll 1$; see the discussion in Section 13.2.3), then we can use the Taylor series in Eq. (1.2) to make an approximation to $v(t)$, to leading order in t. (The leading-order term is the smallest power of t with a nonzero coefficient.) To first order in x, Eq. (1.2) gives $e^x \approx 1 + x$. If we let x be $-bt/m$, then we see that Eq. (1.3) can be written as

$$v(t) \approx -\frac{mg}{b}\left(1 - \left(1 - \frac{bt}{m}\right)\right)$$
$$\approx -gt. \tag{1.4}$$

This answer makes sense, because the drag force is negligible at the start (because v, and hence bv, is very small), so we essentially have a freely falling body with acceleration g downward. And $v(t) = -gt$ is the standard expression in that case (see the introduction to Chapter 2). This successful check of a limiting case makes us have a little more faith that Eq. (1.3) is actually correct.

If we mistakenly had, say, $-2mg/b$ as the coefficient in Eq. (1.3), then we would have obtained $v(t) \approx -2gt$ in the small-t limit, which is incorrect. So we would know that we needed to go back and check over our work. Although it isn't obvious that an extra factor of 2 in Eq. (1.3) is incorrect, it *is* obvious that it is incorrect in the limiting $v(t) \approx -2gt$ result. As mentioned above, your intuition about limiting cases is generally much better than your intuition about generic values of the parameters.

We can also consider the limit of large t (or rather, large bt/m). In this limit, $e^{-bt/m}$ is essentially zero, so the $v(t)$ in Eq. (1.3) becomes (there's no need for a Taylor series in this case)

$$v(t) \approx -\frac{mg}{b}. \tag{1.5}$$

This is the "terminal velocity" that the ball approaches as time goes on. Its value makes sense, because it is the velocity for which the total force (gravitational plus air drag), $-mg - bv$, equals zero. And zero force means constant velocity. Mathematically, the velocity never quite reaches the value of $-mg/b$, but it gets extremely close as t becomes large.

Whenever you derive approximate answers as we just did, you gain something and you lose something. You lose some truth, of course, because your new answer is an approximation and therefore technically not correct (although the error becomes arbitrarily small in the appropriate limit). But you gain some aesthetics. Your new answer is invariably much cleaner (often involving only one term), and that makes it a lot easier to see what's going on.

In the above beach-ball example, we checked limiting cases of an answer that was correct. This whole process is more useful (and a bit more fun) when you check limiting cases of an answer that is *incorrect* (as in the case of the erroneous coefficient of $-2mg/b$ we mentioned

---
[4]Calculus is required if you want to *derive* a Taylor series. However, if you just want to *use* a Taylor series (which is what we will do in this book), then algebra is all you need. So although some Taylor-series manipulations might look a bit scary, there's nothing more than algebra involved.

above). When this happens, you gain the unequivocal information that your answer is wrong (assuming that your incorrect answer doesn't just happen to give the correct result in a certain limit, by pure luck). However, rather than leading you into despair, this information is something you should be quite happy about, considering that the alternative is to carry on in a state of blissful ignorance. Once you know that your answer is wrong, you can go back through your work and figure out where the error is (perhaps by checking limiting cases at various intermediate stages to narrow down where the error could be). Personally, if there's any way I'd like to discover that my answer is garbage, this is it. So you shouldn't check limiting cases (and units) because you're being told to, but rather because you *want* to.

1.2 List of strategies

This section contains a list of all the problem-solving strategies I can think of. *The list is long, so there is certainly no need to memorize it.* It would be a step backward if you spent your time worrying about covering all of the strategies, when you should instead be thinking about actually solving a problem. The best way to use this list is to read through it now, and then occasionally refer back to it, especially if you get stuck.

You will inevitably apply many of the strategies without even trying to. But others in the list might seem like meaningless gibberish for now; we're not applying them to any problems here, so there isn't much context. However, if you refer back to the list when solving problems, a given strategy will mean much more if it helps you solve a problem.

Different people think differently, of course. Some strategies might work for you, while others might not. In the end, there's no overall magic bullet for solving all problems. It just comes down to practice and doing lots of problems. But the strategies listed below should help make the practice more efficient. We've divided them into five categories: (1) Getting started, (2) Solving the problem, (3) Troubleshooting, (4) Finishing up, and (5) Looking ahead.

1.2.1 Getting started

The following nine strategies will help you get started on a problem. They don't require too much thinking; they're standard mechanical things that you can do on auto-pilot.

1. Read the problem slowly and carefully

There's no better way to waste time than to read a problem quickly in an effort to save time. If you miss a piece of the given information, you'll end up just spinning your wheels, trying to solve an unsolvable problem.

There is famous statement about the existence of known knowns (things that we know we know), known unknowns (things that we know we don't know), and unknown unknowns (things that we don't know we don't know). Leaving firmly aside who made the statement and why, you might wonder about the fourth permutation: the unknown knowns. What might those be? Well, one thing that certainly falls into this category is the information you miss when you read a problem too quickly. The information is certainly known, but you just don't know that you know it!

2. Identify the things you know, and the things you are trying to find

Identifying the known quantities enables you to see what you have to work with. And identifying the unknown quantities enables you to see what you're aiming for, which gives you some guidance in thinking about what physical principles you should consider (Strategy 10 below). Of course, as mentioned in Strategy 1, identifying the things you know requires reading the problem carefully!

The "knowns and unknowns" reference in Strategy 1 is relevant here too. We mentioned there that you want to avoid unknown knowns. You also want to avoid unknown unknowns.

1.2. LIST OF STRATEGIES

These are things you're going to need to find, but you don't know yet that you need to look for them. It's much easier, of course, to reach a destination if you know what that destination is. So do your best to make sure you know what all of the unknowns are. Basically, try to make sure everything is known – even if it's (an) unknown!

3. Draw a picture

Draw a nice *big* picture, one where you can label everything clearly. Make a note of which quantities you know, and which quantities you're trying to find. Although a small set of mechanics problems involve doing only some math, the vast majority involve a setup that you really need to visualize in order to get anywhere. A picture makes things much more concrete.

4. Draw free-body diagrams

A special kind of picture is a free-body diagram. This is a picture where you draw all of the external forces acting on a given object. As with a general picture, make a note of the known and unknown quantities. Free-body diagrams are absolutely critical when solving problems involving forces. More precisely, they are necessary, and nearly sufficient. That is, many problems are impossible if you don't draw the free-body diagrams, and trivial if you do. Often the only thing that remains to be done after drawing the diagrams is to solve some $F = ma$ equations by doing some math. The physics is all contained in the diagrams. See the introduction to Chapter 4 for further discussion of free-body diagrams.

5. Strip the problem down to its basics

Some problems are posed as idealized "toy model" problems, for example a point mass colliding with a uniform stick with negligible thickness. Other problems deal with more realistic setups that you might encounter in the real world, for example two skaters colliding and grabbing on to each other. When dealing with the latter type, the first thing you should do is strip the problem down to its basics. If possible, reduce the problem to point masses, sticks, massless strings, etc. Many real-life problems that look different at first glance end up being the same when reduced to the underlying toy model. So when you solve the toy-model version, you're actually solving a more general problem, which is a good thing.

Of course, you need to be careful that your toy model mimics the original setup correctly. For example, simplifying an object to a point mass works fine if you're using forces, but not necessarily if you're using torques. Your goal is to simplify the setup as much as possible without changing the physics. It takes some thought not to go too far, but this thought process is helpful in solving the problem. It helps you decide which aspects of the problem are important, and which aspects are irrelevant. This in turn helps you decide which physical principles you need to use (Strategy 10 below). Along these lines, if the original real-life setup is one for which you have some physical intuition, remember to use it when you start dealing with the toy model!

6. Choose wisely your coordinate system or reference frame

There are always only a couple of reasonable coordinate systems and reference frames to choose from, but a particular choice may greatly simplify things. For example, when dealing with an inclined plane, choosing tilted axes (parallel and perpendicular to the plane) is often helpful. And when dealing with circular motion, it is of course usually best to work with polar coordinates. And when two (or more) objects are moving with respect to each other, it is often helpful to analyze the setup in a new reference frame (the CM frame, or perhaps a frame moving along with one of the objects).

Part of choosing a coordinate system involves choosing the positive directions for the coordinate axes. This is completely your choice, but you must remember that once you pick a convention, you must stick with it. It's fine to let downward be positive for a falling object; just don't forget later on in your solution that you've made that choice.

7. Identify the initial and final states of the system

This is especially important in problems involving conservation principles (conservation of energy, momentum, angular momentum). In many cases, you can ignore the specifics of what happens during a process and simply equate the initial and final values of a particular quantity.

Another class of problems is "initial condition" problems. If you've calculated a general expression for, say, an object's position involving some unknown parameters, you can determine the values of these parameters by invoking the initial conditions (usually the initial position and velocity).

8. Identify the constraints

Is an object constrained to lie on a plane? Or travel in a circle? Or move with constant velocity? Is the system static? In the end, a constraint means that you have one fewer unknown than you otherwise might have thought. For example, if an object lies on a plane inclined at angle θ, then its coordinates are related by $y = x\tan\theta$. So if you choose x as your unknown, then y is determined.

9. Convert numbers to letters, so that you can solve things symbolically

This strategy is extremely helpful and very simple to apply. It is discussed in depth in Section 1.1.1 above, where its many benefits are noted.

1.2.2 Solving the problem

Having taken the above mechanical steps, it's now time to start thinking. There's no sure-fire way to guarantee that you'll solve every problem you encounter, but the following five strategies will certainly help.

10. Identify the physical principles involved

Think about what physical principle(s) will allow you to solve the problem. The most fundamental principles in mechanics are $F = ma$ and $\tau = I\alpha$ (or more accurately $F = dp/dt$ and $\tau = dL/dt$), and conservation of E, p, and L. A given problem can invariably be solved in multiple ways. For example, since conservation of energy can be derived from $F = ma$, any problem that can be solved with conservation of energy can also be solved with $F = ma$, although the latter solution may be more cumbersome.

In addition to the overarching fundamental principles listed above, there are many other physical principles/facts that you may need to use. For example, the radial acceleration is v^2/r (Eq. (3.7)); $v_y = 0$ at the highest point in projectile motion; the energy of an object that is both translating and rotating consists of two terms (Eq. (7.8)); Hooke's law for a spring is $F = -kx$ (Eq. (10.1)); Newton's law of gravitation is an inverse-square law (Eq. (11.1)); and so on.

11. Convert physical statements into mathematical equations

Having identified the relevant physical principles, you must now convert them into mathematical equations. For example, having noted that the horizontal speed in projectile motion is constant, you need to write down $x = (v_0\cos\theta)t$, or something equivalent. Or having decided that you will use $F = ma$ to solve a problem, you need to explicitly write down the F_x (and maybe F_y and F_z) equations, which may involve breaking vectors into their components. Or having decided to use conservation of energy, you need to determine what kinds of energy are involved, and then equate the initial total energy with the final total energy.

1.2. LIST OF STRATEGIES

12. Think initially in terms of physical statements, rather than equations

It is important to *first* think about the physical principles, and *then* think about how you can express them with equations. Don't just write down a bunch of equations and look for ways to plug things into them. *The initial goal when attacking a problem isn't to write down the correct equation; rather, it's to say the correct thing in words.* If you proceed by blindly writing down all the equations you can think of that seem somewhat relevant, you might end up just going around in circles. You wouldn't try to get to a certain destination by randomly walking around with the hope that you'll eventually stumble upon it. And that strategy doesn't work any better in problem solving!

13. Make sure you have as many facts/equations as unknowns

If you are trying to solve for three unknowns and you have only two equations/facts, then there's no way you're going to be successful. Along the same lines, if you've identified an unknown but haven't incorporated it into any of your equations/facts, then there's no way you're going to be able to solve for it. If you can't think of which additional physical principle to apply to generate the necessary equation, it's helpful to run through all of the given information and think about the implications of each bit.

14. Be organized

Sometimes you can see right away exactly how to solve a problem, in which case you can fly right through it, without much need for organization. But unless you're positive that the solution will be quick, it is critical to be organized about the other strategies in this list, by explicitly writing things out. For example, you should write out the knowns and unknowns, as opposed to just thinking them. And likewise for the physical principles involved, etc. There's no need to write a book, but some brief notes will do wonders in organizing your thoughts.

1.2.3 Troubleshooting

In many cases the preceding strategies are sufficient for solving a problem. But if you get stuck, the following thirteen strategies should be helpful.

The following three strategies are bread-and-butter ones.

15. Reduce the problem to an intermediate one

Equivalently, work backwards. Say to yourself, "I'd be able to get the answer to the problem if I somehow knew the quantity A. And I'd be able to get A if I somehow knew B." And so on. Eventually you'll hit a quantity that you can figure out from the given information. For example, you can find the distance x traveled by a projectile if you somehow know the time t (because $x = (v_0 \cos\theta)t$). So the problem reduces to finding t. And you can find t by (among other ways) noting that at the top of the projectile motion (after time $t/2$), the y component of the velocity is zero, so $v_0 \sin\theta - g(t/2) = 0$.

As an analogy, if you can't remember how to get to a certain destination, you're still in luck if you remember that it's just north of a park, which you remember is a few blocks down a certain street from a statue, which you remember is around the corner from a school, which you remember how to get to.

16. Exaggerate/change the parameters to understand their influence

This is basically the same as checking limiting cases of your final answer (Strategy 28 below, discussed in detail in Section 1.1.3). However, there is no need to wait until you obtain your final answer (or even an intermediate result) to take advantage of this extremely useful strategy. Your intuition about extreme cases is much better than your intuition about normal scenarios, so you should use it. Once you see that a certain parameter influences the result, you can hone in on how exactly this influence comes about. This can then lead you to the relevant physical principle (Strategy 10).

17. Think about how the various quantities (known or unknown) are related

The task of Strategy 10 is to identify the relevant physical principles. This will yield relations among the various quantities. If you've missed some of the principles, it might be possible to figure out what they are by thinking about how the various quantities relate. For example, consider a mass on the end of a spring, and let's say you pull the mass a distance d away from its equilibrium position and then let go. It is intuitively clear that the larger d is, the larger the mass's speed v will be when it passes through the equilibrium position during the resulting oscillatory motion. If your goal is to find v, the preceding qualitative statement might help lead you to the useful physical principle of energy conservation, which will then allow you to write down a quantitative mathematical equation.

The following three strategies are quick checks.

18. Check that you have incorporated all of the given information

Part of the task of Strategy 2 is to identify everything that you know. When immersed in a problem, it's easy to forget some of this information, and this will likely make the problem unsolvable. So double check that for every given piece of information, you've either incorporated it or declared it to be irrelevant.

19. Check your math

Check over your algebra, of course. It's good to do at least a cursory check after each step. If you eventually hit a roadblock, go back and do a more careful check through all the steps.

20. Check the signs in all equations

In some sense this is just a subcase of the preceding strategy of checking your math. But often when people check through algebra, they fixate on the numerical values of the various terms and neglect the signs. So if you're stuck, just do a quick check where you ignore the numerical values and look only at the signs, just to make sure that at least those are correct. This check should be very quick. Pay special attention to the initial equation that you wrote down. A common mistake is to have an incorrect sign right from the start (for example, having the wrong sign in a vector component), which won't show up as an algebra mistake.

The following three strategies involve building on other knowledge.

21. Think of similar problems you know how to do

Try to reduce the problem (all, or part of it) to a previously solved problem. There are only so many types of problems in introductory mechanics, so odds are that if you've done a good number of problems, they should start looking familiar. How is the present problem similar to an old one, and how is it different?

You might wonder whether someone becomes an expert problem solver by being brilliant, or by solving a zillion problems, which has the effect of making any new problem look vaguely familiar. Elite athletes, chess players, debaters, comedians, etc., rely on recognizing familiar situations that they know how to react to. You can argue about what percentage of their strategy/action is based on this reaction. But you can't argue with the fact that a huge arsenal of familiar situations, built up from endless hours of practice, is a necessary condition for elite status in pretty much anything.

1.2. LIST OF STRATEGIES

22. Think of a real-life setup that behaves the same way

This strategy in is the same spirit as the preceding one. We continually gather physical intuition from everyday life, so in a sense we're always practicing physics without even knowing it. For example, if you need to push a large object, say, a car, then you know that you should lean forward with your feet behind you, as opposed to pushing with your body upright. We have much more physical intuition about Newtonian mechanics than we do about other subfields of physics (electricity and magnetism, relativity, quantum mechanics), so you should use it to your advantage when solving mechanics problems!

Of course, there are times when your intuition might lead you astray, sometimes due to the fact that certain observations dominate others. For example, based on observations of skidding and general braking in a car or on a bike, you might think that friction always slows things down. It's easy to forget that when you accelerate from rest (often a more gentle acceleration than braking), friction is what speeds you up. You won't go anywhere if you're on ice!

23. Solve a simpler problem

If you can't get anywhere, if never hurts to solve a simpler version of the problem, to get a feel for what's going on. A particular example of this (which appears more often in math than in physics) is a problem that involves a large number or a general number N. In such problems, you definitely want to try solving things for the case of $N = 1, 2, 3$. Once you see what's going on for small numbers, it's much easier to generalize to an arbitrary number N.

The following four strategies involve helping your brain get going.

24. Explain (or imagine explaining) the solution to someone else

This strategy might seem a little silly, but it's really just a way of forcing yourself to organize your thoughts and proceed slowly. I assume I'm not the only person who has checked over an incorrect solution multiple times in my head, only to make the same mistake each time. It's very easy to repeatedly slide over a mistake or faulty assumption in the confines of one's own head. This can often be remedied by explaining your solution to someone else. And in most cases, you will see the mistake even before the other person says anything. But in the event that there's no one else around to lend an ear, a little talking to yourself never hurt.

25. Imagine your teacher explaining the solution

This strategy might also seem a little silly, but it does help sometimes, due to the fact that you're a human and not a machine. When floundering with a problem, it's easy to lose confidence and give up, even if you don't consciously know that you're giving up. If you imagine your teacher (or another student you look up to) explaining the solution, then since you expect them to be able to solve the problem with confidence, you just might end up solving it yourself. Occasionally it's better to ditch the "I think I can" mantra for "I *know* someone *else* can." Hey, if it gets the job done...

26. When you can't think of anything to do, just do something!

If all else fails, just start trying some random things. It can't hurt. You might hit something that gets you on track. A reasonable analogy is the unrealistic scenario in which you're lost in the woods with an infinite supply of food, and with zero chance that anyone else is going to help you. It doesn't do any good to just sit there. If you can't think of a reason to head in any particular direction, you should just head in *some* direction. Maybe you'll hit a stream that heads somewhere. Of course, after the fact, you'll probably see why you should have known the stream was there in the first place (a gulley between two hills, the dripping wet moose that walked past you, etc.). But that's knowledge you can use the next time you get stuck.

27. Set the problem aside and go for a walk

If nothing is working and you're truly at an impasse, take a break from the problem. It's easy to get stuck in a rut in your thinking, where you keep making the same mistakes and your mind keeps heading from a correct thought to an incorrect or unhelpful one. If you take a break and let your mind wander a little, it might semi-randomly jump to a helpful thought.

1.2.4 Finishing up

Assuming you've produced an answer to a problem, the following three strategies are ones you should always apply, as double checks on your answer.

28. Check limits and special cases

This strategy is immensely helpful. It is discussed in depth in Section 1.1.3, where its many benefits are explained.

29. Check units

This strategy is also very helpful and is discussed in depth in Section 1.1.2.

30. Check the rough size of numerical values

If your final answer involves actual numbers, be sure to check that they make sense. You can do this by checking that the rough "order of magnitude" (the nearest power of 10) is plausible. It's quite possible that you dropped a factor of 10 somewhere in the calculation. Or maybe you dropped one of the parameters when going from one line of the math to the next (although checking units is often a safeguard against this, too).

1.2.5 Looking ahead

After solving a problem, you can use the following four strategies to build upon what you've done and better prepare yourself for future problems.

31. Review the solution

After you're finished with a problem, go back and carefully review the entire solution. First think about the big-picture idea(s). Then think not only about what each step was, but also *why* you performed it. That is, how did each step logically follow from the previous one, as opposed to just being a random step? This review is extremely helpful in making the solution sink in, and it usually takes only a minute or two. You get a lot of bang for your buck, timewise.

Even if the solution *did* sink in first time around, there is another benefit to a careful review. It's often not enough just to know something; you also need to *know that you know it*. That way, you can confidently apply it to a future problem. If a certain tool is in your arsenal but you don't know that it's there, it doesn't help you much. (Think of all the memories that reside in your brain but that you'll never ever think of again, because you don't know that you have them.) A careful review of something will let you know that you know it.

32. Analyze where/why you went wrong

This strategy is actually relevant *while* you are solving the problem. As mentioned in the preface, it is critical that you never just read a solution straight through, unless you've already solved the problem. Just read enough to get a hint to get started. If you *do* need to read a little bit to get a hint, then before you use the hint to move on with the problem, think about your previous train of thought. Where exactly did you get stuck? What would you have needed to realize to get unstuck? How will you modify your thinking in the future so that you don't hit the same roadblock?

In short, it doesn't hurt to obsess a little about where you went wrong and what you can do better. If you get schooled on the basketball court with a crossover dribble that leads to the game-winning layup, you're probably going to obsess for a while about what to do differently next time. That's a good thing. And it works the same way with problem solving. And for that matter, anything else you're trying to get good at.

33. Think of another approach

Just because you solved a problem once, that doesn't mean you can't solve it again. There are often many ways to solve a given problem, and you can learn a lot by working through another solution. Furthermore, most of what you learn actually comes from the things you do wrong, so if you happened to have breezed through your first solution, you might not have learned much. But if you work through a second solution and have to struggle here and there, you'll significantly increase what you take away.

34. Think of a variation or extension

Try to make up a similar problem. You could vary the parameters, of course. But if you've solved the problem symbolically instead of with numbers (as you should always do!), then you've actually already solved the problem for any values of the parameters. So there's nothing new there. So what can you change? Perhaps add new forces and/or objects, or change the shape of an object, or allow something that was fixed to be moveable, or change the direction of the motion, or relax some of the given information and see what the most general motion is, or generalize from 1-D to 2-D, etc. Thinking about variations will not only solidify the problem you just solved, but also make you much more prepared for new problems that come your way.

As mentioned at the start of this section, the above list of strategies is long, so you certainly shouldn't try to memorize it. Just refer back to it now and then.

1.3 How to use this book

The preface contained some advice on the proper use of the solutions in this book. We'll repeat some of that advice here and also give a few more pointers for using this book.

- Read the introduction to each chapter, to become familiar with the material.

- Solve the foundational problems in a given chapter first, to make sure that you have all the tools you will need. These problems are the first few in each chapter.

- Solve lots of problems.

- Don't look at the solution to a problem too soon. If you do need to look at it, don't just read it straight through. Read it line by line until you get a hint to get going again. (Have a piece of paper handy, to cover up the rest of the solution.) Then set it aside and solve the problem on your own. Repeat as necessary.

- When solving the multiple-choice questions, be sure to *fully commit* to an answer before checking to see if it's correct. Don't just make a reasonable guess and then cross your fingers. Think hard until you're sure of your answer. If it turns out to be wrong, then solve the question again, without looking at the explanation in the solution.

The most important piece of advice is the fourth bullet point above, which also appeared in the boxed paragraph in the preface. But it's so important that one more appearance, now in bold, can't hurt:

Unless you have already completely solved a given problem,

> **Don't just read through the solution!**

If you read through a solution without first solving the problem, you will gain essentially nothing from it.

1.4 Multiple-choice questions

As mentioned above: In the multiple-choice questions, be sure to fully commit to an answer before checking to see if it is correct.

1.1. If the task of a given problem is to find a certain length, which of the following quantities could be the answer? (The ℓ, v, a, t, and m in this and the following two questions are given quantities with the dimensions of length, velocity, acceleration, time, and mass.)

(a) at (b) mvt (c) $\sqrt{a\ell}$ (d) v/t (e) v^2/a

1.2. If the task of a given problem is to find a certain time, which of the following quantities could be the answer?

(a) a/t (b) mv/ℓ (c) v^2/a (d) $\sqrt{\ell/a}$ (e) $\sqrt{v/a}$

1.3. If the task of a given problem is to find a certain force (with units kg m/s^2), which of the following quantities could be the answer?

(a) mv^2 (b) mat (c) mv/t (d) mv/ℓ (e) v^2/ℓ

1.4. One mile per hour equals how many meters per second? (There are 1609 meters in a mile.)

(a) 0.04 (b) 0.45 (c) 1 (d) 2.2 (e) 27

In the following seven questions, don't solve things from scratch. Just use dimensional analysis.

1.5. A block rests on an inclined plane with coefficient of friction μ (which is dimensionless). Let $\theta_{\max}$ be the largest angle of inclination for which the block doesn't slide down. Which of the following is true?

(a) $\theta_{\max}$ is larger on the moon than on the earth.

(b) $\theta_{\max}$ is larger on the earth than on the moon.

(c) $\theta_{\max}$ is the same on the earth and the moon.

1.6. A mass m oscillates back and forth on a spring with spring constant k (with units kg/s^2). If the amplitude (the maximum displacement) is A, which of the following quantities is the maximum speed the mass achieves as it passes through the equilibrium point?

(a) $\dfrac{kA}{m}$ (b) $\dfrac{kA^2}{m}$ (c) $\sqrt{\dfrac{kA}{m}}$ (d) $\sqrt{\dfrac{kA^2}{m}}$ (e) $\sqrt{mkA^2}$

1.7. A bucket of water with mass density ρ (with units kg/m^3) has a small hole in it, at a depth h below the surface. Assuming that the viscosity of the water is negligible, which of the following quantities is the speed of the water as it exits the hole?

(a) $\sqrt{2gh}$ (b) $\sqrt{2\rho g h}$ (c) $\sqrt{2g/h}$ (d) $\sqrt{2h/g}$ (e) $\sqrt{2gh/\rho}$

1.8. The increase in pressure ΔP (force per area) as you descend in a lake depends on your depth h, the density of water ρ, and g. Which of the following quantities is ΔP?

(a) $\rho g/h$ (b) $\rho g h$ (c) $\rho^2 g h$ (d) $\rho g^2 h$ (e) $\rho g h^3$

1.4. MULTIPLE-CHOICE QUESTIONS

1.9. The drag force F_d on a sphere moving slowly through a viscous fluid depends on the viscosity of the fluid η (with units kg/(m s)), the radius R, and the speed v. Which of the following quantities is F_d?

(a) $6\pi\eta R/v$ (b) $6\pi\eta/Rv$ (c) $6\pi\eta Rv$ (d) $6\pi\eta R^2 v$ (e) $6\pi\eta R^2 v^2$

1.10. The drag force F_d on a sphere moving quickly through a nonviscous fluid depends on the density of the fluid ρ, the radius R, and the speed v. Which of the following quantities is F_d proportional to?

(a) ρv (b) $\rho R v$ (c) $\rho R v^2$ (d) $\rho R^2 v$ (e) $\rho R^2 v^2$

1.11. The Schwarzschild radius R_S of a black hole depends on its mass m, the speed of light c, and the gravitation constant G (with units $m^3/(kg\,s^2)$). Which of the following quantities is R_S?

(a) $\dfrac{2G}{mc^2}$ (b) $\dfrac{2Gm}{c^2}$ (c) $\dfrac{2Gm}{c^3}$ (d) $\dfrac{2c^2}{Gm}$ (e) $\dfrac{2c^3}{Gm}$

In the remaining questions, don't solve things from scratch. Just check special cases.

1.12. The plane in Fig. 1.2 is inclined at an angle θ, and two vectors are drawn. One vector is perpendicular to the plane, and its horizontal and vertical components are shown. The other vector is horizontal, and its components parallel and perpendicular to the plane are shown. Which of the following angles equal(s) θ? (Circle all that apply.)

(a) A (b) B (c) C (d) D

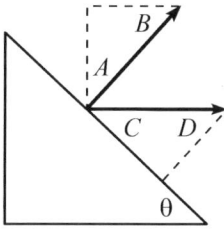

Figure 1.2

1.13. Two massless strings support a mass m as shown in Fig. 1.3. Which of the following quantities is the tension (that is, force) T in each string?

(a) $\dfrac{mg}{2}$ (b) $\dfrac{mg\sin\theta}{2}$ (c) $\dfrac{mg\cos\theta}{2}$ (d) $\dfrac{mg}{2\sin\theta}$ (e) $\dfrac{mg}{2\cos\theta}$

1.14. A block slides down a plane inclined at angle θ. What should the coefficient of kinetic friction μ be so that the block slides with constant velocity?

(a) 1 (b) $\sin\theta$ (c) $\cos\theta$ (d) $\tan\theta$ (e) $\cot\theta$

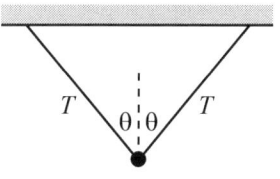

Figure 1.3

1.15. Consider the "endcap" of the sphere shown in Fig. 1.4, obtained by slicing the sphere with a vertical plane perpendicular to the plane of the paper. Which of the following expressions is the volume of the cap?

(a) $\pi R^3 \big(4/3 - (2/3)\sin\theta\big)$

(b) $\pi R^3 \big((2/3)\sin\theta\big)$

(c) $\pi R^3 \big(2/3 - (2/3)\cos\theta + \sin\theta\big)$

(d) $\pi R^3 \big(2/3 + (1/3)\cos^3\theta - \cos\theta\big)$

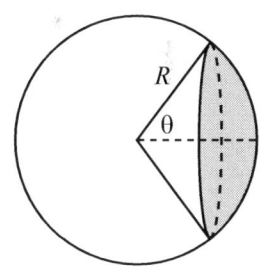

Figure 1.4

1.16. Consider the line described by $ax + by + c = 0$. Which of the following expressions is the distance from this line to the point (x_0, y_0)?

(a) $\dfrac{bx_0 + ay_0 + c}{\sqrt{a^2 + b^2}}$ (b) $\dfrac{ax_0 + by_0 + c}{\sqrt{a^2 + b^2}}$ (c) $\dfrac{ax_0 + by_0}{\sqrt{a^2 + b^2}}$ (d) $\dfrac{ax_0 + by_0 + c}{\sqrt{a^2 + b^2 + c^2}}$

1.17. A person throws a ball with a given speed v (at the optimal angle for the following task) toward a wall of height h. Which of the following quantities is the maximum distance the person can stand from the wall and still be able to throw the ball over the wall?

(a) $\dfrac{gh^2}{v^2}$ (b) $\dfrac{v^2}{g}$ (c) $\dfrac{v^4}{g^2 h}$ (d) $\sqrt{\dfrac{v^2 h}{g}}$ (e) $\dfrac{v^2}{g}\sqrt{1 - \dfrac{2gh}{v^2}}$ (f) $\dfrac{v^2/g}{1 + 2gh/v^2}$

1.5 Problems

1.1. Furlongs per fortnight squared

Convert $g = 9.8 \text{ m/s}^2$ into furlongs/fortnight^2. A fortnight is two weeks, a furlong is 220 yards, and there are 1.09 yards in a meter.

1.2. Miles per gallon

The efficiency of a car is commonly rated in miles per gallon, the dimensions of which are length per volume, or equivalently inverse area. What is the value of this area for a car that gets 30 miles per gallon? What exactly is the physical interpretation of this area? *Note*: There are 3785 milliliters (cubic centimeters) in a gallon, and 1609 meters in a mile.

1.3. Painting a funnel

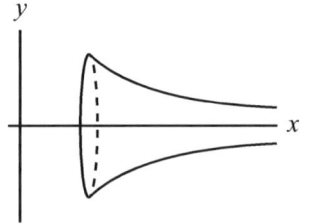

Figure 1.5

Consider the curve $y = 1/x$, from $x = 1$ to $x = \infty$. Rotate this curve around the x axis to create a funnel-like surface of revolution, as shown in Fig. 1.5. By slicing up the funnel into disks with radii $r = 1/x$ and thickness dx (and hence volume $(\pi r^2)\, dx$) stacked side by side, we see that the volume of the funnel is

$$V = \int_1^\infty \frac{\pi}{x^2}\, dx = -\frac{\pi}{x}\bigg|_1^\infty = \pi, \tag{1.6}$$

which is finite. The surface area, however, involves the circumferential area of the disks, which is $(2\pi r)\, dx$ multiplied by a $\sqrt{1 + y'^2}$ factor accounting for the tilt of the area. The surface area of the funnel is therefore

$$A = \int_1^\infty \frac{2\pi \sqrt{1 + y'^2}}{x}\, dx \;>\; \int_1^\infty \frac{2\pi}{x}\, dx, \tag{1.7}$$

which is infinite. (The square root factor is irrelevant for the present purposes.) Since the volume is finite but the area is infinite, it appears that you can fill up the funnel with paint but you can't paint it. However, we then have a problem, because filling up the funnel with paint implies that you can certainly paint the *inside* surface. But the inside surface is the same as the outside surface, because the funnel has no thickness. So we should be able to paint the outside surface too. What's going on here? Can you paint the funnel or not?

1.4. Planck scales

Three fundamental physical constants are Planck's constant, $\hbar = 1.05 \cdot 10^{-34}$ kg m^2/s; the gravitational constant, $G = 6.67 \cdot 10^{-11}$ m^3/(kg s^2); and the speed of light, $c = 3.0 \cdot 10^8$ m/s. These constants can be combined to yield quantities with dimensions of length, time, and mass (known as the *Planck length*, etc.). Find these three combinations and the associated numerical values.

1.5. Capillary rise

If the bottom end of a narrow tube is placed in a cup of water, the surface tension of the water causes the water to rise up in the tube. The height h of the column of water depends on the surface tension γ (with dimensions of force per length), the radius of the tube r, the mass density of the water ρ, and g. Is it possible to determine from dimensional analysis alone how h depends on these four quantities? Is it possible if we invoke the fact that h is proportional to γ? (This is believable; doubling the surface tension γ should double the height, because the surface tension is what is holding up the column of water.)

1.6. Fluid flow

Poiseuille's equation gives the flow rate Q (volume per time) of a fluid in a pipe, in the case where viscous drag is important. How does Q depend on the following four quantities: the pressure difference ΔP (force per area) between the ends of the pipe, the radius R and length L of the pipe, and the viscosity η (with units of kg/(m s)) of the fluid? To

1.6. MULTIPLE-CHOICE ANSWERS

answer this, you will need to invoke the fact that Q is inversely proportional to L. (This is believable; doubling the length of the pipe will double the effect of friction between the fluid and the walls, and thereby halve the flow rate.)

1.7. 1-D collision

If a mass M moving with velocity V collides head-on elastically with a mass m that is initially at rest, it can be shown (see Problem 6.3) that the final velocities are given by

$$V_M = \frac{(M-m)V}{M+m} \quad \text{and} \quad v_m = \frac{2MV}{M+m}. \tag{1.8}$$

Check the $M = m$, $M \ll m$, and $M \gg m$ limits of these expressions.

1.8. Atwood's machine

Consider the Atwood's machine in Fig. 1.6, consisting of three masses and two frictionless pulleys. It can be shown that the acceleration of m_2, with upward taken to be positive, is given by (just accept this)

$$a_2 = -g \frac{4m_2 m_3 + m_1(m_2 - 3m_3)}{4m_2 m_3 + m_1(m_2 + m_3)}. \tag{1.9}$$

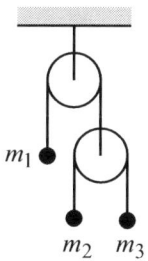

Figure 1.6

Find a_2 for the following special cases:

(a) $m_1 = 2m_2 = 2m_3$

(b) m_2 much larger than both m_1 and m_3

(c) m_2 much smaller than both m_1 and m_3

(d) $m_1 \gg m_2 = m_3$

(e) $m_1 = m_2 = m_3$

1.9. Dropped ball

In Section 1.1.4, we looked at limiting cases of the velocity, given in Eq. (1.3), of a beach ball dropped from rest. Let's now look at the height of the ball. If the ball is dropped from rest at height h, and if the drag force from the air takes the form $F_d = -bv$, then it can be shown that the ball's height as a function of time equals

$$y(t) = h - \frac{mg}{b}\left(t - \frac{m}{b}\left(1 - e^{-bt/m}\right)\right). \tag{1.10}$$

Find an approximate expression for $y(t)$ in the limit where t is very small (or more precisely, in the limit where $bt/m \ll 1$).

1.6 Multiple-choice answers

1.1. [e] This is the only choice with dimensions of length.

1.2. [d] This is the only choice with dimensions of time.

1.3. [c] This is the only choice with units of kg m/s².

1.4. [b] There are $60 \cdot 60 = 3600$ seconds in an hour, so one mile per hour equals

$$1 \frac{\text{mile}}{\text{hour}} = \frac{1609 \text{ meters}}{3600 \text{ seconds}} = 0.447 \text{ m/s}. \tag{1.11}$$

A common automobile speed of, say, 60 mph is therefore about 27 m/s. The inverse relation, going from m/s to mph, is 1 m/s = 2.24 mph.

REMARK: We see that a meter per second is larger than a mile per hour, by a factor of slightly more than 2. This factor of about 2.2 is easy to remember, because it happens to be essentially the same as the conversion factor between kilograms and pounds: 1 kg weighs 2.2 pounds. (Note that since a kilogram is a unit of mass, and a pound is a unit of weight, we used the word "weighs" here, instead of "equals.")

1.5. $\boxed{c}$ The only things that $\theta_{\max}$ can possibly depend on are the mass m of the block, the relevant acceleration due to gravity g, and the coefficient of friction μ. But $\theta_{\max}$ is dimensionless, and there is no way to form a dimensionless quantity involving g or m. So $\theta_{\max}$ can depend only on μ (which is dimensionless). Therefore, since $\theta_{\max}$ can't depend on g, it is the same on the earth and the moon.

1.6. $\boxed{d}$ This is the only choice with units of m/s. The answer must involve the ratio k/m, to get rid of the kg units. And it must involve $\sqrt{k}$ in the numerator to produce the desired single power of seconds in the denominator. Choice (d) additionally produces the desired single power of meters in the numerator. If you want to solve the problem for real, you can quickly use conservation of energy (discussed in Chapter 5) to say that $mv^2/2 = kA^2/2$.

1.7. $\boxed{a}$ This is the only choice with units of m/s. Note that the answer can't involve ρ, because there would then be no way to get rid of the units of kg.

REMARKS: This speed of $\sqrt{2gh}$ is the same as the speed of a dropped ball after it has fallen a height h, assuming that air drag can be neglected. (This speed can't depend on the mass m of the ball for the same dimensional reason.) If air drag *is* included, then this introduces a new parameter which involves units of kg, so now the speed of the ball can (and does) depend on m. A metal ball falls faster than a styrofoam ball.

There is one issue we've glossed over. The size of the hole introduces another length scale ℓ (which you can take to be the radius or diameter or whatever). This means that there are now an infinite number of possible answers. Any expression of the form $\sqrt{2gh^{n+1}/\ell^n}$ has the correct units. However, assuming that the viscosity is negligible, it can be shown that the size of the hole doesn't matter. That is, $n = 0$. At any rate, choice (a) is certainly the only correct choice among the given options.

1.8. $\boxed{b}$ Let's be a little more systematic here than in the previous few questions. Since the units of force are kg m/s², the units of pressure (force per area) are kg/(m s²). So the units of the various quantities are:

$$\Delta P : \frac{\text{kg}}{\text{m s}^2}, \qquad h : \text{m}, \qquad \rho : \frac{\text{kg}}{\text{m}^3}, \qquad g : \frac{\text{m}}{\text{s}^2}. \qquad (1.12)$$

Our goal is to create ΔP from the other three quantities. By looking at the powers of kg and s in ΔP, we quickly see that the answer must be proportional to ρg. This then means that we need one power of h, to produce the correct power of m.

1.9. $\boxed{c}$ The units of the various quantities are:

$$F_d : \frac{\text{kg m}}{\text{s}^2}, \qquad \eta : \frac{\text{kg}}{\text{m s}}, \qquad R : \text{m}, \qquad v : \frac{\text{m}}{\text{s}}. \qquad (1.13)$$

Our goal is to create F_d from the other three quantities, and we quickly see that the simple product $\eta R v$ gets the job done. A detailed calculation is required to generate the numerical coefficient of 6π.

1.10. $\boxed{e}$ The units of the various quantities are:

$$F_d : \frac{\text{kg m}}{\text{s}^2}, \qquad \rho : \frac{\text{kg}}{\text{m}^3}, \qquad R : \text{m}, \qquad v : \frac{\text{m}}{\text{s}}. \qquad (1.14)$$

Our goal is to create F_d from the other three quantities. By looking at the powers of kg and s in F_d, we see that we need one power of ρ and two powers of v. This then implies that we need two powers of R, to produce the correct power of m. The actual numerical coefficient in F_d depends on the specifics of the surface of the sphere (how rough it is).

1.6. MULTIPLE-CHOICE ANSWERS

1.11. [b] The units of the various quantities are:

$$R_S : \text{m}, \quad m : \text{kg}, \quad c : \frac{\text{m}}{\text{s}}, \quad G : \frac{\text{m}^3}{\text{kg s}^2}. \tag{1.15}$$

Our goal is to create the Schwarzschild radius R_S from the other three quantities. Since R_S doesn't involve units of kg, we see that the answer must involve the product Gm (raised to some power). And since R_S also doesn't involve units of seconds, the answer must involve the quotient G/c^2 (raised to some power). We quickly see that Gm/c^2 makes the units of meters work out correctly. The factor of 2 requires a detailed calculation.

REMARK: The Schwarzschild radius of an object is the radius with the property that if you shrink the object down to that radius (keeping the mass the same), it will be a black hole. That is, not even light can escape from it. The R_S for the sun is about 3 km, and the R_S for the earth is about 1 cm. These objects are (of course) not black holes, because R_S is smaller than the radius of the object. However, note that R_S is proportional to m, which in turn is proportional to the cube of the radius r of the object, for a given density ρ. This means that for any given ρ, if r is increased, m grows faster than r. So eventually R_S will become as large as r, and the object will be a black hole. Since $m = (4\pi r^3/3)\rho$, you can quickly show that the critical radius at which this occurs is $r_{\text{crit}} = \sqrt{3c^2/8\pi G\rho}$. If $\rho = 1000\,\text{kg/m}^3$ (the density of water), then $r_{\text{crit}} \approx 4 \cdot 10^{11}$ m, which is about three times the radius of the earth's orbit.

1.12. [A,C] You can figure out the angles by doing some geometry, but it's much easier to just check the special case where θ is very small or very close to 90°. Fig. 1.7 shows the case where θ is small. In setups like this, every angle is either θ or $90° - \theta$ (or 90°). So all of the small angles in the figure must be θ, as shown.

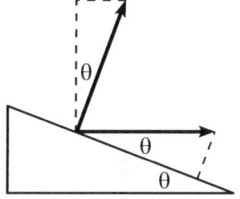

Figure 1.7

REMARK: Even if you want to work out the geometry to determine the angles, it would be silly to pass up the quick and easy double check of small or large θ. Of course, once you get used to doing this quick check, you'll realize that there's not much need to work through the geometry in the first place. A corollary of this is that you should never draw anything close to a 45-45-90 triangle (as we purposely did in the statement of this question), because in that case you can't tell which are the big angles and which are the small angles!

1.13. [e] Intuitively, the tension must go to infinity when $\theta \to 90°$, that is, when the strings approach being horizontal. Imagine pulling the top ends of the strings outward, to try to make the strings horizontal. This will take a large (infinite) force, because the mass will always sag a little in the middle. If the strings were exactly horizontal, then they would have no vertical force component to balance the mg weight. Choice (e) is the only one that goes to infinity in the $\theta \to 90°$ limit.

Some additional reasoning: (a) is incorrect because T should depend on θ, (b) is incorrect because T shouldn't be zero when $\theta = 0$, (c) is incorrect because T shouldn't be zero when $\theta = 90°$, and (d) is incorrect because T shouldn't be infinite when $\theta = 0$.

1.14. [d] The coefficient of friction must be very small in the $\theta \to 0$ limit, otherwise the block won't move. And it must be very large in the $\theta \to \pi/2$ limit, otherwise the block will keep accelerating. The $\mu = \tan\theta$ choice is the only one that satisfies these conditions.

1.15. [d] The volume must be zero when $\theta = 0$; this rules out (a). And it must be $(2/3)\pi R^3$ when $\theta = \pi/2$ (half of the whole sphere); this rules out (c). And it must be $(4/3)\pi R^3$ when $\theta = \pi$ (the whole sphere); this rules out (b). So the answer must be (d).

1.16. [b] For a horizontal line (with $a = 0$), the distance can't depend on x_0; this rules out (a). The answer must depend on c, because c affects the position (the height) of the line; this rules out (c). Furthermore, in the $c \to \infty$ limit, the distance should go to infinity. But choice (d) approaches 1. So the answer must be (b).

Another bit of reasoning: if the point (x_0, y_0) lies on the line, that is, if $ax_0 + by_0 + c = 0$, then the distance is zero. This implies that the correct answer must be (b) or (d).

Alternatively, you can eliminate all of the wrong answers by explicitly using the fact that in the case of a horizontal line (with $a = 0$, so the line is given by $y = -c/b$), the distance is $y_0 - (-c/b)$.

1.17. $\boxed{e}$ All of the possible answers have the correct units, so we'll have to figure things out by looking at special cases. Let's look at each choice in turn:

Choice (a) is incorrect, because the answer shouldn't be zero for $h = 0$. Also, it shouldn't grow with g. And furthermore it shouldn't be infinite for $v \to 0$.

Choice (b) is incorrect, because the answer should depend on h.

Choice (c) is incorrect, because the answer shouldn't be infinite for $h = 0$.

Choice (d) is incorrect, because the answer shouldn't be zero for $h = 0$.

Choice (e) can't be ruled out, and it happens to be the correct answer.

Choice (f) is incorrect, because there should be no possible distance when $h \to \infty$ (you can't throw the ball over an infinitely high wall). But this answer gives zero in this limit. This reasoning actually gets rid of all the answers except the correct one in one fell swoop. There are certainly cases for which there is no distance from which it is possible to throw the ball over the wall (for example, if h or g is very large, or v is very small). So any expression that gives a real result for all values of the parameters cannot be correct. Choice (e) is the only one that correctly gives an imaginary (and hence nonphysical) result in certain cases.

1.7 Problem solutions

1.1. Furlongs per fortnight squared

The systematic way of doing the conversion is to trade certain units for certain other units, by multiplying by 1 in the appropriate form. We want to trade meters for furlongs, and seconds for fortnights. This can be done as follows:

$$g = 9.8 \frac{\text{m}}{\text{s}^2} \left(\frac{1.09 \text{ yard}}{1 \text{ m}} \cdot \frac{1 \text{ furlong}}{220 \text{ yard}} \right) \left(\frac{60 \text{ s}}{1 \text{ min}} \cdot \frac{60 \text{ min}}{1 \text{ hr}} \cdot \frac{24 \text{ hr}}{1 \text{ day}} \cdot \frac{14 \text{ day}}{1 \text{ fortnight}} \right)^2$$
$$= 7.1 \cdot 10^{10} \frac{\text{furlongs}}{\text{fortnight}^2}. \tag{1.16}$$

All we've done here is multiply by 1 six times, so we haven't changed the value. The right-hand side still equals g; it's just that it's now expressed in different units. You can see that the m, yard, s, min, hr, and day units all cancel (don't forget that the second set of fractions is squared), so we're left with only furlongs and fortnights (squared), as desired.

REMARKS: This numerical result of $7.1 \cdot 10^{10}$ is very large. The reason for this is the following. The $g = 9.8 \text{ m/s}^2$ expression tells us that after one second, a falling body will be traveling at a speed of 9.8 m/s. So the question we want to ask is "If a body has been falling for one fortnight, what will its speed be, as expressed in furlongs per fortnight?" (We'll ignore the fact that a body certainly can't freefall for two weeks on the earth!) The numerical answer to this question is large for *two* reasons, consistent with the fact that the fortnight is *squared* in the expression for g. First, a fortnight is a long time, so the body will be moving very fast after falling for all this time. Second, because the body is moving so fast, it will (if it were to continue to travel with that speed) travel a very large distance after another lengthy time of one fortnight. And this distance is the numerical value of the speed when expressed in furlongs per fortnight. A competing effect is that since a furlong is larger than a meter by a factor of 220, the numerical result in Eq. (1.16) is decreased by this factor. But this effect is washed out by the squared larger affect of the lengthy fortnight.

There are many conversions that you can just do in your head. If someone asks you to convert 1 minute into seconds, you know that the answer is simply 60 seconds. There is no need to multiply 1 min by $(60 \text{ s})/(1 \text{ min})$ to cancel the minutes and be left with only seconds. But for more complicated conversions, you can systematically multiply by 1 in the appropriate form.

1.7. PROBLEM SOLUTIONS

Note the word "appropriate" in the previous sentence. You can't just blindly multiply by 1; you need to think about which units you're trying to cancel out. If you unwisely multiply 1 min by (1 min)/(60 s), which still equals 1, then you will end up with $(1\,\text{min})^2/(60\,\text{s})$. This does indeed equal 1 minute, but it isn't very informative. No one will know what you're talking about if you say to take a $(5\,\text{min}^2)/(60\,\text{s})$ break!

1.2. Miles per gallon

30 miles per gallon equals

$$\frac{30\,\text{miles}}{1\,\text{gallon}} = \frac{30 \cdot 1609\,\text{m}}{3785\,\text{cm}^3} = \frac{30 \cdot 1609\,\text{m}}{3785\,(10^{-2}\,\text{m})^3} = \frac{1}{7.84 \cdot 10^{-8}\,\text{m}^2}. \qquad (1.17)$$

This tiny area of $7.84 \cdot 10^{-8}\,\text{m}^2$ corresponds to a square with side length $2.8 \cdot 10^{-4}\,\text{m}$, or about 0.3 millimeters. For comparison, a common diameter for the pencil lead in a mechanical pencil is 0.5 or 0.7 millimeters.

What does this area actually have to do with a car that gets 30 miles per gallon? Note that Eq. (1.17) can alternatively be written as

$$(7.84 \cdot 10^{-8}\,\text{m}^2)(30\,\text{miles}) = 1\,\text{gallon}. \qquad (1.18)$$

What this says is that if we have a narrow tube of gasoline running parallel to the road, with a cross-sectional area of $7.84 \cdot 10^{-8}\,\text{m}^2$, and if our car gobbles up the gasoline in the tube as it travels along, then it will gobble up one gallon every 30 miles, which is exactly what it needs to operate. This interpretation of the area gives you an intuitive sense of how much gasoline you're using; think of a long pencil lead running parallel to the road. If you instead want to think in terms of a small unit of volume, you can work with drops. In medicine, the unit of one "drop" equals 1/20 of a milliliter. You can show that you burn about one drop of gasoline for every two feet you travel (assuming 30 miles per gallon).

1.3. Painting a funnel

It is true that the volume of the funnel is finite, and that you can fill it up with paint. It is also true that the surface area is infinite, but you actually *can* paint it.

The apparent paradox arises from essentially comparing apples and oranges. In our case we are comparing *volumes* (which are three dimensional) with *areas* (which are two dimensional). When someone says that the funnel can't be painted, he is saying that it would take an infinite *volume* of paint to cover it. But the fact that the surface *area* is infinite does *not* imply that it takes an infinite *volume* of paint to cover it. To be sure, if we try to paint the funnel with a given fixed thickness of paint, then we would indeed need an infinite volume of paint. But in this case, if we look at very large values of x where the funnel has negligible thickness, we would essentially have a tube of paint with a fixed radius, extending to $x = \infty$, with the funnel taking up a negligible volume at the center of the tube. This tube certainly has an infinite volume.

But what if we paint the funnel with a decreasing thickness of paint, as x gets larger? For example, if we make the thickness be proportional to $1/x$, then the volume of paint is proportional to $\int_1^\infty (1/x)(1/x)\,dx$, which is finite. (The first $1/x$ factor here comes from the $2\pi r$ factor in the area, and the second $1/x$ factor comes from the thickness of the paint. We have ignored the $\sqrt{1 + y'^2}$ factor, which goes to 1 for large x.) In this manner, we can indeed paint the funnel. To sum up, you buy paint by the gallon, not by the square meter. And a gallon of paint can cover an infinite area, as long as you make the thickness go to zero fast enough. The moral of this problem, therefore, is to not mix up things with different units!

1.4. Planck scales

FIRST SOLUTION: Since only G and $\hbar$ involve units of kilograms (one in the numerator and one in the denominator), it's fairly easy to see what the three desired combinations are. For example, the Planck length and time must involve the product $\hbar G$, because this eliminates the kilograms. You should check that the expressions below in Eq. (1.20) all have the correct units (the subscript P is for Planck). Using the numerical values,

$$\hbar = 1.05 \cdot 10^{-34} \frac{\text{kg m}^2}{\text{s}},$$
$$G = 6.67 \cdot 10^{-11} \frac{\text{m}^3}{\text{kg s}^2},$$
$$c = 3.0 \cdot 10^8 \frac{\text{m}}{\text{s}}, \qquad (1.19)$$

we obtain the three Planck scales:

$$\text{Length}: \quad \ell_P = \sqrt{\frac{\hbar G}{c^3}} = 1.6 \cdot 10^{-35} \text{ m},$$
$$\text{Time}: \quad t_P = \sqrt{\frac{\hbar G}{c^5}} = 5.4 \cdot 10^{-44} \text{ s},$$
$$\text{Mass}: \quad m_P = \sqrt{\frac{\hbar c}{G}} = 2.2 \cdot 10^{-8} \text{ kg}. \qquad (1.20)$$

Another such quantity is the Planck energy:

$$\text{Energy}: \quad E_P = \sqrt{\frac{\hbar c^5}{G}} = 2.0 \cdot 10^9 \text{ J}. \qquad (1.21)$$

SECOND SOLUTION: If you want to be more systematic, you can write down a general expression of the form $\hbar^\alpha G^\beta c^\gamma$ and then solve for the values of the three exponents that make the overall units be correct. For example, if we want to find the Planck length, then we need the overall units to be meters, so we want

$$\left(\frac{\text{kg m}^2}{\text{s}}\right)^\alpha \left(\frac{\text{m}^3}{\text{kg s}^2}\right)^\beta \left(\frac{\text{m}}{\text{s}}\right)^\gamma = \text{m}. \qquad (1.22)$$

Equating separately the powers of kg, m, and s on the two sides of the equation gives a system of three equations in three unknowns:

$$\begin{aligned} \text{kg}: & \quad \alpha - \beta = 0, \\ \text{m}: & \quad 2\alpha + 3\beta + \gamma = 1, \\ \text{s}: & \quad -\alpha - 2\beta - \gamma = 0. \end{aligned} \qquad (1.23)$$

The first equation gives $\alpha = \beta$. The third equation then gives $\gamma = -3\alpha$. And the second equation then gives $\alpha = 1/2$. So $\beta = 1/2$ and $\gamma = -3/2$. This agrees with the result for the Planck length in Eq. (1.20). The Plank time and mass proceed similarly. However, even though this method will always get the job done, it was certainly quicker in this problem to just fiddle around with the units by noting, for example, that the Planck length and time must involve the product $\hbar G$.

REMARK: The physical significance of all these scales is that they are the scales at which the quantum effects of gravity can no longer be ignored. Said in another way, they are the scales at which the four known forces in nature (gravitational, electromagnetic, weak, strong) all become roughly equal. This equality should be contrasted with the fact that at, say, atomic length scales, the electrical force between two protons is vastly larger (about 10^{36} times larger) than the gravitational force.

To get a sense of how small the planck length is, the size of an atomic nucleus is on the order of a femtometer (10^{-15} m). The Planck length is therefore 10^{20} times smaller. So the Planck length is to the nuclear size as the nuclear size is to 100 km. And to get a sense of how large the Planck energy E_P is, the energies presently probed at the CERN collider are only on the order of $10^{-15} E_P$.

1.7. PROBLEM SOLUTIONS

1.5. Capillary rise

Since the units of force are kg m/s^2, the units of surface tension (force per length) are kg/s^2. So the units of the various quantities are:

$$h : \text{m}, \quad \gamma : \frac{\text{kg}}{\text{s}^2}, \quad r : \text{m}, \quad \rho : \frac{\text{kg}}{\text{m}^3}, \quad g : \frac{\text{m}}{\text{s}^2}. \tag{1.24}$$

Our goal is to create h from the other four quantities. We need to get rid of the units of kg and s, and we quickly see that the combination $\gamma/\rho g$ accomplishes this. It leaves us with m^2 in the numerator. But now we have a problem, because although simply dividing by r will give us the desired units of meters, there are many other possibilities that work too. In fact, any expression of the form $(\gamma/\rho g)^n/r^{2n-1}$ has units of meters. The issue here is that we have four unknowns (the powers of each of γ, r, ρ, and g), but only three equations (the facts that the overall powers of kg, m, and s must be 0, 1, and 0, respectively). So the system is under-determined; we can't uniquely solve for the four unknowns.

However, if we use the additional fourth piece of information that the power of γ is 1, then this tells us that the value of n in the above general expression is 1. So we can now say that

$$h \propto \frac{\gamma}{\rho g r}. \tag{1.25}$$

The actual result has a numerical factor of 2 in the numerator, but it takes a little more effort to show that. Additionally, h depends on the contact angle θ between the water and the tube, at the top of the meniscus; this brings in a factor of $\cos\theta$ (so the correct result is $h = 2\gamma\cos\theta/\rho g r$). But θ is a dimensionless quantity, so we can't say anything about it by using dimensional analysis.

1.6. Fluid flow

Since the units of force are kg m/s^2, the units of pressure (force per area) are $\text{kg}/(\text{m s}^2)$. So the units of the various quantities are:

$$Q : \frac{\text{m}^3}{\text{s}}, \quad \Delta P : \frac{\text{kg}}{\text{m s}^2}, \quad R : \text{m}, \quad L : \text{m}, \quad \eta : \frac{\text{kg}}{\text{m s}}. \tag{1.26}$$

Our goal is to create Q from the other four quantities, with the condition that there is one power of L in the denominator. As in Problem 1.5, we can't solve for four unknowns (the powers of ΔP, R, L, and η) with only three pieces of information (the required powers of kg, m, and s). This fact about L is the necessary fourth piece of information.

To produce the units of Q, we need to get rid of the kg and have one power of s in the denominator. We quickly see that the quotient $\Delta P/\eta$ accomplishes this. This expression has no m's, but we need an m^3 in the numerator of Q. Since we are told that there is one power of L in the denominator, we must have four powers of R in the numerator. So the desired expression is

$$Q \propto \frac{\Delta P R^4}{\eta L}. \tag{1.27}$$

The actual result has a numerical factor of $\pi/8$ out front, but it requires a detailed calculation to show that.

REMARK: The R^4 dependence in Q is stronger than the R^2 dependence you might expect by simply considering the fact that the cross-sectional area of the pipe is proportional to R^2. What happens is that the wider the pipe, the faster the average speed of the fluid (for a given ΔP). So we have a faster fluid flowing through a wider pipe. The fourth power of R isn't obvious, though.

This fourth power implies that if the pipe's radius is decreased to, say, 0.8 of what it was, then the flow rate (with ΔP held constant) is decreased to $(0.8)^4 \approx 0.41$ of what it was. In other words, a 20% reduction in radius produces a nearly 60% reduction in flow rate, which is more than you might naively expect. This fact is highly relevant, for example, when dealing with plaque buildup in arteries.

1.7. 1-D collision

The given expressions are

$$V_M = \frac{(M-m)V}{M+m} \quad \text{and} \quad v_m = \frac{2MV}{M+m}. \quad (1.28)$$

If $M = m$ (for example, two identical billiard balls) then we have

$$V_M = 0 \quad \text{and} \quad v_m = V. \quad (1.29)$$

The mass that was initially moving ends up at rest, and the the mass that was initially at rest ends up moving with whatever velocity the other mass had. This result is familiar to pool players; barring any effects of spin, the cue ball ends up at rest in a head-on collision.

If $M \ll m$ (for example, a marble bouncing off a bowling ball) then we can ignore the M's in the expressions in Eq. (1.28). (More precisely, we can ignore the additive M's; see the discussion following Eq. (1.1).) This yields

$$V_M \approx \frac{0-m}{0+m}V = -V \quad \text{and} \quad v_m \approx \frac{2M}{0+m}V \approx 0. \quad (1.30)$$

In this case, M is basically a ball bouncing backward off a unmoveable brick wall.

If $M \gg m$ (for example, a bowling ball colliding with a marble) then we can ignore the m's in the expressions in Eq. (1.28). This yields

$$V_M \approx \frac{M-0}{M+0}V = V \quad \text{and} \quad v_m \approx \frac{2M}{M+0}V = 2V. \quad (1.31)$$

In this case the bowling ball M plows forward with the same velocity V, as expected. But interestingly the marble m picks up *twice* this velocity.

REMARK: This result of $2V$ for the speed of m in the $M \gg m$ limit isn't so obvious in the given lab frame, but it's fairly easy to understand if you imagine riding along with M. In the reference frame where M is at rest, m comes flying in with speed V and then bounces off with essentially the same speed V (because M is a brick wall in the $M \gg m$ limit). So the final relative speed of M and m is V. But we must now shift back to the lab frame and remember that M is still plowing forward with speed V (its speed hardly changes during the collision). This means that m is moving forward with speed V *relative* to M, which itself is moving forward with speed V relative to the lab frame. So m is moving with speed $V + V = 2V$ relative to the lab frame, as desired. In terms of energy (the subject of Chapter 5), the transfer of energy from the large object to the small object occurs via the energy stored in the elasticity of the balls as they deform during the collision.

The fact that you can use a large object moving with speed V to make a small object move faster than V is the basic principle behind a whip. Initially the thick heavy part of the whip is moving with a given speed, and eventually the thin light part at the end moves with a much larger speed. A whip is a continuous object, whereas the above problem involved two discrete balls, but see Problem 6.14 for a discussion of how to transition from the discrete case to the continuous case.

Whip-like motions are ubiquitous in sports. Examples include throwing a baseball, throwing a frisbee™, shooting a hockey puck, and kicking a football. In all cases, you start by moving a large object, and you end up with a small object that is moving much faster. There's a reason for the "Put your body into it" mantra. A baseball pitch starts with a relatively slow motion of the body/torso/shoulder and ends with a much faster motion of the forearm/hand/fingers. The transfer of energy occurs via the energy stored in the elasticity of tendons and ligaments. Incidentally, the fastest you can make any part of your body move (relative to your center of mass) is your fingers when throwing something.

1.8. Atwood's machine

The given expression for a_2 is

$$a_2 = -g\frac{4m_2m_3 + m_1(m_2 - 3m_3)}{4m_2m_3 + m_1(m_2 + m_3)}. \quad (1.32)$$

1.7. PROBLEM SOLUTIONS

(a) If $m_1 = 2m_2 = 2m_3 \equiv 2m$, then a_2 becomes

$$a_2 = -g\frac{4m^2 + (2m)(m-3m)}{4m^2 + (2m)(m+m)} = 0. \quad (1.33)$$

In this case, m_2 and m_3 balance m_1.

(b) If m_2 is very large, then we can ignore the $m_1 m_3$ terms in Eq. (1.32). This gives

$$a_2 \approx -g\frac{4m_2 m_3 + m_1(m_2 - 0)}{4m_2 m_3 + m_1(m_2 - 0)} = -g. \quad (1.34)$$

In this case, m_2 is simply in freefall.

(c) If m_2 is very small, then we can ignore the m_2 terms in Eq. (1.32). This gives

$$a_2 \approx -g\frac{0 + m_1(0 - 3m_3)}{0 + m_1(0 + m_3)} = 3g. \quad (1.35)$$

In this case, m_2 accelerates upward at $3g$. To understand this factor of 3, you can convince yourself that if m_1 and m_3 both freefall a distance ℓ, then m_2 must rise up a distance 3ℓ. This then implies that the acceleration of m_2 is 3 times the freefall acceleration g of m_1 and m_3. You may want to wait to do this until we discuss Atwood's machines (in particular, the topic of conservation of string) in Chapter 4.

(d) If $m_1 \gg m_2 = m_3 \equiv m$, then we can ignore the $m_2 m_3$ terms in Eq. (1.32). This gives

$$a_2 \approx -g\frac{0 + m_1(m - 3m)}{0 + m_1(m + m)} = g. \quad (1.36)$$

In this case, m_1 is in freefall, so m_2 and m_3 accelerate upward at g.

(e) If $m_1 = m_2 = m_3 \equiv m$, then Eq. (1.32) gives

$$a_2 = -g\frac{4m^2 + m(m-3m)}{4m^2 + m(m+m)} = -\frac{g}{3}. \quad (1.37)$$

In this case, a_2 is correctly negative, but the factor of 1/3 isn't obvious; we would have to solve the problem for real to derive that.

1.9. Dropped ball

As with the beach ball's velocity $v(t)$ given in Eq. (1.3), the position $y(t)$ in Eq. (1.10) is a somewhat complicated expression, so it's hard to feel too confident about its validity by just looking at it. But if we can verify that it gives the correct answer in a particular limit, then we'll feel much better about it.

In the limit of small bt/m, we can use the Taylor series $e^{-x} \approx 1 - x + x^2/2$ to produce an approximate expression for $y(t)$, to leading order in t. We obtain (as you can verify, being careful with all the minus signs!)

$$y(t) = h - \frac{mg}{b}\left[t - \frac{m}{b}\left(1 - \left(1 - \frac{bt}{m} + \frac{1}{2}\left(\frac{bt}{m}\right)^2 - \cdots\right)\right)\right]$$
$$\approx h - \frac{gt^2}{2}. \quad (1.38)$$

This answer is expected, because we essentially have a freely falling body at the start (v is small, so there is hardly any drag force), which implies that the distance fallen is the standard $gt^2/2$. (This is covered in Chapter 2, so just take it on faith for now. But the $gt^2/2$ term probably looks familiar to you anyway.) Note that in obtaining the leading-order term (the smallest power of t with a nonzero coefficient) in $y(t)$, we needed to go to second order in the Taylor series for e^{-x}, whereas we needed to go only to first order in obtaining the expression for $v(t)$ in Eq. (1.4).

REMARK: We can also look at the limit of large bt/m. In this case, $e^{-bt/m}$ is essentially zero, so the $y(t)$ in Eq. (1.10) becomes (there's no need for a Taylor series in this case)

$$y(t) \approx h - \frac{mgt}{b} + \frac{m^2 g}{b^2}. \tag{1.39}$$

Apparently, after a long time, m^2g/b^2 is the distance that our ball lags behind another ball that started out already at the terminal velocity $-mg/b$, because that ball has $y(t) = h - (mg/b)t$. (The terminal velocity is in fact $-mg/b$; see the discussion in Section 1.1.4.) You can verify that the quantity m^2g/b^2 does indeed have dimensions of length, using the fact that the original expression for the drag force, $F_d = -bv$, tells us that b has units of N/(m/s), or equivalently kg/s. This m^2g/b^2 result is by no means obvious, so our check of the large bt/m limit doesn't do anything to make us feel better about the original expression in Eq. (1.10). But that's fine; when checking limiting cases, the result is either that we feel better about our answer, or we get an interesting result that we didn't expect.

Chapter 2

Kinematics in 1-D

As mentioned in the preface, this book should not be thought of as a textbook. The introduction to each chapter is brief and is therefore no substitute for an actual textbook. You will most likely want to have a textbook on hand when reading the introductions.

2.1 Introduction

In this chapter and the next, we won't be concerned with the forces that cause an object to move in the particular way it is moving. We will simply take the motion as given, and our goal will be to relate positions, velocities, and accelerations as functions of time. Our objects can be treated like point particles; we will not be concerned with what they are actually made of. This is the study of *kinematics*. In Chapter 4 we will move on to *dynamics*, where we will deal with mass, force, energy, momentum, etc.

Velocity and acceleration

In one dimension, the *average* velocity and acceleration over a time interval Δt are given by

$$v_{\text{avg}} = \frac{\Delta x}{\Delta t} \quad \text{and} \quad a_{\text{avg}} = \frac{\Delta v}{\Delta t}. \tag{2.1}$$

The *instantaneous* velocity and acceleration at a particular time t are obtained by letting the interval Δt become infinitesimally small. In this case we write the "Δ" as a "d," and the instantaneous v and a are given by

$$v = \frac{dx}{dt} \quad \text{and} \quad a = \frac{dv}{dt}. \tag{2.2}$$

In calculus terms, v is the derivative of x, and a is the derivative of v. Equivalently, v is the slope of the x vs. t curve, and a is the slope of the v vs. t curve. In the case of the velocity v, you can see how this slope arises by taking the limit of $v = \Delta x/\Delta t$, as Δt becomes very small; see Fig. 2.1. The smaller Δt is, the better the slope $\Delta x/\Delta t$ approximates the actual slope of the tangent line at the given point P.

In 2-D and 3-D, the velocity and acceleration are vectors. That is, we have a separate pair of equations of the form in Eq. (2.2) for each dimension; the x components are given by $v_x = dx/dt$ and $a_x = dv_x/dt$, and likewise for the y and z components. The velocity and acceleration are also vectors in 1-D, although in 1-D a vector can be viewed simply as a number (which may be positive or negative). In any dimension, the *speed* is the magnitude of the velocity, which means the absolute value of v in 1-D and the length of the vector **v** in 2-D and 3-D. So the speed is a positive number by definition. The units of velocity and speed are m/s, and the units of acceleration are m/s^2.

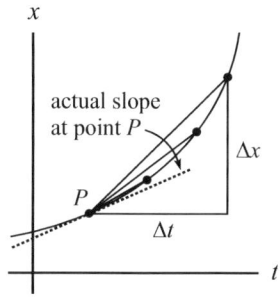

Figure 2.1

Displacement as an area

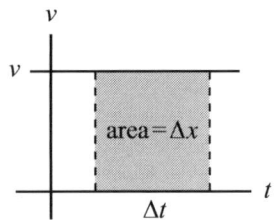

Figure 2.2

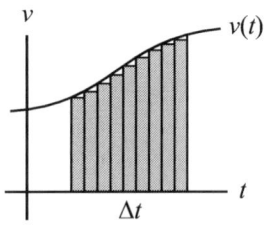

Figure 2.3

If an object moves with constant velocity v, then the displacement Δx during a time Δt is $\Delta x = v\Delta t$. In other words, the displacement is the area of the region (which is just a rectangle) under the v vs. t "curve" in Fig. 2.2. Note that the *displacement* (which is Δx by definition), can be positive or negative. The *distance* traveled, on the other hand, is defined to be a positive number. In the case where the displacement is negative, the v vs. t line in Fig. 2.2 lies below the t axis, so the (signed) area is negative.

If the velocity varies with time, as shown in Fig. 2.3, then we can divide time into a large number of short intervals, with the velocity being essentially constant over each interval. The displacement during each interval is essentially the area of each of the narrow rectangles shown. In the limit of a very large number of very short intervals, adding up the areas of all the thin rectangles gives exactly the total area under the curve; the areas of the tiny triangular regions at the tops of the rectangles become negligible in this limit. So the general result is:

- *The displacement (that is, the change in x) equals the area under the v vs. t curve.*

Said in a more mathematical way, the displacement equals the time integral of the velocity. This statement is equivalent (by the fundamental theorem of calculus) to the fact that v is the time derivative of x.

All of the relations that hold between x and v also hold between v and a. In particular, the change in v equals the area under the a vs. t curve. And conversely, a is the time derivative of v. This is summarized in the following diagram:

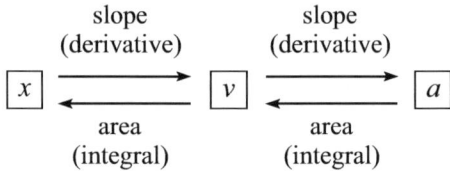

Motion with constant acceleration

For motion with constant acceleration a, we have

$$a(t) = a,$$
$$v(t) = v_0 + at,$$
$$x(t) = x_0 + v_0 t + \frac{1}{2}at^2, \tag{2.3}$$

where x_0 and v_0 are the initial position and velocity at $t = 0$. The above expressions for $v(t)$ and $x(t)$ are correct, because $v(t)$ is indeed the derivative of $x(t)$, and $a(t)$ is indeed the derivative of $v(t)$. If you want to derive the expression for $x(t)$ in a graphical manner, see Problem 2.1.

The above expressions are technically all you need for any setup involving constant acceleration, but one additional formula might make things easier now and then. If an object has a displacement d with constant acceleration a, then the initial and final velocities satisfy

$$v_f^2 - v_i^2 = 2ad. \tag{2.4}$$

See Problem 2.2 for a proof. If you know three out of the four quantities v_f, v_i, a, and d, then this formula quickly gives the fourth. In the special case where the object starts at rest (so $v_i = 0$), we have the simple result, $v_f = \sqrt{2ad}$.

Falling bodies

Perhaps the most common example of constant acceleration is an object falling under the influence of only gravity (that is, we'll ignore air resistance) near the surface of the earth. The constant nature of the gravitational acceleration was famously demonstrated by Galileo. (He mainly rolled balls down ramps instead of dropping them, but it's the same idea.) If we take the positive y axis to point upward, then the acceleration due to gravity is $-g$, where $g = 9.8\,\mathrm{m/s^2}$. After every second, the velocity becomes more negative by $9.8\,\mathrm{m/s}$; that is, the downward speed increases by $9.8\,\mathrm{m/s}$. If we substitute $-g$ for a in Eq. (2.3) and replace x with y, the expressions become

$$a(t) = -g,$$
$$v(t) = v_0 - gt,$$
$$y(t) = y_0 + v_0 t - \frac{1}{2} g t^2, \tag{2.5}$$

For an object dropped from rest at a point we choose to label as $y = 0$, Eq. (2.5) gives $y(t) = -gt^2/2$.

In some cases it is advantageous to choose the positive y axis to point downward, in which case the acceleration due to gravity is g (with no minus sign). In any case, it is always a good idea to take g to be the *positive* quantity $9.8\,\mathrm{m/s^2}$, and then throw in a minus sign by hand if needed, because working with quantities with minus signs embedded in them can lead to confusion.

The expressions in Eq. (2.5) hold only in the approximation where we neglect air resistance. This is generally a good approximation, as long as the falling object isn't too light or moving too quickly. Throughout this book, we will ignore air resistance unless stated otherwise.

2.2 Multiple-choice questions

2.1. If an object has negative velocity and negative acceleration, is it slowing down or speeding up?

(a) slowing down

(b) speeding up

2.2. The first figure below shows the a vs. t plot for a certain setup. The second figure shows the v vs. t plot for a different setup. The third figure shows the x vs. t plot for a yet another setup. Which of the twelve labeled points correspond(s) to zero acceleration? Circle all that apply. (To repeat, the three setups have nothing to do with each other. That is, the v plot is *not* the velocity curve associated with the position in the x plot. etc.)

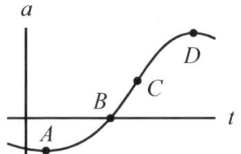 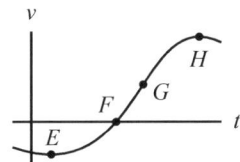 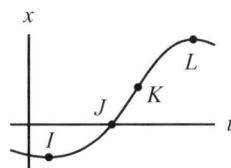

2.3. If the acceleration as a function of time is given by $a(t) = At$, and if $x = v = 0$ at $t = 0$, what is $x(t)$?

(a) $\dfrac{At^2}{2}$ (b) $\dfrac{At^2}{6}$ (c) At^3 (d) $\dfrac{At^3}{2}$ (e) $\dfrac{At^3}{6}$

2.4. Under what condition is the average velocity (which is defined to be the total displacement divided by the time) equal to the average of the initial and final velocities, $(v_i + v_f)/2$?

(a) The acceleration must be constant.

(b) It is true for other motions besides constant acceleration, but not for all possible motions.

(c) It is true for all possible motions.

2.5. Two cars, with initial speeds of $2v$ and v, lock their brakes and skid to a stop. Assume that the deceleration while skidding is independent of the speed. The ratio of the distances traveled is

(a) 1   (b) 2   (c) 4   (d) 8   (e) 16

2.6. You start from rest and accelerate with a given constant acceleration for a given distance. If you repeat the process with twice the acceleration, then the time required to travel the same distance

(a) remains the same

(b) is doubled

(c) is halved

(d) increases by a factor of $\sqrt{2}$

(e) decreases by a factor of $\sqrt{2}$

2.7. A car travels with constant speed v_0 on a highway. At the instant it passes a stationary police motorcycle, the motorcycle accelerates with constant acceleration and gives chase. What is the speed of the motorcycle when it catches up to the car (in an adjacent lane on the highway)? *Hint*: Draw the v vs. t plots on top of each other.

(a) v_0 (b) $3v_0/2$ (c) $2v_0$ (d) $3v_0$ (e) $4v_0$

2.8. You start from rest and accelerate to a given final speed v_0 after a time T. Your acceleration need not be constant, but assume that it is always positive or zero. If d is the total distance you travel, then the range of possible d values is

(a) $d = v_0 T/2$

(b) $0 < d < v_0 T/2$

(c) $v_0 T/2 < d < v_0 T$

(d) $0 < d < v_0 T$

(e) $0 < d < \infty$

2.9. You are driving a car that has a maximum acceleration of a. The magnitude of the maximum deceleration is also a. What is the maximum distance that you can travel in time T, assuming that you begin and end at rest?

(a) $2aT^2$ (b) aT^2 (c) $aT^2/2$ (d) $aT^2/4$ (e) $aT^2/8$

2.10. A golf club strikes a ball and sends it sailing through the air. Which of the following choices best describes the sizes of the position, speed, and acceleration of the ball at a moment in the middle of the strike? ("Medium" means a non-tiny and non-huge quantity, on an everyday scale.)

(a) x is tiny, v is medium, a is medium

(b) x is tiny, v is medium, a is huge

(c) x is tiny, v is huge, a is huge

(d) x is medium, v is medium, a is medium

(e) x is medium, v is medium, a is huge

2.11. Which of the following answers is the best estimate for the time it takes an object dropped from rest to fall a vertical mile (about 1600 m)? Ignore air resistance, as usual.

(a) 5 s   (b) 10 s   (c) 20 s   (d) 1 min   (e) 5 min

2.3. PROBLEMS

2.12. You throw a ball upward. After half of the time to the highest point, the ball has covered

   (a) half the distance to the top
   (b) more than half the distance
   (c) less than half the distance
   (d) It depends on how fast you throw the ball.

2.13. A ball is dropped, and then another ball is dropped from the same spot one second later. As time goes on while the balls are falling, the distance between them (ignoring air resistance, as usual)

   (a) decreases
   (b) remains the same
   (c) increases and approaches a limiting value
   (d) increases steadily

2.14. You throw a ball straight upward with initial speed v_0. How long does it take to return to your hand?

   (a) $v_0^2/2g$ (b) v_0^2/g (c) $v_0/2g$ (d) v_0/g (e) $2v_0/g$

2.15. Ball 1 has mass m and is fired directly upward with speed v. Ball 2 has mass $2m$ and is fired directly upward with speed $2v$. The ratio of the maximum height of Ball 2 to the maximum height of Ball 1 is

   (a) 1   (b) $\sqrt{2}$ (c) 2   (d) 4   (e) 8

2.3 Problems

The first three problems are foundational problems.

2.1. Area under the curve

At $t = 0$ an object starts with position x_0 and velocity v_0 and moves with constant acceleration a. Derive the $x(t) = x_0 + v_0 t + at^2/2$ result by finding the area under the v vs. t curve (without using calculus).

2.2. A kinematic relation

Use the relations in Eq. (2.3) to show that if an object moves through a displacement d with constant acceleration a, then the initial and final velocities satisfy $v_f^2 - v_i^2 = 2ad$.

2.3. Maximum height

If you throw a ball straight upward with initial speed v_0, it reaches a maximum height of $v_0^2/2g$. How many derivations of this result can you think of?

2.4. Average speeds

   (a) If you ride a bike up a hill at 10 mph, and then down the hill at 20 mph, what is your average speed?
   (b) If you go on a bike ride and ride for half the time at 10 mph, and half the time at 20 mph, what is your average speed?

2.5. Colliding trains

Two trains, A and B, travel in the same direction on the same set of tracks. A starts at rest at position d, and B starts with velocity v_0 at the origin. A accelerates with acceleration a, and B decelerates with acceleration $-a$. What is the maximum value of v_0 (in terms of d and a) for which the trains don't collide? Make a rough sketch of x vs. t for both trains in the case where they barely collide.

2.6. Ratio of distances

Two cars, A and B, start at the same position with the same speed v_0. Car A travels at constant speed, and car B decelerates with constant acceleration $-a$. At the instant when B reaches a speed of zero, what is the ratio of the distances traveled by A and B? Draw a reasonably accurate plot of x vs. t for both cars.

You should find that your answer for the ratio of the distances is a nice simple number, independent of any of the given quantities. Give an argument that explains why this is the case.

2.7. How far apart?

An object starts from rest at the origin at time $t = -T$ and accelerates with constant acceleration a. A second object starts from rest at the origin at time $t = 0$ and accelerates with the same a. How far apart are they at time t? Explain the meaning of the two terms in your answer, first in words, and then also with regard to the v vs. t plots.

2.8. Ratio of odd numbers

An object is dropped from rest. Show that the distances fallen during the first second, the second second, the third second, etc., are in the ratio of $1 : 3 : 5 : 7 \ldots$.

2.9. Dropped and thrown balls

A ball is dropped from rest at height h. Directly below on the ground, a second ball is simultaneously thrown upward with speed v_0. If the two balls collide at the moment the second ball is instantaneously at rest, what is the height of the collision? What is the relative speed of the balls when they collide? Draw the v vs. t plots for both balls.

2.10. Hitting at the same time

A ball is dropped from rest at height h. Another ball is simultaneously thrown downward with speed v from height $2h$. What should v be so that the two balls hit the ground at the same time?

2.11. Two dropped balls

A ball is dropped from rest at height $4h$. After it has fallen a distance d, a second ball is dropped from rest at height h. What should d be (in terms of h) so that the balls hit the ground at the same time?

2.4 Multiple-choice answers

2.1. $\boxed{b}$ The object is speeding up. That is, the magnitude of the velocity is increasing. This is true because the negative acceleration means that the change in velocity is negative. And we are told that the velocity is negative to start with. So it might go from, say, -20 m/s to -21 m/s a moment later. It is therefore speeding up.

REMARKS: If we had said that the object had negative velocity and positive acceleration, then it would be slowing down. Basically, if the sign of the acceleration is the same as (or the opposite of) the sign of the velocity, then the object is speeding up (or slowing down).

A comment on terminology: The word "decelerate" means to slow down. The word "accelerate" means in a colloquial sense to speed up, but as a physics term it means (in 1-D) to either speed up *or* slow down, because acceleration can be positive or negative. More generally, in 2-D or 3-D it means to change the velocity in any general manner (magnitude and/or direction).

2.2. $\boxed{B,E,H,K}$ Point B is where a equals zero in the first figure. Points E and H are where the slope (the derivative) of the v vs. t plot is zero; and the slope of v is a. Point K is where the slope of the x vs. t plot is *maximum*. In other words, it is where v is maximum. But the slope of a function is zero at a maximum, so the slope of v (which is a) is zero at K.

REMARK: In calculus terms, K is an *inflection point* of the x vs. t curve. It is a point where the slope is maximum. Equivalently, the derivative of the slope is zero. Equivalently again, the second derivative is zero. In the present case, the tangent line goes from lying below the x vs. t curve to lying above it; the slope goes from increasing to decreasing as it passes through its maximum value.

2.3. $\boxed{e}$ Since the second derivative of $x(t)$ equals $a(t)$, we must find a function whose second derivative is At. Choice (e) is satisfies this requirement; the first derivative equals $At^2/2$, and then the second derivative equals At, as desired. The standard $At^2/2$ result is valid only for a *constant* acceleration a. Note that all of the choices satisfy $x = v = 0$ at $t = 0$.

REMARK: If we add on a constant C to $x(t)$, so that we now have $At^3/6 + C$, then the $x = 0$ initial condition isn't satisfied, even though $a(t)$ is still equal to At. Similarly, if we add on a linear term Bt, then the $v = 0$ initial condition isn't satisfied, even though $a(t)$ is again still equal to At. If we add on quadratic term Dt^2, then although the $x = v = 0$ initial conditions are satisfied, the second derivative is now not equal to At. Likewise for any power of t that is 4 or higher. So not only is the $At^3/6$ choice the only correct answer among the five given choices, it is the only correct answer, period. Formally, the integral of a (which is v) must take the form of $At^2/2 + B$, where B is a constant of integration. And the integral of v (which is x) must then take the form of $At^3/6 + Bt + C$, where C is a constant of integration. The initial conditions $x = v = 0$ then quickly tell us that $C = B = 0$.

2.4. $\boxed{b}$ The statement is at least true in the case of constant acceleration, as seen by looking at the v vs. t plot in Fig. 2.4(a). The area under the v vs. t curve is the distance traveled, and the area of the trapezoid (which corresponds to constant acceleration) is the same as the area of the rectangle (which corresponds to constant velocity $(v_i + v_f)/2$). Equivalently, the areas of the triangles above and below the $(v_i + v_f)/2$ line are equal. If you want to work things out algebraically, the displacement is

$$d = v_i t + \frac{1}{2}at^2 = \frac{1}{2}(2v_i + at)t = \frac{1}{2}\big(v_i + (v_i + at)\big)t = \frac{1}{2}(v_i + v_f)t. \qquad (2.6)$$

The average velocity d/t is therefore equal to $(v_i + v_f)/2$, as desired.

The statement is certainly not true in all cases; a counterexample is shown in Fig. 2.4(b). The distance traveled (the area under the curve) is essentially zero, so the average velocity is essentially zero and hence not equal to $(v_i + v_f)/2$.

However, the statement *can* be true for motions without constant acceleration, as long as the area under the v vs. t curve is the same as the area of the rectangle associated with velocity $(v_i + v_f)/2$, as shown in Fig. 2.4(c). For the curve shown, this requirement is the same as saying that the areas of the two shaded regions are equal.

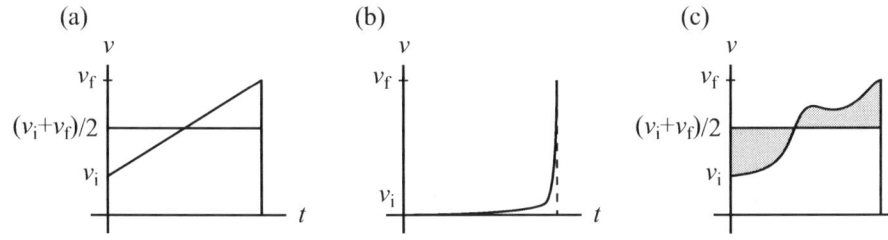

Figure 2.4

2.5. $\boxed{c}$ The final speed is zero in each case, so the $v_f^2 - v_i^2 = 2ad$ relation in Eq. (2.4) gives $0 - v_i^2 = 2(-a)d$, where a is the magnitude of the (negative) acceleration. So $d = v_i^2/2a$. Since this is proportional to v_i^2, the car with twice the initial speed has four times the stopping distance.

Alternatively, the distance traveled is $d = v_i t - at^2/2$, where again a is the magnitude of the acceleration. Since the car ends up at rest, the $v(t) = v_i - at$ expression for the velocity tells us that $v = 0$ when $t = v_i/a$. So

$$d = v_i\left(\frac{v_i}{a}\right) - \frac{1}{2}a\left(\frac{v_i}{a}\right)^2 = \frac{v_i^2}{2a}, \qquad (2.7)$$

in agreement with the relation obtained via Eq. (2.4).

Alternatively again, we could imagine reversing time and accelerating the cars from rest. Using the fact that one time is twice the other (since $t = v_i/a$), the relation $d = at^2/2$ immediately tells us that twice the time implies four times the distance.

Alternatively yet again, the factor of 4 quickly follows from the v vs. t plot shown in Fig. 2.5. The area under the diagonal line is the distance traveled, and the area of the large triangle is four times the area of the small lower-left triangle, because all four of the small triangles have the same area.

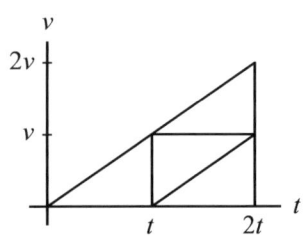

Figure 2.5

REMARK: When traveling in a car, the safe distance (according to many sources) to keep between your car and the car in front of you is dictated by the "three-second rule" (in good weather). That is, your car should pass, say, a given tree at least three seconds after the car in front of you passes it. This rule involves *time*, but it immediately implies that the minimum following *distance* is proportional to your speed. It therefore can't strictly be correct, because we found above that the stopping distance is proportional to the *square* of your speed. This square behavior means that the three-second rule is inadequate for sufficiently high speeds. There are of course many other factors involved (reaction time, the nature of the road hazard, the friction between the tires and the ground, etc.), so the exact formula is probably too complicated to be of much use. But if you take a few minutes to observe some cars and make some rough estimates of how drivers out there are behaving, you'll find that many of them are following at astonishingly unsafe distances, by any measure.

2.6. $\boxed{\text{e}}$ The distance traveled is given by $d = at^2/2$, so $t = \sqrt{2d/a}$. Therefore, if a is doubled then t decreases by a factor of $\sqrt{2}$.

REMARK: Since $v = at$ for constant acceleration, the speeds in the two given scenarios (label them S_1 and S_2) differ by a factor of 2 at any given time. So if at all times the speed in S_2 is twice the speed in S_1, shouldn't the time simply be halved, instead of decreased by the factor of $\sqrt{2}$ that we just found? No, because although the S_1 distance is only $d/2$ when the S_2 distance reaches the final value of d, it takes S_1 less time to travel the remaining $d/2$ distance, because its speed increases as time goes on. This is shown in Fig. 2.6. The area under each v vs. t curve equals the distance traveled. Compared with S_1, S_2's final speed is $\sqrt{2}$ times larger, but its time is $1/\sqrt{2}$ times smaller. So the areas of the two triangles are the same.

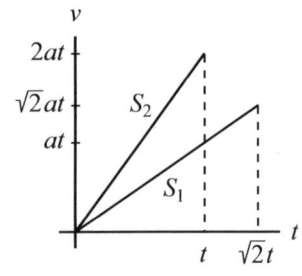

Figure 2.6

2.7. $\boxed{\text{c}}$ The area under a v vs. t curve is the distance traveled. The car's curve is the horizontal line shown in Fig. 2.7, and the motorcycle's curve is the tilted line. The two vehicles will have traveled the same distance when the area of the car's rectangle equals the area of the motorcycle's triangle. This occurs when the triangle has twice the height of the rectangle, as shown. (The area of a triangle is half the base times the height.) So the final speed of the motorcycle is $2v_0$. Note that this result is independent of the motorcycle's (constant) acceleration. If the acceleration is small, then the process will take a long time, but the speed of the motorcycle when it catches up to the car will still be $2v_0$.

Alternatively, the position of the car at time t is $v_0 t$, and the position of the motorcycle is $at^2/2$. These two positions are equal when $v_0 t = at^2/2 \Longrightarrow at = 2v_0$. But the motorcycle's speed is at, which therefore equals $2v_0$ when the motorcycle catches up to the car.

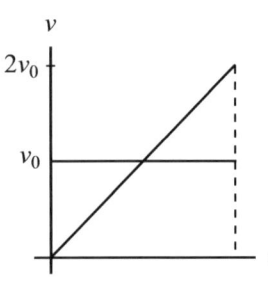

Figure 2.7

2.8. $\boxed{\text{d}}$ A distance of essentially zero can be obtained by sitting at rest for nearly all of the time T, and then suddenly accelerating with a huge acceleration to speed v_0. Approximately zero distance is traveled during this acceleration phase. This is true because Eq. (2.4) gives $d = v_0^2/2a$, where v_0 is a given quantity and a is huge.

Conversely, a distance of essentially $v_0 T$ can be obtained by suddenly accelerating with a huge acceleration to speed v_0, and then coasting along at speed v_0 for nearly all of the time T.

2.4. MULTIPLE-CHOICE ANSWERS

These two cases are shown in the v vs. t plots in Fig. 2.8. The area under the curve (which is the distance traveled) for the left curve is approximately zero, and the area under the right curve is approximately the area of the whole rectangle, which is $v_0 T$. This is the maximum possible distance, because an area larger than the $v_0 T$ rectangle would require that the v vs. t plot extend higher than v_0, which would then require a negative acceleration (contrary to the stated assumption) to bring the final speed back down to v_0.

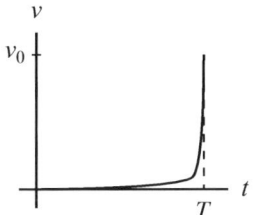
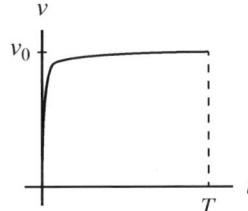

Figure 2.8

2.9. [d] The maximum distance is obtained by having acceleration a for a time $T/2$ and then deceleration $-a$ for a time $T/2$. The v vs. t plot is shown in Fig. 2.9. The distance traveled during the first $T/2$ is $a(T/2)^2/2 = aT^2/8$. Likewise for the second $T/2$, because the two triangles have the same area, and the area under a v vs. t curve is the distance traveled. So the total distance is $aT^2/4$.

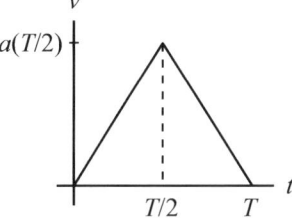

Alternatively, we see from the triangular plot that the average speed is half of the maximum v, which gives $v_{\text{avg}} = (aT/2)/2 = aT/4$. So the total distance traveled is $v_{\text{avg}} T = (aT/4)T = aT^2/4$.

Figure 2.9

Note that the triangle in Fig. 2.9 does indeed yield the maximum area under the curve (that is, the maximum distance traveled) subject to the given conditions, because (1) the triangle is indeed a possible v vs. t plot, and (2) velocities above the triangle aren't allowed, because the given maximum a implies that it would either be impossible to accelerate from zero initial speed to such a v, or impossible to decelerate to zero final speed from such a v.

2.10. [b] The distance x is certainly tiny, because the ball is still in contact with the club during the (quick) strike. The speed v is medium, because it is somewhere between the initial speed of zero and the final speed (on the order of 100 mph); it would be exactly half the final speed if the acceleration during the strike were constant. The acceleration is huge, because (assuming constant acceleration to get a rough idea) it is given by v/t, where v is medium and t is tiny (the strike is very quick).

REMARK: In short, the ball experiences a very large a for a very small t. The largeness and smallness of these quantities cancel each other and yield a medium result for the velocity $v = at$ (again, assuming constant a). But in the position $x = at^2/2$, the two factors of t win out over the one factor of a, and the result is tiny. These results (tiny x, medium v, and huge a) are consistent with Eq. (2.4), which for the present scenario says that $v^2 = 2ax$.

2.11. [c] From $d = at^2/2$ we obtain (using $g = 10\,\text{m/s}^2$)

$$1600\,\text{m} = \frac{1}{2}(10\,\text{m/s}^2)t^2 \implies t^2 = 320\,\text{s}^2 \implies t \approx 18\,\text{s}. \tag{2.8}$$

So 20 s is the best answer. The speed at this time is $gt \approx 10 \cdot 20 = 200\,\text{m/s}$, which is about 450 mph (see Multiple-Choice Question 1.4). In reality, air resistance is important, and a terminal velocity is reached. For a skydiver in a spread-eagle position, the terminal velocity is around 50 m/s.

2.12. [b] Let the time to the top be t. Since the ball decelerates on its way up, it moves faster in the first $t/2$ time span than in the second $t/2$. So it covers more than half the distance in the first $t/2$.

REMARK: If you want to find the exact ratio of the distances traveled in the two $t/2$ time spans, it is easiest to imagine dropping the ball instead of firing it upward; the answer is the same. In the upper $t/2$ of the motion, the ball falls $g(t/2)^2/2$, whereas in the total time t the ball falls $gt^2/2$. The ratio of these distances is 1 to 4, so the distance in the upper $t/2$ is 1/4 of the total, which means that the distance in the lower $t/2$ is 3/4 of the total. The ratio of the distances traveled in the two $t/2$ time spans is therefore 3 to 1. This also quickly follows from drawing a v vs. t plot like the one in Fig. 2.5.

2.13. $\boxed{d}$ If T is the time between the dropping of each ball (which is one second here), then the first ball has a speed of gT when the second ball is dropped. At a time t later than this, the speeds of the two balls are $g(t+T)$ and gt. So the difference in speeds is always gT. That is, the second ball always sees the first ball pulling away with a relative speed of gT. The separation therefore increases steadily at a rate gT.

This result ignores air resistance. In reality, the objects will reach the same terminal velocity (barring any influence of the first ball on the second), so the distance between them will approach a constant value. The real-life answer is therefore choice (c).

2.14. $\boxed{e}$ The velocity as a function of time is given by $v(t) = v_0 - gt$. Since the velocity is instantaneously zero at the highest point, the time to reach the top is $t = v_0/g$. The downward motion takes the same time as the upward motion (although it wouldn't if we included air resistance), so the total time is $2v_0/g$. Note that choices (a) and (b) don't have the correct units; choice (a) is the maximum height.

2.15. $\boxed{d}$ From general kinematics (see Problem 2.3), or from conservation of energy (the subject of Chapter 5), or from dimensional analysis, the maximum height is proportional to v^2/g (it equals $v^2/2g$). The v^2 dependence implies that the desired ratio is $2^2 = 4$. The difference in the masses is irrelevant.

2.5 Problem solutions

Although this was mentioned many times in the preface and in Chapter 1, it is worth belaboring the point: Don't look at the solution to a problem (or a multiple-choice question) too soon. If you do need to look at it, read it line by line until you get a hint to get going again. If you read through a solution without first solving the problem, you will gain essentially nothing from it!

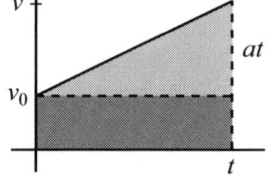

Figure 2.10

2.1. Area under the curve

The v vs. t curve, which is simply a tilted line in the case of constant acceleration, is shown in Fig. 2.10. The slope of the line equals the acceleration a, which implies that the height of the triangular region is at, as shown. The area under the v vs. t curve is the distance traveled. This area consists of the rectangle with area $t \cdot v_0$ and the triangle with area $(1/2) \cdot t \cdot at$. So the total area is $v_0 t + at^2/2$. To find the present position $x(t)$, we must add the initial position, x_0, to the distance traveled. The present position is therefore $x(t) = x_0 + v_0 t + at^2/2$, as desired.

2.2. A kinematic relation

FIRST SOLUTION: Our strategy will be to eliminate t from the equations in Eq. (2.3) by solving for t in the second equation and plugging the result into the third. This gives

$$\begin{aligned}
x &= x_0 + v_0 \left(\frac{v - v_0}{a}\right) + \frac{1}{2}a\left(\frac{v - v_0}{a}\right)^2 \\
&= x_0 + \frac{1}{a}(v_0 v - v_0^2) + \frac{1}{a}\left(\frac{v^2}{2} - v v_0 + \frac{v_0^2}{2}\right) \\
&= x_0 + \frac{1}{2a}(v^2 - v_0^2).
\end{aligned} \qquad (2.9)$$

2.5. PROBLEM SOLUTIONS

Hence $2a(x - x_0) = v^2 - v_0^2$. But $x - x_0$ is the displacement d. Changing the notation, $v \to v_f$ and $v_0 \to v_i$, gives the desired result, $2ad = v_f^2 - v_i^2$. A quick corollary is that if d and a have the same (or opposite) sign, then v_f is larger (or smaller) than v_i. You should convince yourself that this makes sense intuitively.

SECOND SOLUTION: A quicker derivation is the following. The displacement equals the average velocity times the time, by definition. The time is $t = (v - v_0)/a$, and the average velocity is $v_{avg} = (v + v_0)/2$, where this second expression relies on the fact that the acceleration is constant. (The first expression does too, because otherwise we wouldn't have a unique a in the denominator.) So we have

$$d = v_{avg} t = \left(\frac{v + v_0}{2}\right)\left(\frac{v - v_0}{a}\right) = \frac{v^2 - v_0^2}{2a}. \qquad (2.10)$$

Multiplying by $2a$ gives the desired result.

2.3. Maximum height

The solutions I can think of are listed below. Most of them use the fact that the time to reach the maximum height is $t = v_0/g$, which follows from the velocity $v(t) = v_0 - gt$ being zero at the top of the motion. The fact that the acceleration is constant also plays a critical role in all of the solutions.

1. Since the acceleration is constant, the average speed during the upward motion equals the average of the initial and final speeds. So $v_{avg} = (v_0 + 0)/2 = v_0/2$. The distance equals the average speed times the time, so $d = v_{avg} t = (v_0/2)(v_0/g) = v_0^2/2g$.

2. Using the standard expression for the distance traveled, $d = v_0 t - gt^2/2$, we have

$$d = v_0\left(\frac{v_0}{g}\right) - \frac{g}{2}\left(\frac{v_0}{g}\right)^2 = \frac{v_0^2}{2g}. \qquad (2.11)$$

3. If we imagine reversing time (or just looking at the downward motion, which takes the same time), then the ball starts at rest and accelerates downward at g. So we can use the simpler expression $d = gt^2/2$, which quickly gives $d = g(v_0/g)^2/2 = v_0^2/2g$.

4. The kinematic relation $v_f^2 - v_i^2 = 2ad$ from Eq. (2.4) gives $0^2 - v_0^2 = 2(-g)d \Longrightarrow d = v_0^2/2g$. We have been careful with the signs here; if we define positive d as upward, then the acceleration is negative.

5. The first three of the above solutions have graphical interpretations (although perhaps these shouldn't count as separate solutions). The v vs. t plots associated with these three solutions are shown in Fig. 2.11. The area under each curve, which is the distance traveled, equals $v_0^2/2g$.

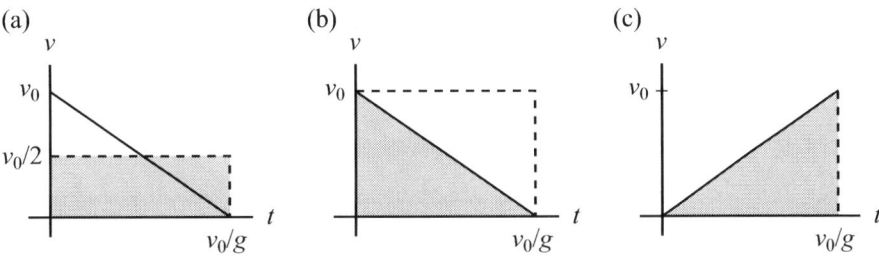

Figure 2.11

6. We can also use conservation of energy to solve this problem. Even though we won't discuss energy until Chapter 5, the solution is quick enough to state here. The

initial kinetic energy $mv_0^2/2$ gets completely converted into the gravitational potential energy mgd at the top of the motion (because the ball is instantaneously at rest at the top). So $mv_0^2/2 = mgd \implies d = v_0^2/2g$.

2.4. **Average speeds**

(a) Let the length of the hill be ℓ, and define $v \equiv 10$ mph. Then the time up the hill is ℓ/v, and the time down is $\ell/2v$. Your average speed is therefore

$$v_{\text{avg}} = \frac{d_{\text{total}}}{t_{\text{total}}} = \frac{2\ell}{\ell/v + \ell/2v} = \frac{2}{3/2v} = \frac{4v}{3} = 13.3 \text{ mph}. \tag{2.12}$$

(b) Let $2t$ be the total time of the ride, and again define $v \equiv 10$ mph. Then during the first half of the ride, you travel a distance vt. And during the second half, you travel a distance $(2v)t$. Your average speed is therefore

$$v_{\text{avg}} = \frac{d_{\text{total}}}{t_{\text{total}}} = \frac{vt + 2vt}{2t} = \frac{3v}{2} = 15 \text{ mph}. \tag{2.13}$$

REMARK: This result of 15 mph is simply the average of the two speeds, because you spend the *same amount of time* traveling at each speed. This is *not* the case in the scenario in part (a), because you spend longer (twice as long) traveling uphill at the slower speed. So that speed matters more when taking the average. In the extreme case where the two speeds differ greatly (in a multiplicative sense), the average speed in the scenario in part (a) is very close to twice the smaller speed (because the downhill time can be approximated as zero), whereas the average speed in the scenario in part (b) always equals the average of the two speeds. For example, if the two speeds are 1 and 100 (ignoring the units), then the answers to parts (a) and (b) are, respectively,

$$v_{\text{avg}}^{(a)} = \frac{2\ell}{\ell/1 + \ell/100} = \frac{200}{101} = 1.98,$$
$$v_{\text{avg}}^{(b)} = \frac{1 \cdot t + 100 \cdot t}{2t} = \frac{101}{2} = 50.5. \tag{2.14}$$

2.5. **Colliding trains**

The positions of the two trains are given by

$$x_A = d + \frac{1}{2}at^2 \quad \text{and} \quad x_B = v_0 t - \frac{1}{2}at^2. \tag{2.15}$$

These are equal when

$$d + \frac{1}{2}at^2 = v_0 t - \frac{1}{2}at^2 \implies at^2 - v_0 t + d = 0$$
$$\implies t = \frac{v_0 \pm \sqrt{v_0^2 - 4ad}}{2a}. \tag{2.16}$$

The trains *do* collide if there is a real solution for t, that is, if $v_0^2 > 4ad \implies v_0 > 2\sqrt{ad}$. The relevant solution is the "$-$" root. The "$+$" root corresponds to the case where the trains "pass through" each other and then meet up again a second time.

The trains *don't* collide if the roots are imaginary, that is, if $v_0 < 2\sqrt{ad}$. So the maximum value of v_0 that avoids a collision is $2\sqrt{ad}$. In the cutoff case where $v_0 = 2\sqrt{ad}$, the trains barely touch, so it's semantics as to whether you call that a "collision."

Note that $\sqrt{ad}$ correctly has the units of velocity. And in the limit of large a or d, the cutoff speed $2\sqrt{ad}$ is large, which makes intuitive sense.

A sketch of the x vs. t curves for the $v_0 = 2\sqrt{ad}$ case is shown in Fig. 2.12. If v_0 is smaller than $2\sqrt{ad}$, then the bottom curve stays lower (because its initial slope at the origin is v_0),

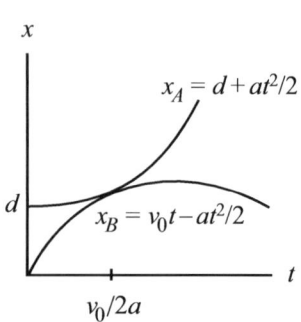

Figure 2.12

2.5. PROBLEM SOLUTIONS

so the curves don't intersect. If v_0 is larger than $2\sqrt{ad}$, then the bottom curve extends higher, so the curves intersect twice.

REMARKS: As an exercise, you can show that the location where the trains barely collide in the $v_0 = 2\sqrt{ad}$ case is $x = 3d/2$. And the maximum value of x_B is $2d$. If B has normal friction brakes, then it will of course simply stop at this maximum value and not move backward as shown in the figure. But in the hypothetical case of a jet engine with reverse thrust, B would head backward as the curve indicates.

In the $a \to 0$ limit, B moves with essentially constant speed v_0 toward A, which is essentially at rest, initially a distance d away. So the time is simply $t = d/v_0$. As an exercise, you can apply a Taylor series to Eq. (2.16) to produce this $t = d/v_0$ result. A Taylor series is required because if you simply set $a = 0$ in Eq. (2.16), you will obtain the unhelpful result of $t = 0/0$.

2.6. Ratio of distances

The positions of the two cars are given by

$$x_A = v_0 t \quad \text{and} \quad x_B = v_0 t - \frac{1}{2}at^2. \qquad (2.17)$$

B's velocity is $v_0 - at$, and this equals zero when $t = v_0/a$. The positions at this time are

$$x_A = v_0\left(\frac{v_0}{a}\right) = \frac{v_0^2}{a} \quad \text{and} \quad x_B = v_0\left(\frac{v_0}{a}\right) - \frac{1}{2}a\left(\frac{v_0}{a}\right)^2 = \frac{v_0^2}{2a}. \qquad (2.18)$$

The desired ratio is therefore $x_A/x_B = 2$. The plots are shown in Fig. 2.13. Both distances are proportional to v_0^2/a, so large v_0 implies large distances, and large a implies small distances. These make intuitive sense.

The only quantities that the ratio of the distances can depend on are v_0 and a. But the ratio of two distances is a dimensionless quantity, and there is no non-trivial combination of v_0 and a that gives a dimensionless result. Therefore, the ratio must simply be a number, independent of both v_0 and a.

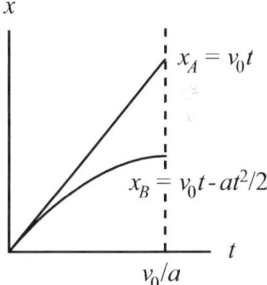

Figure 2.13

Note that it is easy to see from a v vs. t graph why the ratio is 2. The area under A's velocity curve (the rectangle) in Fig. 2.14 is twice the area under B's velocity curve (the triangle). And these areas are the distances traveled.

2.7. How far apart?

At time t, the first object has been moving for a time $t+T$, so its position is $x_1 = a(t+T)^2/2$. The second object has been moving for a time t, so its position is $x_2 = at^2/2$. The difference is

$$x_1 - x_2 = aTt + \frac{1}{2}aT^2. \qquad (2.19)$$

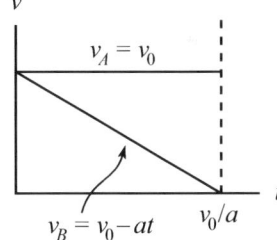

Figure 2.14

The second term here is the distance the first object has already traveled when the second object starts moving. The first term is the relative speed, aT, times the time. The relative speed is always aT because this is the speed the first object has when the second object starts moving. And from that time onward, both speeds increase at the same rate (namely a), so the objects always have the same relative speed. In summary, from the second object's point of view, the first object has a head start of $aT^2/2$ and then steadily pulls away with relative speed aT.

The v vs. t plots are shown in Fig. 2.15. The area under a v vs. t curve is the distance traveled, so the difference in the distances is the area of the shaded region. The triangular region on the left has an area equal to half the base times the height, which gives $T(aT)/2 = aT^2/2$. And the parallelogram region has an area equal to the horizontal width times the height, which gives $t(aT) = aTt$. These terms agree with Eq. (2.19).

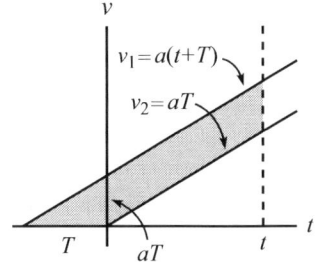

Figure 2.15

2.8. Ratio of odd numbers

This general result doesn't depend on the 1-second value of the time interval, so let's replace 1 second with a general time t. The *total* distances fallen after times of $0, t, 2t, 3t, 4t$, etc., are

$$0, \quad \frac{1}{2}gt^2, \quad \frac{1}{2}g(2t)^2, \quad \frac{1}{2}g(3t)^2, \quad \frac{1}{2}g(4t)^2, \quad \text{etc.} \quad (2.20)$$

The distances fallen *during* each interval of time t are the differences between the above distances, which yield

$$\frac{1}{2}gt^2, \quad 3 \cdot \frac{1}{2}gt^2, \quad 5 \cdot \frac{1}{2}gt^2, \quad 7 \cdot \frac{1}{2}gt^2, \quad \text{etc.} \quad (2.21)$$

These are in the desired ratio of $1 : 3 : 5 : 7\ldots$. Algebraically, the difference between $(nt)^2$ and $((n+1)t)^2$ equals $(2n+1)t^2$, and the $2n+1$ factor here generates the odd numbers.

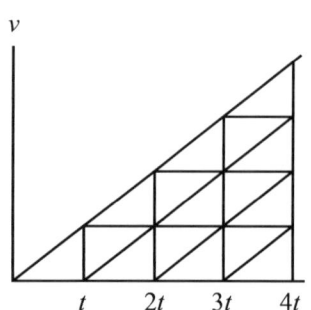

Figure 2.16

Geometrically, the v vs. t plot is shown in Fig. 2.16. The area under the curve (a tilted line in this case) is the distance traveled, and by looking at the number of (identical) triangles in each interval of time t, we quickly see that the ratio of the distances traveled in each interval is $1 : 3 : 5 : 7\ldots$.

2.9. Dropped and thrown balls

The positions of the two balls are given by

$$y_1(t) = h - \frac{1}{2}gt^2 \quad \text{and} \quad y_2(t) = v_0 t - \frac{1}{2}gt^2. \quad (2.22)$$

These are equal (that is, the balls collide) when $h = v_0 t \Longrightarrow t = h/v_0$. The height of the collision is then found from either of the y expressions to be $y_c = h - gh^2/2v_0^2$. This holds in any case, but we are given the further information that the second ball is instantaneously at rest when the collision occurs. Its speed is $v_0 - gt$, so the collision must occur at $t = v_0/g$. Equating this with the above $t = h/v_0$ result tells us that v_0 must be given by $v_0^2 = gh$. Plugging this into $y_c = h - gh^2/2v_0^2$ gives $y_c = h/2$.

The two velocities are given by $v_1(t) = -gt$, and $v_2(t) = v_0 - gt$. The difference of these is v_0. This holds for *all* time, not just at the moment when the balls collide. This is due to the fact that both balls are affected by gravity in exactly the same way, so the initial relative speed (which is v_0) equals the relative speed at any other time. This is evident from the v vs. t plots in Fig. 2.17. The upper line is v_0 above the lower line for all values of t.

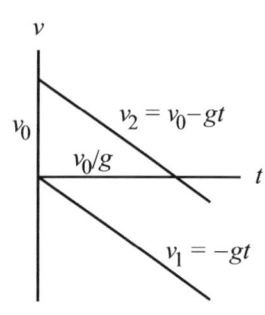

Figure 2.17

2.10. Hitting at the same time

The time it takes the first ball to hit the ground is given by

$$\frac{gt_1^2}{2} = h \quad \Longrightarrow \quad t_1 = \sqrt{\frac{2h}{g}}. \quad (2.23)$$

The time it takes the second ball to hit the ground is given by $vt_2 + gt_2^2/2 = 2h$. We could solve this quadratic equation for t_2 and then set the result equal to t_1. But a much quicker strategy is to note that since we want t_2 to equal t_1, we can just substitute t_1 for t_2 in the quadratic equation. This gives

$$v\sqrt{\frac{2h}{g}} + \frac{g}{2}\left(\frac{2h}{g}\right) = 2h \quad \Longrightarrow \quad v\sqrt{\frac{2h}{g}} = h \quad \Longrightarrow \quad v = \sqrt{\frac{gh}{2}}. \quad (2.24)$$

In the limit of small g, the process will take a long time, so it makes sense that v should be small. Note that without doing any calculations, the consideration of units tells us that the answer must be proportional to $\sqrt{gh}$.

2.5. PROBLEM SOLUTIONS

REMARK: The intuitive interpretation of the above solution is the following. If the second ball were dropped from rest, it would be at height h when the first ball hits the ground at time $\sqrt{2h/g}$ (after similarly falling a distance h). The second ball therefore needs to be given an initial downward speed v that causes it to travel an extra distance of h during this time. But this is just what the middle equation in Eq. (2.24) says.

2.11. Two dropped balls

The total time it takes the first ball to fall a height $4h$ is given by $gt^2/2 = 4h \implies t = 2\sqrt{2h/g}$. This time may be divided into the time it takes to fall a distance d (which is $\sqrt{2d/g}$), plus the remaining time it takes to hit the ground, which we are told is the same as the time it takes the second ball to fall a height h (which is $\sqrt{2h/g}$). Therefore,

$$2\sqrt{\frac{2h}{g}} = \sqrt{\frac{2d}{g}} + \sqrt{\frac{2h}{g}} \implies 2\sqrt{h} = \sqrt{d} + \sqrt{h} \implies d = h. \qquad (2.25)$$

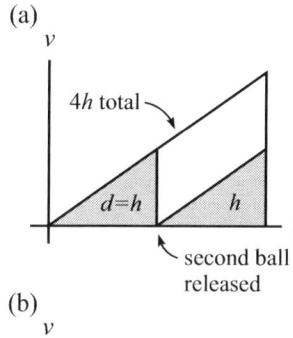
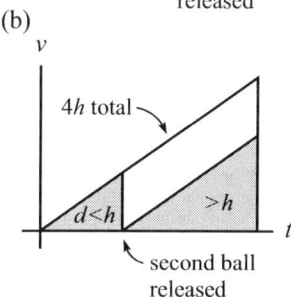

REMARK: Graphically, the process is shown in the v vs. t plot in Fig. 2.18(a). The area of the large triangle is the distance $4h$ the first ball falls. The right small triangle is the distance h the second ball falls, and the left small triangle is the distance d the first ball falls by the time the second ball is released. If, on the other hand, the second ball is released too soon, after the first ball has traveled a distance d that is less than h, then we have the situation shown in Fig. 2.18(b). The second ball travels a distance that is larger than h (assuming it can fall into a hole in the ground) by the time the first ball travels $4h$ and hits the ground. In other words (assuming there is no hole), the second ball hits the ground first. Conversely, if the second ball is released too late, then it travels a distance that is smaller than h by the time the first ball hits the ground. This problem basically boils down to the fact that freefall distances fallen are proportional to t^2 (or equivalently, to the areas of triangles), so twice the time means four times the distance.

Figure 2.18

Chapter 3

Kinematics in 2-D (and 3-D)

3.1 Introduction

In this chapter, as in the previous chapter, we won't be concerned with the actual forces that cause an object to move the way it is moving. We will simply take the motion as given, and our goal will be to relate positions, velocities, and accelerations as functions of time. However, since we are now dealing with more general motion in two and three dimensions, we will give one brief mention of forces:

Motion in more than one dimension

Newton's second law (for objects with constant mass) is $\mathbf{F} = m\mathbf{a}$, where $\mathbf{a} \equiv d\mathbf{v}/dt$. This law (which is the topic of Chapter 4) is a *vector* equation. (See Appendix A in Section 13.1 for a review of vectors.) So it really stands for three different equations: $F_x = ma_x$, $F_y = ma_y$, and $F_z = ma_z$. In many cases, these three equations are "decoupled," that is, the x equation has nothing to do with what is going on in the y and z equations, etc. In such cases, we simply have three copies of 1-D motion (or two copies if we're dealing with only two dimensions). So we just need to solve for the three *independent* motions along the three coordinate axes.

Projectile motion

The classic example of independent motions along different axes is projectile motion. Projectile motion is the combination of two separate linear motions. The horizontal motion doesn't affect the vertical motion, and vice versa. Since there is no acceleration in the horizontal direction (ignoring air resistance), the projectile moves with constant velocity in the x direction. And since there is an acceleration of $-g$ in the vertical direction, we can simply copy the results from the previous chapter (in particular, Eq. (2.3) with $a_y = -g$) for the motion in the y direction. We therefore see that if the initial position is (X, Y) and the initial velocity is (V_x, V_y), then the acceleration components

$$a_x = 0 \quad \text{and} \quad a_y = -g \quad (3.1)$$

lead to velocity components

$$v_x(t) = V_x \quad \text{and} \quad v_y(t) = V_y - gt \quad (3.2)$$

and position components

$$x(t) = X + V_x t \quad \text{and} \quad y(t) = Y + V_y t - \frac{1}{2}gt^2. \quad (3.3)$$

Projectile motion is completely described by these equations for the velocity and position components.

3.1. INTRODUCTION

Standard projectile results

The initial velocity **V** of a projectile is often described in terms of the initial speed v_0 (we'll use a lowercase v here, since it looks a little nicer) and the launch angle θ with respect to the horizontal. From Fig. 3.1, the initial velocity components are then $V_x = v_0 \cos\theta$ and $V_y = v_0 \sin\theta$, so the velocity components in Eq. (3.2) become

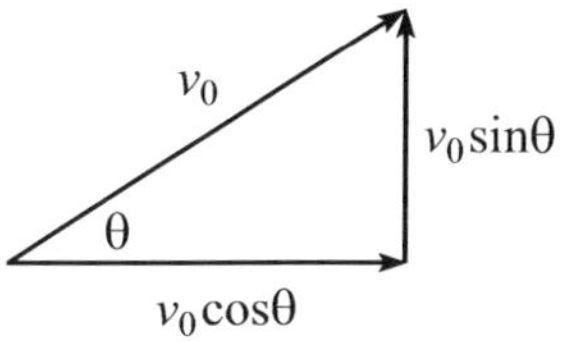

$$v_x(t) = v_0 \cos\theta \quad \text{and} \quad v_y(t) = v_0 \sin\theta - gt, \tag{3.4}$$

and the positions in Eq. (3.3) become (assuming that the projectile is fired from the origin, so that $(X, Y) = (0, 0)$)

$$x(t) = (v_0 \cos\theta)t \quad \text{and} \quad y(t) = (v_0 \sin\theta)t - \frac{1}{2}gt^2. \tag{3.5}$$

A few results that follow from these expressions are that the time to the maximum height, the maximum height attained, and the total horizontal distance traveled are given by (see Problem 3.1)

$$t_{\text{top}} = \frac{v_0 \sin\theta}{g}, \quad y_{\text{max}} = \frac{v_0^2 \sin^2\theta}{2g}, \quad x_{\text{max}} = \frac{2v_0^2 \sin\theta \cos\theta}{g} = \frac{v_0^2 \sin 2\theta}{g}. \tag{3.6}$$

The last of these results holds only if the ground is level (more precisely, if the projectile returns to the height from which it was fired). As usual, we are ignoring air resistance.

Figure 3.1

Motion along a plane

If an object slides down a frictionless plane inclined at angle θ, the acceleration down the plane is $g \sin\theta$, because the component of **g** (the downward acceleration due to gravity) that points along the plane is $g \sin\theta$; see Fig. 3.2. (There is no acceleration perpendicular to the plane because the normal force from the plane cancels the component of the gravitational force perpendicular to the plane. We'll discuss forces in Chapter 4.) Even though the motion appears to take place in 2-D, we really just have a (tilted) 1-D setup. We effectively have "freefall" motion along the tilted axis, with the acceleration due to gravity being $g \sin\theta$ instead of g. If $\theta = 0$, then the $g \sin\theta$ acceleration along the plane equals 0, and if $\theta = 90°$ it equals g (downward), as expected.

More generally, if a projectile flies through the air above an inclined plane, the object's acceleration (which is the downward-pointing vector **g**) can be viewed as the sum of its components along any choice of axes, in particular the $g \sin\theta$ acceleration along the plane and the $g \cos\theta$ acceleration perpendicular to the plane. This way of looking at the downward **g** vector can be very helpful when solving projectile problems involving inclined planes. See Section 13.1.5 in Appendix A for further discussion of vector components.

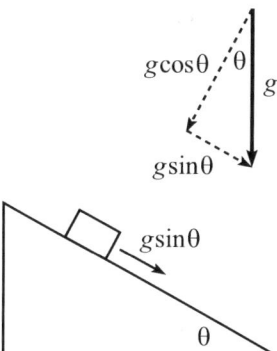

Figure 3.2

Circular motion

Another type of 2-D motion is circular motion. If an object is moving in a circle of radius r with speed v at a given instant, then the (inward) radial component of the acceleration vector **a** equals (see Problem 3.2(a))

$$a_r = \frac{v^2}{r}. \tag{3.7}$$

This radially inward acceleration is called the *centripetal* acceleration. If additionally the object is speeding up or slowing down as it moves around the circle, then there is also a tangential component of **a** given by (see Problem 3.2(b))

$$a_t = \frac{dv}{dt}. \tag{3.8}$$

This tangential component is the more intuitive of the two components of the acceleration; it comes from the change in the speed v, just as in the simple case of 1-D motion. The a_r component is the less intuitive one; it comes from the change in the *direction* of **v**. Remember that the acceleration $\mathbf{a} \equiv d\mathbf{v}/dt$ involves the rate of change of the entire vector **v**, not just the magnitude $v \equiv |\mathbf{v}|$. A vector can change because its magnitude changes or because its direction changes (or both). The former change is associated with a_t, while the latter is associated with a_r.

It is sometimes convenient to work with the *angular frequency* ω (also often called the *angular speed* or *angular velocity*), which is defined to be the rate at which the angle θ around the circle (measured in radians) is swept out. That is, $\omega \equiv d\theta/dt$. If we multiply both sides of this equation by the radius r, we obtain $r\omega = d(r\theta)/dt$. But $r\theta$ is simply the distance s traveled along the circle,[1] so the right-hand side of this equation is ds/dt, which is just the tangential speed v. Hence $r\omega = v \implies \omega = v/r$. In terms of ω, the radial acceleration can be written as

$$a_r = \frac{v^2}{r} = \frac{(r\omega)^2}{r} = \omega^2 r. \qquad (3.9)$$

Similarly, we can define the *angular acceleration* as $\alpha \equiv d\omega/dt \equiv d^2\theta/dt^2$. If we multiply through by r, we obtain $r\alpha = d(r\omega)/dt$. But from the preceding paragraph, $r\omega$ is the tangential speed v. Therefore, $r\alpha = dv/dt$. And since the right-hand side of this equation is just the tangential acceleration, we have

$$a_t = r\alpha. \qquad (3.10)$$

We can summarize most of the results in the previous two paragraphs by saying that the "linear" quantities (distance s, speed v, tangential acceleration a_t) are related to the angular quantities (angle θ, angular speed ω, angular acceleration α) by a factor of r:

$$s = r\theta, \qquad v = r\omega, \qquad a_t = r\alpha. \qquad (3.11)$$

However, the radial acceleration a_r doesn't fit into this pattern.

3.2 Multiple-choice questions

3.1. A bullet is fired horizontally from a gun, and another bullet is simultaneously dropped from the same height. Which bullet hits the ground first? (Ignore air resistance, the curvature of the earth, etc.)

   (a) the fired bullet

   (b) the dropped bullet

   (c) They hit the ground at the same time.

3.2. A projectile is fired at an angle θ with respect to level ground. Is there a point in the motion where the velocity is perpendicular to the acceleration?

   Yes   No

3.3. A projectile is fired at an angle θ with respect to level ground. Does there exist a θ such that the maximum height attained equals the total horizontal distance traveled?

   Yes   No

3.4. Is the following reasoning correct? If the launch angle θ of a projectile is increased (while keeping v_0 the same), then the initial v_y velocity component increases, so the time in the air increases, so the total horizontal distance traveled increases.

   Yes   No

---

[1]This is true by the definition of a radian. If you take a piece of string with a length of one radius and lay it out along the circumference of a circle, then it subtends an angle of one radian, by definition. So each radian of angle is worth one radius of distance. The total distance s along the circumference is therefore obtained by multiplying the number of radians (that is, the number of "radiuses") by the length of the radius.

3.2. MULTIPLE-CHOICE QUESTIONS

3.5. A ball is thrown at an angle θ with speed v_0. A second ball is simultaneously thrown straight upward from the point on the ground directly below the top of the first ball's parabolic motion. How fast should this second ball be thrown if you want it to collide with the first ball?

(a) $v_0/2$ (b) $v_0/\sqrt{2}$ (c) v_0 (d) $v_0 \cos\theta$ (e) $v_0 \sin\theta$

3.6. A wall has height h and is a distance ℓ away. You wish to throw a ball over the wall with a trajectory such that the ball barely clears the wall at the top of its parabolic motion. What initial speed is required? (Don't solve this from scratch, just check special cases. See Problem 3.10 for a quantitative solution.)

(a) $\sqrt{2gh}$

(b) $\sqrt{4gh}$

(c) $\sqrt{g\ell^2/2h}$

(d) $\sqrt{2gh + g\ell^2/2h}$

(e) $\sqrt{4gh + g\ell^2/2h}$

3.7. Two balls are thrown with the same speed v_0 from the top of a cliff. The angles of their initial velocities are θ above and below the horizontal, as shown in Fig. 3.3. How much farther along the ground does the top ball hit than the bottom ball? *Hint*: The two trajectories have a part in common. No calculations necessary!

(a) $2v_0^2/g$

(b) $2v_0^2 \sin\theta/g$

(c) $2v_0^2 \cos\theta/g$

(d) $2v_0^2 \sin\theta \cos\theta/g$

(e) $2v_0^2 \sin^2\theta \cos^2\theta/g$

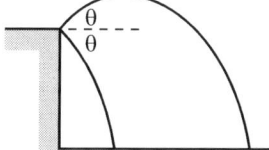

Figure 3.3

3.8. A racecar travels in a horizontal circle at constant speed around a circular banked track. A side view is shown in Fig. 3.4. (The triangle is a cross-sectional slice of the track; the car is heading into the page at the instant shown.) The direction of the racecar's acceleration is

(a) horizontal rightward

(b) horizontal leftward

(c) downward along the plane

(d) upward perpendicular to the plane

(e) The acceleration is zero.

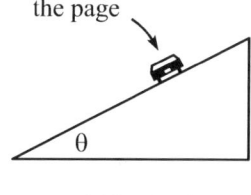

Figure 3.4

3.9. Which one of the following statements is *not* true for *uniform* (constant speed) circular motion?

(a) **v** is perpendicular to **r**.

(b) **v** is perpendicular to **a**.

(c) **v** has magnitude $R\omega$ and points in the **r** direction.

(d) **a** has magnitude v^2/R and points in the negative **r** direction.

(e) **a** has magnitude $\omega^2 R$ and points in the negative **r** direction.

3.10. A car travels around a horizontal circular track, *not* at constant speed. The acceleration vectors at five different points are shown in Fig. 3.5 (the four nonzero vectors have equal length). At which of these points is the car's speed the largest?

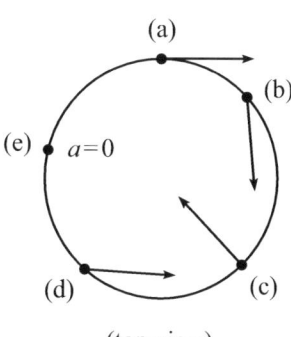

Figure 3.5

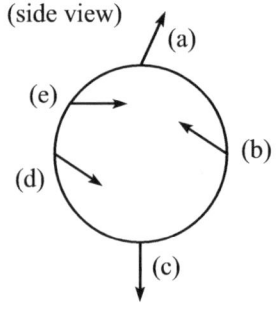

Figure 3.6

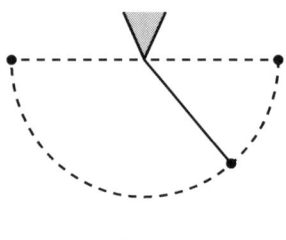

(side view)

Figure 3.7

3.11. A bead is given an initial velocity and then circles indefinitely around a frictionless vertical hoop. Only one of the vectors in Fig. 3.6 is a possible acceleration vector at the given point. Which one?

3.12. A pendulum is released from rest at an angle of 45° with respect to the vertical, as shown below. Which vector shows the direction of the initial acceleration?

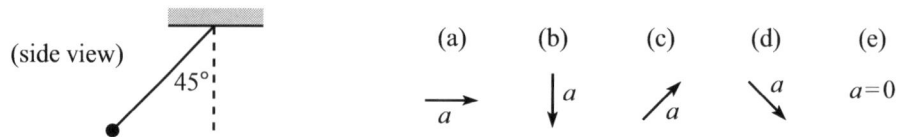

3.13. A pendulum swings back and forth between the two horizontal positions shown in Fig. 3.7. The acceleration is vertical (g downward) at the highest points, and is also vertical (upward) at the lowest point.

   (a) There is at least one additional point where the acceleration is vertical.
   (b) There is at least one point where the acceleration is horizontal.
   (c) There is at least one point where the acceleration is zero.
   (d) None of the above

3.3 Problems

The first three problems are foundational problems.

3.1. A few projectile results

On level ground, a projectile is fired at angle θ with speed v_0. Derive the expressions in Eq. (3.6). That is, find (a) the time to the maximum height, (b) the maximum height attained, and (c) the total horizontal distance traveled.

3.2. Radial and tangential accelerations

(a) If an object moves in a circle at constant speed v (uniform circular motion), show that the acceleration points radially inward with magnitude $a_r = v^2/r$. Do this by drawing the position and velocity vectors at two nearby times and then making use of similar triangles.

(b) If the object speeds up or slows down as it moves around in the circle, then the acceleration also has a tangential component. Show that this component is given by $a_t = dv/dt$.

3.3. Radial and tangential accelerations, again

A particle moves in a circle, not necessarily at constant speed. Its coordinates are given by $(x, y) = (R\cos\theta, R\sin\theta)$, where $\theta \equiv \theta(t)$ is an arbitrary function of t. Take two time derivatives of these coordinates to find the acceleration vector, and then explain why the result is consistent with the a_r and a_t magnitudes derived in Problem 3.2.

3.4. Movie replica

(a) A movie director wants to shoot a certain scene by building a detailed replica of the actual setup. The replica is 1/100 the size of the real thing. In the scene, a person jumps from rest from a tall building (into a net, so it has a happy ending). If the director films a tiny doll being dropped from the replica building, by what factor should the film be sped up or slowed down when played back, so that the falling person looks realistic to someone watching the movie? (Assume that the motion is essentially vertical.)

3.3. PROBLEMS

(b) The director now wants to have a little toy car zoom toward a cliff in the replica (with the same scale factor of 1/100) and then sail over the edge down to the ground below (don't worry, the story has the driver bail out in time). Assume that the goal is to have the movie viewer think that the car is traveling at 50 mph before it goes over the cliff. As in part (a), by what factor should the film be sped up or slowed down when played back? What should the speed of the toy car be as it approaches the cliff in the replica?

3.5. Doubling gravity

A ball is thrown with speed v at an angle θ with respect to the horizontal ground. At the highest point in the motion, the strength of gravity is somehow magically doubled. What is the total horizontal distance traveled by the ball?

3.6. Ratio of heights

From the standard $d = gt^2/2$ expression for freefall from rest, we see that if the falling time is doubled, the falling distance is quadrupled. Use this fact to find the ratio of the height of the top of projectile motion (point A in Fig. 3.8) to the height where the projectile would be if gravity were turned off (point B in the figure). Two suggestive distances are drawn.

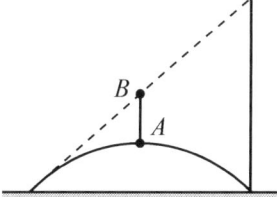

Figure 3.8

3.7. Hitting horizontally

A ball is thrown with speed v_0 at an angle θ with respect to the horizontal. It is thrown from a point that is a distance ℓ from the base of a cliff that has a height also equal to ℓ. What should θ and v_0 be so that the ball hits the corner of the cliff moving horizontally, as shown in Fig. 3.9?

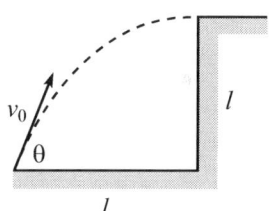

Figure 3.9

3.8. Projectile and tube

A projectile is fired horizontally with speed v_0 from the top of a cliff of height h. It immediately enters a fixed tube with length x, as shown in Fig. 3.10. There is friction between the projectile and the tube, the effect of which is to make the projectile decelerate with constant acceleration $-a$ (a is a positive quantity here). After the projectile leaves the tube, it undergoes normal projectile motion down to the ground.

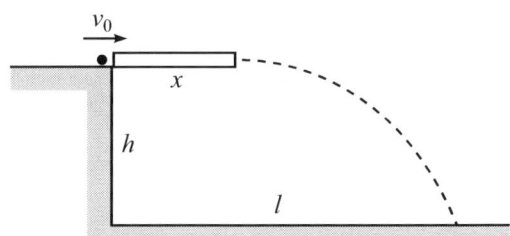

Figure 3.10

(a) What is the total horizontal distance (call it ℓ) that the projectile travels, measured from the base of the cliff? Give your answer in terms of x, h, v_0, g, and a.

(b) What value of x yields the maximum value of ℓ?

3.9. Car in the mud

A wheel is stuck in the mud, spinning in place. The radius is R, and the points on the rim are moving with speed v. Bits of the mud depart from the wheel at various random locations. In particular, some bits become unstuck from the rim in the upper left quadrant, as shown in Fig. 3.11. What should θ be so that the mud reaches the maximum possible height (above the ground) as it flies through the air? What is this maximum height? You may assume $v^2 > gR$.

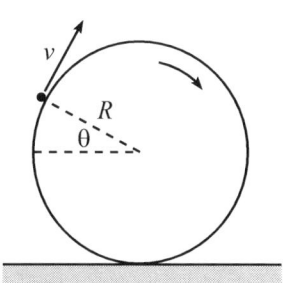

Figure 3.11

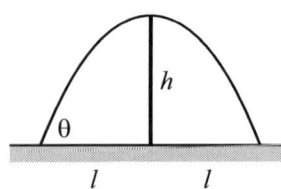

Figure 3.12

Figure 3.13

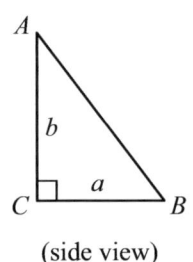

(side view)

Figure 3.14

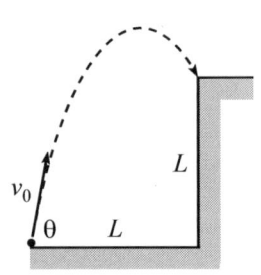

Figure 3.15

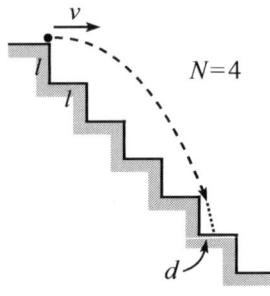

Figure 3.16

3.10. **Clearing a wall**

(a) You wish to throw a ball to a friend who is a distance 2ℓ away, and you want the ball to just barely clear a wall of height h that is located halfway to your friend, as shown in Fig. 3.12. At what angle θ should you throw the ball?

(b) What initial speed v_0 is required? What value of h (in terms of ℓ) yields the minimum v_0? What is the value of θ in this minimum case?

3.11. **Bounce throw**

A person throws a ball with speed v_0 at a 45° angle and hits a given target. How much quicker does the ball get to the target if the person instead throws the ball with the same speed v_0 but at the angle that makes the trajectory consist of two identical bumps, as shown in Fig. 3.13? (Assume unrealistically that there is no loss in speed at the bounce.)

3.12. **Maximum bounce**

A ball is dropped from rest at height h. At height y, it bounces elastically (that is, without losing any speed) off a board. The board is inclined at the angle (which happens to be 45°) that makes the ball bounce off horizontally. In terms of h, what should y be so that the ball hits the ground as far off to the side as possible? What is the horizontal distance in this optimal case?

3.13. **Falling along a right triangle**

In the vertical right triangle shown in Fig. 3.14, a particle falls from A to B either along the hypotenuse, or along the two legs (lengths a and b) via point C. There is no friction anywhere.

(a) What is the time (call it t_H) if the particle travels along the hypotenuse?

(b) What is the time (call it t_L) if the particle travels along the legs? Assume that at point C there is an infinitesimal curved arc that allows the direction of the particle's motion to change from vertical to horizontal without any change in speed.

(c) Verify that $t_H = t_L$ when $a = 0$.

(d) How do t_H and t_L compare in the limit $b \ll a$?

(e) Excluding the $a = 0$ case, what triangle shape yields $t_H = t_L$?

3.14. **Throwing to a cliff**

A ball is thrown at an angle θ up to the top of a cliff of height L, from a point a distance L from the base, as shown in Fig. 3.15.

(a) As a function of θ, what initial speed causes the ball to land right at the edge of the cliff?

(b) There are two special values of θ for which you can check your result. Check these.

3.15. **Throwing from a cliff**

A ball is thrown with speed v at angle θ (with respect to horizontal) from the top of a cliff of height h. How far from the base of the cliff does the ball land? (The ground is horizontal below the cliff.)

3.16. **Throwing on stairs**

A ball is thrown horizontally with speed v from the floor at the top of some stairs. The width and height of each step are both equal to ℓ.

(a) What should v be so that the ball barely clears the corner of the step that is N steps down? Fig. 3.16 shows the case where $N = 4$.

(b) How far along the next step (the distance d in the figure) does the ball hit?

3.3. PROBLEMS

(c) What is d in the limit $N \to \infty$?

(d) Find the components of the ball's velocity when it grazes the corner, and then explain why their ratio is consistent with your answer to part (c).

3.17. Bullet and sphere

A bullet is fired horizontally with speed v_0 from the top of a fixed sphere with radius R, as shown in Fig. 3.17. What is the minimum value of v_0 for which the bullet doesn't touch the sphere after it is fired? (*Hint*: Find y as a function of x for the projectile motion, and also find y as a function of x for the sphere near the top where x is small; you'll need to make a Taylor-series approximation. Then compare your two results.) For the v_0 you just found, where does the bullet hit the ground?

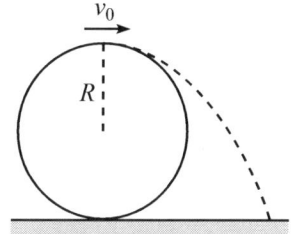

Figure 3.17

3.18. Throwing on an inclined plane

You throw a ball from a plane inclined at angle θ. The initial velocity is perpendicular to the plane, as shown in Fig. 3.18. Consider the point P on the trajectory that is farthest from the plane. For what angle θ does P have the same height as the starting point? (For the case shown in the figure, P is higher.) Answer this in two steps:

(a) Give a continuity argument that explains why such a θ should in fact exist.

(b) Find θ. In getting a handle on where (and when) P is, it is helpful to use a tilted coordinate system and to isolate what is happening in the direction perpendicular to the plane.

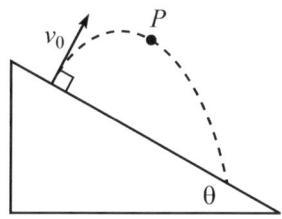

Figure 3.18

3.19. Ball landing on a block

A block is fired up along a frictionless plane inclined at angle β, and a ball is simultaneously thrown upward at angle θ (both β and θ are measured with respect to the horizontal). The objects start at the same location, as shown in Fig. 3.19. What should θ be in terms of β if you want the ball to land on the block at the instant the block reaches its maximum height on the plane? (An implicit equation is fine.) What is θ if β equals $45°$? (You might think that we've forgotten to give you information about the initial speeds, but it turns out that you don't need these to solve the problem.)

3.20. g's in a washer

A typical front-loading washing machine might have a radius of 0.3 m and a spin cycle of 1000 revolutions per minute. What is the acceleration of a point on the surface of the drum at this spin rate? How many g's is this equivalent to?

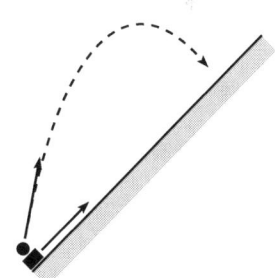

Figure 3.19

3.21. Acceleration after one revolution

A car starts from rest on a circular track with radius R and then accelerates with constant tangential acceleration a_t. At the moment the car has completed one revolution, what angle does the total acceleration vector make with the radial direction? You should find that your answer doesn't depend on a_t or R. Explain why you don't have to actually solve the problem to know this.

3.22. Equal acceleration components

An object moves in a circular path of radius R. At $t = 0$, it has speed v_0. From this point on, the magnitudes of the radial and tangential accelerations are arranged to be equal at all times.

(a) As functions of time, find the speed and the distance traveled.

(b) If the tangential acceleration is positive (that is, if the object is speeding up), there is special value for t. What is it, and why is it special?

50   CHAPTER 3. KINEMATICS IN 2-D (AND 3-D)

3.23. **Horizontal acceleration**

A bead is at rest at the top of a fixed frictionless hoop of radius R that lies in a vertical plane. The bead is given an infinitesimal push so that it slides down and around the hoop. Find all the points on the hoop where the bead's acceleration is horizontal. (We haven't covered conservation of energy yet, but use the fact that the bead's speed after it has fallen through a height h is given by $v = \sqrt{2gh}$.)

3.4 Multiple-choice answers

3.1. $\boxed{c}$ This setup is perhaps the most direct example of the independence of horizontal and vertical motions, under the influence of only gravity. The horizontal motion doesn't affect the vertical motion, and vice versa. The gravitational force causes both objects to have the same acceleration g downward, so their heights are both given by $h - gt^2/2$. The fired bullet might travel a mile before it hits the ground, but will still take a time of $t = \sqrt{2h/g}$, just like the dropped ball.

3.2. $\boxed{\text{Yes}}$ At the highest point in the projectile motion, the velocity is sideways, and the acceleration is (always) downward.

REMARK: Since **a** (which is **g** $= -g\hat{\mathbf{y}}$) is perpendicular to **v** at the top of the motion, the component of **a** in the direction of **v** is zero. But this component is what causes a change in the speed (this is just the $a_t = dv/dt$ statement). So $dv/dt = 0$ at the top of the motion. This makes sense because on the way up, the speed decreases (from a tilted v_0 to a horizontal $v_0 \cos\theta$); **a** has a component in the negative **v** direction. And on the way down, the speed increases (from a horizontal $v_0 \cos\theta$ to a tilted v_0); **a** has a component in the positive **v** direction. So at the top of the motion, the speed must be neither increasing nor decreasing. That is, $dv/dt = 0$. On the other hand, the vertical v_y component of the velocity steadily decreases (at a rate of $-g$) during the entire flight, from $v_0 \sin\theta$ to $-v_0 \sin\theta$.

3.3. $\boxed{\text{Yes}}$ If θ is very small, then the projectile barely climbs above the ground, so the total horizontal distance traveled is much larger than the maximum height. In the other extreme where θ is close to $90°$, the projectile goes nearly straight up and down, so the maximum height is much larger than the total horizontal distance. By continuity, there must exist an intermediate angle for which the maximum height equals the total horizontal distance. As an exercise, you can show that this angle is given by $\tan\theta = 4 \implies \theta \approx 76°$.

3.4. $\boxed{\text{No}}$ The reasoning is not valid for all θ. For all θ, the reasoning is correct up until the last "so." The time t in the air does indeed increase as θ increases (it equals $2v_0 \sin\theta/g$). However, an additional consequence of increasing θ is that the v_x velocity component (which equals $v_0 \cos\theta$) *decreases*. The total horizontal distance equals $v_x t$, so there are competing effects: increasing t vs. decreasing v_x. If we invoke the standard result that the maximum distance is obtained when $\theta = 45°$ (see the solution to Problem 3.1), we see that for $\theta < 45°$, the increase in t wins and the distance increases; but for $\theta > 45°$, the decrease in v_x wins and the distance decreases.

3.5. $\boxed{e}$ The vertical velocity component of the first ball is $v_0 \sin\theta$. If the second ball is thrown with this speed, then it will always have the same height as a function of time as the first ball. The balls will therefore collide when the first ball's horizontal position coincides with the second ball's (at the top of the parabolic motion).

REMARK: The initial location of the second ball on the ground is actually irrelevant. As long as it is thrown simultaneously with speed $v_0 \sin\theta$, it can be thrown from *any* point below the parabolic motion of the first ball, and the balls will still always have the same heights at any moment. They will therefore collide when the first ball's horizontal position coincides with the second ball's. If the collision occurs during the second half of the parabolic motion, the balls will be on their way down.

3.6. $\boxed{d}$ The answer certainly depends on ℓ, because the speed must be very large if ℓ is very large. So choices (a) and (b) are ruled out. Alternatively, these two choices can be ruled out by noting that in the $h \to 0$ limit, you must throw the ball infinitely fast. This is true because you must throw the ball at a very small angle; so if the speed weren't large, the initial vertical velocity would be very small, which means that the top of the parabolic motion would occur too soon.

In the $\ell \to 0$ limit, you are throwing the ball straight up. And the initial speed in this case is the standard $v = \sqrt{2gh}$. (This can be derived in many ways, for example by using Eq. (2.4).) So the answer must be (d).

3.7. $\boxed{d}$ When the top ball returns to the initial height (the height of the cliff), its velocity (both magnitude and direction) will be the same as the initial velocity of the bottom ball (speed v_0 at an angle θ below the horizontal). So the trajectory from that point onward will look exactly the same as the entire trajectory of the bottom ball. So the difference in the trajectories is just the symmetric parabola that lies above the initial height. And from Eq. (3.6) we know that the horizontal distance traveled in this part is $2v_0^2 \sin\theta \cos\theta/g$.

REMARK: The first three choices can be eliminated by checking limiting cases. The answer must be zero in both the $\theta = 0$ case (the trajectories are the same) and the $\theta = 90°$ case (both balls travel vertically and hence have the same horizontal distance of zero). So the answer must be (d) or (e). But it takes the above reasoning to show that (d) is correct.

3.8. $\boxed{b}$ The acceleration has magnitude v^2/r and points toward the center of the circular motion. Since the car is traveling in a horizontal circle, the radial direction is to the left. So the acceleration is horizontal leftward.

REMARK: As long as we are told that the racecar is undergoing uniform (constant speed) circular motion, there is no need to know anything about the various forces acting on the car (which happen to be gravity, normal, and friction; we'll discuss forces in Chapter 4). The acceleration for uniform circular motion, no matter what the cause of the motion, points radially inward with magnitude v^2/r, period.

3.9. $\boxed{c}$ The velocity $\mathbf{v}$ points in the tangential, not radial, direction. The other four statements are all true.

REMARK: If you want to consider *non-uniform* (that is, changing speed) circular motion, then only statement (a) is always true. The acceleration can now have a tangential component, which ruins (b), (d), and (e). And statement (c) is still incorrect.

3.10. $\boxed{c}$ The radial component of the acceleration has magnitude $a_r = v^2/r$. So the largest speed v corresponds to the $\mathbf{a}$ with the largest (inward) radial component, which is choice (c).

REMARK: Note that the tangential component of the acceleration, which is $a_t = dv/dt$, has nothing to do with the *instantaneous* value of v, which is what we're concerned with in this question. A large a_t component (as in choices (a), (b), and (d)) does not imply a large v. On the other hand, a zero a_r component (as in choices (a) and (e)) implies a zero v.

3.11. $\boxed{d}$ The acceleration is the vector sum of the radially inward $a_r = v^2/r$ component and the tangentially downward $a_t = g \sin\theta$ component, where $\theta = 0$ corresponds to the top of the hoop. (This is just the component of g that points in the tangential direction.) Only choice (d) satisfies both of these properties. Choice (e) is the trickiest. The acceleration can't be horizontal there, because both a_r and a_t have downward components. There *is* a point in each lower quadrant where the acceleration is horizontal, because in the bottom half of the circle, a_r has an *upward* component which can cancel the downward component of a_t at two particular points.

REMARK: Since the radial $a_r = v^2/r$ component always points radially *inward*, the acceleration vector in any arbitrary circular motion can never have a radially *outward* component. This immediately rules out choices (a) and (c). The borderline case occurs when $a_r = 0$, that is, when $\mathbf{a}$ points tangentially. In this case $v = 0$, so the bead is instantaneously at rest. But any nonzero speed at all will cause an inward a_r component.

3.12. $\boxed{d}$ The tangential component of the acceleration is $a_t = g \sin 45°$. And the radial component is $a_r = v^2/r = 0$, since $v = 0$ at the start. No matter where the pendulum is released *from rest*, the initial acceleration is always tangential (or zero, if it is "released" when hanging vertically), because $a_r = 0$ when $v = 0$.

3.13. $\boxed{b}$ Since the acceleration is negative vertical at the highest points and positive vertical at the lowest point, by continuity it must have zero vertical component somewhere in between. That is, it must be horizontal somewhere in between.

The acceleration is never vertical (except at the highest and lowest points), because the $\mathbf{a}_r$ and $\mathbf{a}_t$ vectors either both have rightward components, or both have leftward components, which means that the x component of the total acceleration vector $\mathbf{a}$ is nonzero. This is consistent (if we invoke $F = ma$) with the fact that the tension in the tilted string has a nonzero horizontal component (except at the highest points where the tension is zero and the lowest point where the string is vertical).

3.5 Problem solutions

3.1. A few projectile results

(a) The components of the velocity and position are given in Eqs. (3.4) and (3.5). At the highest point in the motion, v_y equals zero because the projectile is instantaneously moving horizontally. So Eq. (3.4) gives the time to the highest point as $t_{\text{top}} = v_0 \sin \theta / g$.

(b) FIRST SOLUTION: Plugging t_{top} into Eq. (3.5) gives the maximum height as

$$y_{\max} = v_0 \sin\theta \left(\frac{v_0 \sin\theta}{g}\right) - \frac{1}{2} g \left(\frac{v_0 \sin\theta}{g}\right)^2 = \frac{v_0^2 \sin^2\theta}{2g}. \qquad (3.12)$$

This is just the $v_0^2/2g$ result from Problem 2.3, with v_0 replaced with the vertical component of the velocity, $v_0 \sin \theta$.

SECOND SOLUTION: We can imagine reversing time (or equivalently, looking at the second half of the motion), in which case the motion is equivalent to (at least as far as the y motion is concerned) an object dropped from rest. We know that the time it takes to reach the ground is $t_{\text{top}} = v_0 \sin\theta/g$, so the distance is $g t_{\text{top}}^2/2 = g(v_0 \sin\theta/g)^2/2 = v_0^2 \sin^2\theta/(2g)$, in agreement with Eq. (3.12).

(c) FIRST SOLUTION: Because the ground is level, the up and down parts of the motion are symmetrical, so the total time t in the air is twice the time to the top, that is, $t = 2 t_{\text{top}} = 2 v_0 \sin\theta/g$. From Eq. (3.5) the total horizontal distance traveled is then

$$x_{\max} = v_0 \cos\theta \left(\frac{2 v_0 \sin\theta}{g}\right) = \frac{2 v_0^2 \sin\theta \cos\theta}{g}. \qquad (3.13)$$

SECOND SOLUTION: The total time in the air can be determined by finding the value of t for which $y(t) = 0$. From Eq. (3.5), we quickly obtain $t = 2 v_0 \sin\theta/g$, as we found in the first solution. (A second value of t that makes $y = 0$ in Eq. (3.5) is $t = 0$, of course, because the projectile is on the ground at the start.) Note that if the ground isn't level, then the up and down parts of the motion are *not* symmetrical. So this alternative method of finding the total time (by finding the time for which y takes on a particular value) must be used.

3.5. PROBLEM SOLUTIONS

REMARK: If we use the double-angle formula $\sin 2\theta = 2\sin\theta\cos\theta$, the expression for $x_{\max}$ in Eq. (3.13) can alternatively be written as $x_{\max} = v_0^2 \sin 2\theta / g$. Since $\sin 2\theta$ achieves its maximum value when $2\theta = 90°$, this form makes it immediately clear that for a given speed v_0, the maximum horizontal distance is achieved when $\theta = 45°$. This maximum distance is v_0^2/g. Note that consideration of units tells us that the maximum distance must be proportional to v_0^2/g. But a calculation is necessary to show that the multiplicative factor is 1.

The $v_0^2 \sin 2\theta/g$ form of $x_{\max}$ makes it clear (although it is also clear from the $2v_0^2 \sin\theta\cos\theta/g$ form) that the distance is symmetric on either side of $45°$. That is, $46°$ yields the same distance as $44°$, and $80°$ yields the same distance as $10°$, etc.

3.2. Radial and tangential accelerations

(a) The position and velocity vectors at two nearby times are shown in Fig. 3.20. Their differences, $\Delta \mathbf{r} \equiv \mathbf{r}_2 - \mathbf{r}_1$ and $\Delta \mathbf{v} \equiv \mathbf{v}_2 - \mathbf{v}_1$, are shown in Fig. 3.21. (See Figs. 13.9 and 13.10 in Appendix A for a comment on this $\Delta \mathbf{v}$.) The angle between the $\mathbf{v}$'s is the same as the angle between the $\mathbf{r}$'s, because each $\mathbf{v}$ makes a right angle with the corresponding $\mathbf{r}$. Therefore, the triangles in Fig. 3.21 are similar, and we have

$$\frac{|\Delta \mathbf{v}|}{v} = \frac{|\Delta \mathbf{r}|}{r}, \quad (3.14)$$

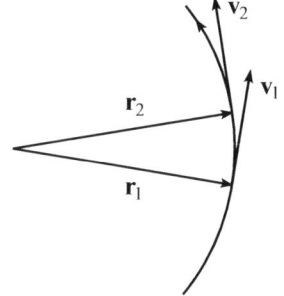

Figure 3.20

where $v \equiv |\mathbf{v}|$ and $r \equiv |\mathbf{r}|$. Our goal is to obtain an expression for $a \equiv |\mathbf{a}| \equiv |\Delta \mathbf{v}/\Delta t|$. The Δt here suggests that we should divide Eq. (3.14) through by Δt. This gives (using $v \equiv |\mathbf{v}| \equiv |\Delta \mathbf{r}/\Delta t|$)

$$\frac{1}{v}\left|\frac{\Delta \mathbf{v}}{\Delta t}\right| = \frac{1}{r}\left|\frac{\Delta \mathbf{r}}{\Delta t}\right| \implies \frac{|\mathbf{a}|}{v} = \frac{|\mathbf{v}|}{r} \implies a = \frac{v^2}{r}, \quad (3.15)$$

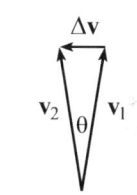

as desired. We have assumed that Δt is infinitesimal here, which allows us to convert the above quotients into derivatives.

The direction of the acceleration vector $\mathbf{a}$ is radially inward, because $\mathbf{a} \equiv d\mathbf{v}/dt$ has the same direction as $d\mathbf{v}$ (or $\Delta \mathbf{v}$), which points radially inward (leftward) in Fig. 3.21 (in the limit where θ is very small).

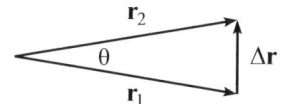

Figure 3.21

REMARK: The $a = v^2/r$ result involves the *square* of v. That is, v matters twice in a. The physical reason for this is the following. The first effect is that the larger v is, the larger the $\Delta \mathbf{v}$ is for a given angle θ in Fig. 3.21 (because the triangle is larger). The second effect is that the larger v is, the faster the object moves around in the circle, so the larger the angle θ is (and hence the larger the $\Delta \mathbf{v}$ is) for a given time Δt. Basically, if v increases, then the $\mathbf{v}$ triangle in Fig. 3.21 gets both taller and wider. Each of these effects is proportional to v, so for a given time Δt the change in velocity $\Delta \mathbf{v}$ is proportional to v^2, as we wanted to show.

(b) If the speed isn't constant, then the radial component a_r still equals v^2/r, because the velocity triangle in Fig. 3.21 becomes the triangle shown in Fig. 3.22 (for the case where the speed increases, so that $\mathbf{v}_2$ is longer than $\mathbf{v}_1$). The lower part of this triangle is exactly the same as the triangle in Fig. 3.21 (or at least it would be, if we had drawn $\mathbf{v}_1$ with the same length). So all of the preceding reasoning carries through, leading again to $a_r = v^2/r$. The v here could technically be either v_1 or v_2. But in the $\Delta t \to 0$ limit, the angle θ goes to zero, and both v_1 and v_2 are equal to the instantaneous speed v.

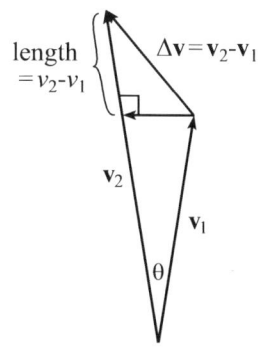

Figure 3.22

To obtain the tangential component a_t, we can use the upper part of the triangle in Fig. 3.22. This is a right triangle in the $\Delta t \to 0$ limit, and it tells us that a_t (which is the vertical component in Fig. 3.22) is

$$a_t = \frac{v_2 - v_1}{\Delta t} \equiv \frac{\Delta v}{\Delta t} \longrightarrow \frac{dv}{dt}, \quad (3.16)$$

as desired.

REMARK: A word about the placement of absolute value signs ("| |"): The tangential component of **a** is $a_t = dv/dt \equiv d|\mathbf{v}|/dt$, while the complete vector **a** is $\mathbf{a} \equiv d\mathbf{v}/dt$, which has magnitude $a = |\mathbf{a}| = |d\mathbf{v}|/dt$ (which equals $a = \sqrt{a_r^2 + a_t^2}$). The placement of the absolute value signs is critical, because $d|\mathbf{v}|$, which is the *change in the magnitude* of the velocity vector, is *not* equal to $|d\mathbf{v}|$, which is the *magnitude of the change* in the velocity vector. The former is associated with the left leg of the right triangle in Fig. 3.22, while the latter is associated with the hypotenuse. The disparity between $d|\mathbf{v}|$ and $|d\mathbf{v}|$ is most obvious in the case of uniform circular motion, where we have $d|\mathbf{v}| = 0$ and $|d\mathbf{v}| \neq 0$; the speed is constant, but the velocity is not. (The one exception to the $d|\mathbf{v}| \neq |d\mathbf{v}|$ statement occurs when the speed is instantaneously zero, so that $a_r = 0$. Since the acceleration is only tangential in this case, we have $d|\mathbf{v}| = |d\mathbf{v}|$.)

3.3. Radial and tangential accelerations, again

When taking the time derivatives, we must be careful to use the chain rule and the product rule. Starting with $(x, y) = R(\cos\theta, \sin\theta)$, the velocity is found by taking one time derivative:

$$(\dot{x}, \dot{y}) = R(-\dot{\theta}\sin\theta, \dot{\theta}\cos\theta), \tag{3.17}$$

where the $\dot{\theta}$'s come from the chain rule, because θ is a function of t. Another time derivative (using the chain rule again, along with the product rule) yields the acceleration:

$$(\ddot{x}, \ddot{y}) = R(-\ddot{\theta}\sin\theta - \dot{\theta}^2\cos\theta, \ddot{\theta}\cos\theta - \dot{\theta}^2\sin\theta). \tag{3.18}$$

If we group the $\ddot{\theta}$ terms together, and likewise the $\dot{\theta}^2$ terms, we find

$$(\ddot{x}, \ddot{y}) = R\ddot{\theta}(-\sin\theta, \cos\theta) + R\dot{\theta}^2(-\cos\theta, -\sin\theta). \tag{3.19}$$

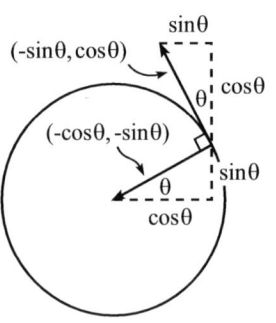

Figure 3.23

The first vector here is the tangential acceleration vector, because the magnitude is $R\ddot{\theta} \equiv R\alpha = a_t$, where we have used the fact that $(-\sin\theta, \cos\theta)$ is a unit vector. And this vector $(-\sin\theta, \cos\theta)$ points in the tangential direction, as shown in Fig. 3.23.

The second vector in Eq. (3.19) is the radial acceleration vector, because the magnitude is $R\dot{\theta}^2 \equiv R\omega^2 = R(v/R)^2 = v^2/R = a_r$, where we have used the fact that $(-\cos\theta, -\sin\theta)$ is a unit vector. And this vector $(-\cos\theta, -\sin\theta)$ points radially inward.

UNITS: The units of all of the components in Eq. (3.19) are all correctly m/s^2, because both $\ddot{\theta}$ and $\dot{\theta}^2$ have units of $1/s^2$. If you forgot to use the chain rule and omitted the $\dot{\theta}$'s, the units wouldn't work out.

LIMITS: If $\ddot{\theta} = 0$ (uniform circular motion), then the first vector in Eq. (3.19) is zero, so the acceleration is only radial; this is correct. And if $\dot{\theta} = 0$ (the object is instantaneously at rest), then the second vector in Eq. (3.19) is zero, so the acceleration is only tangential; this is also correct.

3.4. Movie replica

(a) If the movie were played back at normal speed, the person would appear to fall unnaturally fast. To see why, let's say that the doll falls for 1 s in the replica, which means that it falls a distance of 4.9 m (from $d = gt^2/2$, and we're neglecting air resistance, as usual). Since the scale factor is 100, someone watching the movie would think that the person falls 490 m in 1 s. This would look very strange, because it is far too large a distance; an object should fall only 4.9 m in 1 s. An object would fall 490 m in 1 s on a planet that has $g = 980$ m/s^2, but not on the earth.

How long does it take something to fall 490 m on the earth? From $d = gt^2/2$, we obtain $t = 10$ s. So the answer to this problem is that the movie should be slowed down by a factor of 10 when played back, so that the movie watcher sees a 490 m fall take the correct time of 10 s.

REMARK: The factor of 10 here arises because it is the square root of the scale factor, 100. Mathematically, the 10 comes from the fact that the units of g involve s^2, which leads to the t^2 in the expression $d = gt^2/2$. A factor of 10 in the time then leads to the desired factor of

3.5. PROBLEM SOLUTIONS

$10^2 = 100$ in the distance. But more intuitively, the factor of 10 arises in the following way. If we don't scale the time at all, then as we saw above, we basically end up on a planet with $g = 980 \text{ m/s}^2$. If we then slow down the movie by a factor of 10, this decreases g by a factor of 10^2, because it causes the falling object to take *10 times as long* to fall a given distance, at which point it is going only *1/10 as fast*. So the factor of 10 matters *twice*. The rate of change $\Delta v/\Delta t$ of the velocity (that is, the acceleration) is therefore only $(1/10)/10 = 1/100$ of the $g = 980 \text{ m/s}^2$ value that it would have been, so we end up with the desired g value of 9.8 m/s^2.

(b) The vertical motion must satisfy the same conditions as in part (a), because it is unaffected by the horizontal motion. So we again need to slow down the playback by a factor of 10.

We want the speed of the car to appear to be 50 mph. This means that if the movie weren't slowed down by the required factor of 10, then the car's speed in the movie would appear to be 500 mph. Due to the scale factor of 100, this would then require the speed in the replica to be 5 mph. So 5 mph is the desired answer.

Said in an equivalent way, starting from the replica: A speed of 5 mph in the replica translates to a speed of 500 mph in an unmodified movie, due to the scale factor of 100. But then slowing down the movie by a factor of 10 brings the apparent speed down to the desired 50 mph.

3.5. Doubling gravity

The horizontal distance traveled during the upward part of the motion is half the total distance of normal projectile motion. Therefore, from Eq. (3.6) the distance traveled during the upward motion is $(v_0^2/g)\sin\theta\cos\theta$. We must now find the horizontal distance traveled during the downward motion. Since the horizontal velocity is constant throughout the entire motion, we just need to get a handle on the time of the downward part.

The time of the upward part is given by $h = gt_u^2/2$, where h is the maximum height, because we can imagine running time backwards, in which case the ball is dropped from rest, as far as the vertical motion is concerned. (The height h equals $(v_0^2/2g)\sin^2\theta$ from Eq. (3.6), but we won't need to use this.) The time of the downward part is given by $h = (2g)t_d^2/2$, because gravity is doubled. Therefore $t_d = t_u/\sqrt{2}$. So the horizontal distance traveled during the downward motion is $1/\sqrt{2}$ times the horizontal distance traveled during the upward motion. The total distance is therefore

$$d = \frac{v_0^2 \sin\theta \cos\theta}{g}\left(1 + \frac{1}{\sqrt{2}}\right). \tag{3.20}$$

3.6. Ratio of heights

The given $gt^2/2$ expression holds for an object dropped from rest. But it is also a valid expression for the distance fallen relative to where the object would be if gravity were turned off. Mathematically, this is true because in Eq. (3.3), $Y + V_y t$ is the height of the object in the absence of gravity, and $gt^2/2$ is what is subtracted from this.

The two vertical distances we drew in Fig. 3.8 are the distances fallen relative to the zero-gravity line of motion (the dotted line). Since the time of the entire projectile motion is twice the time to the top, the ratio of these two vertical distances is $2^2 = 4$ (hence the d and $4d$ labels in Fig. 3.24). This figure contains two similar right triangles, with one being twice the size of the other. Comparing the vertical legs of each triangle tells us that $4d/(h+d) = 2$. This yields $h = d$, which means that point A has half the height of point B. The answer to the problem is therefore $1/2$.

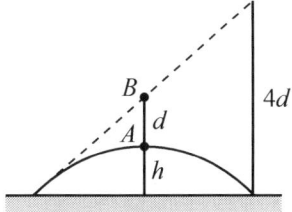

Figure 3.24

3.7. Hitting horizontally

If t is the time of flight, then we have three unknowns: t, v_0, and θ. And we have three facts:

- The vertical speed is zero at the top, so $v_0 \sin\theta - gt = 0 \implies t = v_0 \sin\theta/g$.
- The horizontal distance is ℓ, so $(v_0 \cos\theta)t = \ell$.
- The vertical distance is ℓ, so $(v_0 \sin\theta)t - gt^2/2 = \ell$.

Plugging the value of t from the first fact into the other two gives

$$\frac{v_0^2 \sin\theta \cos\theta}{g} = \ell \quad \text{and} \quad \frac{v_0^2 \sin^2\theta}{2g} = \ell. \tag{3.21}$$

Dividing the second of these equations by the first gives

$$\frac{\sin\theta}{2\cos\theta} = 1 \implies \tan\theta = 2 \implies \theta \approx 63.4°. \tag{3.22}$$

(What we've done here is basically find the firing angle of a projectile that makes the maximum height be half of the total range.) This value of θ corresponds to a 1-2-$\sqrt{5}$ right triangle, which yields $\sin\theta = 2/\sqrt{5}$. The second of the equations in Eq. (3.21) (the first would work just as well) then gives

$$\frac{v_0^2}{2g}\left(\frac{2}{\sqrt{5}}\right)^2 = \ell \implies v_0 = \sqrt{\frac{5g\ell}{2}}. \tag{3.23}$$

3.8. Projectile and tube

(a) From the standard $v_f^2 = v_i^2 + 2ad$ formula, the speed of the projectile when it exits the tube is $v = \sqrt{v_0^2 - 2ax}$. You can also obtain this by using $v = v_0 - at$, where t is found by solving $v_0 t - at^2/2 = x$. This quadratic equation has two solutions, of course. You want the "$-$" root. (What is the meaning of the "$+$" root?)

Since the projectile motion has zero initial v_y, the time to reach the ground is given by $gt^2/2 = h \implies t = \sqrt{2h/g}$. The horizontal distance traveled in the air is vt, but we must add on the length of the tube to get the total distance ℓ. So we have

$$\ell = x + vt = x + \sqrt{v_0^2 - 2ax}\sqrt{\frac{2h}{g}}. \tag{3.24}$$

(b) Looking at the two terms in Eq. (3.24), we see that we have competing effects of x. Increasing x increases ℓ by having the projectile motion start farther to the right. But increasing x also decreases the projectile motion's initial speed and hence its horizontal distance. Maximizing the ℓ in Eq. (3.24) by taking the derivative with respect to x gives

$$0 = \frac{d\ell}{dx} = 1 + \frac{1}{2}\frac{-2a}{\sqrt{v_0^2 - 2ax}}\sqrt{\frac{2h}{g}}$$

$$\implies \sqrt{v_0^2 - 2ax} = a\sqrt{\frac{2h}{g}}$$

$$\implies x = \frac{v_0^2}{2a} - \frac{ah}{g}. \tag{3.25}$$

Both terms here correctly have dimensions of length. This result for the optimal value of x is smaller than $v_0^2/2a$, as it should be, because otherwise the projectile would reach zero speed inside the tube (since $v = \sqrt{v_0^2 - 2ax}$) and never make it out. However, we aren't quite done, because there are two cases to consider. To see why, note that if $a = v_0\sqrt{g/2h}$, then the x in Eq. (3.25) equals zero. So Eq. (3.25) is

3.5. PROBLEM SOLUTIONS

applicable only if $a \leq v_0 \sqrt{g/2h}$. If $a \geq v_0 \sqrt{g/2h}$, then Eq. (3.25) yields a negative value of x, which isn't physical. The optimal x in the $a \geq v_0 \sqrt{g/2h}$ case is therefore simply $x = 0$. Physically, if a is large then any nonzero value of x will hurt ℓ (by slowing down the initial projectile speed) more than it will help (by adding on the "head start" distance of x). Mathematically, in this case the extremum of ℓ occurs at the boundary of the allowed values of x (namely, $x = 0$), as opposed to at a local maximum; a zero derivative therefore isn't relevant.

LIMITS: If $a \to 0$ then Eq. (3.25) gives $x \to \infty$, which is correct. We can make the tube be very long, and the projectile will keep sliding along; the projectile motion at the end is largely irrelevant. We also see that the optimal x increases with v_0 and g, and decreases with a and h. You should convince yourself that these all make sense.

3.9. Car in the mud

At the instant the mud leaves the wheel, it is at height $R + R \sin \theta$ above the ground. It then rises an extra height of $h = (v \cos \theta)^2 / 2g$ during its projectile motion. This is true because the initial v_y is $v \cos \theta$ (because the initial velocity makes an angle of $90° - \theta$ with respect to the horizontal), so the time to the highest point is $(v \cos \theta)/g$; plugging this into $h = (v \cos \theta)t - (1/2)gt^2$ gives $h = (v \cos \theta)^2 / 2g$. The total height above the ground at the highest point is therefore

$$H = R + R \sin \theta + \frac{v^2 \cos^2 \theta}{2g}. \quad (3.26)$$

Maximizing this by taking the derivative with respect to θ gives

$$R \cos \theta - \frac{v^2}{g} \sin \theta \cos \theta = 0. \quad (3.27)$$

Since $\cos \theta$ is a factor of this equation, we see that there are two solutions. One is $\cos \theta = 0 \implies \theta = \pi/2$, which corresponds to the top of the wheel. This is the maximum height if $v^2 < gR$, because in this case the best the mud can do is stay in contact with the wheel the whole time. (After learning about forces in Chapter 4, you can show that if $v^2 < gR$ then the normal force between the wheel and the mud is always nonzero, so the mud will never fly off the wheel.) But we are assuming $v^2 > gR$, so we want the other solution,

$$\sin \theta = \frac{gR}{v^2}. \quad (3.28)$$

Note that gR/v^2 is less than 1 if $v^2 > gR$, so such a θ does indeed exist. (For $v^2 = gR$ we correctly obtain $\theta = \pi/2$.) Plugging Eq. (3.28) into Eq. (3.26) gives a maximum height of

$$H_{\max} = R + R \sin \theta + \frac{v^2}{2g}(1 - \sin^2 \theta)$$

$$= R + R\left(\frac{gR}{v^2}\right) + \frac{v^2}{2g}\left(1 - \left(\frac{gR}{v^2}\right)^2\right)$$

$$= R + \frac{gR^2}{2v^2} + \frac{v^2}{2g}. \quad (3.29)$$

All three terms here correctly have dimensions of length. This result is valid if $v^2 \geq gR$. If $v^2 \leq gR$, then the maximum height (at the top of the wheel) is $2R$.

LIMITS: If v is large (more precisely, if $v^2 \gg gR$), then Eq. (3.29) gives $H_{\max} \approx R + v^2/2g$. From Eq. (3.28) the mud leaves the wheel at $\theta \approx 0$ (the side point) and travels vertically an extra height of $v^2/2g$, which is the standard height achieved in vertical projectile motion. If $v^2 \approx gR$, then Eq. (3.29) gives $H_{\max} \approx 2R$. The mud barely leaves the wheel at the top, so the maximum height is simply $2R$. It doesn't make sense to take the small-v limit (more precisely, $v^2 \ll gR$) of Eq. (3.29), because this result assumes $v^2 > gR$.

3.10. Clearing a wall

(a) The unknowns in this problem are v_0 and θ. Problem 3.1(b) gives the maximum height of a projectile, so we want

$$h = \frac{v_0^2 \sin^2 \theta}{2g}. \tag{3.30}$$

And Problem 3.1(c) gives the range of a projectile, so we want

$$2\ell = \frac{2v_0^2 \sin\theta \cos\theta}{g}. \tag{3.31}$$

Dividing Eq. (3.30) by Eq. (3.31) gives

$$\frac{h}{\ell} = \frac{\sin\theta}{2\cos\theta} \implies \tan\theta = \frac{2h}{\ell}. \tag{3.32}$$

This means that you should pretend that the wall is twice as tall as it is, and then aim for the top of that imaginary wall.

LIMITS: If $h \to 0$ then $\theta \to 0$. And if $h \to \infty$ then $\theta \to 90°$. Both of these limits make sense.

(b) FIRST SOLUTION: If $\tan\theta = 2h/\ell$, then drawing a right triangle with legs of length ℓ and $2h$ tells us that $\sin\theta = 2h/\sqrt{4h^2 + \ell^2}$. So Eq. (3.30) gives

$$h = \frac{v_0^2}{2g} \cdot \frac{4h^2}{4h^2 + \ell^2} \implies v_0^2 = g\left(\frac{4h^2 + \ell^2}{2h}\right). \tag{3.33}$$

(Eq. (3.31) would give the same result.) We want to minimize this function of h. Setting the derivative equal to zero gives (ignoring the denominator of the derivative, since we're setting the result equal to zero)

$$0 = 2h(8h) - (4h^2 + \ell^2) \cdot 2 \implies 0 = 4h^2 - \ell^2 \implies h = \frac{\ell}{2}. \tag{3.34}$$

If h takes on this value, then Eq. (3.32) gives $\tan\theta = 2(\ell/2)/\ell = 1 \implies \theta = 45°$. You should convince yourself why this result is consistent with the familiar fact that $\theta = 45°$ gives the maximum range of a projectile.

LIMITS: The v_0 in Eq. (3.33) correctly goes to infinity as $h \to \infty$. It also goes to infinity as $h \to 0$. This makes sense because $h \approx 0$ corresponds to a nearly horizontal "line drive." If the speed weren't large, then the ball would quickly hit the ground (since the initial v_y would be very small). Note that since $v_0 \to \infty$ for both $h \to \infty$ and $h \to 0$, a continuity argument implies that v_0 achieves a minimum for some intermediate value of h. But it takes a little work to show that this h equals $\ell/2$.

SECOND SOLUTION: A somewhat quicker way of obtaining v_0 (without first obtaining θ in Eq. (3.32)) is the following. Let t be the time to the top of the motion. Then the horizontal component of $\mathbf{v}_0$ is ℓ/t, and the vertical component is gt (because v_y is zero at the top). From the second half of the motion, we also know that $gt^2/2 = h \implies t^2 = 2h/g$. So

$$v_0^2 = v_x^2 + v_y^2 = \frac{\ell^2}{t^2} + g^2 t^2 = \frac{\ell^2}{2h/g} + g^2 \frac{2h}{g} = g\left(\frac{\ell^2 + 4h^2}{2h}\right), \tag{3.35}$$

in agreement with Eq. (3.33).

3.5. PROBLEM SOLUTIONS

3.11. Bounce throw

From Eq. (3.6) we know that the total horizontal distance traveled for a 45° throw is v_0^2/g. (As always, we are ignoring air resistance, which is actually a pretty lousy approximation for thrown balls.) For the bounce throw, each of the two bumps has a horizontal span of $v_0^2/2g$. The throwing speed is still v_0, so we want the $\sin 2\theta$ factor in Eq. (3.6) to be $1/2$. Hence $2\theta = 30° \implies \theta = 15°$. The (constant) horizontal component of the velocity is therefore $v_0 \cos 15°$ for the entire duration of the bounce throw, as opposed to $v_0 \cos 45°$ for the original throw. Since the time of flight is inversely proportional to the horizontal speed, the total time for the bounce throw is $(v_0 \cos 45°)/(v_0 \cos 15°) \approx 0.73$ as long as the total time for the original throw.

REMARKS: Since $\cos 15° = \cos(45° - 30°)$, you can use the trig sum formula for cosine to show that the exact answer to this problem is $2/(\sqrt{3} + 1)$, which can be written as $\sqrt{3} - 1$.

In real life, air resistance makes the trajectory be nonparabolic, and there is also an abrupt decrease in speed at the bounce, due to friction with the ground. But it is still possible for a bounce throw to take less time than the no-bounce throw. This is particularly relevant in baseball games. If a player is making a long throw from the outfield (or even from third base to first base if the player is off balance and the throwing speed is low), then a bounce throw is desirable. The advantage of throwing in a more direct line can outweigh the disadvantage of the loss in speed at the bounce. But the second bump in the throw needs to be relatively small.

3.12. Maximum bounce

The time it takes the ball to fall the distance $h - y$ to the board is given by $gt^2/2 = h - y \implies t = \sqrt{2(h-y)/g}$. The speed at this time is $v = gt = \sqrt{2g(h-y)}$. (This is just the standard $v_f = \sqrt{2ad}$ result that follows from Eq. (2.4) when $v_i = 0$.) Since the collision is elastic, this is also the horizontal speed v_x right after the bounce.

The time to fall the remaining distance y to the ground after the horizontal bounce is given by $gt^2/2 = y \implies t = \sqrt{2y/g}$. The horizontal distance traveled is then

$$d = v_x t = \sqrt{2g(h-y)} \sqrt{\frac{2y}{g}} = 2\sqrt{y(h-y)}. \tag{3.36}$$

Our goal is therefore to maximize the function $hy - y^2$. Setting the derivative equal to zero gives $0 = h - 2y \implies y = h/2$. So the board should be at the halfway point. The desired horizontal distance is then $d = 2\sqrt{(h/2)(h - h/2)} = h$.

LIMITS: The distance d in Eq. (3.36) goes to zero for both $y = 0$ and $y = h$. These limits make sense. In the first case, the board is on the ground, so there is no time after the collision for the ball to travel any horizontal distance. In the second case, the board is located right where the ball is released, so the horizontal speed after the "collision" is zero, and the ball falls straight down.

3.13. Falling along a right triangle

(a) If θ is the inclination angle of the hypotenuse, then the component of the gravitational acceleration along the hypotenuse is $g \sin \theta$, where $\sin \theta = b/\sqrt{a^2 + b^2}$. Using $d = at^2/2$ (this a is the acceleration, not the length of the lower leg!), the time to travel along the hypotenuse is given by

$$\sqrt{a^2 + b^2} = \frac{1}{2}\left(g \frac{b}{\sqrt{a^2 + b^2}}\right) t_H^2 \implies t_H = \sqrt{\frac{2(a^2 + b^2)}{gb}}. \tag{3.37}$$

LIMITS: If $a \to \infty$ or $b \to \infty$ or $b \to 0$, then $t_H \to \infty$, which makes sense.

(b) The time to fall to C is given by $gt_1^2/2 = b \implies t_1 = \sqrt{2b/g}$. The speed at C is then $gt_1 = \sqrt{2gb}$. The particle then travels along the bottom leg of the triangle at this constant speed, which takes a time of $t_2 = a/\sqrt{2gb}$. The total time is therefore

$$t_L = t_1 + t_2 = \sqrt{\frac{2b}{g}} + \frac{a}{\sqrt{2gb}}. \tag{3.38}$$

LIMITS: Again, if $a \to \infty$ or $b \to \infty$ or $b \to 0$, then $t_L \to \infty$.

(c) If $a = 0$ then both t_H and t_L reduce to $\sqrt{2b/g}$.

(d) If $b \ll a$ then $t_H \approx \sqrt{2}a/\sqrt{gb}$ (the b^2 term in Eq. (3.37) is negligible), and $t_L \approx a/\sqrt{2gb}$ (the first term in Eq. (3.38) is negligible). So in this limit we have

$$t_H \approx 2t_L. \tag{3.39}$$

This makes sense for the following reason. In the journey along the legs, the particle is moving at the maximum speed of $\sqrt{2gb}$ for essentially the entire time. But in the journey along the hypotenuse, the particle has the same maximum speed (as you can show with kinematics; this also follows quickly from conservation of energy, discussed in Chapter 5), and the average speed is half of the maximum speed, because the acceleration is constant.

(e) Setting the t_H in Eq. (3.37) equal to the t_L in Eq. (3.38) gives

$$\sqrt{\frac{2(a^2+b^2)}{gb}} = \sqrt{\frac{2b}{g}} + \frac{a}{\sqrt{2gb}} \implies 2\sqrt{a^2+b^2} = 2b + a$$
$$\implies 4(a^2+b^2) = 4b^2 + 4ba + a^2 \implies 3a^2 = 4ba$$
$$\implies a = \frac{4b}{3}. \tag{3.40}$$

So the two times are equal if we have a 3-4-5 right triangle, with the bottom leg being the longer one.

REMARKS: Without doing any calculations, the following continuity argument demonstrates that there must exist a triangle shape for which t_H equals t_L. We found in part (d) that $t_H > t_L$ (by a factor of 2) when $b \ll a$. But $t_H < t_L$ when $a \ll b$. This can be seen by noting that $t_H = t_L$ when $a = 0$ and that a appears only at second order in the expression for t_H in Eq. (3.37), but at first order in the expression for t_L in Eq. (3.38). When a is small, the second-order a^2 term is much smaller than the first-order a term, making t_H smaller than t_L. The preceding $t_H > t_L$ and $t_H < t_L$ inequalities imply, by continuity, that there must exist some relation between a and b for which the two times are equal.

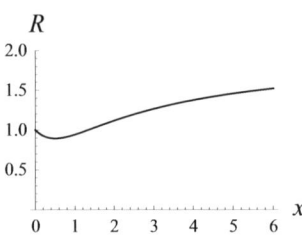

Figure 3.25

Note that since the times are equal in both the 3-4-5 case and the $a = 0$ case, the ratio $R \equiv t_H/t_L$ must achieve an extremum (it's a minimum) somewhere in between. As an exercise, you can show that this occurs when $a = b/2$. The associated minimum is $R = 2/\sqrt{5} \approx 0.89$. The plot of R vs. $x \equiv a/b$ is shown in Fig. 3.25. In terms of x, you can show that $R(x) = 2\sqrt{1+x^2}/(2+x)$. The value $x \equiv a/b = 0$ corresponds to a tall thin triangle (with $R = 1$), and $x \equiv a/b = \infty$ corresponds to a wide squat triangle (with $R = 2$). You should think physically about the competing effects that make t_H larger than t_L (that is, $R > 1$) in some (most) cases, but smaller in others.

3.14. **Throwing to a cliff**

(a) If t is the time to hit the edge of the cliff, then the standard expressions for the horizontal and vertical positions yield

$$L = (v_0 \cos\theta)t,$$
$$L = (v_0 \sin\theta)t - \frac{gt^2}{2}. \tag{3.41}$$

3.5. PROBLEM SOLUTIONS

Solving for t in the first equation and plugging the result into the second equation gives

$$L = v_0 \sin\theta \left(\frac{L}{v_0 \cos\theta}\right) - \frac{g}{2}\left(\frac{L}{v_0 \cos\theta}\right)^2 = L\tan\theta - \frac{gL^2}{2v_0^2 \cos^2\theta}. \quad (3.42)$$

Solving for v_0 yields

$$v_0 = \sqrt{\frac{gL}{2\cos\theta(\sin\theta - \cos\theta)}}. \quad (3.43)$$

(b) If $\theta \to 90°$ then $\cos\theta \to 0$, so $v_0 \to \infty$. This makes sense, because the ball is essentially thrown straight up. The horizontal component of the velocity is very small, so the ball needs to spend a very long time in the air. It must therefore have a very large initial speed.

If $\theta \to 45°$ then $\sin\theta \to \cos\theta$, so $v_0 \to \infty$. This also makes sense, because the ball is aimed right at the corner of the cliff, so if it isn't thrown infinitely fast, it will have time to fall down relative to the "zero-gravity" straight-line path and hence hit below the corner.

REMARK: Since the speed v_0 goes to infinity for both $\theta \to 45°$ and $\theta \to 90°$, by continuity it must achieve a minimum value for some angle in between. As an exercise, you can show that v_0 is minimum when $\theta = 3\pi/8 = 67.5°$, which happens to be exactly halfway between $45°$ and $90°$.

3.15. Throwing from a cliff

With $y = 0$ taken to correspond to the base of the cliff, the height of the ball as a function of time is $y(t) = h + (v\sin\theta)t - gt^2/2$. The ball hits the ground when this equals zero, which gives

$$\frac{g}{2}t^2 - (v\sin\theta)t - h = 0 \implies t = \frac{v\sin\theta + \sqrt{v^2\sin^2\theta + 2gh}}{g}, \quad (3.44)$$

where we have chosen the "+" root because t must be positive. (The "−" root corresponds to the negative time at which the parabolic motion would hit $y = 0$ if it were extended backward through the cliff.) The desired horizontal position at this time is

$$x = (v\cos\theta)t = \frac{v\cos\theta}{g}\left(v\sin\theta + \sqrt{v^2\sin^2\theta + 2gh}\right). \quad (3.45)$$

LIMITS: If $\theta = \pi/2$ then $x = 0$, of course, because the ball is thrown straight up. If $\theta = 0$ then $x = v\sqrt{2h/g}$. This is correct because the ball is fired horizontally, so the time to fall the height h is the standard $\sqrt{2h/g}$. And since the horizontal speed is always v, the horizontal distance is $v\sqrt{2h/g}$. If $h = 0$ then $x = (2v^2/g)\sin\theta\cos\theta$, which is the standard projectile range on flat ground. If $h \to \infty$ then we can ignore the $v^2\sin^2\theta$ under the square root, and we end up with

$$x \approx \frac{v^2 \sin\theta\cos\theta}{g} + (v\cos\theta)\sqrt{\frac{2h}{g}}. \quad (3.46)$$

The first term here is the horizontal distance traveled by the time the ball reaches the highest point in its motion. The second term is the horizontal distance traveled during the time it takes to fall a height h from the highest point (because $\sqrt{2h/g}$ is the time it takes to fall a height h).

However, the reason why Eq. (3.46) isn't exact is that after falling a distance h from the highest point, the ball hasn't quite reached the ground, because there is still an extra distance to fall, corresponding to the initial gain in height from the top of the cliff to the highest point. From Eq. (3.6) this height is $(v^2/2g)\sin^2\theta$. But the ball is traveling so fast (assuming h is large) during this last stage of the motion that it takes a negligible time to fall through the last $(v^2/2g)\sin^2\theta$ interval and hit the ground. The speed is essentially $gt = g\sqrt{2h/g} = \sqrt{2gh}$ at this point, so the additional time is approximately $(v^2/2g)\sin^2\theta/\sqrt{2gh}$. You can show by using a Taylor series that the next term in the approximation to the time in Eq. (3.44) would yield this tiny additional time. (You will want to factor the $2gh$ out of the square root, to put it in the standard $\sqrt{1+\epsilon}$ form.) Further corrections from additional terms in the Taylor series correspond to the fact that the speed isn't constant during this final tiny time.

3.16. **Throwing on stairs**

(a) When the ball reaches the corner of the Nth step, it has traveled a distance of $N\ell$ both sideways and downward. So if t is the time in the air, then looking at the horizontal distance gives $vt = N\ell$, and looking at the vertical distance gives $gt^2/2 = N\ell$. Solving for t in the first of these equations and plugging the result into the second yields

$$\frac{g}{2}\left(\frac{N\ell}{v}\right)^2 = N\ell \implies v = \sqrt{\frac{N\ell g}{2}}. \tag{3.47}$$

Limits: Big N, ℓ, or g implies big v, as expected.

(b) As we noted above, the time needed to fall the distance $N\ell$ to the corner of the Nth step is given by

$$\frac{1}{2}gt_N^2 = N\ell \implies t_N = \sqrt{\frac{2N\ell}{g}}. \tag{3.48}$$

Likewise, the total time needed to fall to the $(N+1)$st step is $t_{N+1} = \sqrt{2(N+1)\ell/g}$. The difference in these times is the time of flight from the Nth corner to the $(N+1)$st step. So the horizontal distance along the $(N+1)$st step is (using the result for v from part (a))

$$d = v(t_{N+1} - t_N) = \sqrt{\frac{N\ell g}{2}}\left(\sqrt{\frac{2(N+1)\ell}{g}} - \sqrt{\frac{2N\ell}{g}}\right)$$
$$= \ell\left(\sqrt{N(N+1)} - N\right). \tag{3.49}$$

(c) We need to apply the Taylor series $\sqrt{1+\epsilon} \approx 1 + \epsilon/2$ to the above result for d. If we take out a factor of $\sqrt{N^2}$ from the square root, it will take the requisite $\sqrt{1+\epsilon}$ form. We then have

$$d = N\ell\left(\sqrt{1+\frac{1}{N}} - 1\right) \approx N\ell\left(\left(1+\frac{1}{2N}\right) - 1\right) = \frac{\ell}{2}. \tag{3.50}$$

(d) The x component of the velocity is (always) $v_x = v = \sqrt{N\ell g/2}$. The y component of the velocity at the corner of the Nth step is $v_y = gt_N = g\sqrt{2N\ell/g} = \sqrt{2N\ell g}$. The ratio of these components is $v_y/v_x = 2$.

This is consistent with the result in part (c), because for large N, the ball is moving very fast when it grazes the corner, so it travels essentially in a straight line in going to the next step; there's basically no time to accelerate and have the trajectory bend. From part (c), we know that the ball goes down a distance ℓ, and sideways a distance $\ell/2$. These distances imply (in the straight-line-trajectory approximation) that the ratio of the velocity components is $v_y/v_x = \ell/(\ell/2) = 2$, in agreement with the ratio obtained directly from the components calculated at the corner.

Units: Note that this ratio of 2 is independent of ℓ, g, and N. It can't depend on g due to the seconds in g's units. And then it can't depend on ℓ due to the meters in ℓ's units. The argument that eliminates N is a little tricker, but it is short: The ratio of the velocity components can't depend on N, because we could simply turn each step into many little ones. N therefore increases, but the velocity components are the same; the ball has no clue that we subdivided the steps, so the velocity components can't change.

3.5. PROBLEM SOLUTIONS

3.17. Bullet and sphere

FIRST SOLUTION: If we take the origin to be the center of the sphere, then the projectile motion is given by

$$x(t) = v_0 t \quad \text{and} \quad y(t) = R - \frac{1}{2}gt^2. \tag{3.51}$$

Plugging the t from the first equation into the second gives

$$y = R - \frac{gx^2}{2v_0^2}. \tag{3.52}$$

If a point on the sphere (which is a 2-D circle for our purposes) is a distance x to the right of the center, then by the Pythagorean theorem it is a distance $\sqrt{R^2 - x^2}$ above the center. So the y coordinate is $y = \sqrt{R^2 - x^2}$. Taking the R^2 out of the square root and using $\sqrt{1 + \epsilon} \approx 1 + \epsilon/2$ gives the approximate expression for y as a function of (small) x:

$$y = R\sqrt{1 - \frac{x^2}{R^2}} \approx R\left(1 - \frac{x^2}{2R^2}\right) = R - \frac{x^2}{2R}. \tag{3.53}$$

Comparing Eqs. (3.52) and (3.53), we see that the projectile motion matches up with the circle (at least for small x) if

$$\frac{g}{2v_0^2} = \frac{1}{2R} \implies v_0 = \sqrt{gR}. \tag{3.54}$$

If v_0 takes on this value, you can quickly show that the y value in Eq. (3.52) satisfies $x^2 + y^2 \geq R^2$. That is, the projectile motion always lies *outside* the circle, which is intuitively reasonable. So the bullet does indeed avoid touching the sphere if $v_0 = \sqrt{gR}$.

The bullet hits the ground when $y = -R$ (because we defined the origin to be the center of the sphere). Using the expression for y in Eq. (3.52) with $v_0^2 = gR$, we find the desired distance along the ground to be

$$R - \frac{x^2}{2R} = -R \implies x_g = 2R. \tag{3.55}$$

UNITS: Note that considerations of units tells us that x_g must be proportional to R. In general it depends on v_0, but we've specifically chosen v_0 to equal $\sqrt{gR}$; and x_g can't depend on g, due to the seconds in g. But it takes a calculation to show that there is a factor of 2 out front.

LIMITS: Large R or large g implies large v_0, which makes sense. Large R implies large x_g, which also makes sense.

SECOND SOLUTION: A quicker way of finding v_0 is to use the fact that the radial acceleration is given by $a = v^2/r$, where r is the radius of the circle that the parabolic trajectory instantaneous matches up with. But the acceleration of the bullet is also of course just g, because it is undergoing freefall projectile motion (assuming that it isn't touching the sphere). So if we want the radius of the instantaneous circle to be R, then we need

$$\frac{v_0^2}{R} = g \implies v_0 = \sqrt{gR}. \tag{3.56}$$

3.18. Throwing on an inclined plane

(a) If θ is very small (so that the plane is nearly horizontal), then the point P is essentially at the top of the ball's parabolic motion. So P is certainly higher than the starting point.

In the other extreme where θ is close to $90°$ (so that the plane is nearly vertical), the ball starts out moving essentially horizontally. So all later points in the motion (including P) are lower than the starting point.

Therefore, by continuity there must exist a θ between 0 and $90°$ for which P has the same height as the starting point.

(b) We'll calculate the special value of θ by finding the time at which the ball returns to the starting height, and also the time at which the ball is at point P farthest from the plane. We'll then set these two times equal to each other.

The ball is thrown at an angle of $90° - \theta$ with respect to horizontal, so the vertical component of the initial velocity is $v_0 \cos\theta$ (as opposed to the usual $v_0 \sin\theta$). The time to the top of the parabolic motion is therefore $v_0 \cos\theta/g$. The time to fall back down to the initial height is twice this, or $2v_0 \cos\theta/g$.

The time to reach the point P farthest from the plane can be found in the following quick manner, by using tilted axes. The acceleration components parallel and perpendicular to the plane are $g \sin\theta$ and $g \cos\theta$, respectively. We aren't concerned with the first of these; the motions in the two tilted directions are independent, and we are concerned only with the motion perpendicular to the plane. The $g \cos\theta$ acceleration perpendicular to the plane tells us that the time to reach P equals $v_0/g \cos\theta$. This is true because the velocity component perpendicular to the plane at the start is just v_0, while the velocity component perpendicular to the plane at P is zero (by definition, since P is the farthest point from the plane). Basically, someone living in the tilted-axis world would think that the acceleration due to gravity is $g \cos\theta$ "downward" toward the plane, while there is also an additional mysterious force causing an acceleration of $g \sin\theta$ "rightward."

The time to return to the initial height is the same as the time to reach P if

$$\frac{2v_0 \cos\theta}{g} = \frac{v_0}{g \cos\theta} \implies \cos^2\theta = \frac{1}{2} \implies \theta = 45°. \qquad (3.57)$$

The corresponding trajectory is sketched in Fig. 3.26. You can show that P is a distance $v_0^2/\sqrt{2}g$ from the plane in this case.

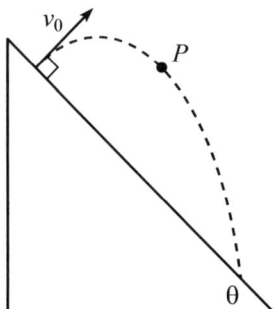

Figure 3.26

3.19. Ball landing on a block

FIRST SOLUTION: Let the initial speed of the block be u, and let the initial speed of the ball be v. Our strategy will be to (1) equate the times when the ball hits the plane and when the block reaches its maximum height, and then (2) equate the distances along the plane at this time.

Since the slope of the plane is $\tan\beta$, the ball hits the plane when its coordinates satisfy $y/x = \tan\beta$. Using $y = (v \sin\theta)t - gt^2/2$ and $x = (v \cos\theta)t$, this becomes

$$\frac{(v \sin\theta)t - gt^2/2}{(v \cos\theta)t} = \tan\beta \implies t_{\text{hit}} = \frac{2v \cos\theta}{g}(\tan\theta - \tan\beta). \qquad (3.58)$$

The block reaches its maximum height at $t_{\max} = u/(g \sin\beta)$, because the acceleration downward along the plane is $g \sin\beta$, so this is the time when the speed is zero. Equating $t_{\max}$ with t_{hit} gives

$$\frac{u}{g \sin\beta} = \frac{2v \cos\theta}{g}(\tan\theta - \tan\beta) \implies u = 2v \sin\beta \cos\theta(\tan\theta - \tan\beta). \qquad (3.59)$$

Now let's demand that the distances along the plane are equal at this time. The ball's distance along the plane is $x/\cos\beta = (v \cos\theta)t/\cos\beta$. And the block's distance is $ut - (g \sin\beta)t^2/2$, because it undergoes motion with constant acceleration $g \sin\theta$ pointing down the plane. Equating these distances, canceling a factor of t, and using $t = u/(g \sin\beta)$ from above, we obtain

$$\frac{v \cos\theta}{\cos\beta} = u - \frac{1}{2}(g \sin\beta)\frac{u}{g \sin\beta} \implies v \cos\theta = \frac{u \cos\beta}{2}. \qquad (3.60)$$

This makes sense; the constant horizontal speed of the ball (the left-hand side) correctly equals the average horizontal speed of the block (the right-hand side); the block starts with

3.5. PROBLEM SOLUTIONS

$u_x = u\cos\theta$ and ends up with $u_x = 0$. Plugging the u from Eq. (3.59) into Eq. (3.60) gives

$$v\cos\theta = v\sin\beta\cos\theta(\tan\theta - \tan\beta)\cdot\cos\beta$$
$$\implies 1 = \sin\beta\cos\beta(\tan\theta - \tan\beta)$$
$$\implies \tan\theta = \tan\beta + \frac{1}{\sin\beta\cos\beta}. \tag{3.61}$$

This is the desired implicit equation that gives θ in terms of β. If $\beta = 45°$ we have

$$\tan\theta = 1 + \frac{1}{\frac{1}{\sqrt{2}}\cdot\frac{1}{\sqrt{2}}} = 3 \implies \theta \approx 71.6°. \tag{3.62}$$

LIMITS: If $\beta \to 90°$ then Eq. (3.61) gives $\theta \to 90°$; this makes sense because θ must be at least as large as β. If $\beta \to 0$ then $\theta \to 90°$; this makes sense because for a given u, the process takes a very long time in the $\beta \to 0$ limit (because the block hardly slows down on the plane). So the ball must spend a very long time in the air. It must therefore be thrown very fast, which means that θ must be very close to $90°$ so that the horizontal speed is essentially equal to the given speed u.

SECOND SOLUTION: In this solution, we'll work with axes parallel and perpendicular to the plane. The objects have the same acceleration along the plane, namely $-g\sin\beta$. Therefore, if their initial velocity components along the plane are equal, then these velocity components will always be equal. And since the objects start at the same place, equal velocity components implies equal positions along the plane. That is, the ball will always be "above" the block (in the tilted reference frame), which implies that the ball will land on the block. Since the angle between the ball's firing angle and the plane is $\theta - \beta$, equating the initial velocity components along the plane gives

$$u = v\cos(\theta - \beta). \tag{3.63}$$

This condition guarantees that the ball will land on the block, but let's now also demand that the landing happens when the block reaches its maximum height on the plane. As in the first solution, the time for the block to reach its maximum height (where it is instantaneously at rest) is $t_{\max} = u/(g\sin\beta)$, because $g\sin\beta$ is the (magnitude of the) acceleration along the plane.

We claim that the time for the ball to return to the plane is $t_{\text{hit}} = 2\cdot v\sin(\theta - \beta)/g\cos\beta$. (You can verify by using the sum formula for sine that this agrees with Eq. (3.58).) This is true because $v\sin(\theta - \beta)$ is the initial velocity perpendicular to the plane, and $g\cos\beta$ is the (magnitude of the) acceleration perpendicular to the plane. We effectively live in a tilted world where gravity has strength $g' = g\cos\beta$ in the "upward" (tilted) direction, so the time to the "top" of the motion is the standard $v_{y'}/g'$, where $v_{y'} = v\sin(\theta - \beta)$ is the initial "upward" velocity. The total time is twice this; hence the above expression for t_{hit}. There is also a mysterious "sideways" acceleration $g\sin\beta$ along the plane in our tilted world, but this doesn't come into play when calculating the time to return to the plane.

We want $t_{\max}$ and t_{hit} to be equal:

$$t_{\max} = t_{\text{hit}} \implies \frac{u}{g\sin\beta} = \frac{2v\sin(\theta - \beta)}{g\cos\beta}. \tag{3.64}$$

Substituting the u from Eq. (3.63) into Eq. (3.64) gives

$$\frac{v\cos(\theta - \beta)}{g\sin\beta} = \frac{2v\sin(\theta - \beta)}{g\cos\beta} \implies 1 = 2\tan(\theta - \beta)\tan\beta. \tag{3.65}$$

You can use the sum formula for tan to verify that this leads to the same value of $\tan\theta$ we found in Eq. (3.61).

3.20. g's in a washer

1000 revolutions per minute equals 16.7 revolutions per second. A point on the surface of the drum moves a distance of $2\pi r = 2\pi(0.3\,\text{m}) = 1.88\,\text{m}$ during one revolution, so the speed of such a point is $v = (1.88\,\text{m})(16.7\,\text{s}^{-1}) \approx 31.4\,\text{m/s}$. The radial acceleration is therefore $a_r = v^2/r = (31.4\,\text{m/s})^2/(0.3\,\text{m}) = 3300\,\text{m/s}^2$. Since one g is about $10\,\text{m/s}^2$, this acceleration is equivalent to about 330 g's. That's huge!

REMARK: Since 31.4 m/s equals about 70 miles per hour, the results in this problem carry over to the spinning wheels on a car moving at a good highway clip. Note that it is irrelevant that the center of the wheel is moving down the road, as opposed to remaining stationary like the washing machine. In the reference frame moving along with the car, the wheel is simply spinning in place, so we have the same setup as with the washing machine. Adding on the constant velocity of the car to transform to the reference frame of the ground doesn't affect the acceleration of a point on the rim, because the derivative of a constant velocity is zero.

3.21. Acceleration after one revolution

The tangential acceleration is always a_t, and the radial acceleration is v^2/R. So our goal is to find the value of v after one revolution. The time it takes to complete one revolution is given by $a_t t^2/2 = 2\pi R \implies t = \sqrt{4\pi R/a_t}$. The speed after one revolution is therefore $v = a_t t = \sqrt{4\pi R a_t}$. (This also follows from the standard $v = \sqrt{2ad}$ kinematic result.) The radial acceleration is then $a_r = v^2/R = 4\pi a_t$, which is more than 12 times a_t (a surprisingly large factor). The angle that the acceleration vector **a** makes with the radial direction is given by $\tan\theta = a_t/a_r = a_t/(4\pi a_t) = 1/4\pi$. This angle is about $4.5°$, which means that **a** points only slightly away from the radial direction.

The angle doesn't depend on a_t or R, because an angle is a dimensionless quantity, and it is impossible to form a dimensionless quantity from a_t (which has units of m/s^2) and R (which has units of m).

REMARK: The angle θ that **a** makes with the radial direction starts off at $90°$ (when $v = 0 \implies a_r = 0$) and approaches zero after a long time (when $v \to \infty \implies a_r \to \infty$). From the above reasoning, you can show that the general result for θ is $\tan\theta = R/2d$, where d is the distance traveled around the circle. The angle that **a** makes with the radial direction therefore equals, for example, $45°$ when the car has traveled a distance $d = R/2$, which corresponds to $28.6°$ around the circle (half of a radian).

3.22. Equal acceleration components

(a) The acceleration components are $a_r = v^2/R$ and $a_t = dv/dt$. We are told that $|dv/dt| = v^2/R$. Let's assume for now that dv/dt is positive (so that the object is speeding up), in which case we can ignore the absolute value operation. Separating variables in $dv/dt = v^2/R$ and integrating gives

$$\int_{v_0}^{v} \frac{dv'}{v'^2} = \int_0^t \frac{dt'}{R} \implies -\frac{1}{v}\bigg|_{v_0}^{v} = \frac{t}{R}$$

$$\implies \frac{1}{v_0} - \frac{1}{v} = \frac{t}{R} \implies v(t) = \frac{1}{\frac{1}{v_0} - \frac{t}{R}}. \quad (3.66)$$

If dv/dt were negative (so that the object were slowing down), then $|dv/dt|$ would be equal to $-dv/dt$. The negative sign would carry through the above calculation, and we would end up with $v(t) = 1/(1/v_0 + t/R)$.

The distance (arclength) traveled, s, equals the integral of v. That is, $s = \int v\,dt$.

Assuming that dv/dt is positive, this gives

$$s(t) = \int_0^t \frac{dt}{\frac{1}{v_0} - \frac{t}{R}} = -R \ln\left(\frac{1}{v_0} - \frac{t}{R}\right)\Big|_0^t$$

$$= -R\left[\ln\left(\frac{1}{v_0} - \frac{t}{R}\right) - \ln\left(\frac{1}{v_0}\right)\right]$$

$$= -R \ln\left(\frac{1/v_0 - t/R}{1/v_0}\right) = -R \ln\left(1 - \frac{v_0 t}{R}\right). \tag{3.67}$$

If dv/dt is negative, the distance comes out to be $s(t) = R \ln(1 + v_0 t/R)$.

LIMITS: Using the Taylor approximation $\ln(1 - \epsilon) \approx -\epsilon$, we find that if t is small then $s(t)$ for the $dv/dt > 0$ case behaves like $s(t) \approx -R(-v_0 t/R) = v_0 t$. This is correct, because the object hasn't had any time to change its speed. The $s(t)$ for the $dv/dt < 0$ case also correctly reduces to $v_0 t$.

(b) In the case where dv/dt is positive, the special value of t is $T = R/v_0$. At this time, both the v in Eq. (3.66) and the s in Eq. (3.67) go to infinity. After this time, the stated motion is impossible.

In the case where dv/dt is negative, v goes to zero as $t \to \infty$. But it goes to zero slowly enough so that the distance s diverges (slowly, like a log) as $t \to \infty$.

LIMITS: Small v_0 implies large T, and large R also implies large T. These make intuitive sense.

3.23. Horizontal acceleration

Let θ be the angular position below the horizontal. Then the height fallen is $R + R\sin\theta$, which gives a speed of $v = \sqrt{2gh} = \sqrt{2gR(1 + \sin\theta)}$. The radial acceleration is then $a_r = v^2/R = 2g(1 + \sin\theta)$. The tangential acceleration comes from the tangential component of gravity, so it is simply $a_t = g\cos\theta$. The total acceleration is horizontal if the vertical components of a_r and a_t cancel, as shown in Fig. 3.27.[2] These two vertical components are $a_r \sin\theta$ upward and $a_t \cos\theta$ downward. So we want

$$a_r \sin\theta = a_t \cos\theta$$
$$\implies 2g(1 + \sin\theta) \cdot \sin\theta = g\cos\theta \cdot \cos\theta$$
$$\implies 2\sin\theta + 2\sin^2\theta = \cos^2\theta$$
$$\implies 2\sin\theta + 2\sin^2\theta = 1 - \sin^2\theta$$
$$\implies 3\sin^2\theta + 2\sin\theta - 1 = 0$$
$$\implies (3\sin\theta - 1)(\sin\theta + 1) = 0$$
$$\implies \sin\theta = \frac{1}{3}, \tag{3.68}$$

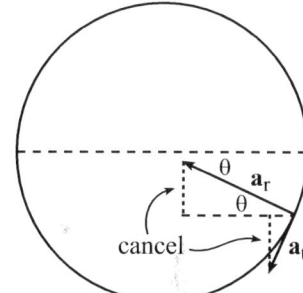

Figure 3.27

which gives $\theta \approx 19.5°$ or $\theta \approx 160.5°$. The positions corresponding to these two angles are shown in Fig. 3.28.

There is also another root of the above quadratic equation, namely $\sin\theta = -1 \implies \theta = -90°$. This corresponds to the top of the hoop, where $a_r = a_t = 0$. So the acceleration does indeed have a zero vertical component there. But it's semantics as to whether or not the zero vector is "horizontal."

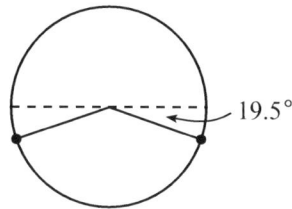

Figure 3.28

You can also solve this problem by considering the forces on the bead; see Problem 5.22.

---

[2] A common mistake is to say that the vertical component of a_r should cancel the downward acceleration g due to gravity. This isn't correct, because part of the gravitational force (the radial component) is already "included" in the radial acceleration. So you'd be double counting this component of gravity.

Chapter 4

F=ma

4.1 Introduction

Newton's laws

In the preceding two chapters, we dealt with *kinematics*. We took the motions of objects as given and then looked at positions, velocities, and accelerations as functions of time. We weren't concerned with the forces that caused the objects' motions. We will now deal with *dynamics*, where the goal is to understand *why* objects move the way they do. This chapter and the following ones will therefore be concerned with force, mass, energy, momentum, etc.

The motion of any object is governed by Newton's three laws:

- First Law: "A body moves with constant velocity (which may be zero) unless acted on by a force."

  If you think hard about this law, it seems a bit circular because we haven't defined what a force is. But if you think harder, there is in fact some content there. See Section 3.1 in Morin (2008) for a discussion of this.

- Second Law: "The rate of change of the momentum of a body equals the force acting on the body."

  In cases where the mass of the body doesn't change (we'll deal with the more general case in Chapter 6), this law becomes

  $$\mathbf{F} = m\mathbf{a}. \tag{4.1}$$

  This is a vector equation, so it is really three equations, namely $F_x = ma_x$, $F_y = ma_y$, and $F_z = ma_z$.

- Third Law: "Given two bodies A and B, if A exerts a force on B, then B exerts an equal and opposite force on A."

  As we'll see in Chapter 6, this law basically postulates conservation of momentum. There is a great deal of physical content in this law; it says that things don't happen in isolation by magic. Instead, if an object feels a force, then there must be another object somewhere feeling the opposite force.

The second law, $\mathbf{F} = m\mathbf{a}$, is the one we'll get the most mileage out of. The unit of force is called a newton (N), and from $F = ma$ we see that $1\,\text{N} = 1\,\text{kg m/s}^2$. A few types of forces (gravitational, tension, normal, friction, and spring) come up again and again, so let's take a look at each of these in turn.

Gravity

The gravitational force on an object near the surface of the earth is proportional to the mass of the object. More precisely, the force is mg downward, where $g = 9.8 \text{ m/s}^2$. This mg force is a special case of the more general gravitational force we will encounter in Chapter 11. Substituting mg for the force F in Newton's second law quickly gives $mg = ma \implies a = g$. That is, all objects fall with the same acceleration (in the absence of air resistance).

Tension

If you pull on a rope, the rope pulls back on you with the same force, by Newton's third law. The magnitude of this force is called the tension in the rope. As far as the direction of the tension goes, the question, "At a given interior point in the rope, which way does the tension point?" can't be answered. What we *can* say is that the tension at the given point pulls leftward on the point just to its right, and pulls rightward on the point just to its left. Equivalently, the tension pulls leftward on someone holding the right end of the rope, and it pulls rightward on someone holding the left end of the rope.

Normal force

Whereas a tension arises from a material resisting being stretched, a normal force arises from a material resisting being compressed. If you push leftward on the right face of a wooden block, then the normal force from the block pushes rightward on your hand. And similarly, the leftward force from your hand is itself a normal force pushing on the block. In the case where you push inward on the ends of an object shaped like a rod, people sometimes say that there is a "negative tension" in the rod, instead of calling it a normal force at the ends. But whatever name you want to use, the rod pushes back on you at the ends. If instead of a rigid rod we have a flexible rope, then the rope can support a tension, but not a normal force.

Friction

The friction force between two objects is extremely complicated on a microscopic scale. But fortunately we don't need to understand what is going on at that level to get a rough handle on friction forces. To a good approximation under most circumstances, we can say the following things about kinetic friction (where two objects are moving with respect to each other) and static friction (where two objects are at rest with respect to each other).

- KINETIC FRICTION: If there is slipping between two objects, then to a good approximation under non-extreme conditions, the friction force is proportional to the normal force between the objects, with the constant of proportionality (called the *coefficient of kinetic friction*) labeled as μ_k:

$$F_k = \mu_k N. \tag{4.2}$$

  The friction force is independent of the contact area and relative speed. The direction of the friction force on a given object is opposite to the direction of the velocity of that object relative to the other object.

- STATIC FRICTION: If there is no slipping between two objects, then to a good approximation under non-extreme conditions, the *maximum value* of the friction force is proportional to the normal force between the objects, with the constant of proportionality (called the *coefficient of static friction*) labeled as μ_s:

$$F_s \leq \mu_s N. \tag{4.3}$$

  As in the kinetic case, the friction force is independent of the contact area. Note well the *equality* in Eq. (4.2) and the *inequality* in Eq. (4.3). Equation (4.3) gives only an *upper limit* on the static friction force. If you push on an object with a force that is smaller than $\mu_s N$, then the static friction force is exactly equal and opposite to your force, and the object stays at rest.[1] But if you increase your force so that it exceeds $\mu_s N$, then the maximum friction force isn't enough to keep the object at rest. So it will move, and the friction force will abruptly drop to the kinetic value of $\mu_k N$. (It turns out that μ_k is always less than or equal to μ_s; see Problem 4.1 for an explanation why.)

The expressions in Eqs. (4.2) and (4.3) will of course break down under extreme conditions (large normal force, high relative speed, pointy shapes). But they work fairly well under normal conditions. Note that the coefficients of kinetic and static friction, μ_k and μ_s, are properties of *both* surfaces together. A single surface doesn't have a coefficient of friction. What matters is how two surfaces interact.

Spring force

To a good approximation for small displacements, the restoring force from a spring is proportional to the stretching distance. That is, $F = -kx$, where k is the *spring constant*. A large value of k means a stiff spring; a small value means a weak spring. The reason why the $F = -kx$ relation is a good approximation for small displacements in virtually any system is explained in Problem 10.1.

If x is positive then the force F is negative; and if x is negative then F is positive. So $F = -kx$ does indeed describe a restoring force, where the spring always tries to bring x back to zero. The $F = -kx$ relation, known as *Hooke's law*, breaks down if x is too large, but we'll assume that it holds for the setups we're concerned with.

The tension and normal forces discussed above are actually just special cases of spring forces. If you stand on a floor, the floor acts like a very stiff spring. The matter in the floor compresses a tiny amount, exactly the amount that makes the upward "spring" force (which we call a normal force in this case) be equal to your weight.

We'll always assume that our springs are massless. Massive springs can get very complicated because the force (the tension) will in general vary throughout the spring. In a massless spring, the force is the same everywhere in it. This follows from the reasoning in Problem 4.3(a).

Centripetal force

The centripetal force is the force that keeps an object moving in a circle. Since we know from Eq. (3.7) that the acceleration for uniform (constant speed) circular motion points radially inward with magnitude $a = v^2/r$, the centripetal force likewise points radially inward with magnitude $F = ma = mv^2/r$. This force might be due to the tension in a string, or the friction force acting on a car's tires as it rounds a corner, etc. Since $v = r\omega$, we can also write F as $mr\omega^2$.

The term "centripetal" isn't the same type of term as the above "gravity," "tension," etc. descriptors, because the latter terms describe the *type* of force, whereas "centripetal" simply describes the *direction* of the force (radially inward). The word "centripetal" is therefore more like the words "eastward" or "downward," etc. For example, we might say, "The downward force is due to gravity," or "The centripetal force is due to the tension in a string." The centripetal force is *not* a magical special new kind of force. It is simply one of the standard forces (or a combination of them) that points radially inward, and whose magnitude we know always equals mv^2/r.

Free-body diagrams

The "**F**" in Newton's second law in Eq. (4.1) is the *total* force on an object, so it is important to determine what all the various forces are. The best way to do this is to draw a picture. The picture of an object that shows all of the forces acting on it is called a *free-body diagram*. More precisely,

---

[1]The friction force certainly can't be *equal* to $\mu_s N$ in this case, because if you push rightward with a very small force, then the leftward (incorrect) $\mu_s N$ friction force would cause there to be a nonzero net force, which would hurl the object leftward back toward you!

4.1. INTRODUCTION

a free-body diagram shows all of the *external* forces (that is, forces due to other objects) acting on a given object. There are undoubtedly also *internal* forces acting within the object; each atom might be pushing or pulling on the atom next to it. But these internal forces cancel in pairs (by Newton's third law), so they don't produce any acceleration of the object. Only external forces can do that. (We'll assume we have a rigid object, so that the distance between any two given points remains fixed.)

In simple cases (for example, ones involving only one force), you can get away with not drawing a free-body diagram. But in more complicated cases (for example, ones involving forces pointing in various directions), a diagram is absolutely critical. A problem is often hopeless without a diagram, but trivial with one.

The length of a force vector is technically a measure of the magnitude of the force. But when drawing a free-body diagram, the main point is just to indicate all the forces that exist. In general we don't yet know the relative sizes, so it's fine to give all the vectors the same length (unless it's obvious that a certain force is larger than another). Also, the exact location of each force vector isn't critical (at least in this chapter), although the most sensible thing to do is to draw the vector near the place where the force acts. However, when we discuss torque in Chapter 7, the location of the force *will* be important.

Note that due to Newton's third law, for every force vector that appears in the free-body diagram for one object, there is an opposite force vector that appears in the free-body diagram for another object. An example involving two blocks on a table is shown in Fig. 4.1. If a person applies a force F to the left block, then the two free-body diagrams are shown (assume there is no friction from the table). Note that the force pushing the right block rightward is *only* the normal force between the blocks, and *not* the applied force F. True, the N force wouldn't exist without the F force, but the right block feels only the N force; it doesn't care about the original cause of N.

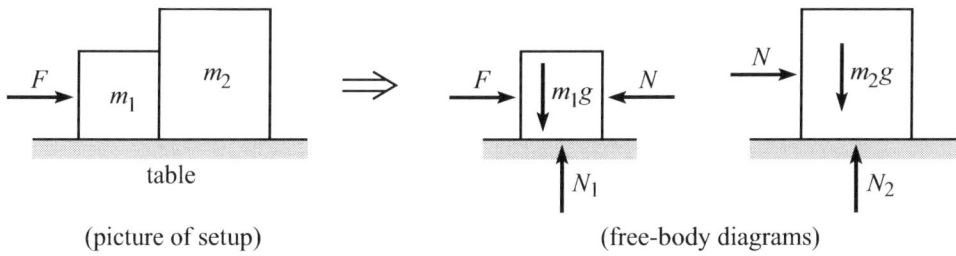

Figure 4.1

If we happened to be concerned also with the free-body diagram for the person applying the force F, then we would draw a force F acting leftward, along with a downward mg gravitational force and an upward normal force from the ground (and also probably a rightward friction force from the ground). Likewise, the free-body diagram for the table would involve gravity and the N_1 and N_2 normal forces pointing downward, along with upward normal forces from the ground at the bases of the legs. But if we're concerned only with the blocks, then all of this other information is irrelevant.

If the direction of the acceleration of an object is known, it is often helpful to draw the acceleration vector in the free-body diagram. But you should be careful to draw this vector with a dotted line or something similar, so that you don't mistake it for a force. Remember that although F equals ma, the quantity ma is *not* a force; see the last section of this introduction.

Atwood's machines

The name *Atwood's machine* is the term used for any system of pulleys, strings, and masses. Although a subset of these systems is certainly very useful in everyday life (a "block and tackle" enables you to lift heavy objects; see Problem 4.4), the main reason for all the Atwood's problems in this chapter is simply that they're good practice for drawing free-body diagrams and applying $F = ma$. An additional ingredient in solving any Atwood's problem is the so-called "conservation of string" relation. This is the condition that the length of any given string doesn't change. This constrains the motion of the various masses and pulleys. A few useful Atwood's facts that come up again and again are derived in Problem 4.3. In this chapter we will assume that all strings and pulleys are massless.

The four forces

Having discussed many of the forces we see in everyday life, we should make at least a brief mention of where all these forces actually come from. There are four known fundamental forces in nature:

- GRAVITATIONAL: Any two masses attract each other gravitationally. We are quite familiar with the gravitational force due to the earth. The gravitational force between everyday-sized objects is too small to observe without sensitive equipment. But on the planetary scale and larger, the gravitational force dominates the other three forces.

- ELECTROMAGNETIC: The single word "electromagnetic" is indeed the proper word to use here, because the electric and magnetic forces are two aspects of the same underlying force. (In some sense, the magnetic force can be viewed as a result of combining the electric force with special relativity.) Virtually all everyday forces have their origin in the electric force. For example, a tension in a string is due to the electric forces holding the molecules together in the string.

- WEAK: The weak force is responsible for various nuclear processes; it isn't too important in everyday life.

- STRONG: The strong force is responsible for holding the protons and neutrons together in a nucleus. Without the strong force, matter as we know it wouldn't exist. But taking the existence of matter for granted, the strong force doesn't show up much in everyday life.

ma is not a force!

Newton's second law is "F equals ma," which says that ma equals a force. Does this imply that ma *is* a force? Absolutely not. What the law says is this: Write down the sum of all the forces on an object, and also write down the mass times the acceleration of the object. The law then says that these two quantities have the same value. This is what a physical law does; it says to take two things that aren't obviously related, and then demand that they are equal. In a simple freefall setup, the $F = ma$ equation is $mg = ma$, which tells us that a equals g. But just because $a = g$, this doesn't mean that the two side of $mg = ma$ represent the same type of thing. The left side is a force, the right side is a mass times an acceleration. So when drawing a free-body diagram, you should *not* include ma as one of the forces. If you do, you will end up double counting things. However, as mentioned above, it is often helpful to indicate the acceleration of the object in the free-body diagram. Just be careful to distinguish this from the forces by drawing it with a dotted line.

4.2 Multiple-choice questions

4.1. Two people pull on opposite ends of a rope, each with a force F. The tension in the rope is

(a) $F/2$ (b) F (c) $2F$

4.2. MULTIPLE-CHOICE QUESTIONS

4.2. You accelerate the two blocks in Fig. 4.2 by pushing on the bottom block with a force F. The top block moves along with the bottom block. What force directly causes the top block to accelerate?

(a) the normal force between the blocks

(b) the friction force between the blocks

(c) the gravitational force on the top block

(d) the force you apply to the bottom block

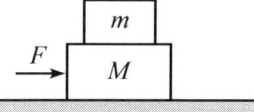

Figure 4.2

4.3. Three boxes are pushed with a force F across a frictionless table, as shown in Fig. 4.3. Let N_1 be the normal force between the left two boxes, and let N_2 be the normal force between the right two boxes. Then

(a) $F = N_1 = N_2$

(b) $F + N_1 = N_2$

(c) $F > N_1 = N_2$

(d) $F < N_1 < N_2$

(e) $F > N_1 > N_2$

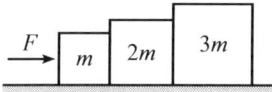

Figure 4.3

4.4. Two blocks with masses 2 kg and 1 kg lie on a frictionless table. A force of 3 N is applied as shown in Fig. 4.4. What is the normal force between the blocks?

(a) 0 (b) 0.5 N (c) 1 N (d) 2 N (e) 3 N

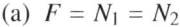

Figure 4.4

4.5. In the system shown in Fig. 4.5, the ground is frictionless, the blocks have mass m and $2m$, and the string connecting them is massless. If you accelerate the system to the right, as shown, the tension is the same everywhere throughout the string connecting the masses because

(a) the string is massless

(b) the ground is frictionless

(c) the ratio of the masses is 2 to 1

(d) the acceleration of the system is nonzero

(e) The tension is the same throughout any string; no conditions are necessary.

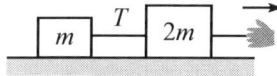

Figure 4.5

4.6. You are in a plane accelerating down a runway during takeoff, and you are holding a pendulum (say, a shoe hanging from a shoelace). The string of the pendulum

(a) hangs straight downward

(b) hangs downward and forward, because the net force on the pendulum must be zero

(c) hangs downward and forward, because the net force must be nonzero

(d) hangs downward and backward, because the net force must be zero

(e) hangs downward and backward, because the net force must be nonzero

4.7. When you stand at rest on a floor, you exert a downward normal force on the floor. Does this force cause the earth to accelerate in the downward direction?

(a) Yes, but the earth is very massive, so you don't notice the motion.

(b) Yes, but you accelerate along with the earth, so you don't notice the motion.

(c) No, because the normal force isn't a real force.

(d) No, because you are also pulling on the earth gravitationally.

(e) No, because there is also friction at your feet.

4.8. If you stand at rest on a bench, the bench exerts a normal force on you, equal and opposite to your weight. Which force is related to this normal force by Newton's third law?

(a) the gravitational force from the earth on you

(b) the gravitational force from you on the earth

(c) the normal force from you on the bench

(d) none of the above

4.9. The driver of a car steps on the gas, and the car accelerates with acceleration a. When writing down the horizontal $F = ma$ equation for the car, the "F" acting on the car is

(a) the normal force between the tires and the ground

(b) the friction force between the tires and the ground

(c) the force between the driver's foot and the pedal

(d) the energy obtained by burning the gasoline

(e) the backward friction force that balances the forward ma force

4.10. The static friction force between a car's tires and the ground can do all of the following *except*

(a) speed the car up

(b) slow the car down

(c) change the car's direction

(d) It can do all of the above things.

4.11. A car is traveling forward along a road. The driver wants to arrange for the car's acceleration to point diagonally backward and leftward. The driver should

(a) turn right and accelerate

(b) turn right and brake

(c) turn left and accelerate

(d) turn left and brake

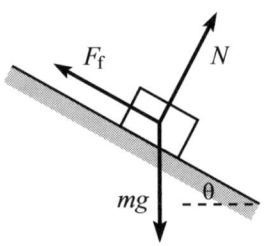

Figure 4.6

4.12. A block is at rest on a plane inclined at angle θ. The forces on it are the gravitational, normal, and friction forces, as shown in Fig. 4.6. These are *not* drawn to scale. Which of the following statements is *always* true, for any θ?

(a) $mg \leq N$ and $mg \leq F_f$

(b) $mg \geq N$ and $mg \geq F_f$

(c) $F_f \not> N$

(d) $F_f + N = mg$

(e) $F_f > N$ if $\mu_s > 1$

4.13. A block sits on a plane, and there is friction between the block and the plane. The plane is accelerated to the right. If the block remains at the same position on the plane, which of the following pictures might show the free-body diagram for the block? (All of the vectors shown are forces.)

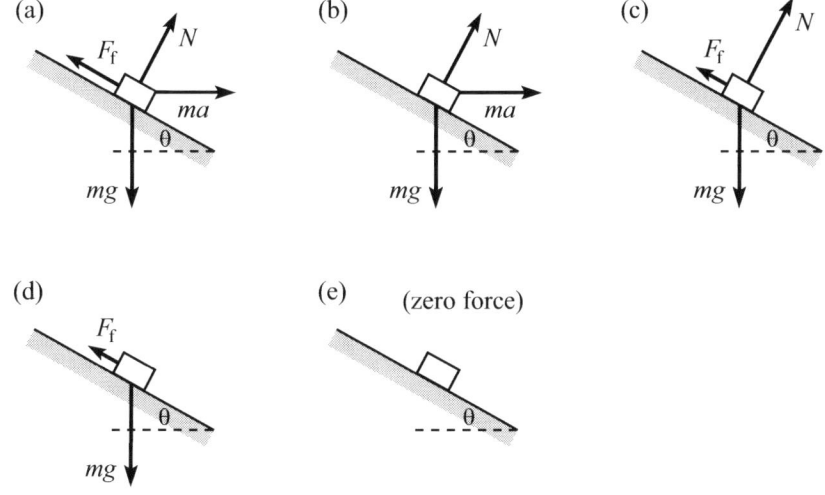

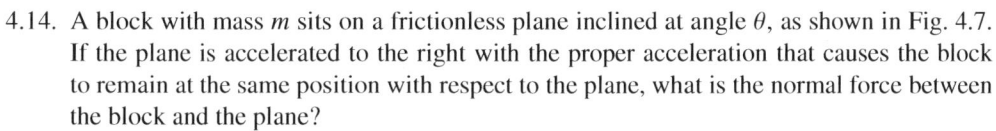

4.14. A block with mass m sits on a frictionless plane inclined at angle θ, as shown in Fig. 4.7. If the plane is accelerated to the right with the proper acceleration that causes the block to remain at the same position with respect to the plane, what is the normal force between the block and the plane?

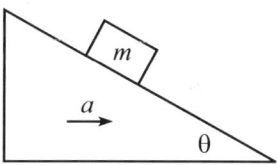

Figure 4.7

(a) mg (b) $mg\sin\theta$ (c) $mg/\sin\theta$ (d) $mg\cos\theta$ (e) $mg/\cos\theta$

4.15. A bead is arranged to move with *constant* speed around a hoop that lies in a vertical plane. The magnitude of the net force on the bead is

(a) largest at the bottom

(b) largest at the top

(c) largest at the side points

(d) the same at all points

4.16. A toy race car travels through a loop-the-loop (a circle in a vertical plane) on a track. Assuming that the speed at the top of the loop is above the threshold to remain in contact with the track, the car's acceleration at the top is

(a) downward and larger than g

(b) downward and smaller than g

(c) zero

(d) upward and smaller than g

(e) upward and larger than g

4.17. A plane in a holding pattern is flying in a horizontal circle at constant speed. Which of the following free-body diagrams best illustrates the various forces acting on the plane at the instant shown? (See Problem 4.20 for a quantitative treatment of this setup.)

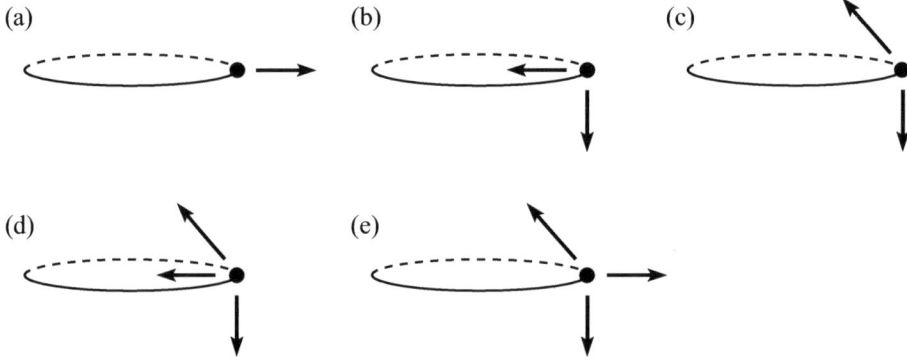

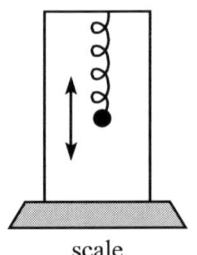

scale

Figure 4.8

4.18. A mass hangs from a spring and oscillates vertically. The top end of the spring is attached to the top of a box, and the box is placed on a scale, as shown in Fig. 4.8. The reading on the scale is largest when the mass is

(a) at its maximum height

(b) at its minimum height

(c) at the midpoint of its motion

(d) All points give the same reading.

4.19. A spring with spring constant k hangs vertically from a ceiling, initially at its relaxed length. You attach a mass m to the end and bring it down to a position that is $3mg/k$ below the initial position. You then let go. What is the upward acceleration of the mass right after you let go?

(a) 0 (b) g (c) $2g$ (d) $3g$ (e) $4g$

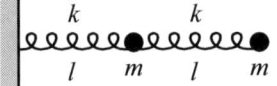

Figure 4.9

4.20. Two springs both have spring constant k and relaxed length zero. They are each stretched to a length ℓ and then attached to two masses and a wall, as shown in Fig. 4.9. The masses are simultaneously released. Immediately afterward, the magnitudes of the accelerations of the left and right masses are, respectively,

(a) $2k\ell/m$ and $k\ell/m$

(b) $k\ell/m$ and $2k\ell/m$

(c) $k\ell/m$ and $k\ell/m$

(d) 0 and $2k\ell/m$

(e) 0 and $k\ell/m$

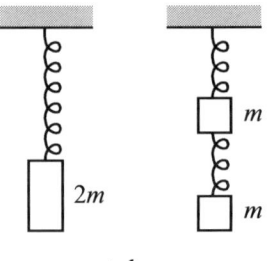

not drawn to scale

Figure 4.10

4.21. A mass $2m$ suspended from a given spring causes it to stretch relative to its relaxed length. The mass and the spring are then each cut into two identical pieces and connected as shown in Fig. 4.10. Is the bottom of the lower mass higher than, lower than, or at the same height as the bottom of the original mass? (This one takes a little thought.)

(a) higher

(b) lower

(c) same height

4.3. PROBLEMS

4.22. What is the conservation-of-string relation for the Atwood's machine shown in Fig. 4.11? All of the accelerations are defined to be positive upward.

(a) $a_3 = -2(a_1 + a_2)$
(b) $a_3 = -(a_1 + a_2)$
(c) $a_3 = -(a_1 + a_2)/2$
(d) $a_3 = -(a_1 + a_2)/4$
(e) $a_3 = -2a_2$

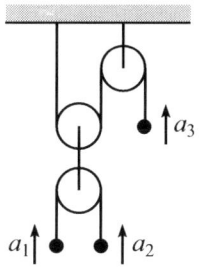

Figure 4.11

4.23. What is the conservation-of-string relation for the Atwood's machine shown in Fig. 4.12? All of the accelerations are defined to be positive upward.

(a) $a_3 = -a_1 - a_2$
(b) $a_3 = -2a_1 - 2a_2$
(c) $a_3 = -4a_2$
(d) $2a_3 = -a_1 - a_2$
(e) $4a_3 = -a_1 - a_2$

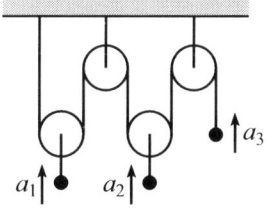

Figure 4.12

4.24. What is the conservation-of-string relation for the Atwood's machine shown in Fig. 4.13? All of the accelerations are defined to be positive upward.

(a) $a_1 = a_3$
(b) $a_1 = -a_3$
(c) $a_2 = -(a_1 + a_3)/2$
(d) $a_2 = -(a_1 + a_3)$
(e) $a_2 = -2(a_1 + a_3)$

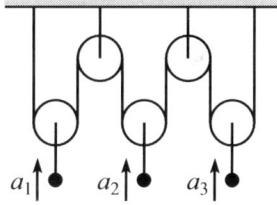

Figure 4.13

4.3 Problems

The first five problems are foundational problems.

4.1. Coefficients of friction

Explain why the coefficient of static friction, μ_s, must always be at least as large as the coefficient of kinetic friction, μ_k.

4.2. Cutting a spring in half

A spring has spring constant k. If it is cut in half, what is the spring constant of each of the resulting shorter springs?

4.3. Useful Atwood's facts

In the Atwood's machine shown in Fig. 4.14(a), the pulleys and strings are massless (as we will assume in all of the Atwood's problems in this chapter). Explain why (a) the tension is the same throughout the long string, as indicated, (b) the tension in the bottom string is twice the tension in the long string, as indicated, and (c) the acceleration of the right mass is negative twice the acceleration of the left mass.

Also, in Fig. 4.14(b), explain why (d) the acceleration of the left mass equals negative the average of the accelerations of the right two masses.

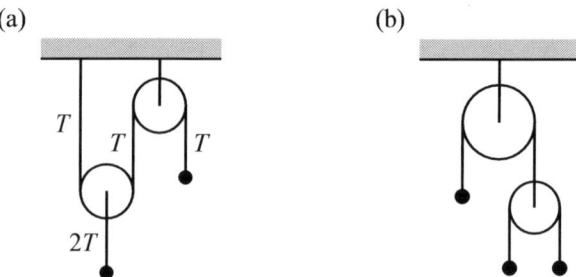

Figure 4.14

4.4. Block and tackle

(a) What force on the rope must be exerted by the person in Fig. 4.15(a) in order to hold up the block, or equivalently to move it upward at constant speed? The rope wraps twice around the top of the top pulley and the bottom of the bottom pulley. (Assume that the segment of rope attached to the center of the top pulley is essentially vertical.)

(b) Now consider the case where the person (with mass m) stands on the block, as shown in Fig. 4.15(b). What force is now required?

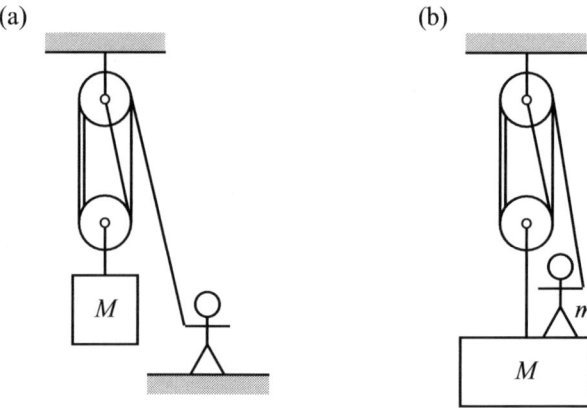

Figure 4.15

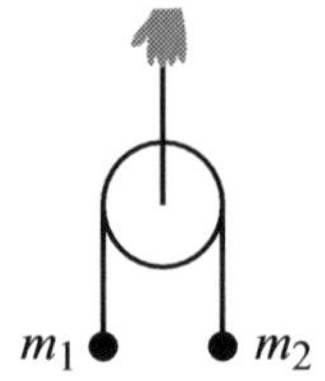

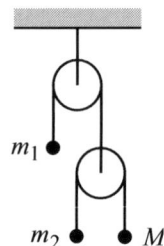

Figure 4.16

4.5. Equivalent mass

In Fig. 4.16 you support the pulley system, with your hand at rest. If you have your eyes closed and think that you are instead supporting a single mass M at rest, what is M in terms of m_1 and m_2? Is M simply equal to $m_1 + m_2$?

The following eight problems involve Atwood's machines. This large number of Atwood's problems shouldn't be taken to imply that they're terribly important in physics (they're not). Rather, they are included here because they provide good practice with $F = ma$.

4.6. Atwood's 1

Consider the Atwood's machine shown in Fig. 4.17. The masses are held at rest and then released. In terms of m_1 and m_2, what should M be so that m_1 doesn't move? What relation must hold between m_1 and m_2 so that such an M exists?

Figure 4.17

4.3. PROBLEMS

4.7. Atwood's 2

Consider the Atwood's machine shown in Fig. 4.18. Masses of m and $2m$ lie on a frictionless table, connected by a string that passes around a pulley. The pulley is connected to another mass of $2m$ that hangs down over another pulley, as shown. Find the accelerations of all three masses.

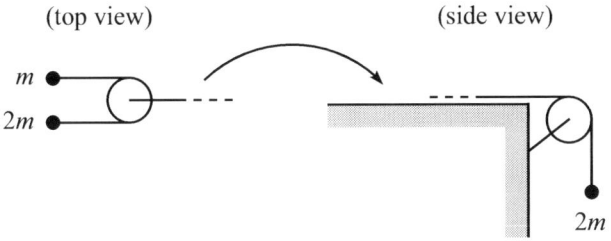

Figure 4.18

4.8. Atwood's 3

Consider the Atwood's machine shown in Fig. 4.19, with masses m, m, and $2m$. Find the acceleration of the mass $2m$.

4.9. Atwood's 4

In the Atwood's machine shown in Fig. 4.20, both masses are m. Find their accelerations.

4.10. Atwood's 5

Consider the triple Atwood's machine shown in Fig. 4.21. What is the acceleration of the rightmost mass? *Note*: The math isn't as bad as it might seem at first. You should take advantage of the fact that many of your $F = ma$ equations look very similar.

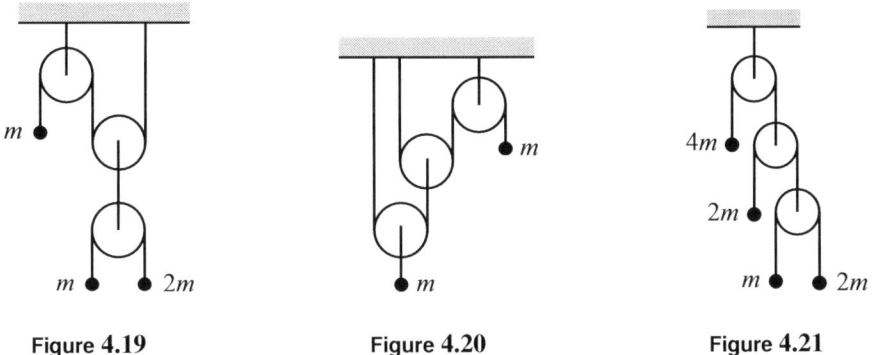

Figure 4.19   **Figure 4.20**   **Figure 4.21**

4.11. Atwood's 6

Consider the Atwood's machine shown in Fig. 4.22. The middle mass is glued to the long string. Find the accelerations of all three masses, and also the tension everywhere in the long string.

4.12. Atwood's 7

In the Atwood's machine shown in Fig. 4.23, both masses are m. Find their accelerations.

4.13. Atwood's 8

In the Atwood's machine shown in Fig. 4.24, both masses are m. The two strings that touch the center of the left pulley are both attached to its axle. Find the accelerations of the masses.

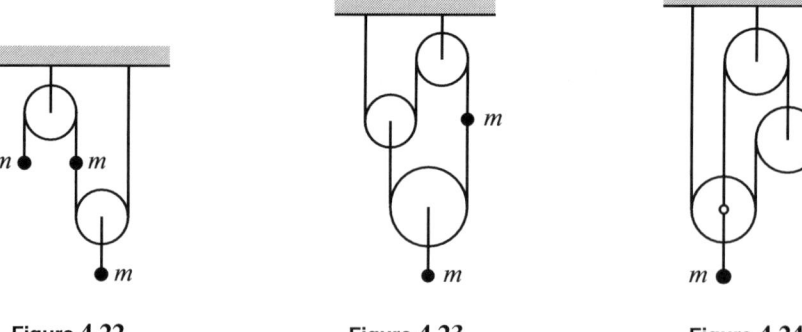

Figure 4.22   Figure 4.23   Figure 4.24

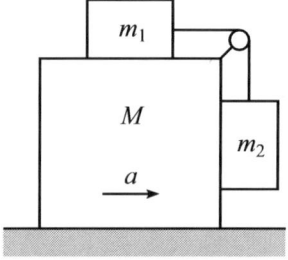

(side view)

Figure 4.25

4.14. No relative motion

All of the surfaces in the setup in Fig. 4.25 are frictionless. You push on the large block and give it an acceleration a. For what value of a is there no relative motion among the masses?

4.15. Slipping blocks

A block with mass m sits on top of a block with mass $2m$ which sits on a table. The coefficients of friction (both static and kinetic) between all surfaces are $\mu_s = \mu_k = 1$. A string is connected to each mass and wraps halfway around a pulley, as shown in Fig. 4.26. You pull on the pulley with a force of $6mg$.

(a) Explain why the bottom block must slip with respect to the table. *Hint*: Assume that it doesn't slip, and show that this leads to a contradiction.

(b) Explain why the top block must slip with respect to the bottom block. (Same hint.)

(c) What is the acceleration of your hand?

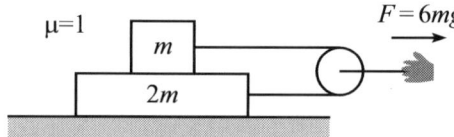

Figure 4.26

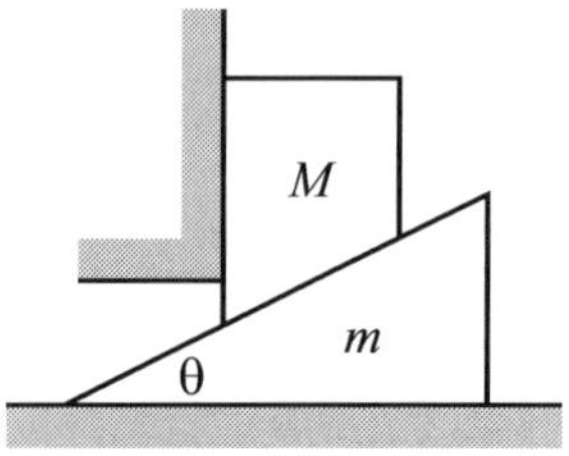

4.16. Block and wedge

A block with mass M rests on a wedge with mass m and angle θ, which lies on a table, as shown in Fig. 4.27. All surfaces are frictionless. The block is constrained to move vertically by means of a wall on its left side. What is the acceleration of the wedge?

4.17. Up and down a plane

A block with mass m is projected up along the surface of a plane inclined at angle θ. The initial speed is v_0, and the coefficients of both static and kinetic friction are equal to 1. The block reaches a highest point and then slides back down to the starting point.

(a) Show that in order for the block to in fact slide back down (instead of remaining at rest at the highest point), θ must be greater than $45°$.

(b) Assuming that $\theta > 45°$, find the times of the up and down motions.

(c) Assuming that $\theta > 45°$, is the total up and down time longer or shorter than the total time it would take (with the same initial v_0) if the plane were frictionless? Or does the answer to this question depend on what θ is? (The solution to this gets a little messy.)

Figure 4.27

4.3. PROBLEMS

4.18. Rope in a tube

A rope is free to slide frictionlessly inside a circular tube that lies flat on a horizontal table. In part (a) of Fig. 4.28, the rope moves at constant speed. In part (b) of the figure, the rope is at rest, and you pull on its right end to give it a tangential acceleration. What is the direction of the net force on the rope in each case?

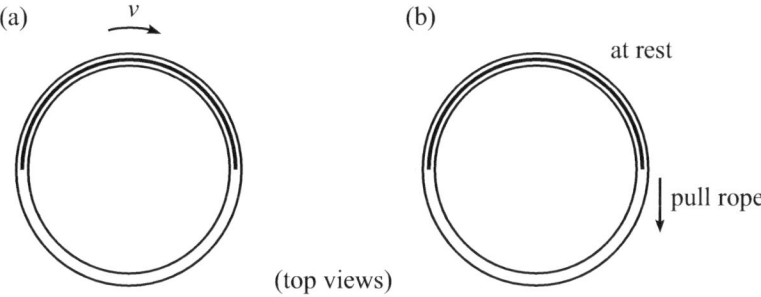

Figure 4.28

4.19. Circling bucket

You hold the handle of a bucket of water and swing it around in a vertical circle, keeping your arm straight. If you swing it around fast enough, the water will stay inside the bucket, even at the highest point where the bucket is upside down. What, roughly, does "fast enough" here mean? (You can specify the maximum time of each revolution.) Make whatever reasonable assumptions you want to make for the various parameters involved. You can work in the approximation where the speed of the bucket is roughly constant throughout the motion.

4.20. Banking an airplane

A plane in a holding patter flies at speed v in a horizontal circle of radius R. At what angle should the plane be banked so that you don't feel like you are getting flung to the side in your seat? At this angle, what is your apparent weight (that is, what is the normal force from the seat)?

4.21. Breaking and turning

You are driving along a horizontal straight road that has a coefficient of static friction μ with your tires. If you step on the brakes, what is your maximum possible deceleration? What is it if you are instead traveling with speed v around a bend with radius of curvature R?

4.22. Circle of rope

A circular loop of rope with radius R and mass density λ (kg/m) lies on a frictionless table and rotates around its center, with all points moving at speed v. What is the tension in the rope? *Hint*: Consider the net force on a small piece of rope that subtends an angle $d\theta$.

4.23. Cutting the string

A mass m is connected to the end of a massless string of length ℓ. The top end of the string is attached to a ceiling that is a distance ℓ above the floor. Initial conditions have been set up so that the mass swings around in a horizontal circle, with the string always making an angle θ with respect to the vertical, as shown in Fig. 4.29. If the string is cut, what horizontal distance does the mass cover between the time the string is cut and the time the mass hits the floor?

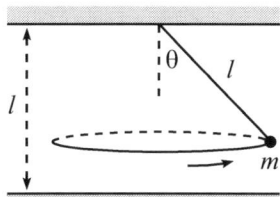

Figure 4.29

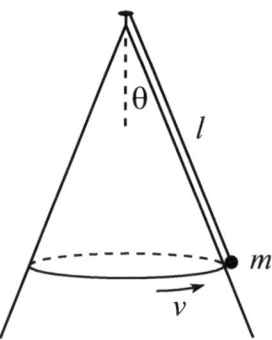

Figure 4.30

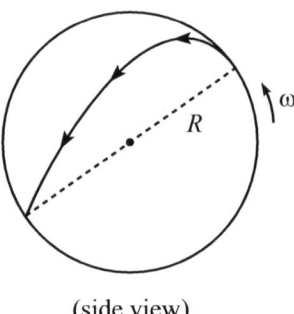

(side view)

Figure 4.31

4.24. **Circling around a cone**

A mass m is attached by a massless string of length ℓ to the tip of a frictionless cone, as shown in Fig. 4.30. The half-angle at the vertex of the cone is θ. If the mass moves around in a horizontal circle at speed v on the cone, find (a) the tension in the string, (b) the normal force from the cone, and (c) the maximum speed v for which the mass stays in contact with the cone.

4.25. **Penny in a dryer**

Consider a clothes dryer with radius R, which spins with angular frequency ω. (In other words, consider a cylinder that spins with angular frequency ω around its axis, which is oriented horizontally.) A small object, such as a penny, is in the dryer. The penny rotates along with the dryer and gets carried upward, but eventually loses contact with the dryer and sails through the air (as clothes do in a dryer), eventually coming in contact with the dryer again. Assume for simplicity that the coefficient of friction is very large, so that the penny doesn't slip with respect to the dryer as long as the normal force is nonzero.

Let's assume that you want the trajectory of the penny to look like the one shown in Fig. 4.31, starting and ending at diametrically opposite points. In order for this to happen, where must the penny lose contact with the dryer? (Give the angle θ with respect to the vertical.) What must ω be, in terms of g and R?

4.4 Multiple-choice answers

4.1. $\boxed{b}$ The tension is simply F. A common error is to double F (because there are two people pulling) and say that the tension is $2F$. This is incorrect, because every piece of the rope pulls on the piece to its right with a force F, and also on the piece to its left with a force F. This common value is the tension.

REMARK: If you try to change the setup by replacing one person with a wall, then the wall still pulls on the rope with a force F (assuming that the other person still does), so the setup hasn't actually changed at all. If you instead remove one person (say, the left one) and replace her with nothing, then the setup has certainly changed. The remaining (right) person will accelerate the rope rightward, and the tension will vary over the length (assuming that the rope has mass). Points closer to the left end don't have as much mass to their left that they need to accelerate, so the tension is smaller there.

4.2. $\boxed{b}$ Friction is the horizontal force that acts on the top block.

REMARK: The friction force wouldn't exist without the normal force between the blocks, which in turn wouldn't exist without the gravitational force on the top block. But these forces are vertical and therefore can't directly cause the horizontal acceleration of the top block. Likewise, the friction force wouldn't exist if you weren't pushing on the bottom block. But your force isn't what directly causes the acceleration of the top block; if the surfaces are greased down, then you can push on the bottom block all you want, and the top block won't move.

4.3. $\boxed{e}$ The three boxes all have the same acceleration; call it a. Then the force F equals $F = (6m)a$ because this is the force that accelerates all three boxes, which have a total mass of $6m$. Similarly, $N_1 = (5m)a$ because N_1 is the force that accelerates the right two boxes. And $N_2 = (3m)a$ because N_2 is the force that accelerates only the right box. Therefore, $F > N_1 > N_2$. As a double check, the net force on the middle block is $N_1 - N_2 = 5ma - 3ma = 2ma$, which is correctly $(2m)a$.

4.4. $\boxed{c}$ The acceleration of the system is $a = F/m = (3\,\text{N})/(3\,\text{kg}) = 1\,\text{m/s}^2$. The normal force N on the 1 kg block is what causes this block to accelerate at $1\,\text{m/s}^2$, so N must be given by $N = ma = (1\,\text{kg})(1\,\text{m/s}^2) = 1\,\text{N}$.

4.5. $\boxed{a}$ The massless nature of the string implies that the tension is the same everywhere throughout it, because if the tension varied along the length, then there would exist a massless piece that had a net force acting on it, yielding infinite acceleration. Conversely, if the string had mass, then the tension would have to vary along it, so that there would be a net force on each little massive piece, to cause the acceleration (assuming the acceleration is nonzero). So none of the other answers can be the reason why the tension is the same everywhere along the string.

4.6. $\boxed{e}$ The pendulum is accelerating forward (as is everything else in the plane), so there must be a forward net force on it. If the pendulum hangs *downward* and *backward*, then the tension force on the pendulum's mass is *upward* and *forward*. The upward component cancels the gravitational force (the weight), and the forward (uncanceled) component is what causes the forward acceleration.

REMARK: If the pendulum's mass is m, then from the above reasoning, the vertical component of the tension is mg, and the horizontal component is ma (where a is the acceleration of the plane and everything in it). So if the string makes an angle θ with the vertical, then $\tan\theta = ma/mg \implies a = g\tan\theta$. You can use this relation to deduce your acceleration from a measurement of the angle θ. Going in the other direction, a typical plane might have a takeoff acceleration of around 2.5 m/s^2, in which case we can deduce what θ is: $\tan\theta = a/g \approx 1/4 \implies \theta \approx 15°$. For comparison, a typical car might be able to go from 0 to 60 mph (27 m/s) in 7 seconds, which implies an acceleration of about 4 m/s^2 and an angle of $\theta = 22°$.

A more extreme case is a fighter jet taking off from an aircraft carrier. With the help of a catapult, the acceleration can be as large as $3g \approx 30 \text{ m/s}$. A pendulum in the jet would therefore hang at an angle of $\tan^{-1}(3) \approx 72°$, which is more horizontal than vertical. In the accelerating reference frame of the jet (accelerating frames are the subject of Chapter 12, so we're getting ahead of ourselves here), the direction of the hanging pendulum defines "downward." So the pilot effectively lives in a world where gravity points diagonally downward and backward at an angle of 72° with respect to the vertical. The direction of the jet's forward motion along the runway is therefore nearly *opposite* to this "downward" direction (as opposed to being roughly perpendicular to downward in the case of a passenger airplane). The jet pilot will therefore have the sensation that he is flying *upward*, even though he is actually moving horizontally along the runway. A possible dangerous consequence of this sensation is that if it is nighttime and there are minimal visual cues, the pilot may mistakenly try to correct this "error" by turning downward, causing the jet to crash into the ocean.

4.7. $\boxed{d}$ The normal force from you on the earth is equal and opposite to the gravitational force from you on the earth. (Yes, you pull on the earth, just as it pulls on you.) So the net force on the earth is zero, and it therefore doesn't accelerate.

REMARK: Similarly, *you* also don't accelerate, because the normal force from the earth on you is equal and opposite to the gravitational force from the earth on you. So the net force on you is zero. This wouldn't be the case if instead of standing at rest, you jump upward. You are now accelerating upward (while your feet are in contact with the ground). And consistent with this, the upward normal force from the earth on you is *larger* than the downward gravitational force from the earth on you. So the net force on you is upward. In contrast, after you leave the ground the normal force drops to zero, so the downward gravitational force is all there is, and you accelerate downward.

4.8. $\boxed{c}$ Newton's third law says that the forces that *two* bodies exert on *each other* are equal in magnitude and opposite in direction. The given force is the (upward) normal force from the bench on you. The two objects here are the *bench* and *you*, so the force that is related by the third law must be the (downward) normal force from you on the bench. Similarly, choices (a) and (b) are a third-law pair.

REMARK: The earth isn't relevant at all in the third-law statement concerning you and the bench. Of course, the gravitational force from the earth on you (that is, your weight) *is* related to the given normal force from the bench on you, but this relation does *not* involve the third law. It involves the *second* law. More precisely, the second law says that since you are at rest (and hence not accelerating), the total force on you must be zero. So the downward force from the earth on you (your weight) must be equal and opposite to the upward normal force from the bench on you. Note that since three objects (earth, you, bench) were mentioned in the preceding statement, there is no way

4.9. [b] The friction force is the horizontal force that makes the car accelerate; you won't go anywhere on ice. The other choices are incorrect because: (a) the normal force is vertical, (c) the force applied to the pedal is an internal force within the car, and only external forces appear in $F = ma$ (and besides, the force on the pedal is far smaller than the friction between the tires and the ground), (d) energy isn't a force, and (e) the friction force doesn't point backward, and ma isn't a force! (See the discussion on page 72.)

4.10. [d] When you step on the gas, the friction force speeds you up (if you are on ice, you won't go anywhere). When you hit the brakes, the friction force slows you down (if you are on ice, you won't slow down). And when you turn the steering wheel, the friction force is the centripetal force that causes you to move in the arc of a circle as you change your direction (if you are on ice, your direction won't change).

REMARK: Note that unless you are skidding (which rarely happens in everyday driving), the friction force between the tires and the ground is static and not kinetic. The point on a tire that is instantaneously in contact with the ground is instantaneously at rest. (The path traced out by a point on a rolling wheel is known as a *cycloid*, and the speed of the point is zero where it touches the ground.) If this weren't the case (that is, if we were perpetually skidding in our cars), then we would need to buy new tires every week, and we would be listening constantly to the sound of screeching tires.

4.11. [d] Braking will yield a backward component of the acceleration, and turning left will yield a leftward component. This leftward component is the centripetal acceleration for the circular arc (at least locally) that the car is now traveling in.

REMARK: Depending on what you do with the gas pedal, brake, and steering wheel, the total acceleration vector (or equivalently, the total force vector) can point in any horizontal direction. The acceleration can have a forward or backward component, depending on whether you are stepping on the gas or the brake. And it can have a component to either side if the car is turning. The relative size of these components is arbitrary, so the total acceleration vector can point in any horizontal direction.

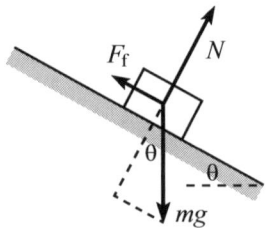

Figure 4.32

4.12. [b] The total acceleration (and hence total force) is zero. In Fig. 4.32 we have broken the $m\mathbf{g}$ force into its components parallel are perpendicular to the plane, and we have drawn the forces to scale. Zero net force perpendicular to the plane gives $N = mg \cos\theta$, which implies $mg \geq N$. And zero net force along the plane gives $F_f = mg \sin\theta$, which implies $mg \geq F_f$. The $\theta \to 0$ limit gives a counterexample to choices (a), (c), and (e), because $F_f \to 0$ when $\theta \to 0$. Choice (d) would be almost true if we were talking about vectors (the correct statement would be $\mathbf{F}_f + \mathbf{N} + m\mathbf{g} = 0$). But we're dealing with magnitudes here, so choice (d) is equivalent to $\sin\theta + \cos\theta = 1$, which isn't true.

4.13. [c] The gravitational force exists, as does the normal force (to keep the block from falling though the plane). The friction force might or might not exist (and it might point in either direction). So the correct answer must be (a), (b), or (c). But ma isn't a force, so we're left with choice (c) as the only possibility. If you want to draw the acceleration vector in a free-body diagram, you must draw it differently (say, with a dotted line) to signify that it isn't a force!

REMARK: Since we are assuming that the block remains at the same position on the plane, it has a nonzero a_x but a zero a_y. Given a_x, the (positive) horizontal component of $\mathbf{N}$ plus the (positive or negative or zero) horizontal component of $\mathbf{F}_f$ must equal ma_x. ($\mathbf{F}_f$ will point down along the plane if a_x is larger than a certain value; as an exercise, you can determine this value.) And zero a_y means that the upward components of $\mathbf{N}$ and $\mathbf{F}_f$ must balance the downward $m\mathbf{g}$ force.

4.4. MULTIPLE-CHOICE ANSWERS

4.14. $\boxed{e}$ Since the plane is frictionless, the only forces acting on the block are gravity and the normal force, as shown in Fig. 4.33. Since the block's acceleration is horizontal, the vertical component of the normal force must equal mg, to yield zero net vertical force. The right triangle then implies that $N = mg/\cos\theta$.

LIMITS: If $\theta = 0$ then $N = mg$. This makes sense, because the plane is horizontal and the normal force simply needs to balance the weight mg. If $\theta \to 90°$ then $N \to \infty$. This makes sense, because a needs to be huge to keep the block from falling.

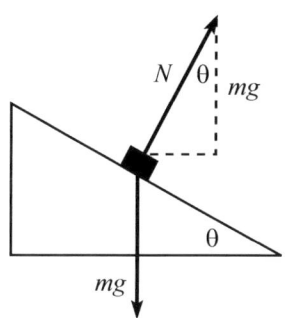

Figure 4.33

REMARK: Consider instead the case where there is friction between the block and the plane, and where the entire system is *static*. Then the normal force takes on the standard value of $N = mg\cos\theta$. This is most easily derived from the fact that there is no acceleration perpendicular to the plane, which then implies that N is equal to the component of gravity perpendicular to the plane; see Fig. 4.34. This should be contrasted with the original setup, where mg was equal to a component (the vertical component) of N. In any case, the normal force doesn't just magically turn out to be $mg\cos\theta$ or $mg/\cos\theta$ or whatever. It (or any other force) is determined by applying $F = ma$.

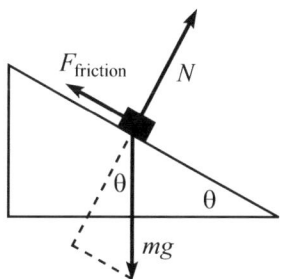

Figure 4.34

4.15. $\boxed{d}$ The magnitude of the radial acceleration is v^2/R, which is the same at all points because v is constant. And the tangential acceleration is always zero because again v is constant. So the net force always points radially inward with constant magnitude mv^2/R. The vertical orientation of the hoop is irrelevant in this question, given that the speed is constant. The answer would be the same if the hoop were horizontal.

REMARK: Be careful not to confuse the *total* force on the bead with the *normal* force from the hoop. The normal force *does* depend on the position of the bead. It is largest at the bottom of the hoop, because there the radial $F = ma$ equation (with inward taken to be positive for both N and a) is

$$N_{\text{bot}} - mg = \frac{mv^2}{R} \implies N_{\text{bot}} = \frac{mv^2}{R} + mg. \tag{4.4}$$

At the top of the hoop, the radial $F = ma$ equation (with inward again taken to be positive for both N and a) is

$$N_{\text{top}} + mg = \frac{mv^2}{R} \implies N_{\text{top}} = \frac{mv^2}{R} - mg. \tag{4.5}$$

If this is negative (if v is small), it just means that the normal force actually points radially outward (that is, upward).

4.16. $\boxed{a}$ In the threshold case where the car barely doesn't stay in contact with the track at the top, the normal force N is zero, so the car is in freefall (while moving sideways). The downward acceleration is therefore g. If the speed is above the threshold value, then N is nonzero. The total downward force, $F = mg + N$, is therefore larger than the mg due to gravity. The downward acceleration, which is $F/m = g + N/m$, is therefore larger than g.

REMARK: It isn't necessary to mention the v^2/R expression for the centripetal acceleration in this problem. But if you want to write down the radial $F = ma$ equation at the top of the loop, you can show as an exercise that the minimum speed required to barely maintain contact with the track at the top of the loop (that is, to make $N \geq 0$) is $v = \sqrt{gR}$.

4.17. $\boxed{c}$ Gravity acts downward. The force from the air must have an upward component to balance gravity (because there is zero vertical acceleration), and also a radially inward component to provide the nonzero centripetal acceleration. So the net force from the air points in a diagonal direction, upward and leftward. The gravitational and air forces are the only two forces, so the correct answer is (c).

REMARK: Choices (d) and (e) are incorrect because although they have the correct gravitational and air forces, they have an incorrect additional horizontal force. Remember that $ma = mv^2/r$ is *not* a force (see the discussion on page 72), so (d) can't be correct. Choice (e) would be correct if we were working in an accelerating frame and using fictitious forces. But we won't touch accelerating frames until Chapter 12.

4.18. $\boxed{b}$ At the bottom of the motion, the upward force from the spring on the mass is maximum (because the spring is stretched maximally there), which means that the downward force from the spring on the box is maximum (because the spring exerts equal and opposite forces at its ends). This in turn means that the upward force from the scale on the box is maximum (because the net force on the box is always zero, because it isn't accelerating). And this force is the reading on the scale.

4.19. $\boxed{c}$ The forces on the mass are the spring force, which is $k\,\Delta y = k(3mg/k) = 3mg$ upward, and the gravitational force, which is mg downward. The net force is therefore $2mg$ upward, so the acceleration is $2g$ upward.

4.20. $\boxed{e}$ The left mass feels forces from both springs, so it feels equal and opposite forces of $k\ell$. Its acceleration is therefore zero. The right mass feels only the one force of $k\ell$ from the right spring (directed leftward). Its acceleration is therefore $k\ell/m$ leftward. The amount that the left spring is stretched is completely irrelevant as far as the right mass goes. Also, as far as the left mass goes, it is irrelevant that the far ends of the springs are attached to different things (the immovable wall and the movable right mass), at least right at the start.

4.21. $\boxed{a}$ Let the stretching, relative to the equilibrium position, of the spring in the original scenario be 2ℓ. Then the given information tells us that half of the spring stretches by ℓ when its tension is $2mg$ (because the whole spring stretches by 2ℓ when its tension is $2mg$, and the tension is the same throughout). Therefore, the top half of the spring in the second scenario is stretched by ℓ, because it is holding up a total mass of $2m$ below it (this mass is split into two pieces, but that is irrelevant as far as the top spring is concerned). But the bottom half of the spring is stretched by only $\ell/2$, because it is holding up only a mass m below it (and $x \propto F$ by Hooke's law). The total stretch is therefore $\ell + \ell/2 = 3\ell/2$. This is less than 2ℓ, so the desired answer is "higher."

REMARK: This answer of "higher" can be made a little more believable by looking at some limiting cases. Consider a more general version of the second scenario, where we still cut the spring into equal pieces, but we now allow for the two masses to be unequal (although they must still add up to the original mass). Let the top and bottom masses be labeled m_t and m_b, and let the original mass be M. In the limit where $m_t = 0$ and $m_b = M$, we simply have the original setup, so the answer is "same height." In the limit where $m_t = M$ and $m_b = 0$, we have a mass M hanging from a shorter spring. But a shorter spring has a larger spring constant (see Problem 4.2), which means that it stretches less. So the answer is "higher." Therefore, since the $m_t = M/2$ and $m_b = M/2$ case presented in the problem lies between the preceding two cases with answers of "same height" and "higher," it is reasonable to expect that the answer to the original problem is "higher." You can also make a similar argument by splitting the original mass into two equal pieces, but now allowing the spring to be cut into unequal pieces.

4.22. $\boxed{b}$ $(a_1 + a_2)/2$ is the acceleration of the bottom pulley, because the average height of the bottom two masses always stays the same distance below the bottom pulley. And the acceleration of the bottom pulley equals the acceleration of the middle pulley, which in turn equals $-a_3/2$. This is true because if the middle pulley goes up by d, then $2d$ of string disappears above it, which must therefore appear above m_3; so m_3 goes down by $2d$. Putting all this together yields $a_3 = -(a_1 + a_2)$. See the solution to Problem 4.3 for more discussion of these concepts.

4.23. $\boxed{b}$ If the left mass goes up by y_1, there is $2y_1$ less string in the segments above it. Likewise, if the middle mass goes up by y_2, there is $2y_2$ less string in the segments above it. All of this missing string must appear above the right mass, which therefore goes down by $2y_1 + 2y_2$. So $y_3 = -2y_1 - 2y_2$. Taking two time derivatives gives choice (b).

4.24. $\boxed{d}$ If the left mass goes up by y_1, then $2y_1$ worth of string disappears from the left region. Similarly, if the right mass goes up by y_3, then $2y_3$ worth of string disappears from the right region. This $2y_1 + 2y_3$ worth of string must appear in the middle region. It gets divided evenly between the two segments there, so the middle mass goes down by $y_1 + y_3$. Hence $y_2 = -(y_1 + y_3)$. Taking two time derivatives gives choice (d).

4.5 Problem solutions

4.1. Coefficients of friction

Assume, in search of a contradiction, that μ_k is larger than μ_s. Imagine pushing a block that rests on a surface. If the applied force is smaller than $\mu_s N$ (such as the force indicated by the point A on the scale in Fig. 4.35), then nothing happens. The static friction exactly cancels the applied force, and the block just sits there.

However, if the applied force lies between $\mu_s N$ and $\mu_k N$ (such as the force indicated by the point B), then we have problem. On one hand, the applied force exceeds $\mu_s N$ (which is the maximum static friction force, by definition), so the block should move. But on the other hand, the applied force is smaller than the kinetic friction force, $\mu_k N$, so the block *shouldn't* move. Basically, as soon as the block moves even the slightest infinitesimal amount (at which point the kinetic friction force becomes the relevant force), it will decelerate and stop because the kinetic friction force wins out over the applied force. So it actually never moves at all, even for the applied force represented by point B. This means, by the definition of μ_s, that $\mu_s N$ is actually located higher than point B. The above contradiction (where the block both does move and doesn't move) will arise unless $\mu_s N \geq \mu_k N$. So we conclude that it must be the case that $\mu_s \geq \mu_k$. There then exists no point B below $\mu_k N$ and above $\mu_s N$.

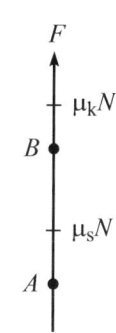

This situation with $\mu_k > \mu_s$ leads to a contradiction

Figure 4.35

In the real world, μ_s is rarely larger than twice μ_k. In some cases the two are essentially equal, with μ_s being a hair larger.

4.2. Cutting a spring in half

Take the half-spring and stretch it a distance x. Our goal is to find the force F that it exerts on something attached to an end; the spring constant is then given by $k_{1/2} = F/x$ (ignoring the minus sign in Hooke's law; we'll just deal with the magnitude). Imagine taking two half-springs that are each stretched by x, lying along the same line, and attaching the right end of one to the left end of the other. We are now back to our original spring with spring constant k. And it is stretched by a distance $2x$, so the force it exerts is $k(2x)$. But this is also the force F that each of the half-springs exerts, because we didn't change anything about these springs when we attached them together. So the desired value of $k_{1/2}$ is $k_{1/2} = F/x = k(2x)/x = 2k$.

REMARK: The same type of reasoning shows (as you can verify) that if we have a spring and then cut off a piece with a length that is a factor f times the original (so $f = 1/2$ in the above case), then the spring constant of the new piece is $1/f$ times the spring constant of the original. Actually, the reasoning works only in the case of rational numbers f. (It works with f of the form $f = 1/N$, where N is an integer. And you can show with similar reasoning that it also works with f of the form $f = N$, which corresponds to attaching N springs together in a line. Combining these two results yields all of the rational numbers.) But any real number is arbitrarily close to a rational number, so the result is true for any factor f.

Basically, a shorter spring is a stiffer spring, because for a given total amount of stretching, a centimeter of a shorter spring must stretch more than a centimeter of a longer spring. So the tension in the former centimeter (which is the same as the tension throughout the entire shorter spring) is larger than the tension in the latter centimeter.

4.3. Useful Atwood's facts

(a) Assume, in search of a contradiction, that the tension varies throughout the string. Then there exists a segment of the string for which the tension is different at the two ends. This means that there is a nonzero net force on the segment. But the string is massless, so the acceleration of this segment must be infinite. Since this can't be the case, the tension must in fact be the same throughout the string. In short, any massless object must always have zero net force acting on it. (Ignoring photons and such!)

REMARK: If there is friction between a string and a pulley, and if the pulley has a nonzero moment of inertia (a topic covered in Chapter 7), then the tension in the string will vary, even if the string is massless. But the statement, "Any massless object must always have zero net force acting on it," still holds; there is now a nonzero friction force from the pulley acting on a segment of the string touching it. So the net force on the segment is still zero.

(b) The reasoning here is similar to the reasoning in part (a). Since the left pulley is massless, the net force on it must be zero. If we draw a free-body diagram as shown in Fig. 4.36, then we see that two T's protrude from the top of the box. So the downward tension from the bottom string must be $2T$, to make the net force be zero. (By the same reasoning, the tension in the short string above the right pulley is also $2T$.) We'll use this result many times in the Atwood's problems in this chapter.

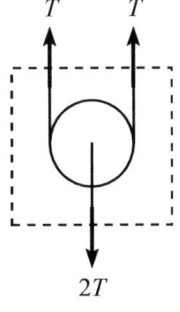

Figure 4.36

REMARK: In the above free-body diagram, the top string does indeed get counted twice, as far as the forces go. In a modified scenario where we have two people pulling upward on the ends of the string, each with a force T, as shown in Fig. 4.37, their total upward force is of course $2T$. And the pulley can't tell the difference between this scenario and the original one.

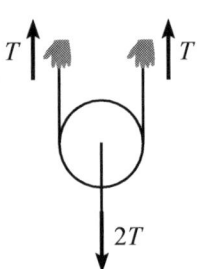

Figure 4.37

(c) Imagine that the left mass (and hence left pulley) goes up by a distance d. Then a length d of string disappears from each of the two segments above the pulley, as shown in Fig. 4.38. So a total length $2d$ of string disappears from these two segments. This string has to go somewhere, so it appears above the right mass. The right mass therefore goes down by $2d$, as shown. So the downward displacement of the right mass is always twice the upward displacement of the left mass. Taking two derivatives of this relation tells us that downward acceleration of the right mass is always twice the upward acceleration of the left mass. (Equivalently, the general relation $d = at^2/2$ holds, so the ratio of the accelerations must be the same as the ratio of the displacements.) This is the so-called "conservation of string" statement for this setup.

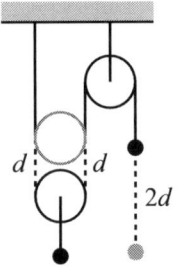

Figure 4.38

The same reasoning applies if the left mass instead goes down. In any case, the sign of the right mass's acceleration is the negative of the sign of the left mass's acceleration.

REMARKS: When thinking about conservation-of-string statements, it is often helpful to imagine cutting out pieces of string in some parts of the setup and then splicing them into other parts. This isn't what actually happens, of course; the string just slides around like a snake. But if you take a photo at two different times during the motion, the splicing photo will look the same as the actual photo. In short, while it is often hard to visualize what is happening as the string *moves*, it is generally much easier to imagine the string as *having moved*.

Conservation of string by itself doesn't determine the motion of the masses. We still need to apply $F = ma$ to find out how the masses actually move. If you grab the masses and move them around in an arbitrary manner while always making sure that the strings stay taut, then the conservation-of-string condition will be satisfied. But the motion will undoubtedly not be the same as the motion where the masses are acted on by only gravity. There is an infinite number of possible motions consistent with conservation of string, but only one of these motions is also consistent with all of the $F = ma$ equations. And conversely, the $F = ma$ equations alone don't determine the motion; the conservation-of-string relation is required. If the strings aren't present, the motion will certainly be different; all the masses will be in freefall!

(d) The average height of the right two masses always remains a constant distance below the right pulley (because the right string keeps the same length). So $y_p = (y_2 + y_3)/2 + C$. Taking two derivatives of this relation gives $a_p = (a_2 + a_3)/2$. But the downward (or upward) acceleration of the right pulley (which we just showed equals the average of the accelerations of the right two masses) equals the upward

4.5. PROBLEM SOLUTIONS

(or downward) acceleration of the left mass (because the left string keeps the same length), as desired.

A few of the Atwood's-machine problems in this chapter contain some unusual conservation-of-string relations, but they all involve the types of reasoning in parts (c) and (d) of this problem.

4.4. Block and tackle

(a) Let T be the tension in the long rope (the tension is the same throughout the rope, from the reasoning in Problem 4.3(a)). Since the bottom pulley is massless, the reasoning in Problem 4.3(b) tells us that the tension in the short rope attached to the block is $4T$ (there would now be four T's protruding from the top of the dashed box in Fig. 4.36). We want this $4T$ tension to balance the Mg weight of the block, so the person must pull on the rope with a force $T = Mg/4$. (The angle of the rope to the person doesn't matter.)

(b) The free-body diagram for the bottom part of the setup is shown in Fig. 4.39. Five tensions protrude from the top of the dashed box, and the two weights Mg and mg protrude from the bottom. The net force must be zero if the setup is at rest (or moving with constant speed), so the tension must equal $T = (M + m)g/5$. This is the desired downward force exerted by the person on the rope.

LIMITS: In the case where $M = 0$ (so the person is standing on a massless platform), he must pull down on the rope with a force equal to one fifth his weight, if he is to hoist himself up. In this case, his mg weight is balanced by an upward $mg/5$ tension force from the rope he is holding, plus an upward $4mg/5$ normal force from the platform (which basically comes from the $4mg/5$ tension in the short rope attached to the platform).

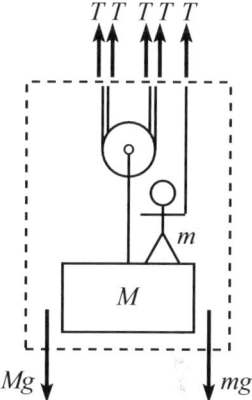

Figure 4.39

4.5. Equivalent mass

With the various parameters defined as in Fig. 4.40, the two $F = ma$ equations are (with upward taken as positive for the left mass, and downward positive for the right)

$$T - m_1 g = m_1 a \qquad \text{and} \qquad m_2 g - T = m_2 a. \tag{4.6}$$

We have used the fact that since your hand (and hence the pulley) is held at constant height, the accelerations of the masses are equal in magnitude and opposite in direction. We can solve for T by multiplying the first equation by m_2, the second by m_1, and then subtracting them. This eliminates the acceleration a, and we obtain $T = 2m_1 m_2 g/(m_1 + m_2)$. The upward force you apply equals the tension $2T$ in the upper string; see Problem 4.3(b) for the explanation of the $2T$. This force of $2T$ equals the weight Mg of a single mass M if

$$M = \frac{4m_1 m_2}{m_1 + m_2}. \tag{4.7}$$

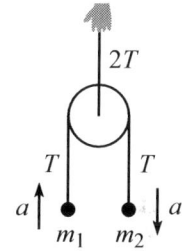

Figure 4.40

This result is *not* equal to the sum of the masses, $m_1 + m_2$. In the special case where the masses are equal ($m_1 = m_2 \equiv m$), the equivalent mass does simply equal $2m$. But in general, M isn't equal to the sum.

LIMITS: In the limit where m_1 is very small, we can ignore the m_1 in the denominator of Eq. (4.7) (but not in the numerator; see the discussion in Section 1.1.3), which yields $M \approx 4m_1$. So the combination of a marble and a bowling ball looks basically like four marbles (this limit makes it clear that M can't be equal to $m_1 + m_2$ in general). This can be seen fairly intuitively: the bowling ball is essentially in freefall downward, so the marble accelerates upward at g. The tension T must therefore be equal to $2m_1 g$ to make the net upward force on the marble be $m_1 g$. The tension $2T$ in the upper string is then $2(2m_1 g)$, which yields $M = 4m_1$.

If we want to solve for a in Eq. (4.6), we can simply add the equations. The result is

$$a = g\frac{m_2 - m_1}{m_2 + m_1}. \tag{4.8}$$

This acceleration makes sense, because a net force of $m_2g - m_1g$ pulls down on the right side, and this force accelerates the total mass of $m_1 + m_2$. Various limits check correctly: if $m_2 = m_1$ then $a = 0$; if $m_2 \gg m_1$ then $a \approx g$; and if $m_2 \ll m_1$ then $a \approx -g$ (in Fig. 4.40 we defined positive a to be upward for m_1).

4.6. **Atwood's 1**

If m_1 is at rest, the tension in the string supporting it must be $m_1 g$. From the reasoning in Problem 4.3(b), the tension in the lower string is then $m_1 g/2$. The $F = ma$ equations for the two lower masses are therefore (with upward taken to be positive for m_2 and downward positive for M)

$$m_2 : \quad \frac{m_1 g}{2} - m_2 g = m_2 a,$$
$$M : \quad Mg - \frac{m_1 g}{2} = Ma. \tag{4.9}$$

We have used the fact that the accelerations of m_2 and M have the same magnitude, because the bottom pulley doesn't move if m_1 is at rest. Equating the two resulting expressions for a from the above two equations gives

$$\frac{m_1 g}{2m_2} - g = g - \frac{m_1 g}{2M} \implies \frac{m_1}{m_2} - 4 = -\frac{m_1}{M}$$
$$\implies M = \frac{m_1 m_2}{4m_2 - m_1}. \tag{4.10}$$

In order for a physical M to exist, we need the denominator of M to be positive. So we need $m_2 > m_1/4$.

Alternatively, you can solve this problem by solving for m_2 in Eq. (4.7) in the solution to Problem 4.5 and then relabeling the masses appropriately.

REMARK: If m_2 is smaller than $m_1/4$, then even an infinitely large M won't keep m_1 from falling. The reason for this is that the best-case scenario is where M is so large that it is essentially in freefall. So m_2 gets yanked upward with acceleration g, which means that the tension in the string pulling on it is $2m_2 g$ (so that the net upward force is $2m_2 g - m_2 g = m_2 g$). The tension in the upper string is then $4m_2 g$. (We've basically just repeated the reasoning for the small-m_1 limit discussed in the solution to Problem 4.5.) This is the largest the tension can be, so if $4m_2 g$ is smaller than $m_1 g$ (that is, if m_2 is smaller than $m_1/4$), then m_1 will fall downward. Note that if $m_2 \to \infty$ (with m_2 finite), then Eq. (4.10) says that $M = m_1/4$, which is consistent with the preceding reasoning. You can show, as you might intuitively expect, that the smallest sum of m_2 and M that supports a given m_1 is achieved when m_2 and M are both equal to $m_1/2$.

4.7. **Atwood's 2**

From the reasoning in Problem 4.3(b), the tensions in the two strings are T and $2T$, as shown in Fig. 4.41. The $F = ma$ equations are therefore

$$T = ma_1,$$
$$T = (2m)a_2,$$
$$(2m)g - 2T = (2m)a_3. \tag{4.11}$$

We have three equations but four unknowns here: a_1, a_2, a_3, and T. So we need one more equation – the conservation-of-string relation. The average position of the left two masses remains the same distance behind the left pulley, which moves the same distance as the right pulley, and hence right mass. So the conservation-of-string relation is $a_3 = (a_1 + a_2)/2$. This is the same reasoning as in Problem 4.3(d).

The first two of the above $F = ma$ equations quickly give $a_1 = 2a_2$. Plugging this into the conservation-of-string relation gives $a_3 = 3a_2/2$. The second two $F = ma$ equations are

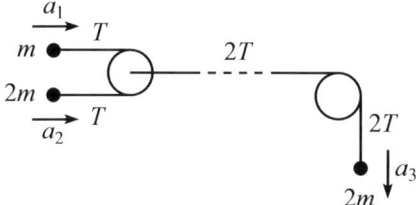

Figure 4.41

then
$$T = (2m)a_2,$$
$$(2m)g - 2T = (2m)(3a_2/2). \tag{4.12}$$

The second equation plus twice the first gives $2mg = 7ma_2 \implies a_2 = 2g/7$. We then have $a_1 = 2a_2 = 4g/7$, and $a_3 = 3a_2/2 = 3g/7$. As a check, a_3 is indeed the average of a_1 and a_2. Additionally, the tension is $T = ma_1 = 4mg/7$.

REMARK: When solving any problem, especially an Atwood's problem, it is important to (1) identify all of the unknowns and (2) make sure that you have as many equations as unknowns, as we did above.

4.8. **Atwood's 3**

If T is the tension in the lowest string, then from Problem 4.3(b) the tensions in the other strings are shown in Fig. 4.42. Let all of the accelerations be defined with upward being positive. Then the three $F = ma$ equations are

$$T - mg = ma_1,$$
$$T - mg = ma_2,$$
$$T - (2m)g = (2m)a_3. \tag{4.13}$$

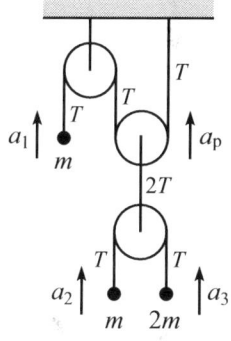

Figure 4.42

The first two of these equations quickly give $a_1 = a_2$.

Now for the conservation-of-string statement. Let a_p be the acceleration of the upper right pulley (which is the same as the acceleration of the lower right pulley). The average height of the two right masses always remains the same distance below this pulley. Therefore $a_p = (a_2 + a_3)/2$. But we also have $a_1 = -2a_p$, because if the pulley goes up a distance d, then a length d of string disappears from both segments above the pulley, so $2d$ of string appears above the left mass. This means that it goes down by $2d$; hence $a_1 = -2a_p$. (We've just redone Problem 4.3(c) and (d) here.) Combining this with the $a_p = (a_2+a_3)/2$ relation gives

$$a_1 = -2\left(\frac{a_2 + a_3}{2}\right) \implies a_1 + a_2 + a_3 = 0. \tag{4.14}$$

Since we know from above that $a_1 = a_2$, we obtain $a_3 = -2a_2$. The last two of the above $F = ma$ equations are then

$$T - mg = ma_2,$$
$$T - (2m)g = (2m)(-2a_2). \tag{4.15}$$

Taking the difference of these equations yields $mg = 5ma_2$, so $a_2 = g/5$. (And this is also a_1.) The desired acceleration of the mass $2m$ is then $a_3 = -2a_2 = -2g/5$. This is negative, so the mass $2m$ goes downward, which makes sense. (If all three masses are equal to m, you can quickly show that $T = mg$ and all three accelerations are zero. Increasing the right mass to $2m$ therefore makes it go downward.)

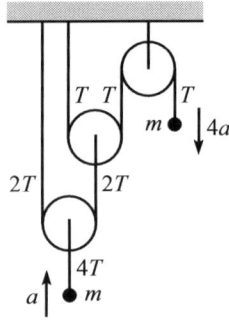

Figure 4.43

4.9. Atwood's 4

Let T be the tension in the string connected to the right mass. Then from the reasoning in Problem 4.3(b), the tensions in the other strings are $2T$ and $4T$, as shown in Fig. 4.43.

The conservation-of-string relation tells us that the accelerations are a and $4a$, as shown. This is true because if the bottom pulley goes up by d, then the middle pulley goes up by $2d$, from the reasoning in Problem 4.3(c). From the same reasoning, the right mass then goes down by $4d$. So the ratio of the distances moved is 4. Taking two time derivatives of this relation tells us that the (magnitudes of the) accelerations are in the same ratio. The $F = ma$ equations are therefore

$$4T - mg = ma,$$
$$mg - T = m(4a). \tag{4.16}$$

The first equation plus four times the second gives $3mg = 17ma \Longrightarrow a = 3g/17$. This is the upward acceleration of the left mass. The downward acceleration of the right mass is then $4a = 12g/17$.

REMARK: Consider the more general case where we have a "tower" of n movable pulleys extending down to the left (so the given problem has $n = 2$). If we define $N \equiv 2^n$, then as an exercise you can show that the upward acceleration of the left mass and the downward acceleration of the right mass are

$$a_{\text{left}} = g\frac{N-1}{N^2+1} \quad \text{and} \quad a_{\text{right}} = g\frac{N^2-N}{N^2+1}. \tag{4.17}$$

These expressions correctly reproduce the above results when $n = 2 \Longrightarrow N = 4$. If $N \to \infty$, we have $a_{\text{left}} \to 0$ and $a_{\text{right}} \to g$. You can think physically about what is going on here; in some sense the system behaves like a lever, where forces and distances are magnified. The above expressions for the a's are also valid when $n = 0 \Longrightarrow N = 1$, in which case both accelerations are zero; we just have two masses hanging over the fixed pulley connected to the ceiling.

4.10. Atwood's 5

Let T be the tension in the bottom string. Then from the reasoning in Problem 4.3(b), the tensions in the other strings are $2T$ and $4T$, as shown in Fig. 4.44. If all of the accelerations are defined with upward being positive, the four $F = ma$ equations are

$$4T - 4mg = 4ma_1,$$
$$2T - 2mg = 2ma_2,$$
$$T - mg = ma_3,$$
$$T - 2mg = 2ma_4. \tag{4.18}$$

The first three of these equations quickly give $a_1 = a_2 = a_3$.

We must now determine the conservation-of-string relation. Let a_m and a_b be the accelerations of the middle and bottom pulleys (with upward being positive). The average position of the bottom two masses stays the same distance below the bottom pulley, so $a_b = (a_3 + a_4)/2$. Similarly, $a_m = (a_2 + a_b)/2$. Substituting the value of a_b from the first of these relations into the second, and using the fact that $a_1 = -a_m$, we have

$$a_1 = -\left(\frac{a_2 + (a_3 + a_4)/2}{2}\right). \tag{4.19}$$

Using $a_1 = a_2 = a_3$, this becomes

$$a_3 = -\left(\frac{a_3 + (a_3 + a_4)/2}{2}\right) \quad \Longrightarrow \quad a_4 = -7a_3. \tag{4.20}$$

The last two of the $F = ma$ equations in Eq. (4.18) are then

$$T - ma = ma_3,$$
$$T - 2mg = 2m(-7a_3). \tag{4.21}$$

Subtracting these equations eliminates T, and we find $a_3 = g/15$. The acceleration of the rightmost mass is then $a_4 = -7a_3 = -7g/15$. This is negative, so this mass goes down. The other three masses all move upward with acceleration $g/15$.

4.11. Atwood's 6

The important thing to note in this problem is that the tension in the long string is *different* above and below the middle mass. The standard fact that we ordinarily use (that the tension is the same everywhere throughout a massless string; see Problem 4.3(a)) holds only if the string is *massless*. If there is a mass attached to the string, then the tensions on either side of the mass will be different (unless the mass is in freefall, as we see from the second $F = ma$ equation below). Let these tensions be T_1 and T_2, as shown in Fig. 4.45. If we label the accelerations as a_1, a_2, and a_3 from left to right (with upward defined to be positive for all), then the $F = ma$ equations are

$$T_1 - mg = ma_1,$$
$$T_1 - T_2 - mg = ma_2,$$
$$2T_2 - mg = ma_3. \tag{4.22}$$

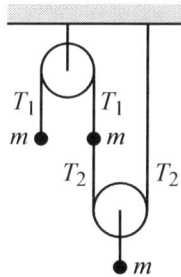

Figure 4.45

Conservation of string quickly gives $a_2 = -a_1$. And it also gives $a_3 = -a_1/2$, from the reasoning in Problem 4.3(c). The above $F = ma$ equations then become

$$T_1 - mg = ma_1,$$
$$T_1 - T_2 - mg = m(-a_1),$$
$$2T_2 - mg = m(-a_1/2). \tag{4.23}$$

We now have three equations in three unknowns (T_1, T_2, and a_1). Solving these by your method of choice gives $a_1 = 2g/9$. Hence $a_2 = -2g/9$ and $a_3 = -g/9$. The tensions turn out to be $T_1 = 11mg/9$ and $T_2 = 4mg/9$.

4.12. Atwood's 7

Let the tension in the upper left part of the string in Fig. 4.46 be T. Then the two other upper parts also have tension T, as shown. Because the left pulley is massless, the net force on it must be zero, so the string below it must have tension $2T$. This then implies the other $2T$ shown. Finally, the tension in the bottom string is $4T$ because the net force on the bottom pulley must be zero. Note that the tension in the long string is allowed to change from T to $2T$ at the mass, because the "massless string" reasoning in Problem 4.3(a) doesn't hold here.

With the accelerations defined as shown in the figure, the $F = ma$ equations for the bottom and top masses are

$$mg - 4T = ma_b,$$
$$mg + 2T - T = ma_t. \tag{4.24}$$

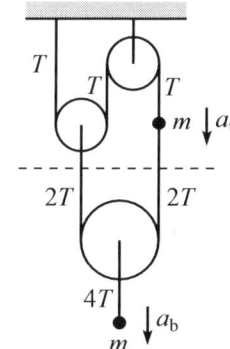

Figure 4.46

We must now determine the conservation-of-string relation. From the reasoning in Problem 4.3(c), the acceleration of the top mass is twice the acceleration of the left pulley (with the opposite sign).

Additionally, if the left pulley goes up by d, then an extra length d of string appears below the dotted line in Fig. 4.46. This is true because d disappears from each of the two parts of the string above the left pulley, but d must also be inserted right below the pulley. So a net length d is left over, and this appears below the dotted line. It gets divided evenly between the two parts of the string touching the bottom pulley, so the bottom pulley (and hence the bottom mass) goes down by $d/2$. So the acceleration of the bottom mass is half the acceleration of the left pulley (with the opposite sign). But from the previous paragraph, the acceleration of the left pulley is half the acceleration of the top mass (with the opposite sign). Putting this all together gives $a_b = -(-a_t/2)/2$, or $a_t = 4a_b$. The above $F = ma$ equations therefore become

$$mg - 4T = ma_b,$$
$$mg + 2T - T = m(4a_b). \quad (4.25)$$

The first equation plus four times the second gives $5mg = 17ma_b \implies a_b = 5g/17$. And then $a_t = 4a_b = 20g/17$. It is perfectly fine that $a_t > g$. In the limit where the top mass is zero, the bottom mass is in freefall, so we have $a_b = g$ and $a_t = 4g$.

4.13. Atwood's 8

Let T be the tension in the string connected to the right mass. Then from Problem 4.3(b), zero net force on the right pulley tells us that the tension in the other long string is $2T$, as shown in Fig. 4.47. And then again from Problem 4.3(b), zero net force on the left pulley tells us that the tension in the bottom short string is $4T$.

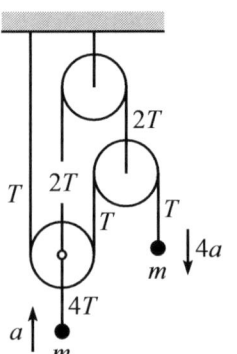

Figure 4.47

Let a be the upward acceleration of the left mass. We claim that the conservation-of-string relation says that the downward acceleration of the right mass is $4a$. This is true for the following reason. If the left pulley (and hence left mass) moves up by a distance d, then a length d of string disappears from each of the two segments of string above it that touch it tangentially. Likewise, if the right pulley moves down by a distance d (and it does indeed move the same distance as the left pulley, because the centers of the two pulleys are connected by a string), then a length d of string disappears from each of the two segments of string below it (assuming for a moment that the right mass doesn't move). A total of $4d$ of string has therefore disappeared (temporarily) from the long string touching the right mass. This $4d$ must end up being inserted above the right mass. So this mass goes down by $4d$, which is four times the distance the left mass goes up, hence the accelerations of a and $4a$ shown in the figure. The $F = ma$ equations for the left and right masses are therefore

$$4T - mg = ma,$$
$$mg - T = m(4a). \quad (4.26)$$

The first equation plus four times the second gives $3mg = 17mg \implies a = 3g/17$. This is the upward acceleration of the left mass. The downward acceleration of the right mass is then $4a = 12g/17$. These accelerations happen to be the same as the ones in Problem 4.9.

4.14. No relative motion

First note that by the following continuity argument, there must exist a value of a for which there is no relative motion among the masses. If a is zero, then m_2 falls and m_1 moves to the right. On the other hand, if a is very large, then intuitively m_1 will drift backward with respect to M, and m_2 will rise (remember that there is no friction anywhere). So for some intermediate value of a, there will be no relative motion.

Now let's find the desired value of a. If there is no relative motion among the masses, then m_2 stays at the same height, which means that the net vertical force on it must be zero. The tension in the string connecting m_1 to m_2 must therefore be $T = m_2 g$. The horizontal $F = ma$ equation on m_1 is then $T = m_1 a \implies m_2 g = m_1 a \implies a = (m_2/m_1)g$. This is the acceleration of the system, and hence the desired acceleration of M.

LIMITS: a is small if m_2 is small (more precisely, if $m_2 \ll m_1$) and large if m_2 is large ($m_2 \gg m_1$). These results make intuitive sense.

4.5. PROBLEM SOLUTIONS

4.15. Slipping blocks

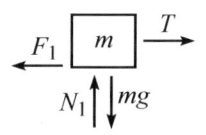
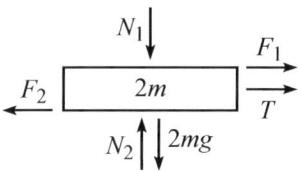

Figure 4.48

(a) The free-body diagrams are shown in Fig. 4.48. The normal forces N_1 (between the blocks) and N_2 (between the bottom block and the table) are quickly found to be $N_1 = mg$ and $N_2 = 3mg$. And the tension in the string is $T = 3mg$, because twice this must equal the applied $6mg$ force (because the net force on the massless pulley must be zero). The friction forces F_1 (between the blocks) and F_2 (between the bottom block and the table) are as yet unknown.

Assume that the bottom block *doesn't* slip with respect to the table. The maximum possible leftward static friction force from the table acting on the bottom block is $\mu_s N_2 = 3mg$. This can cancel out the rightward tension $T = 3mg$ acting on the block. However, there also the friction force F_1 between the blocks. (This is the kinetic friction force $\mu_k N_1 = mg$, because you can quickly show that if the bottom block is at rest, then the top block must slip with respect to the bottom block.) This friction force acts rightward on the bottom block, which means that the net rightward force on the bottom block is nonzero. It will therefore slip with respect to the table. Hence our initial non-slipping assumption was incorrect.

(b) Assume that the top block *doesn't* slip with respect to the bottom block (which we know must be moving, from part (a)). Then the two blocks can be treated like a single block with mass $3m$. The leftward kinetic friction force from the table is $F_2 = \mu_k N_2 = 3mg$. The net force on the effective $3m$ block is therefore $6mg - 3mg = 3mg$ rightward, so the acceleration is $a = g$ rightward. The horizontal $F = ma$ equation for the top block is then

$$T - F_1 = ma \implies 3mg - F_1 = mg \implies F_1 = 2mg. \tag{4.27}$$

But this friction force isn't possible, because it exceeds the maximum possible static friction force between the blocks, which is $\mu_s N_1 = mg$. This contradiction implies that our initial non-slipping assumption must have been incorrect. The blocks therefore slip with respect to each other.

REMARK: If the coefficient of static friction between the blocks were instead made sufficiently large ($\mu_s \geq 2$), then the blocks would in fact move as one effective mass $3m$. The static friction force $2mg$ between the blocks would act backward on m and forward on $2m$, and both blocks would have acceleration g. If the friction force were then suddenly decreased to the kinetic value of $\mu_k N_1 = mg$ relevant to the stated problem, m would accelerate faster than g (because there isn't as much friction holding it back), and $2m$ would accelerate slower than g (because there isn't as much friction pushing it forward). This is consistent with the accelerations we will find below in part (c).

(c) Since we know that all surfaces slip with respect to each other, the friction forces are all kinetic friction forces. Their values are therefore $F_1 = \mu_k N_1 = mg$ and $F_2 = \mu_k N_2 = 3mg$. The $F = ma$ equations for the two blocks are then

$$\begin{aligned} m: \quad & T - F_1 = ma_1 \implies 3mg - mg = ma_1 \\ & \implies a_1 = 2g, \\ 2m: \quad & T + F_1 - F_2 = (2m)a_2 \implies 3mg + mg - 3mg = 2ma_2 \\ & \implies a_2 = g/2. \end{aligned} \tag{4.28}$$

By conservation of string, the average position of the two blocks stays the same distance behind the pulley, and hence also behind your hand. So

$$a_{\text{hand}} = \frac{a_1 + a_2}{2} = \frac{2g + g/2}{2} = \frac{5g}{4}. \tag{4.29}$$

4.16. Block and wedge

Let N be the normal force between the block and the wedge. Then the vertical $F = ma$ equation for the block (with downward taken to be positive) is

$$Mg - N\cos\theta = Ma_M. \tag{4.30}$$

And the horizontal $F = ma$ equation for the wedge (with rightward taken to be positive) is

$$N\sin\theta = ma_m. \tag{4.31}$$

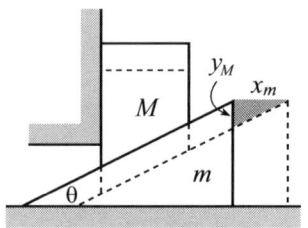

Figure 4.49

We have two equations but three unknowns (N, a_M, a_m), so we need one more equation. This equation is the constraint that the block remains on the wedge at all times. At a later time, let the positions of the block and wedge be indicated by the dotted lines in Fig. 4.49. The block has moved downward a distance y_M, and the wedge has moved rightward a distance x_m. From the triangle that these lengths form (the shaded triangle in the figure), we see that $y_M = x_m \tan\theta$. Taking two time derivatives of this relation gives the desired third equation,

$$a_M = a_m \tan\theta. \tag{4.32}$$

There are various ways to solve the preceding three equations. If we multiply the first by $\sin\theta$ and the second by $\cos\theta$, and then add them, the N terms cancel. If we then use $a_M = a_m \tan\theta$ to eliminate a_M, we obtain

$$Mg\sin\theta = Ma_M \sin\theta + ma_m \cos\theta$$
$$= M(a_m \tan\theta)\sin\theta + ma_m \cos\theta. \tag{4.33}$$

Solving for a_m gives

$$a_m = \frac{Mg\sin\theta\cos\theta}{M\sin^2\theta + m\cos^2\theta}. \tag{4.34}$$

LIMITS: There are many limits we can check. You should verify that all of the following results make sense.

- If $\theta \to 0$ or $\theta \to 90°$, then $a_m \to 0$.
- If $M \ll m$ then $a_m \approx 0$.
- If $M \gg m$ then $a_m \approx g/\tan\theta \implies g \approx a_m \tan\theta$. This is simply Eq. (4.32) with $a_M = g$, because the block is essentially in freefall.
- If $M = m$, then $a_m = g\sin\theta\cos\theta$, which achieves a maximum when $\theta = 45°$. And since the constraining force on the left side of the block equals $N\sin\theta$, which in turn equals ma_m from Eq. (4.31), we see that $\theta = 45°$ necessitates the maximum force by the constraining wall (for the special case where $M = m$).

You can check that various limits for a_M (given in Eq. (4.32)) also work out correctly.

4.17. Up and down a plane

(a) When the block stops (at least instantaneously) at its highest point, the forces along the plane are the $mg\sin\theta$ gravity component downward, and the static friction force F_f upward. We know that $F_f \leq \mu_s N = 1 \cdot mg\cos\theta$. The block will accelerate downward if the gravitational force $mg\sin\theta$ is larger than the maximum possible friction force $mg\cos\theta$. So the block will slide back down if

$$mg\sin\theta > mg\cos\theta \implies \tan\theta > 1 \implies \theta > 45°. \tag{4.35}$$

(b) On the way up the plane, both the gravity component and friction point down the plane, so the force along the plane is $mg\sin\theta + \mu_k mg\cos\theta$ downward. Therefore, since $\mu_k = 1$, the acceleration during the upward motion points down the plane and has magnitude

$$a_u = g(\sin\theta + \cos\theta). \tag{4.36}$$

4.5. PROBLEM SOLUTIONS

The total time for the upward motion is then

$$t_u = \frac{v_0}{a_u} = \frac{v_0}{g(\sin\theta + \cos\theta)}. \tag{4.37}$$

To find the time t_d for the downward motion, we need to find the maximum distance d up the plane that the block reaches. From standard kinematics we have $d = v_0^2/2a_u$. (This can be obtained from either $v_f^2 - v_i^2 = 2ad$, or $d = v_0 t - at^2/2$ with $t = v_0/a$.) During the downward motion, the friction force points up the plane, so the net acceleration points down the plane and has magnitude

$$a_d = g(\sin\theta - \cos\theta). \tag{4.38}$$

Using the value of d we just found, the relation $d = a_d t_d^2/2$ for the downward motion gives

$$\frac{v_0^2}{2a_u} = \frac{a_d t_d^2}{2} \implies t_d = \frac{v_0}{\sqrt{a_u a_d}} = \frac{v_0}{g\sqrt{(\sin\theta+\cos\theta)(\sin\theta-\cos\theta)}}. \tag{4.39}$$

(c) The total time with friction is

$$T_F = t_u + t_d = \frac{v_0}{a_u} + \frac{v_0}{\sqrt{a_u a_d}}, \tag{4.40}$$

where a_u and a_d are given in Eqs. (4.36) and (4.38). Without friction, the acceleration along the plane is simply $g\sin\theta$ downward for both directions of motion, so the total time with no friction is

$$T_{\text{no F}} = \frac{2v_0}{g\sin\theta}. \tag{4.41}$$

(Equivalently, just erase the $\cos\theta$ terms in T_F, since those terms came from the friction.) If we let $s \equiv \sin\theta$, $c \equiv \cos\theta$, and $x \equiv \cot\theta \equiv c/s$, then T_F will be larger than $T_{\text{no F}}$ if

$$\frac{1}{s+c} + \frac{1}{\sqrt{(s+c)(s-c)}} > \frac{2}{s} \implies \frac{1}{1+x} + \frac{1}{\sqrt{1-x^2}} > 2$$

$$\implies \frac{1}{\sqrt{1-x^2}} > 2 - \frac{1}{1+x} \implies \frac{1}{1-x^2} > \frac{(1+2x)^2}{(1+x)^2}$$

$$\implies \frac{1}{1-x} > \frac{(1+2x)^2}{(1+x)}. \tag{4.42}$$

Cross multiplying and simplifying yields

$$4x^3 - 2x > 0 \implies x > \frac{1}{\sqrt{2}}. \tag{4.43}$$

(There is technically also a range of negative solutions to this equation, but x is defined to be a positive number.) However, we also need $x < 1$ for Eq. (4.42) to hold. (If $x > 1$ then our cross multiplication switches the order of the inequality.) So $T_F > T_{\text{no F}}$ if $1 > \cot\theta > 1/\sqrt{2}$, or equivalently if $1 < \tan\theta < \sqrt{2} \implies 45° < \theta < 54.7°$. (Of course, we already knew that $\theta > 45°$ from part (a).) To summarize:

- If $45° < \theta < 54.7°$, then $T_F > T_{\text{no F}}$. That is, the process takes longer with friction.
- If $54.7° \leq \theta$, then $T_{\text{no F}} \geq T_F$. That is, the process takes longer without friction.

The first of these results is clear in the limiting case where θ is only slightly larger than $45°$, because the block will take a very long time to slide back down the plane, since $a_d \approx 0$. In the other extreme where $\theta \to 90°$, we have $T_F = T_{\text{no F}}$ because the friction force vanishes on the vertical plane.

Plots of T_F and T_{noF} for θ values between 45° and 90° are shown in Fig. 4.50. Note that T_F achieves a local minimum. You can show numerically that this minimum occurs at $\theta = 1.34$ radians, which is about 77°. It isn't obvious that there should exist a local minimum. But what happens is that below 77°, T_F is larger than the minimum value because the slowness of the downward motion dominates other competing effects. And above 77°, T_F is larger because the larger distance up the plane dominates other competing effects (this isn't terribly obvious).

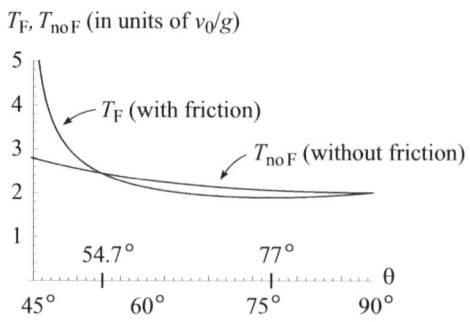

Figure 4.50

4.18. Rope in a tube

(a) FIRST SOLUTION: Each little piece of the rope with tiny mass m has a radially inward acceleration v^2/R, so it feels a radially inward force of mv^2/R (applied by the outer surface of the tube). The sideways components of these forces cancel in pairs in Fig. 4.28(a), so we are left with a net downward force (in the plane of the page).

SECOND SOLUTION: The center of mass (a topic in Chapter 6) of the rope is located somewhere on the y axis at the given instant. It happens to be at radius $2R/\pi$, but that isn't important here. The relevant fact is that the CM travels in a circle. And since $\mathbf{F}_{net} = m\mathbf{a}_{CM}$ (see Chapter 6), we can simply imagine a point mass moving around in a circle. So the acceleration is radially inward (that is, downward in Fig. 4.28(a)) at the given instant.

(b) FIRST SOLUTION: Each little piece of the rope has a tangential acceleration a (but no radial acceleration since v is instantaneously zero), so it feels a tangential force of ma. The vertical (in the plane of the page) components of these forces cancel in pairs, so we are left with a net rightward force in Fig. 4.28(b).

REMARK: Since the rope is being pulled downward, where does this rightward force come from? It comes from the inner surface of the tube, with the important fact being that the force from the right part of the inner circle in Fig. 4.28(b) is larger than the force from the left part. This can be traced to the fact that the tension in the rope is larger closer to the right end that is being pulled; the tension approaches zero at the left end.

SECOND SOLUTION: The CM is initially located on the y axis, and it is accelerated to the right at the given instant (because it will end up moving in a circle once the rope gains some speed). Since $\mathbf{F}_{net} = m\mathbf{a}_{CM}$, the net force is therefore rightward.

If you pull on the rope with a nonzero force while it has a nonzero v, then the total force vector will have both downward and rightward components.

4.19. Circling bucket

Consider a little volume of the water, with mass m. Assuming that the water stays inside the bucket, then at the top of the motion the forces on the mass m are both the downward gravitational force mg and the downward normal force N from the other water in the bucket. So the radial $F = ma$ equation is

$$mg + N = \frac{mv^2}{R}. \tag{4.44}$$

(If you want, you can alternatively consider a little rock at the bottom of an empty bucket; that's effectively the same setup.) If v is large, then N is large. The cutoff case where the water barely stays in the bucket occurs when $N = 0$. The minimum v is therefore given by

$$mg = \frac{mv^2}{R} \implies v_{\min} = \sqrt{gR}. \tag{4.45}$$

If we take R to be, say, 1 m, then this gives $v_{\min} \approx 3$ m/s. The time for each revolution is then $2\pi R/v_{\min} \approx 2$ s. If you swing your arm around with this period of revolution, you'll probably discover that the minimum speed is a lot slower that you would have guessed.

REMARK: The main idea behind this problem is that although at the top of the motion the water is certainly accelerating downward under the influence of gravity, if you accelerate the bucket downward fast enough, then the bucket will maintain contact with the water. So the requirement is that at the top of the motion,[2] you must give the bucket a centripetal (downward) acceleration of at least g. That is, $v^2/R \geq g$, in agreement with Eq. (4.45). If the centripetal acceleration of the bucket is larger than g, then the bucket will need to push downward on the water to keep the water moving along with the bucket. That is, the normal force N will be positive. If the centripetal acceleration of the bucket is smaller than g, then the water will accelerate downward faster than the bucket. That is, the water will leave the bucket.

It isn't necessary to use circular motion to keep the water in the bucket; linear motion works too. If you turn a glass of water upside down and immediately accelerate it straight downward with an acceleration greater than or equal to g, the water will stay in the glass. Of course, you will soon run out of room and smash the glass into the floor! The nice thing about circular motion is that it can go on indefinitely. That's why centrifuges involve circular motion and not linear motion.

4.20. Banking an airplane

FIRST SOLUTION: If you *do* feel like you are getting flung to the side in your seat, then you will need to counter this tendency with some kind of force parallel to the seat, perhaps friction from the seat or a normal force by pushing on the wall, etc. If you *don't* feel like you are getting flung to the side in your seat, then you could just as well be asleep on a frictionless seat, and you would remain at rest on the seat. So the goal of this problem is to determine the banking angle that is consistent with the only forces acting on you being the gravitational force and the normal force from the seat (that is, no friction), as shown in Fig. 4.51.

Let the banking angle be θ. The vertical component of the normal force must be $N_y = mg$, to make the net vertical force be zero. This implies that the horizontal component is $N_x = mg \tan \theta$. The horizontal $F = ma$ equation for the circular motion of radius R is then

$$N_x = \frac{mv^2}{R} \implies mg \tan \theta = \frac{mv^2}{R} \implies \tan \theta = \frac{v^2}{gR}. \tag{4.46}$$

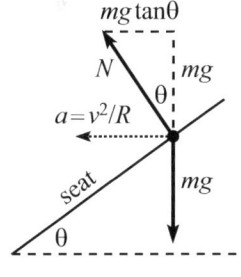

Figure 4.51

Your apparent weight is simply the normal force, because this is the force with which a scale on the seat would have to push up on you. So your weight is

$$N = \sqrt{N_x^2 + N_y^2} = m \sqrt{\left(\frac{v^2}{R}\right)^2 + g^2}. \tag{4.47}$$

LIMITS: If R is very large or v is very small (more precisely, if $v^2 \ll gR$), then $\theta \approx 0$ and $N \approx mg$. (Of course, v can't be too small, or the plane won't stay up!) These limits make sense, because

---
[2] As an exercise, you can show that if the water stays inside the bucket at the highest point, then it will stay inside at all other points too, as you would intuitively expect.

for all you know, you are essentially moving in a straight line. If R is very small or v is very large (more precisely, if $v^2 \gg gR$), then $\theta \approx 90°$ and $N \approx mv^2/R$, which makes sense because gravity is inconsequential in this case.

SECOND SOLUTION: Let's use tilted axes parallel and perpendicular to the seat. If we break the forces and the acceleration into components along these tilted axis, then we see that the force component parallel to the seat is $mg \sin\theta$, and the acceleration component parallel to the seat is $(v^2/R) \cos\theta$, as shown in Fig. 4.52. So the $F = ma$ equation for the motion parallel to the seat is

$$F = ma \implies mg \sin\theta = \frac{mv^2}{R} \cos\theta \implies \tan\theta = \frac{v^2}{gR}, \tag{4.48}$$

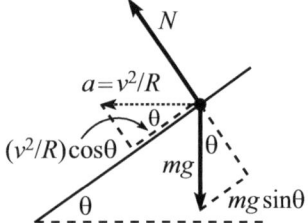

Figure 4.52

in agreement with the result in Eq. (4.46). The $F = ma$ equation for the motion perpendicular to the seat gives us the normal force N:

$$N - mg\cos\theta = \frac{mv^2}{R} \sin\theta \implies N = mg\cos\theta + \frac{mv^2}{R} \sin\theta. \tag{4.49}$$

You can verify that the $\sin\theta$ and $\cos\theta$ values implied by the $\tan\theta = v^2/gR$ relation in Eq. (4.48) make this expression for N reduce to the one we found above in Eq. (4.47).

4.21. **Breaking and turning**

If you brake on a straight road, your acceleration vector points along the road. The friction force satisfies $F_f = ma$. (We'll just deal with magnitudes here, so both F_f and a are positive quantities.) But $F_f \leq \mu N = \mu(mg)$. Therefore,

$$ma = F_f \leq \mu mg \implies a \leq \mu g. \tag{4.50}$$

So your maximum possible deceleration is μg. It makes sense that this should be zero if μ is zero.

If you brake while traveling around a bend, your acceleration vector does *not* point along the road. It has a tangential component a_t pointing along the road (this is the deceleration we are concerned with) and also a radial component $a_r = v^2/R$ pointing perpendicular to the road. These components are shown in Fig. 4.53. The total acceleration vector (and hence also the total friction force vector) points backward and radially inward. Since $\mathbf{F}_f = m\mathbf{a}$, we still have $F_f = ma$, where a is the magnitude of the $\mathbf{a}$ vector, which is now $a = \sqrt{a_t^2 + a_r^2}$. So the $F_f \leq \mu N$ restriction on F_f takes the form

$$m\sqrt{a_t^2 + a_r^2} = F_f \leq \mu mg \implies \sqrt{a_t^2 + (v^2/R)^2} \leq \mu g$$
$$\implies a_t \leq \sqrt{(\mu g)^2 - (v^2/R)^2}. \tag{4.51}$$

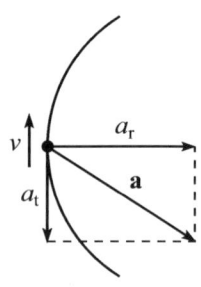

Figure 4.53

LIMITS: This result correctly reduces to $a_t \leq \mu g$ when $R = \infty$, that is, when the road is straight. It also reduces to $a_t \leq \mu g$ when v is very small (more precisely, $v \ll \sqrt{\mu g R}$), because in this case the radial acceleration is negligible; you are effectively traveling on a straight road. If $v = \sqrt{\mu g R}$, then Eq. (4.51) tells us that a_t must be zero. In this case the maximum friction force $\mu N = \mu mg$ is barely large enough to provide the $mv^2/R = m(\mu gR)/R = \mu mg$ force required to keep you going in a circle. Any additional acceleration caused by braking will necessitate a friction force larger than the μN limit.

4.22. **Circle of rope**

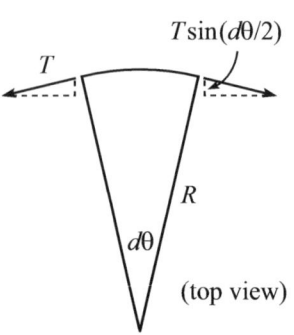

Figure 4.54

The forces on a small piece of rope subtending an angle $d\theta$ are the tensions at its ends. In Fig. 4.54 these tensions point slightly downward; they make an angle of $d\theta/2$ with respect to the horizontal. (This is true because each of the long radial sides of the pie piece in the figure makes an angle of $d\theta/2$ with respect to the vertical, and the tensions are perpendicular to these sides.) If T is the tension throughout the rope, then from the figure the net force on the small piece is $F = 2 \cdot T \sin(d\theta/2)$. This force points downward (radially inward). Using the approximation $\sin x \approx x$ for small x, we see that the net force is $F \approx T\, d\theta$. This force is what causes the centripetal acceleration of the small piece. The length of the piece is $R\, d\theta$, so its mass is $\lambda R\, d\theta$. The radial $F = ma$ equation is therefore

$$F = \frac{mv^2}{R} \implies T\, d\theta = \frac{(\lambda R\, d\theta)v^2}{R} \implies T = \lambda v^2. \tag{4.52}$$

REMARKS: Note that this result of λv^2 is independent of R. This means that if we have an arbitrarily shaped rope (that is, the local radius of curvature may vary), and if the rope is moving along itself (so if the rope were featureless, you couldn't tell that it was actually moving) at constant speed, then the tension equals λv^2 everywhere.

If the rope is stretchable, it will stretch under the tension as it spins around. You should convince yourself that the tension's independence of R implies that if two circles (made of the same material) with different radii have the same v, then the stretching will cause their radii to increase by the same factor.

This λv^2 result comes up often in physics. In addition to being the answer to the present problem, λv^2 is the tension in a rope with density λ if the speed of a traveling wave is v (this is a standard result that can be found in a textbook on waves). And λv^2 is also the tension needed if you have a heap of rope and you grab one end and pull with speed v to straighten it out (see Multiple-Choice Question 6.18). All three of these results appear in a gloriously unified way in the intriguing phenomenon of the "chain fountain."[3]

4.23. Cutting the string

We must first find the speed of the circular motion. The free-body diagram is shown in Fig. 4.55, where we have included the acceleration for convenience. The vertical $F = ma$ equation tells us that $T_y = mg$, because the net vertical force must be zero. The horizontal component of the tension is therefore $T_x = T_y \tan\theta = mg \tan\theta$. So the horizontal $F = ma$ equation is (using the fact that the radius of the circular motion is $r = \ell \sin\theta$)

$$T_x = \frac{mv^2}{r} \implies mg \tan\theta = \frac{mv^2}{\ell \sin\theta} \implies v = \sqrt{g\ell \sin\theta \tan\theta}. \tag{4.53}$$

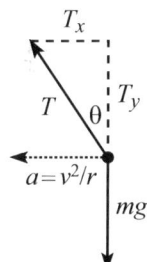

Figure 4.55

After the string is cut, we simply have a projectile problem in which the initial velocity is horizontal (tangential to the circle when the string is cut). The distance down to the floor is $d = \ell - \ell \cos\theta$, so the time to fall to the floor is given by

$$\frac{1}{2}gt^2 = \ell - \ell \cos\theta \implies t = \sqrt{\frac{2\ell(1 - \cos\theta)}{g}}. \tag{4.54}$$

The horizontal distance traveled is therefore

$$x = vt = \sqrt{g\ell \sin\theta \tan\theta} \sqrt{\frac{2\ell(1 - \cos\theta)}{g}}$$
$$= \ell \sqrt{2 \sin\theta \tan\theta (1 - \cos\theta)}. \tag{4.55}$$

LIMITS: If $\theta \approx 0$ then $x \approx 0$ (because both $v \to 0$ and $t \to 0$). And if $\theta \to 90°$ then $x \to \infty$ (because $v \to \infty$). These limits make intuitive sense.

UNITS: Note that x doesn't depend on g. Intuitively, if g is large, then for a given θ the speed v is large (it is proportional to $\sqrt{g}$). But the falling time t is short (it is proportional to $1/\sqrt{g}$). These two effects exactly cancel. The independence of g also follows from dimensional analysis. The distance x must be some function of ℓ, θ, g, and m. But there can't be any dependence on g, which has units of m/s^2, because there would be no way to eliminate the seconds from the units to obtain a pure length. (Likewise for the kilograms in the mass m.)

---
[3]See "Understanding the chain fountain" (by The Royal Society) at http://www.youtube.com/watch?v=-eEi7fO0_O0.

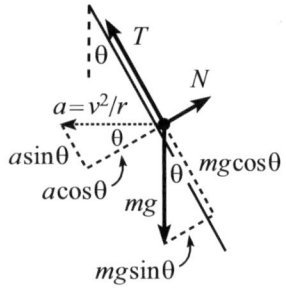

Figure 4.56

4.24. Circling around a cone

(a) The free-body diagram is shown in Fig. 4.56. The net force produces the horizontal centripetal acceleration of $a = v^2/r$, where $r = \ell \sin\theta$. Let's work with axes parallel and perpendicular to the cone. Horizontal and vertical axes would work fine too, but things would be a little messier because the tension T and the normal force N would each appear in both of the $F = ma$ equations, so we would have to solve a system of equations.

The $F = ma$ equation along the cone is (using $a = v^2/\ell \sin\theta$)

$$T - mg\cos\theta = m\left(\frac{v^2}{\ell \sin\theta}\right)\sin\theta \implies T = mg\cos\theta + \frac{mv^2}{\ell}. \quad (4.56)$$

LIMITS: If $\theta \to 0$ then $T \to mg + mv^2/\ell$. We will find below that if contact with the cone is to be maintained, then v must be essentially zero in this case. So we simply have $T \to mg$, which makes sense because the mass is hanging straight down. If $\theta \to \pi/2$ then $T \to mv^2/\ell$, which makes sense because the mass is moving in a circle on a horizontal table.

(b) The $F = ma$ equation perpendicular to the cone is

$$mg\sin\theta - N = m\left(\frac{v^2}{\ell \sin\theta}\right)\cos\theta \implies N = mg\sin\theta - \frac{mv^2}{\ell \tan\theta}. \quad (4.57)$$

LIMITS: If $\theta \to 0$ then $N \to 0 - mv^2/\ell \tan\theta$. Again, we will find below that v must be essentially zero in this case, so we have $N \to 0$, which makes sense because the cone is vertical. If $\theta \to \pi/2$ then $N \to mg$, which makes sense because, as above, the mass is moving in a circle on a horizontal table.

(c) The mass stays in contact with the cone if $N \geq 0$. Using Eq. (4.57), this implies that

$$mg\sin\theta \geq \frac{mv^2}{\ell \tan\theta} \implies v \leq \sqrt{g\ell \sin\theta \tan\theta} \equiv v_{\max}. \quad (4.58)$$

If v equals $v_{\max}$ then the mass is barely in contact with the cone. If the cone were removed in this case, the mass would maintain the same circular motion.

LIMITS: If $\theta \to 0$ then $v_{\max} \to 0$. And if $\theta \to \pi/2$ then $v_{\max} \to \infty$. These limits make intuitive sense.

4.25. Penny in a dryer

The penny loses contact with the dryer when the normal force N becomes zero. If θ is measured with respect to the vertical, then the radially inward component of the gravitational force is $mg\cos\theta$. The radial $F = ma$ equation is therefore $N + mg\cos\theta = mv^2/R$, which implies that the $N = 0$ condition is

$$mg\cos\theta = \frac{mv^2}{R} \implies v^2 = gR\cos\theta. \quad (4.59)$$

This equation tells us how v and θ are related at the point where the penny loses contact. (If $v > \sqrt{gR}$, then there is no solution for θ, so the penny never loses contact.) To solve for θ and v (and hence $\omega = v/R$) individually, we must produce a second equation that relates v and θ. This equation comes from the projectile motion and the condition that the landing point is diametrically opposite.

When the penny loses contact with the dryer, the initial position (relative to the center of the dryer) for the projectile motion is $(R\sin\theta, R\cos\theta)$; remember that θ is measured with respect to the vertical. The initial angle of the velocity is θ upward to the left, as you can check. Therefore, since the initial speed is v, the initial velocity components are $x_x = -v\cos\theta$ and $v_y = v\sin\theta$. The coordinates of the projectile motion are then

$$x(t) = R\sin\theta - (v\cos\theta)t,$$
$$y(t) = R\cos\theta + (v\sin\theta)t - gt^2/2. \quad (4.60)$$

4.5. PROBLEM SOLUTIONS

If the penny lands at the diametrically opposite point, then the final position is the negative of the initial position. So the final time must satisfy

$$R\sin\theta - (v\cos\theta)t = -R\sin\theta \implies (v\cos\theta)t = 2R\sin\theta, \qquad (4.61)$$
$$R\cos\theta + (v\sin\theta)t - gt^2/2 = -R\cos\theta \implies gt^2/2 - (v\sin\theta)t = 2R\cos\theta.$$

Solving for t in the first of these equations and plugging the result into the second gives

$$\frac{g}{2}\left(\frac{2R\sin\theta}{v\cos\theta}\right)^2 - v\sin\theta\left(\frac{2R\sin\theta}{v\cos\theta}\right) = 2R\cos\theta$$

$$\implies \frac{gR\sin^2\theta}{v^2\cos^2\theta} = \frac{\sin^2\theta}{\cos\theta} + \cos\theta$$

$$\implies \frac{gR\sin^2\theta}{v^2\cos^2\theta} = \frac{1}{\cos\theta}$$

$$\implies v^2 = \frac{gR\sin^2\theta}{\cos\theta}. \qquad (4.62)$$

This is the desired second equation that relates v and θ. Equating this expression for v^2 (which guarantees a diametrically opposite landing point) with the one in Eq. (4.59) (which gives the point where the penny loses contact) yields

$$\frac{gR\sin^2\theta}{\cos\theta} = gR\cos\theta \implies \tan^2\theta = 1 \implies \theta = 45°. \qquad (4.63)$$

(And $\theta = -45°$, works too, if the dryer is spinning the other way. Equivalently, the corresponding value of t is negative.) There must be a reason why the angle comes out so nice, but it eludes me.

With this value of θ, Eq. (4.59) gives

$$v = \sqrt{\frac{gR}{\sqrt{2}}} \implies \omega = \frac{v}{R} = \sqrt{\frac{g}{\sqrt{2}R}}. \qquad (4.64)$$

LIMITS: ω grows with g, as expected. The decrease with R isn't as obvious, but it does follow from dimensional analysis.

REMARKS: The relation in Eq. (4.62) describes many different types of trajectories. If v is large, then the corresponding θ is close to $90°$. In this case we have a very tall projectile path that generally lies outside the dryer, so it isn't physical. If v is small, then θ is close to 0. In this case the penny just drops from the top of the dryer down to the bottom; the trajectory lies completely inside the dryer. The cutoff between these two cases (that is, the case where the penny barely stays inside), turns out to be the $\theta = 45°$ case that solves the problem (by also satisfying Eq. (4.59)). You are encouraged to think about why this is true.

As an exercise, you can also produce the second equation relating v and θ (Eq. (4.62), derived from the projectile motion) by using axes that are tilted along the diameter and perpendicular to it. The magnitudes of the accelerations in these directions are $g\cos\theta$ and $g\sin\theta$, respectively, and the penny hits the diameter a distance $2R$ down along it. The math is fairly clean due to the fact that the initial velocity is perpendicular to the diameter. So the solution turns out to be a bit quicker than the one we used above. Problem 3.18(b) involved similar reasoning with the axis perpendicular to the diameter/plane.

Chapter 5

Energy

5.1 Introduction

Work by a constant force

In one dimension, if a constant force F is directed parallel to the line of motion of an object, we define the *work* done by the force on the object to be the force times the displacement:

$$W = F\Delta x. \tag{5.1}$$

Work is a signed quantity, so if the force points opposite to the direction of motion (so that F and Δx have opposite signs), then W is negative. The units of work are

$$[W] = [F][\Delta x] = \frac{\text{kg m}}{\text{s}^2} \cdot \text{m} = \frac{\text{kg m}^2}{\text{s}^2}. \tag{5.2}$$

This combination of units is called a *joule* (J). Why do we define the work as $F\Delta x$ and not, say, $F^3(\Delta x)^2$ or something else? The reason is that the $F\Delta x$ combination of force and displacement happens to appear in a very useful relation, namely the "work-energy" theorem below.

In the more general case where the force (still assumed to be constant) does *not* point along the line of motion, as shown in Fig. 5.1, the work is defined to be

$$W = F\Delta x \cos\theta, \tag{5.3}$$

where θ is the angle between the force $\mathbf{F}$ and the displacement $\Delta \mathbf{x}$ (we'll use $\Delta \mathbf{x}$ to denote a general displacement in space, in any direction). You can write W as $(F\cos\theta)\Delta x$, which corresponds to the entire displacement times the component of the force along the displacement; see Fig. 5.1(a). Or you can write W as $F(\Delta x \cos\theta)$, which corresponds the entire force times the component of the displacement along the force; see Fig. 5.1(b). If you are familiar with the "dot product" (see Section 13.1.6 in Appendix A for the definition of the dot product), you can write W as $W = \mathbf{F} \cdot \Delta\mathbf{x}$ (which tells us that work is a scalar obtained from two vectors), but we won't need to use that form in this book.

Note that our physics definition of work isn't the same as the colloquial definition. If you hold up a heavy object and keep it at rest (or even move it horizontally at constant speed, so that your upward force is perpendicular to the horizontal displacement), then you are doing zero work by our physics definition, whereas you are certainly doing nonzero work in a colloquial sense.

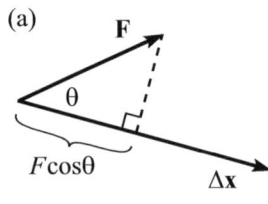
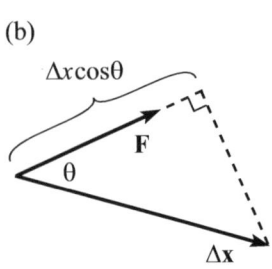

Figure 5.1

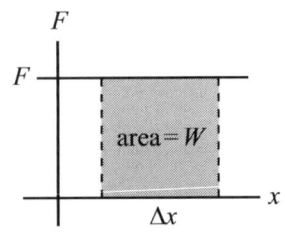

Figure 5.2

Work by a nonconstant force

What is the work done by a nonconstant force? (We'll work in just one dimension here.) In the simple case of a *constant* force, the work done is given in Eq. (5.1) as $F\Delta x$, which equals the area of the region (which is just a rectangle) under the F vs. x "curve" in Fig. 5.2. If the force is

5.1. INTRODUCTION

instead *nonconstant*, that is, if it varies with position as in Fig. 5.3, then we can divide the x axis into a large number of short intervals, with the force being essentially constant over each one. The work done in each interval is essentially the area of each of the narrow rectangles shown. In the limit of a very large number of very short intervals, adding up all the areas of all the thin rectangles gives exactly the total area under the curve. So the general result is: The work done is the area under the F vs. x curve. That is, the work is the integral of the force:

$$W = \int F\, dx. \quad (5.4)$$

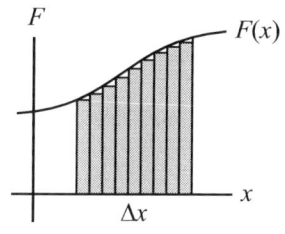

Figure 5.3

The task of Problem 5.1 is to calculate the work done between two given points, x_1 and x_2, by a Hooke's-law spring.

In higher dimensions, Eq. (5.4) becomes $W = \int F\Delta x \cos\theta$. If you want, you can write this in terms of the dot product as $W = \int \mathbf{F} \cdot d\mathbf{x}$. But however you want to write it, you're simply adding up the work done over a large number of tiny intervals.

Work-energy theorem

Define the kinetic energy of a particle as $K \equiv mv^2/2$. Then the *work-energy theorem* states:

- *Work-energy theorem:* The work done on a particle equals the change in kinetic energy of the particle. That is,

$$W = \Delta K. \quad (5.5)$$

See Problem 5.2 for a proof. From $F = ma$, we know that if the force points in the same direction as the velocity of the particle, then the speed increases, so the kinetic energy increases. That is, ΔK is positive. This is consistent with the fact that the work W is positive if the force points in the same direction as the velocity. Conversely, if the force points opposite to the velocity of the particle, then the speed decreases, so the kinetic energy decreases. This is consistent with the fact that the work is negative.

Work-energy theorem (general)

The above statement of the work-energy theorem actually is valid only in the special case where the object has no internal structure, that is, where it is a rigid featureless object. If the object *does* have internal structure (for example, subparts that can move or springs that can be compressed, etc.), then the general form of the work-energy theorem states that the work done equals the change in the *total* energy of the object. This energy comes in the form of not only the kinetic energy K of the object as a whole, but also the energy E_{int} of the internal constituents (this can include both kinetic and potential energies; see below for the definition of potential energy). So the general form of the work-energy theorem can be stated as

$$W = \Delta E_{\text{total}} = \Delta K + \Delta E_{\text{int}}. \quad (5.6)$$

ΔE_{int} includes heat, because heat is a measure of the kinetic energy of molecules vibrating on a microscopic scale.

If an object is deformable, we need to be careful about how we define work. The work is $F\Delta x$ (we'll deal with 1-D here, for simplicity), but if different parts of the object move different amounts, which displacement should we pick as Δx? The correct choice for Δx is the displacement of the point in the object where the force is applied. For contact forces (like pushing or pulling, as opposed to long-range forces like gravity) that don't involve any slipping, we can equivalently say that Δx is the displacement of the thing that is applying the force. If you walk up some stairs, then the stairs do no work on you, because they aren't moving.

When applying Eq. (5.6), the first thing you need to do is define what your *system* is, because your choice of system determines which forces are *external* (which in turn determines the work done) and which changes in energy are *internal*. Multiple-Choice Question 5.2 and Problem 5.5 discuss this issue.

Conservative forces

A *conservative* force can be defined in two equivalent ways:

1. A force is conservative if it does zero total work on an object during a round trip.

2. A force is conservative if the work done between two given points is independent of the path taken.

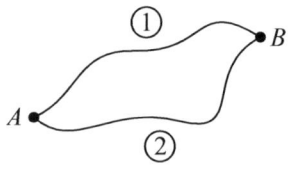

Figure 5.4

You can quickly verify that gravitational and spring forces are conservative, but kinetic friction is not. Kinetic friction always does negative work, because the force always points opposite to the velocity, which means that there can't exist the necessary cancelation of positive and negative contributions to make the total work be zero during a round trip.

The above two definitions are equivalent for the following reason. Let's start with the second definition. From this definition, the work done by a conservative force in going from A to B in Fig. 5.4 is the same along the two paths. Label these works as W. Then the work done in going *from* point B *to* point A along path 2 is $-W$, because the $d\mathbf{x}$ in the $W = \int \mathbf{F} \cdot d\mathbf{x}$ integral changes sign. Therefore the total work done in the round trip from A to B along path 1, and then from B to A along path 2, equals $W + (-W) = 0$, which is consistent with the first definition.

Potential energy

We define the change in the *potential energy* (associated with a given force) of an object to be the negative of the work done by the force on the object:

$$\Delta U \equiv -W. \tag{5.7}$$

(Sometimes the letter V is used instead of U.) For a falling mass, gravity does *positive* work, so ΔU is *negative*, which makes intuitive sense; the object loses gravitational potential energy. Potential energy is defined only for conservative forces, because if a force isn't conservative, then the work is path dependent, so the ΔU between two points isn't well defined. The potential energies associated with two common conservative forces, gravitational and spring, are derived in Problem 5.3.

Note that the definition in Eq. (5.7) deals only with the *change* in U. It makes no sense to ask what U is for an object at a given point; it makes sense only to ask what U is for an object at a given point, *relative* to a given reference point. Only changes in U matter. If one person measures the gravitational U with respect to the floor, and another person measures it with respect to the ceiling, they will have different results for U at any given point. But they will always calculate the same difference between any two given points.

If we combine Eqs. (5.4) and (5.7), we see that ΔU is the negative integral of F. That is, $U(x) - U(x_0) = -\int_{x_0}^{x} F \, dx$. The fundamental theorem of calculus then tells us that F is the negative derivative of U:

$$F(x) = -\frac{dU}{dx}. \tag{5.8}$$

The general result in 3-D is that $\mathbf{F}$ is the negative gradient of U, that is, $\mathbf{F} = -\nabla U$. But we'll stick to one dimension here. Eq. (5.8) gives another explanation of why adding a constant to the potential energy doesn't change the system: Since the derivative of a constant is zero, an additive constant doesn't affect the force. So a particle will move in exactly the same way.

Conservation of energy

If we use $\Delta U \equiv -W$ (which is just a definition) to replace W with $-\Delta U$ in the work-energy theorem, $W = \Delta K$ (which is an actual theorem with content), we obtain

$$-\Delta U = \Delta K \implies \Delta(U + K) = 0. \tag{5.9}$$

This tells us that the quantity $U + K$ is constant (that is, *conserved*) throughout the motion. We call this quantity the total energy: $E \equiv U + K$. In general, U can come from various different

5.1. INTRODUCTION

forces – gravity, spring, electrical, and so on. Note that conservation of energy gives the velocity of an object (in 1-D) as

$$U + K = E \implies U(x) + \frac{mv^2}{2} = E \implies v(x) = \pm\sqrt{\frac{2(E - U(x))}{m}}. \tag{5.10}$$

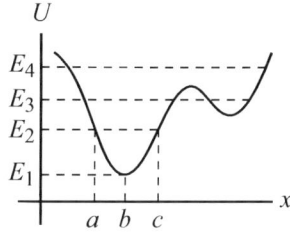

Figure 5.5

Conservation of energy lends itself to an easy graphical visualization. Fig. 5.5 shows an example of a potential-energy function, $U(x)$, relative to a chosen reference point. Various different choices of the total energy E are shown. (While $U(x)$ is determined by the given setup, the total energy E is determined by the initial speed and position of the object, which you are free to choose.) From Eq. (5.10), the object can't exist at values of x for which $U(x) < E$, because v would be imaginary. So, for example, if the energy in Fig. 5.5 is E_1, then the object will sit at rest at $x = b$. If the energy is E_2, then the object will oscillate back and forth between $x = a$ and $x = c$. If the energy is E_3, then the object will oscillate in *either* of two wells (whichever one you put it in at the start). And if the energy is E_4, then the object will oscillate (with a nontrivially varying speed) in one large well.

When using conservation of energy to solve a simple problem (say you want to find the speed of a ball dropped from rest), you can usually get by with saying something like, "the loss in potential energy shows up as kinetic energy," which will allow you to write down $mgh = mv^2/2$. But in more complicated problems, you are strongly advised to be systematic by writing down an $E_{\text{initial}} = E_{\text{final}}$ equation and then dealing with the various terms. For example, if a setup involves an object under the influence of gravity and a spring, then you should immediately write down:

$$K_i + U_i^{\text{grav}} + U_i^{\text{spring}} = K_f + U_f^{\text{grav}} + U_f^{\text{spring}}. \tag{5.11}$$

You can then gradually get a handle on what each term is and then solve for whatever unknown you're trying to solve for. See, for example, Problem 5.9.

Since conservation of energy follows from the work-energy theorem,[1] which in turn follows from $F = ma$ (see Problem 5.2), any problem that you can solve with conservation of energy you can also solve with $F = ma$. However, in many cases a conservation-of-energy approach makes for a much quicker solution, because the $F = ma$ solution would probably involve redoing the derivation of the work-energy theorem in Problem 5.2. This task has already been done once and for all if you use conservation of energy.

Along the same lines (and consistent with Footnote 1), if someone solves a problem by using conservation of energy, and another person uses the work-energy theorem, then they're essentially doing they same thing. A potential-energy term in the former solution will show up as a work term in the latter. See, for example, Problem 5.5.

Conservation of energy is one of the fundamental tools in physics. Two other conservation laws that permeate classical mechanics are conservation of momentum (discussed in Chapter 6) and conservation of angular momentum (discussed in Chapter 8).

Heat

The total energy of an isolated system is always conserved. In standard mechanics, the energy can take three basic forms: potential energy, kinetic energy on a macroscopic scale, and kinetic energy on a microscopic scale. The first two of these are commonly called "mechanical energy" (examples include the $kx^2/2$ potential energy of a spring and the $mv^2/2$ kinetic energy of an object), while the latter is called "thermal energy" or "heat."

Heat is just the sum of the $mv^2/2$ kinetic energies of the tiny molecules moving around inside an object. But the reason we split the total kinetic energy into a macroscopic piece and a microscopic piece is that it is generally easy to write down all the $mv^2/2$ terms for the former, but hopeless for the latter. However, it *is* often easy to get a handle on the *total* heat energy

---

[1]Instead of using the word "follows" here, it might be more accurate to say, "Since conservation of energy *is equivalent to* the work-energy theorem." The only thing separating these two results is the $\Delta U \equiv -W$ definition, which doesn't really have any content, being just a definition.

in a system. In problems in this book, the way that heat arises is from kinetic friction. More precisely, the heat generated equals the magnitude of the work done by kinetic friction.

(An aside, which isn't important for this book: The phrase "work done by kinetic friction" is riddled with subtleties; the exact nature of the friction force needs to be specified before we can determine how much work is done on any given object. See Problem 5.6 in Morin (2008) for a discussion of this. However, for the purpose of calculating the total heat generated by kinetic friction (without caring how it is divided between the objects), it is valid to say that the work equals the kinetic friction force times the distance that one object moves with respect to the other.)

When energy is "lost" to heat, people often say colloquially that energy isn't conserved. What they really mean is that the *mechanical* energy isn't conserved. The *total* energy, including heat, is always conserved. No energy is actually lost; it's just that some of it might change from a macroscopic form to a microscopic from and hence not show up in the overall motion of objects in the system.

Lest you think that there is no way that tiny molecules moving around inside an object can have a substantial amount of energy, remember that there are a lot of them, and that they can be moving very fast, even though you can't see the motion. It turns out that if you increase the temperature of a liter of water from just above freezing to just below boiling, then the energy you need to add is the same as the gravitational potential energy of a 1.5-ton car raised 30 meters, or equivalently the gravitational potential energy of the liter of water raised 40 kilometers, or equivalently the mechanical kinetic energy of the liter of water projected with a speed of 900 m/s! You can verify these claims by using the facts that the amount of energy required to raise one gram of water by one Celsius degree is one calorie, and that there are 4.2 joules in a calorie,

Power

Power is the rate at which work is done. If a force F in 1-D acts on an object for a small time dt, during which the object has a displacement dx, then the small amount of work done is $dW = F\,dx$. So the rate at which work is done (that is, the power) equals

$$P \equiv \frac{dW}{dt} = \frac{F\,dx}{dt} = F\frac{dx}{dt} = Fv. \tag{5.12}$$

In other words, power equals force times velocity. These are all signed quantities, so if the force and velocity have opposite signs, then the power is negative. More generally, in higher dimensions the $dW = \mathbf{F} \cdot d\mathbf{x}$ relation leads to $P = \mathbf{F} \cdot \mathbf{v}$, by the same reasoning. Since the unit of work is the joule, the units of power are joules/second, which are called a *watt* (W).

5.2 Multiple-choice questions

5.1. Which of the following forces can never, under *any* circumstances, do work? (Be careful!)

(a) ~~gravity~~

(b) static friction

(c) ~~kinetic friction~~

(d) ~~tension~~

(e) ~~normal force~~

(f) None of the above; they all can do work.

5.2. If you are driving down a road and you step on the gas, the car accelerates due to the static friction force between the ground and the tires. The car's speed increases, so its kinetic energy increases. Does the static friction force do any work?

Yes   No

5.3. An escalator moves downward at constant speed. You walk up the escalator at this same speed, so that you remain at rest with respect to the earth. Are you doing any work?

Yes   No

5.4. Fill in the blanks: If you walk up some stairs at constant speed, the net work done on your entire body (during some specific time interval) is _____, and the net work done on just your head is _____.

(a) negative, zero

(b) zero, zero

(c) zero, positive

(d) positive, zero

(e) positive, positive

5.5. A spring hangs from a ceiling. It is initially compressed by some distance. A mass is attached to the bottom end and then released from rest. Consider the lowest point in the mass's motion as it bounces up and down. At this point, the spring's potential energy is _____ its initial potential energy.

(a) equal to

(b) larger than

(c) smaller than

(d) The relative size cannot be determined from the given information.

5.6. A block with mass m initially has speed v_0 down a plane inclined at an angle θ. The block is attached to a spring with spring constant k, initially at its relaxed length; see Fig. 5.6. The coefficient of kinetic friction with the plane is μ.

True or false: A method for calculating the position where the block reaches its lowest point on the plane is to find the position where the net force on the block is zero.

T   F

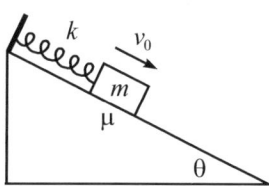

Figure 5.6

5.7. A block with mass m starts from rest and slides down a plane inclined at an angle θ. The coefficient of kinetic friction is μ. Which expression correctly yields the block's speed v after it has traveled a distance d down along the plane, assuming that it does indeed start sliding down? (d is a distance here, so it is a positive quantity.)

(a) $mgd \sin\theta + \mu mgd \cos\theta = \dfrac{mv^2}{2}$

(b) $mgd \sin\theta - \mu mgd \cos\theta = \dfrac{mv^2}{2}$

(c) $-mgd \sin\theta + \mu mgd \cos\theta = \dfrac{mv^2}{2}$

(d) $mgd \cos\theta + \mu mgd \sin\theta = \dfrac{mv^2}{2}$

(e) $mgd \cos\theta - \mu mgd \sin\theta = \dfrac{mv^2}{2}$

5.8. A cart with massless wheels contains sand. The cart starts at rest and then rolls (without any energy loss to friction) down into a valley and then up a hill on the other side. Let the initial height be h_1, and let the final height attained on the other side be h_2. If the cart leaks sand along the way, how does h_2 compare with h_1?

(a) $h_2 < h_1$ (b) $h_2 = h_1$ (c) $h_2 > h_1$

5.3 Problems

The first six problems are foundational problems.

5.1. **Work done by a spring**

If the force from a spring is given by Hooke's-law, $F(x) = -kx$, calculate the work done by the spring in going from x_1 to x_2.

5.2. **Work-energy theorem**

Show that the acceleration a of an object can be written as $a = v\,dv/dx$. Then combine this with $F = ma$ to prove the work-energy theorem: the work done on an object (with no internal structure) equals the change in its kinetic energy.

5.3. **Gravitational and spring U's**

(a) Find the gravitational potential energy of a mass m as a function of y (with upward taken to be positive), measured relative to a given point chosen as $y = 0$.

(b) Find the spring potential energy (for a spring with spring constant k) as a function of x, measured relative to the equilibrium point where $x = 0$.

5.4. **Work in different frames**

An object with mass m is initially at rest, and then a force is applied to it that causes it to undergo a constant rightward acceleration a for a time t.

(a) Calculate the work done on the object, and also the change in kinetic energy. Verify that $W = \Delta K$.

(b) Repeat the same tasks, but now by working in the reference frame moving to the left with constant speed v. (Equivalently, let the object start with speed v instead of starting at rest.)

5.5. **Raising a book**

Assume that you lift a book up a height h at constant speed (so there is no change in kinetic energy). The general work-energy theorem, Eq. (5.6), takes a different form depending on what you pick as your system, because your choice of system determines which forces are external (and thereby do work), and which changes in energy are internal. Write down Eq. (5.6) in the cases where your system is:

(a) the book

(b) the book plus the earth

(c) the book plus the earth plus you

Verify that the three different equations actually say the same thing.

5.6. **Hanging spring**

A massless spring with spring constant k hangs vertically from a ceiling, initially at its relaxed length. A mass m is then attached to the bottom and released.

(a) Calculate the total potential energy U (gravitational and spring) of the system, as a function of the height y (which is negative), relative to the initial position.

(b) Find y_0, the point at which the potential energy is minimum. Make a rough plot of $V(y)$.

(c) Rewrite the potential energy as a function of $z \equiv y - y_0$. Explain why your result shows that a hanging spring can be considered to be a spring in a world without gravity, provided that the new equilibrium point, y_0, is now called the "relaxed" length of the spring.

5.3. PROBLEMS

5.7. A greener world?

In an effort to make the world a greener place by reducing the burning of fossil fuels, consider the following alternative method of producing energy. This brilliant method applies the principle of hydroelectric power to normal solid matter. The plan is to blow up the sides of mountains and convert the loss in potential energy of the falling rock/dirt/etc. into electric power.

Your task is to find (very roughly) the volume of rock that needs to be converted daily to satisfy the entire world's need for power. You will need to look up the values of various things, such as the world's daily energy consumption. Assume that some ingenious method has been devised to capture all of the energy of the falling rock, and assume that on average the rock falls a height of one kilometer.

5.8. Hoisting up

A platform has a rope attached to it which extends vertically upward, over a pulley, and then back down. You stand on the platform. The combined mass of you and the platform is m.

(a) Some friends standing on the ground grab the other end of the rope and hoist you up a height h at constant speed. What is the tension in the rope? How much work do your friends do?

(b) Consider instead the scenario where you grab the other end of the rope and hoist yourself up a height h at constant speed. What is the tension in the rope? How much work do you do?

5.9. Hanging block

A block with mass m is attached to a ceiling by a spring with spring constant k and relaxed length ℓ. Initially, the spring is compressed to a length of $\ell/2$. If the block is released, at what distance below the ceiling will the block be brought to rest (instantaneously, at the lowest point) by the spring?

5.10. Rising on a spring

A spring with spring constant k and relaxed length ℓ stands vertically on the ground, with its bottom end attached to the ground. A mass m is attached to the top and then lowered down to the ground, so that the spring is compressed to zero length. The mass is then released from rest and accelerates vertically upward. What is the maximum height above the ground it reaches? You may assume that $k > mg/\ell$, so that the mass does indeed rise up off the ground. Note that at all times, the mass remains attached to the spring, and the spring remains attached to the ground.

5.11. Jumping onto mattresses

(a) You jump out of a window (with zero initial speed) and do a belly flop onto a mattress. Assume that the mattress can be treated like a spring with spring constant k. If you fall a distance h before hitting the mattress, what is the maximum compression distance of the mattress? What is the maximum force it applies to you? For simplicity, ignore the change in gravitational potential energy during the compression.

(b) Answer the same questions, but now with N identical mattresses stacked on top of each other. Assume that the height fallen before hitting the top mattress is still h. (You will need to use the generalization of the result from Problem 4.2.)

5.12. Bungee jumping 1

A bungee-jump cord has length ℓ and is initially folded back on itself, as shown in Fig. 5.7. The jumper has mass m. After she jumps (or rather, falls) off the platform, she is in freefall for a height ℓ. After that, the spring becomes stretched. Assume that it acts like an ideal

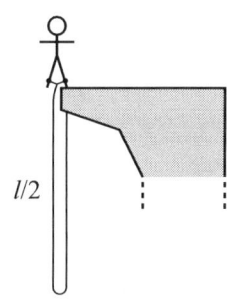

Figure 5.7

spring with a particular spring constant. (This applies only during the stretching motion, of course; the cord can't support compression like a normal spring.) It is observed that the lowest point the jumper reaches is a distance 2ℓ below the platform.

(a) What is the spring constant?

(b) What is the jumper's acceleration at the lowest point?

(c) At what position (specify the distance below the platform) is the jumper's speed maximum? What is this speed?

5.13. Bungee jumping 2

Consider again the setup in Fig. 5.7, with a bungee-jump cord of length ℓ. Assume that the spring constant takes on a particular value k.

(a) What is the lowest point the jumper achieves?

(b) What is the tension in the cord at the lowest point?

(c) If the jumper cuts the cord in half and uses one of the halves for the jump, what is the tension at the lowest point? (You will need to use the result from Problem 4.2.)

The following five problems have a common theme.

5.14. Mass on a spring

A block with mass m is located at position $x = 0$ on a horizontal table. A spring with spring constant k and relaxed length *zero* is connected to it and has its other end anchored at position $x = \ell$, as shown in Fig. 5.8. The coefficient of friction (both static and kinetic) between the block and the table depends on position according to $\mu = Ax$, where A is a constant. Assume that the block is small enough so that it touches the table at essentially only one value of x.

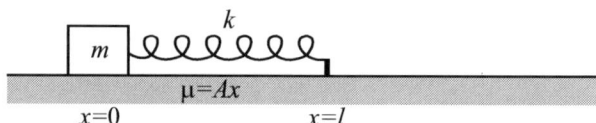

Figure 5.8

(a) The block is released from rest at $x = 0$. Where does it come to rest for the first time? What is the condition on A for which the stopping point is to the right of the $x = \ell$ anchor point? (Assume that the block can somehow pass through the anchor.)

(b) If the stopping point is to the right of the anchor, what is the condition on A for which the block starts moving leftward after it instantaneously comes to rest? In the cutoff case where it barely starts moving again, where is this (first) stopping point?

5.15. Falling with a spring

Consider the system shown in Fig. 5.9, with two equal masses m and a spring with spring constant k. The coefficient of kinetic friction between the left mass and the table is $\mu = 1/4$, and the pulley is frictionless. The system is held with the spring at its relaxed length and then released.

(a) How far does the spring stretch before the masses come to rest?

(b) What is the minimum value of the coefficient of *static* friction for which the system remains at rest once it has stopped?

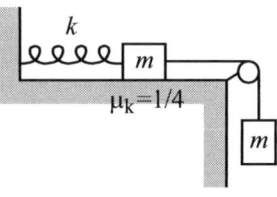

Figure 5.9

5.3. PROBLEMS

(c) If the string is then cut, what is the maximal compression of the spring during the resulting motion?

Note: In parts (a) and (c), you can give the left mass a tiny kick to get it going if the coefficient of static friction happens to be large enough to make this necessary.

5.16. Bead, spring, and rail

Fig. 5.10 shows a setup involving a spring and two angled rails. Assume that the setup is located in deep space, so that you can ignore gravity in this problem. A bead with mass m is constrained to move along one of the rails; this rail has friction, and the coefficient of kinetic friction with the bead is μ. The bead is connected to a spring, the other end of which is constrained to move along a *frictionless* rail. The spring is always perpendicular to this rail (this is a consequence of the facts that the spring is massless and the rail is frictionless). The relaxed length of the spring is *zero*.

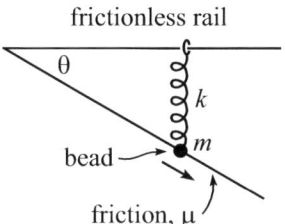

Figure 5.10

The bead starts out at the vertex of the rails (with the spring unstretched at its relaxed length of zero) and is given a kick so that its initial speed is v_0 (so the figure shows a general later time).

(a) Draw the free-body diagram for the bead at a general later time. If x is the distance the bead has traveled along the rail, what are all the forces in terms of x?

(b) How far does the bead travel along the rail before it comes to rest?

(c) Under what condition does the bead start moving again, back toward the vertex? (Assume that the coefficient of static friction is also μ.)

(d) Assuming that the bead does indeed start moving again, what is its speed when it arrives back at the vertex?

5.17. Ring on a pole

A spring with spring constant k and relaxed length *zero* has one end attached to a wall and the other end attached to a ring with mass m. The ring is pulled to the side and is slipped over a vertical pole that is fixed at a distance ℓ from the wall, as shown in Fig. 5.11. The coefficient of friction (both static and kinetic) between the ring and the pole is μ. The ring is held with the spring horizontal and is then released.

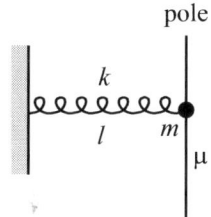

Figure 5.11

(a) Draw the free-body diagram for the ring at a general later time. What is the normal force between the ring and the pole?

(b) How far down the pole does the ring fall before bouncing back up?

(c) What is the cutoff value of μ, below which the ring does indeed:
  i. fall when it is released?
  ii. bounce back up at the bottom of its motion?

5.18. Block on a plane

A block with mass m lies on a plane inclined at angle θ. The coefficient of friction (both kinetic and static) between the block and the plane is $\mu = 1$. A massless spring with spring constant k is placed on the plane, with its lower end held fixed. The block is attached to the top end of the spring (see Fig. 5.12) and then moved down until the spring is compressed a distance ℓ relative to its relaxed length. The block is then released.

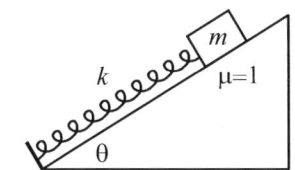

Figure 5.12

(a) For what value of ℓ does the block rise back up exactly to its original position (where the spring is uncompressed)?

(b) What is the condition on θ for which the block then starts to slide back down?

(c) Assuming that the block does indeed slide back down, how far down the plane does it go?

(d) What is the condition on θ for which the block then starts to move back up?

5.19. Tangential acceleration

A bead is initially at rest at the top of a fixed frictionless hoop with radius R that lies in a vertical plane. The bead is then given an infinitesimal push so that it slides down and around the hoop.

(a) What is the speed of the bead after it has fallen through an angle θ (measured relative to the vertical)?

(b) Take the time derivative of your result (don't forget to use the chain rule) to verify that the tangential acceleration dv/dt equals the tangential component of gravity, namely $g\sin\theta$.

5.20. Comparing the tensions

A pendulum with mass m and length ℓ swings back and forth between the two horizontal positions shown on the left in Fig. 5.13. Let the tension in the string as a function of θ be $T_1(\theta)$. The mass is then stopped at an angle θ and held in place with a horizontal rope, as shown on the right in Fig. 5.13. Let the tension in the string (the pendulum's string, not the rope) as a function of θ be $T_2(\theta)$.

(a) Find $T_1(\theta)$ and $T_2(\theta)$. For what θ is $T_1(\theta) = T_2(\theta)$?

(b) Explain with a continuity argument why you can say (without doing any calculations) that there must indeed exist an angle θ for which $T_1(\theta) = T_2(\theta)$.

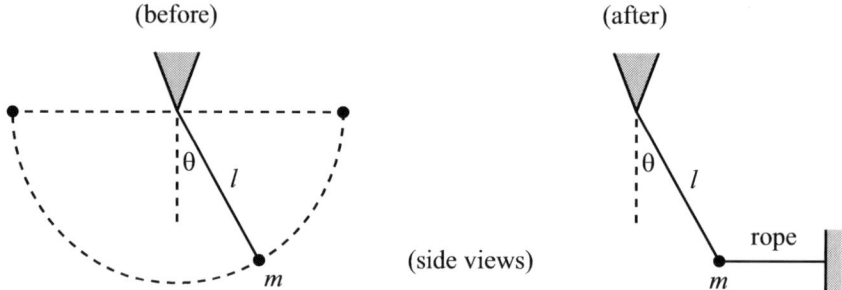

Figure 5.13

5.21. Semicircular tube

A frictionless tube is bent into the shape of a semicircle with radius R. The semicircle is tilted so that its diameter makes a fixed angle θ with respect to the vertical, as shown in Fig. 5.14. A small mass is released from rest at the top of the tube and slides down through it. When the mass leaves the tube, it undergoes projectile motion. Let d be the distance traveled in the projectile motion, up to the time when the mass returns to the height it had when it left the tube. What should θ be so that d is as large as possible?

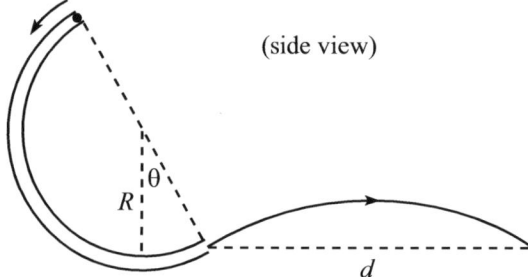

Figure 5.14

5.3. PROBLEMS

5.22. Horizontal force

A bead is constrained to move on a fixed frictionless vertical hoop with radius R. The bead starts at rest at the top and is given an infinitesimal kick. When the bead is at an angle θ below the horizontal, what is the normal force that the hoop applies to the bead? For what θ is the total force on the bead horizontal?

5.23. Maximum vertical normal force

A bead is constrained to move on a fixed frictionless vertical hoop with radius R. The bead is initially at rest at the top. It is then given a kick so that it suddenly acquires a speed v_0; see Fig. 5.15. It then travels around the hoop indefinitely. Let N_y be the vertical component of the bead's normal force on the hoop. (Note that we are talking about the force from the bead on the hoop, and not the hoop on the bead.) Let θ be the angle of the bead's position with respect to the vertical.

(a) What is N_y?

(b) For what θ does N_y achieve its maximum (upward) value? In answering this, you can assume that v_0 is relatively small. Relatively large v_0 is handled in part (c).

(c) There is a certain value of v_0 above which N_y achieves its maximum value at the top of the hoop. What is this value of v_0?

(side view)

Figure 5.15

5.24. Falling stick on a table

A massless stick with length ℓ stands at rest vertically on a table, and a mass m is attached to its top end, as shown in Fig. 5.16. The coefficient of static friction between the stick and the table is μ. The mass is given an infinitesimal kick, and the stick-plus-mass system starts to fall over. At what angle (measured between the stick and the vertical) does the stick start to slip on the table? (Careful, the answer depends on whether μ is larger or smaller than a particular value.)

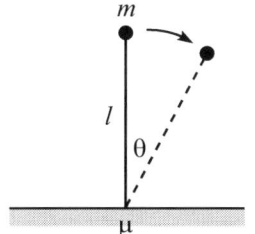

Figure 5.16

5.25. Bead, spring, and hoop

A massless spring with spring constant k and relaxed length *zero* has one end attached to a given point on fixed frictionless *horizontal* hoop of radius R, while the other end is attached to a bead with mass m that is constrained to lie on the hoop. The spring is initially stretched across a diameter, with the bead at rest. The bead is then given an infinitesimal kick, and it gets pulled around the hoop by the spring, as shown in Fig. 5.17.

(a) What is the normal force in the horizontal plane of the hoop (in other words, ignore gravity in this problem) that the hoop exerts on the bead at the moment the bead has gone a quarter of the way around the circle?

(b) Is there a point in the motion where the normal force (in the plane of the hoop) is zero? A simple yes or no, with proper reasoning, is sufficient; you don't have to solve for the normal force as a general function of the angle to answer this question.

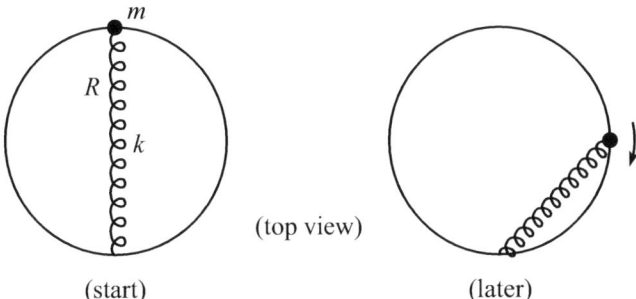

Figure 5.17

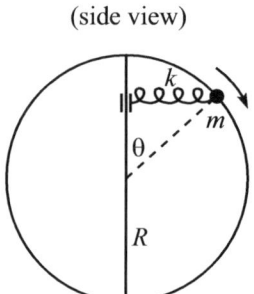

Figure 5.18

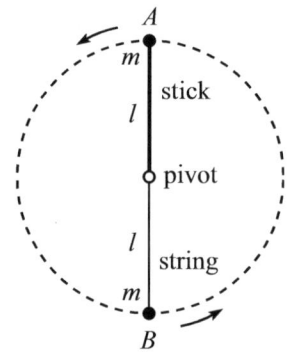

Figure 5.19

5.26. **Bead, spring, hoop, and pole**

A bead with mass m is free to slide on a fixed frictionless vertical hoop with radius R. A pole is located along the vertical diameter. A spring with spring constant k and relaxed length *zero* has one end attached to the mass, while the other end is free to slide without friction along the vertical pole. The bead is initially at rest at the top of the hoop and is given an infinitesimal kick. The situation at a later time is shown in Fig. 5.18. You can assume that the spring is always horizontal (this is a consequence of the facts that the spring is massless and the pole is frictionless).

Assume that before constructing this setup, you determine the value of k by noting that if you hold one end of the spring and let the bead on the other end hang at rest below, the length of the spring is R. This quickly tells you that $k = mg/R$. Given this value of k, at what angle θ (measured with respect to the vertical) is the normal force from the hoop on the bead equal to zero?

5.27. **Entering freefall**

Ball A of mass m is connected to one end of a massless stick with length ℓ, the other end of which is connected to a pivot. The stick is held vertically above the pivot and is then given an infinitesimal kick; see Fig. 5.19. It swings down, and at the bottom of its motion it collides elastically with ball B, also of mass m, which hangs from a *string* with length ℓ and which is initially at rest. Ball B picks up whatever speed ball A had and then swings upward in circular motion.

(a) At what point does ball B leave its circular motion and enter freefall projectile motion?

(b) How high does ball B go in the resulting projectile motion?

5.4 Multiple-choice answers

5.1. $\boxed{f}$ They all can do work. Gravity certainly does work on a falling object. Static friction does work on a book lying on a table if you accelerate the table sideways. Kinetic friction does work in this same setup if you accelerate the table so quickly that the book slips with respect to it. Tension does work if you pull on a block with a piece of rope. And a normal force does work if you push on a book with your hand.

REMARK: Note that in all of these cases, the force has a nonzero component in the direction of the motion, so the product $F\Delta x \cos\theta$ is nonzero. This would *not* be the case for the normal force if we had a block sliding down a *stationary* inclined plane, because the normal force would be perpendicular to the motion. If, on the other had, the plane were moving sideways, then the normal force *would* do nonzero work on the block, because the force would have a nonzero component in the direction of the motion (which now isn't perpendicular to the plane).

5.2. $\boxed{\text{No}}$ The thing applying the force (the ground) doesn't move, so it doesn't do any work.

REMARK: Since no work is done, the general work-energy theorem, Eq. (5.6), says that the total energy of the car doesn't change. But we know that the kinetic energy increases, so what's going on? The car apparently must lose some other form of energy so that the total energy remains constant. And indeed, the car burns gasoline, so the potential energy of the gasoline decreases.

As a more concrete analogy, consider the setup shown in Fig. 5.20. A box has a spring inside it that is compressed against a stationary protrusion from a wall. The box is released, and the spring pushes it to the right. The wall doesn't move, so it doesn't do any work on the box-plus-spring system. The total energy of the box-plus-spring system must therefore remain constant. And indeed, the gain in kinetic energy of the box is exactly canceled by the loss in potential energy of the spring. If the box is opaque so that you can't see inside, you can still conclude that there must be something inside that loses energy.

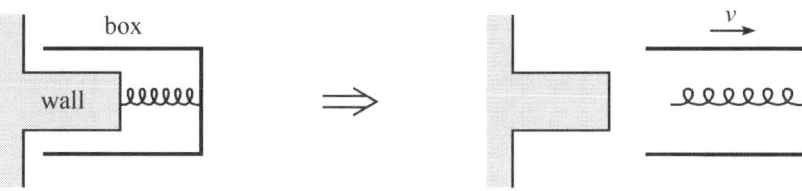

Figure 5.20

If you instead want to consider the box alone to be your system (instead of the box-plus-spring) then the external force acting on the box itself comes from the *right* end of the spring. This end *does* move, so it does do nonzero work on the box. This is consistent with the work-energy theorem, because the energy of the box increases; this energy consists solely of kinetic energy.

5.3. $\boxed{\text{Yes}}$ You do work. Your feet apply a normal force on the steps, and your feet are moving. So there is a nonzero force-times-distance product.

REMARK: The work that you do on the escalator is positive; your feet move downward as they apply a downward force on the escalator. Conversely, the escalator does *negative* work on you; it moves downward as it applies an *upward* force on your feet. Since negative work is done on you, the general work-energy theorem says that your total energy decreases. Your kinetic energy doesn't change (it is always zero; or essentially constant, if we include the roughly constant motion of your legs), nor does your gravitational potential energy change (because you remain at the same height). But your internal chemical potential energy decreases as you use up the dinner you ate the night before.

5.4. $\boxed{a}$ Since you are moving upward, gravity does negative work on your body. The only other force acting on your body as a whole is the force from the stairs, but since the stairs aren't moving, they do no work on you. So the net work comes from only gravity and is therefore negative.

Now consider your head. Gravity again does negative work. The other force acting on your head is the upward force from your neck (equal and opposite to the weight of your head, since your head isn't accelerating). This force *does* do work, because your neck is moving upward. So the net work on your head is zero.

REMARK: These results are consistent with the general work-energy theorem, Eq. (5.6). In the case of your head (which we're assuming is basically a point mass with no internal structure), the zero work is consistent with the zero change in kinetic energy. In the case of your whole body, the negative work is consistent with the fact that your internal energy decreases (your muscles use up chemical potential energy).

5.5. $\boxed{b}$ The total energy, which comes in three forms (kinetic, gravitational U, and spring U), is conserved. There is no net change in the kinetic energy, because it starts and ends at zero (since the mass is instantaneously at rest at the lowest point). Therefore, since the gravitational potential energy decreases, the spring potential energy must increase. What happens is that the spring's U initially decreases as the spring becomes uncompressed, but then U increases and ends up larger than its initial value. So we conclude that the spring ends up stretched by a larger distance than it was initially compressed; this makes intuitive sense.

Alternatively: No net work is done on the mass, because $\Delta K = 0$. Since gravity does positive work, the spring must do negative net work. From Eq. (5.7), the potential energy of the spring therefore increases.

5.6. $\boxed{F}$ The block is instantaneously at rest at the lowest point. The block does *not* come to rest when the net force on it is zero. It comes to rest when the total work done on it equals $-mv_0^2/2$. This follows from the work-energy theorem, because the kinetic energy goes from $mv_0^2/2$ to 0 since the speed goes from v_0 to 0.

REMARK: If you instead want to find the place where the block's speed is maximum, then you *do* want to set the net force equal to zero (so that $a = dv/dt = 0$). But you don't want to do this for the present goal of finding the lowest point. At the lowest point (or rather, a split second before the lowest point is reached, so that we're still dealing with kinetic friction) the force on block is *not* zero. The force points up the plane; we know this because the block is decelerating to zero speed.

Depending on the coefficient of static friction, note that once the block instantaneously stops at the lowest point, the static friction force might be large enough to keep the block from bouncing back up, in which case the total force *is* zero. But there is an infinite number of points where this is the case, so demanding that the total force (including static friction) is zero certainly won't help you find the (unique) lowest point, as this question asks.

5.7. b The component of the gravitational force pointing down along the plane has magnitude $mg \sin\theta$. The block is moving down the plane, so gravity does positive work; this work is $(mg\sin\theta)d$. The kinetic friction force points up along the plane with magnitude $\mu N = \mu(mg\cos\theta)$. The associated work is negative and equals $-(\mu mg \cos\theta)d$. The total work is therefore $mgd\sin\theta - \mu mgd\cos\theta$. Since the block starts at rest, the change in kinetic energy is $mv^2/2 - 0$. The work-energy theorem therefore gives choice (b).

Alternatively: If you want to think in terms of energy instead of work, then the $mgd\sin\theta$ loss in potential energy shows up as $mv^2/2$ kinetic energy plus $\mu mgd \cos\theta$ heat (which is the magnitude of the work done by friction). This yields the same equation, just rearranged.

5.8. b If the cart *doesn't* leak any sand, then it effectively acts like a point object, so conservation of energy tells us that the initial and final potential energies must be equal, because the initial and final kinetic energies are equal (they are both zero). So h_2 must equal h_1.

Now let's take into account the leaked sand. The leaked sand doesn't affect the speed of the cart, because it isn't propelled forward or backward. It just leaks out, so the cart applies no force to the sand, and hence by Newton's third law the sand applies no force to the cart. The cart therefore moves exactly the same way it would move if the sand weren't leaking. So it reaches the same height as it would in the non-leaking case, namely $h_2 = h_1$.

REMARK: Be careful about incorrectly applying conservation of energy. You might say that since the sand comes to rest on the ground somewhere down in the valley, it ends up with less potential energy than it started with; so the cart itself must end up with more potential energy (to conserve the total energy of the system), which means that it must end up at a higher position. This reasoning is incorrect because it doesn't take into account *all* of the energy in the system. The leaked sand has kinetic anergy when it leaves the cart, and this kinetic energy eventually ends up as heat when the sand comes to rest on the ground. The energy is still there, it just takes a different form. But nothing that happens after the sand leaves the cart can affect the cart, of course. If the ground were frictionless and the leaked sand simply slid along the ground, then it would always remain right below the cart. At the end of the process, we would have at height $h_2 = h_1$ both the empty cart and a pile of sand right below it.

If you want to make your head hurt, you can consider the more difficult setup where the cart has *massive* wheels, which we will assume never slip on the ground. This setup requires material from Chapter 7, so you'll need to wait until then to think about it. As a hint, the answer is *not* $h_2 = h_1$.

5.5 Problem solutions

5.1. Work done by a spring

With $F(x) = -kx$, Eq. (5.4) gives the work done by the spring, in going from x_1 to x_2, as

$$W = \int_{x_1}^{x_2} F\, dx = \int_{x_1}^{x_2} (-kx)\, dx = -\frac{1}{2}kx^2 \Big|_{x_1}^{x_2} = \frac{1}{2}kx_1^2 - \frac{1}{2}kx_2^2. \tag{5.13}$$

5.5. PROBLEM SOLUTIONS

This is positive if $x_1^2 > x_2^2$, that is, if $|x_1| > |x_2|$. This makes sense, because if the spring pulls a mass from a larger position x_1 to a smaller position x_2, then both the force and the displacement are negative, so the work is positive.

REMARK: If a mass is attached to the end of the spring, and if you grab the mass and move it at constant speed, then your force exactly cancels the spring force, so your force equals $+kx$. The work you do is therefore exactly opposite to the work the spring does. So $W_{\text{you}} = kx_2^2/2 - kx_1^2/2$. The sign of this makes sense; if you bring the mass from a smaller position x_1 to a larger position x_2, then both your force and the displacement are positive, so your work is positive.

5.2. Work-energy theorem

If we write a as dv/dt and then multiply by 1 in the form of dx/dx, we obtain

$$a = \frac{dv}{dt} = \frac{dx}{dt}\frac{dv}{dx} = v\frac{dv}{dx}, \quad (5.14)$$

as desired. It is indeed legal to perform these operations with infinitesimal quantities. If you feel uneasy about this, you can work with finite quantities like Δx, etc. (for which the manipulations are certainly legal) and then at the end take the limit where all quantities become infinitesimally small.

Plugging $a = v\,dv/dx$ into $F = ma$, and then multiplying by dx and integrating from x_1 (and the corresponding v_1) to x_2 (and the corresponding v_2), gives

$$F = mv\frac{dv}{dx} \implies \int_{x_1}^{x_2} F\,dx = \int_{v_1}^{v_2} mv\,dv. \quad (5.15)$$

The integral on the left-hand side is the work done, by definition. So we have

$$W = \frac{mv^2}{2}\bigg|_{v_1}^{v_2} = \frac{mv_2^2}{2} - \frac{mv_1^2}{2}, \quad (5.16)$$

which is the change in kinetic energy, as desired.

5.3. Gravitational and spring U's

(a) The work done by gravity in moving a mass through a displacement y is

$$W_g = \int_0^y F\,dy = \int_0^y (-mg)\,dy = -mgy. \quad (5.17)$$

The negative sign arises because the gravitational force points downward. If y is positive, then the work done by gravity is negative, because the displacement points upward while the force points downward.

Since ΔU is defined to be $-W$, the gravitational potential energy relative to $y = 0$ is

$$U(y) = mgy. \quad (5.18)$$

This result is valid for both positive and negative y. The higher the object is, the more gravitational potential energy it has, which makes sense because it takes effort to lift up a mass.

(b) From Problem 5.1, the work done by a spring in going from x_1 to x_2 equals $kx_1^2/2 - kx_2^2/2$. If we pick x_1 to be zero and relabel x_2 as x, then the work is $W = -kx^2/2$. Since ΔU is defined to be $-W$, the spring potential energy relative to the equilibrium point at $x = 0$ is

$$U(x) = \frac{1}{2}kx^2. \quad (5.19)$$

This is positive for any nonzero value of x, which makes sense because it takes effort to stretch or compress a spring.

5.4. Work in different frames

(a) The work is $W = Fd$, where the force is $F = ma$ and the distance traveled is $d = at^2/2$. So the work is

$$W = Fd = (ma)\left(\frac{1}{2}at^2\right) = \frac{1}{2}m(at)^2. \tag{5.20}$$

Since the object starts at rest, the change in kinetic energy is simply the final kinetic energy, which is $mv_f^2/2$. But the final speed v_f is just $v_f = at$, so the change in kinetic energy is $m(at)^2/2$. This equals the work W, as it should.

(b) From standard 1-D kinematics, the distance traveled in the reference frame where the object starts with speed v is $d = vt + at^2/2$. The force F is still ma, so the work done in this frame is

$$W = Fd = (ma)\left(vt + \frac{1}{2}at^2\right) = mavt + \frac{1}{2}ma^2t^2. \tag{5.21}$$

The change in kinetic energy is

$$\Delta K = \frac{1}{2}mv_f^2 - \frac{1}{2}mv_i^2 = \frac{1}{2}m(v+at)^2 - \frac{1}{2}mv^2 \tag{5.22}$$
$$= \frac{1}{2}m(v^2 + 2vat + a^2t^2) - \frac{1}{2}mv^2 = mavt + \frac{1}{2}ma^2t^2,$$

which agrees with the work W, as it should.

REMARK: Note that W and ΔK in this frame are larger than W and ΔK in the original frame. So W and ΔK depend on the choice of frame. But given a particular frame, the work-energy theorem tells us that whatever the value of W is, the value of ΔK will be the same.

5.5. Raising a book

(a) SYSTEM = BOOK: Both you and gravity provide external forces. There is no change in the kinetic energy of the book, and also no change in internal energy. So the right-hand side of the general work-energy theorem, $W = \Delta K + \Delta E_{\text{int}}$, equals zero. The work you do is $+mgh$, and the work gravity does is $-mgh$, so Eq. (5.6) becomes

$$W_{\text{you}} + W_{\text{grav}} = 0 \quad \Longrightarrow \quad mgh + (-mgh) = 0. \tag{5.23}$$

(b) SYSTEM = BOOK + EARTH: Now you are the only external force. The gravitational force between the earth and the book is an internal force which is associated with an internal potential energy. (You can think of the book and the earth as being connected by a constant-force "gravitational spring" that is being stretched.) This potential energy $\Delta U_{\text{earth/book}}$ is the usual mgh. This is the change in the internal energy, ΔE_{int}. The general work-energy theorem therefore says

$$W_{\text{you}} = \Delta U_{\text{earth/book}} \quad \Longrightarrow \quad mgh = mgh. \tag{5.24}$$

(c) SYSTEM = BOOK + EARTH + YOU: There is now no external force, so no work is done on the system. The internal energy of the system changes because the earth-book gravitational potential energy increases (by mgh, as in part (b)), and also because *your* potential energy decreases. In order to lift the book, you have to burn some calories from the breakfast you ate. So the general work-energy theorem says

$$0 = \Delta U_{\text{you}} + \Delta U_{\text{earth/book}} \quad \Longrightarrow \quad 0 = -mgh + mgh. \tag{5.25}$$

Actually, the human body isn't 100% efficient, so what really happens is that your potential energy decreases by *more* than mgh, but heat is produced. The sum of these two changes in energy (the decrease in your potential energy plus the increase in heat) equals $-mgh$. In other words, including an amount η of energy in the form of heat, the general work-energy theorem says

$$0 = [\Delta U_{\text{you}} + \text{Heat}] + \Delta U_{\text{earth/book}} \implies 0 = [(-mgh - \eta) + \eta] + mgh. \quad (5.26)$$

Equations (5.23) through (5.26) are all consistent with each other, as they must be. These equations are simply different ways of describing the same setup, based on what we choose to call our system. A term that shows up as work in one equation might show up as a potential energy in another. Basically, when a W term on the left-hand side turns into a ΔU term on the right-hand side as we go from Eq. (5.23) to Eq. (5.25), it picks up a minus sign.

REMARK: If you drop a ball, why does it speed up as it falls? Is it because gravity does work on it (which means that it gains kinetic energy, by the work-energy theorem)? Or is it because it loses potential energy (which means that it gains kinetic energy, by conservation of energy)?

Both of these reasons are valid. They both give the correct result for the speed v after falling a height h. The difference in their language stems from the difference in what is chosen as the system. In the first reason, the system is the ball, and gravity is an external force acting on it. In the second reason, the system is the ball plus the earth. There is no external force on this system, so the general work-energy theorem says that ΔE_{int} doesn't change. That is, energy is conserved. This energy takes the form of kinetic energy of the ball plus gravitational potential energy associated with the earth-ball interaction.

Although both reasons are correct, be sure not to combine them and say that the ball speeds up because gravity does work on it, and *also* that it speeds up some more because it loses potential energy. This double counts the effect of gravity, of course. Equivalently, it uses two different (and hence inconsistent) definitions of the system.

5.6. Hanging spring

(a) The sum of the gravitational and spring potential energies is

$$U(y) = mgy + \frac{1}{2}ky^2. \quad (5.27)$$

Remember that y is negative, so the first of these terms is negative and the second is positive.

(b) The minimum occurs at the value of y for which

$$\frac{dU}{dy} = 0 \implies mg + ky = 0 \implies y_0 = -\frac{mg}{k}. \quad (5.28)$$

The value of the potential energy at this point is $U(y_0) = -m^2g^2/2k$. We also know that $U(y) = 0$ at both $y = 0$ and $y = -2mg/k$. Since $U(y)$ is quadratic function (that is, a parabola), and since we know three points that it passes through, the plot must look like the one shown in Fig. 5.21.

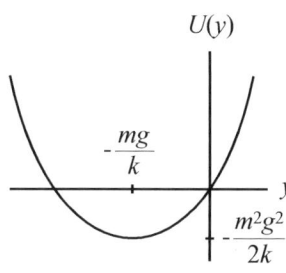

Figure 5.21

REMARK: Since the initial kinetic energy is zero (the mass is released from rest), the total energy of the system is zero. So the line representing the (constant) energy in Fig 5.21 coincides with the y axis. This immediately tells us that the motion bounces back and forth between $y = -2mg/k$ and $y = 0$. (These are the places where $U(y) = E$, so that the kinetic energy is zero; the mass is instantaneously at rest at the extremes of the motion.) So $y = -2mg/k$ is the lowest point in the motion.

(c) To rewrite the potential energy as a function of $z \equiv y - y_0$, we can substitute $y = z + y_0 = z - mg/k$ into $U(y)$. This gives

$$U(z) = mg\left(z - \frac{mg}{k}\right) + \frac{1}{2}k\left(z - \frac{mg}{k}\right)^2$$
$$= \left(mgz - \frac{m^2g^2}{k}\right) + \left(\frac{1}{2}kz^2 - mgz + \frac{1}{2}\frac{m^2g^2}{k}\right)$$
$$= \frac{1}{2}kz^2 - \frac{m^2g^2}{2k}. \tag{5.29}$$

What we've done here is pick a new coordinate system that has its origin at the point $y = -mg/k$ where $U(y)$ in minimum (z is indeed zero when $y = -mg/k$). In terms of z, $U(z)$ is a simple $kz^2/2$ quadratic function, plus an additive constant. But this constant is irrelevant, because only differences in the potential energy matter. (Equivalently $F(z) = -dU/dz$, so an additive constant doesn't affect the force.) We can therefore take the potential energy to be simply $U(z) = kz^2/2$.

The important thing to now note is that there is no mention of gravity in $U(z)$; the constant g doesn't appear. This is why can can consider our hanging spring to be a spring in a world without gravity, with the "relaxed" length corresponding to the equilibrium point at $z = 0$ (that is, $y = y_0$). The underlying reason for this is that the spring force is *linear* in y. So the first mg/k that the spring stretches (after we attach the mass) gives an upward force of $k(mg/k) = mg$ which cancels out gravity, and the remaining part of the stretching gives the net force we're concerned with.

5.7. **A greener world?**

The total power used by the entire world is about $1.5 \cdot 10^{13}$ J/s, or equivalently about $1.3 \cdot 10^{18}$ J/day. Let's assume that the rock is granite with a density of about $2500\,\text{kg/m}^3$. Then if one cubic meter falls one kilometer, the potential energy lost is

$$mgh = (2500\,\text{kg})(10\,\text{m/s}^2)(1000\,\text{m}) = 2.5 \cdot 10^7\,\text{J}. \tag{5.30}$$

The number of these cubic meters needed to yield the $1.3 \cdot 10^{18}$ J each day is therefore about $(1.3 \cdot 10^{18}\,\text{J})/(2.5 \cdot 10^7\,\text{J}) \approx 5 \cdot 10^{10}$. Since one cubic kilometer equals 10^9 cubic meters, the required volume is about $50\,\text{km}^3$. This is equivalent to a cube of side length $3.7\,\text{km}$. That's a lot of rock. Each day. This method of energy production is therefore impractical (in addition to being colossally stupid). At any rate, there is no great urgency in conjuring up new methods of producing energy when we can simply be laying down carpets of solar panels. (New batteries, yes. But new methods, no.)

5.8. **Hoisting up**

(a) Since you are being hoisted up at constant speed, the tension in the rope is mg. Your friends apply this force to the rope and move the rope through a distance h, so the work they do is mgh. This equals your gain in potential energy, as expected.

(b) The tension in the rope is now only $mg/2$, because there are two parts of the rope pulling up on you and the platform; there is the end attached to the platform, and also the end you are holding (see Problem 4.4 for more discussion of this). So you apply a force of $mg/2$ to the rope. But we now seem to have a paradox, because if the work you do is $(mg/2)h$ (since h is again the distance you move), then this doesn't equal your mgh gain in potential energy. We seem to have magically created an energy of $mgh/2$.

The resolution to this paradox is that you actually do mgh work because you apply your force of $mg/2$ over a distance of $2h$ instead of h. This is true because the relevant distance is the distance your hand moves *relative to you* (because that is the only distance the muscles in your arms care about). And your hands will pass over a length $2h$ of the rope by the time you and the platform move up a height h (because relative to the ground, you move up by h and the rope moves down by h).

REMARK: If you are doubtful that it is the *relative* distance that matters, imagine that you have your eyes closed. Then for all you know (since you are moving at constant speed), you might be standing at rest on the ground while pulling a rope downward with a force that is arranged (by whatever means) to be $mg/2$. If a length $2h$ of rope passes through your hands, then you certainly do $(mg/2)(2h) = mgh$ work in this case.

5.9. Hanging block

The initial and final kinetic energies are the same (they both equal zero), so conservation of energy tells us that the sum of the spring and gravitational potential energies must be the same at the start and the end. Let y be the vertical position of the block with respect to the unstretched position of the spring. (So the block starts at $y = \ell/2$ and ends up at a negative value of y.) Conservation of energy gives

$$U_s^i + U_g^i + K^i = U_s^f + U_g^f + K^f$$

$$\implies \frac{1}{2}k\left(\frac{\ell}{2}\right)^2 + mg\left(\frac{\ell}{2}\right) + 0 = \frac{1}{2}ky^2 + mgy + 0$$

$$\implies \frac{1}{2}ky^2 + mgy - \left(\frac{k\ell^2}{8} + \frac{mg\ell}{2}\right) = 0. \tag{5.31}$$

You can solve this equation by using the quadratic formula, or you can just factor it by noting that $y = \ell/2$ must be a solution, because the initial position certainly satisfies conservation of energy with the initial position. We obtain

$$\frac{k}{2}\left(y - \frac{\ell}{2}\right)\left(y + \left(\frac{2mg}{k} + \frac{\ell}{2}\right)\right) = 0. \tag{5.32}$$

(Alternatively, in Eq. (5.31) you can put the quadratic terms on the left and the linear terms on the right, and then $y - \ell/2$ will appear as a common factor.) We are concerned with the $y = -2mg/k - \ell/2$ root. This value of y is the position relative to the end of the relaxed spring, which itself is a distance ℓ below the ceiling. So the desired distance below the ceiling is

$$d = \frac{2mg}{k} + \frac{3\ell}{2}. \tag{5.33}$$

If the block instead started with the spring compressed all the way up to the ceiling, you can show that the $3\ell/2$ here would be replaced by 2ℓ.

REMARK: The above $y = -2mg/k - \ell/2$ root makes sense, because the equilibrium position of the hanging block is $y = -mg/k$, since this is where the spring and gravitational forces balance. And the initial position of $y = \ell/2$ (which is the top of the resulting oscillatory motion) is $mg/k + \ell/2$ above the $y = -mg/k$ point. So the position at the bottom of the oscillatory motion will be a distance $mg/k + \ell/2$ below the $y = -mg/k$ point, which yields $y = -2mg/k - \ell/2$. (The motion is indeed symmetric around the $y = -mg/k$ point, due to the result from Problem 5.6.)

5.10. Rising on a spring

Let y_0 be the maximum displacement that the mass reaches above the relaxed length ℓ of the spring. (So negative y means below the relaxed length ℓ.) The maximum height above the ground is then $\ell + y_0$. The speed is zero at the maximum height, so both the initial and final kinetic energies are zero. Conservation of energy therefore tells us that the initial and final potential energies are equal:

$$\frac{1}{2}k\ell^2 + mg(0) = \frac{1}{2}ky_0^2 + mg(\ell + y_0)$$

$$\implies \frac{1}{2}k(\ell^2 - y_0^2) = mg(\ell + y_0). \tag{5.34}$$

We have measured the gravitational U with respect to the ground. We can choose any origin (it doesn't have to be the relaxed position of the spring), because only differences in U matter. We have grouped the terms as shown in Eq. (5.34) because if we divide both sides by $\ell + y_0$, we can avoid solving a quadratic equation. This gives

$$\frac{1}{2}k(\ell - y_0) = mg \implies y_0 = \ell - \frac{2mg}{k}.\tag{5.35}$$

The maximum height above the ground is therefore

$$h = \ell + y_0 = 2\ell - \frac{2mg}{k}.\tag{5.36}$$

Another solution to the above quadratic equation is $y_0 = -\ell$, because we divided through by $\ell + y_0$. But this is simply the initial position, which we know must be a solution.

LIMITS: If k is large (more precisely, if $k \gg mg/\ell$), then $h \approx 2\ell$. This makes sense because gravity is inconsequential, so we just have an oscillation between $y = -\ell$ and $y = \ell$.

If $k = mg/\ell$, then $h = 0$. This makes sense because the net force ($k\ell$ upward plus mg downward) is zero at the start, so the mass doesn't move. If $k < mg/\ell$, the mass likewise just sits on the ground. The results for y_0 and h in Eqs. (5.35) and (5.36) aren't valid when $k < mg/\ell$ because the $ky_0^2/2$ term in the energy is invalid if $y_0 < -\ell$; the spring can't be compressed more than ℓ.

If $k = 2mg/\ell$, then $h = \ell$. This makes sense because in this case the equilibrium point (where the net force $-ky - mg$ equals zero, which gives $y = -mg/k$) is located at $y = -\ell/2$. And since the oscillation starts a distance $\ell/2$ below this point, it must also extend a distance $\ell/2$ above it, that is, to the point $y = 0 \implies h = \ell$.

REMARK: How high does the mass go if it *isn't* attached to the end of the spring? If the mass doesn't reach a height $h = \ell$, it doesn't matter whether it is attached or not. But if it goes higher than $h = \ell$, the spring simply remains at its relaxed length, and the mass sails up freely in the air. The only change to Eq. (5.34) is that the $ky_0^2/2$ final energy of the spring is now absent. So the maximum y value is $y_0 = k\ell^2/2mg - \ell$. The maximum height above the ground is then $h = \ell + y_0 = k\ell^2/2mg$. This is valid only if $h > \ell$, that is, if $k > 2mg/\ell$. We see that h grows with k and ℓ, and decreases with m and g, as expected.

5.11. **Jumping onto mattresses**

(a) From conservation of energy, your mgh loss in gravitational energy is converted into kinetic energy right before you hit the mattress. This kinetic energy is then converted into $kd^2/2$ spring potential energy at the point of maximum compression, because your speed is instantaneously zero at that point. (As stated in the problem, we are ignoring the addition mgd loss in gravitational potential energy during the compression.) So the maximum compression is given by

$$\frac{1}{2}kd^2 = mgh \implies d = \sqrt{\frac{2mgh}{k}}.\tag{5.37}$$

The upward spring force is kd, so the maximum force, which occurs at the maximum compression, is

$$F = kd = k\sqrt{\frac{2mgh}{k}} = \sqrt{2mghk}.\tag{5.38}$$

LIMITS: The maximum compression increases with m, g, and h. And it decreases with k. These all make sense. The maximum force increases with all four parameters. The increase with k isn't completely obvious, but the main point is that the maximum compression decreases only like $1/\sqrt{k}$. So the maximum force, which is k times this distance, increases like $\sqrt{k}$.

(b) From the solution to Problem 4.2, we know that the spring constant of N springs (each with spring constant k) stacked in a line is k/N. Basically, a longer spring is a weaker spring. So in the present case of N mattresses, we simply need to take the

5.5. PROBLEM SOLUTIONS 125

above results for one mattress and replace every k with k/N. From Eqs. (5.37) and (5.38), the maximum compression and maximum force are therefore

$$d_N = \sqrt{\frac{2mghN}{k}} \quad \text{and} \quad F_N = \sqrt{\frac{2mghk}{N}}. \tag{5.39}$$

Note that the maximum compression is *larger* by a factor of $\sqrt{N}$, while the maximum force is *smaller* by a factor of $1/\sqrt{N}$. This latter fact means that the landing is more gentle in the case of N mattresses. Of course, if N is very large, then the compression distance will be large, so it won't be a good approximation to neglect the change in gravitational potential energy during the compression.

5.12. Bungee jumping 1

(a) The jumper is at rest at both the top and bottom of the motion, so all of the $mg(2\ell)$ loss in gravitational potential energy goes into the potential energy of the spring stretched a distance ℓ. Therefore,

$$mg(2\ell) = \frac{1}{2}k\ell^2 \implies k = \frac{4mg}{\ell}. \tag{5.40}$$

LIMITS: Large m or g means large k, as expected. Small ℓ also means large k, because the spring force, which ranges from zero to $k\ell$, needs to be on the order of mg to slow the jumper down.

(b) At the lowest point in the motion, the spring pulls up with a force $F_s = k\ell = (4mg/\ell)\ell = 4mg$. And gravity pulls down with a force $F_g = mg$. The net force is therefore $3mg$ upward, which implies an acceleration of $3g$ upward.

(c) The speed is maximum when $a = dv/dt = 0$. So we want the total force to be zero. This means that the upward spring force must balance the downward gravitational force. If x is the distance the bungee cord is stretched (so x is defined to be a positive quantity), then we want

$$kx = mg \implies \left(\frac{4mg}{\ell}\right)x = mg \implies x = \frac{\ell}{4}. \tag{5.41}$$

This value of x corresponds to a distance $\ell + \ell/4 = 5\ell/4$ below the platform. Above this point, the downward gravitational force wins, and the speed increases. Below this point, the upward spring force wins, and the speed decreases.

From conservation of energy, the loss in gravitational U goes into spring U plus kinetic energy. So the speed at the point $5\ell/4$ below the platform is given by

$$mg\frac{5\ell}{4} = \frac{1}{2}k\left(\frac{\ell}{4}\right)^2 + \frac{1}{2}mv^2$$

$$\implies \frac{5mg\ell}{4} = \frac{1}{2}\left(\frac{4mg}{\ell}\right)\left(\frac{\ell}{4}\right)^2 + \frac{1}{2}mv^2$$

$$\implies \frac{5mg\ell}{4} = \frac{mg\ell}{8} + \frac{1}{2}mv^2 \implies v = \frac{3}{2}\sqrt{g\ell}. \tag{5.42}$$

For comparison, the speed right when the cord starts stretching (at a distance ℓ below the platform) is just the speed after a distance ℓ in freefall, which is given by $mv^2/2 = mg\ell \implies v = \sqrt{2g\ell} = (1.414)\sqrt{g\ell}$. The maximum speed of $(1.5)\sqrt{g\ell}$ is 6% larger than this.

5.13. **Bungee jumping 2**

(a) Let x be the distance that the spring is stretched at the lowest point (so x is defined to be a positive number). Then the jumper falls a total distance of $\ell + x$. The $mg(\ell + x)$ loss in gravitational potential energy must show up as $kx^2/2$ potential energy in the spring, because there is no kinetic energy at the start or at the lowest point (where the velocity is instantaneously zero). So we have $mg(\ell + x) = kx^2/2$. Solving this quadratic equation for x gives

$$\frac{1}{2}kx^2 - mgx - mg\ell = 0 \implies x = \frac{mg + \sqrt{m^2g^2 + 2mg\ell k}}{k}. \tag{5.43}$$

We have chosen the "+" sign in the quadratic formula because we have defined x to be positive. The "−" root is negative and corresponds to the highest point in the oscillation that would arise if the cord were an actual spring that could be compressed on the way back up.

LIMITS: x correctly goes to zero as $k \to \infty$, and correctly goes to infinity as $k \to 0$. If $\ell \approx 0$ then $x \approx 2mg/k$. You should think about why this makes sense.

(b) Using the result for x in Eq. (5.43), the tension T in the cord at the lowest point is

$$T = kx = mg + \sqrt{m^2g^2 + 2mg\ell k}. \tag{5.44}$$

(c) If the cord is half as long, then from Problem 4.2 we know that the spring constant is twice as large. So the new parameters are given by $\ell' = \ell/2$ and $k' = 2k$. But this means that the new product $\ell'k'$ in Eq. (5.44) is the same as the old product ℓk. So the tension at the lowest point is the same. (Consistent with this, the value of x in Eq. (5.43) is now half as large.) By the same reasoning (see the generalization in the solution to Problem 4.2.), the tension at the lowest point is the same if the cord is one third as long. Or any other factor. The tension at the lowest point is independent of the length of the cord.

REMARK: The stress on your body at the lowest point is proportional to the tension T in the cord. To see why, consider a few cases: If $T = 0$ then you are in freefall, because gravity is the only force acting on you. So there is no stress. If $T = mg$ then you aren't accelerating, so each part of your upside-down body (the cord is attached to your ankles) must support the part below it. This is similar to the case where you are standing on the platform before you jump, when each part of your body must support the part above it (compression now, instead of tension). This is the standard "everyday" stress on your body. If $T = 2mg$ then the net force on you is mg upward, so you are accelerating upward with acceleration g. The tension between different parts of your body is now twice what it was in the $T = mg$ case. Basically, if you close your eyes in the $T = 2mg$ case, then for all you know, you are at rest in a world where gravity is twice as strong. In general, therefore, the stress on your body is proportional to the tension T in the cord. (So the stress isn't proportional to your acceleration, although for large accelerations this distinction isn't important.)

Since the tension T at the lowest point is the same for all lengths of the cord, the stress on your body is also the same for all lengths. And since the stress is what might cause injury if it is too large, we see that changing the length of the cord doesn't make the jump any more jarring or more gentle at the bottom.[2] (Of course, if the cord is too long, then the ground below becomes relevant, in which case things are most certainly not gentle!). Since you are upside down, your eyeballs are especially susceptible to injury. In fact, in any situation involving a

---

[2]In a slightly different setup, this fact is well known to rock climbers. If a climber (leading, not top-roping) is a height ℓ above the last piece of protection (which you can think of as a place where the rope is tied to the wall), and if he falls, then he will fall a distance 2ℓ before the rope starts to slow him down. A climbing rope is stretchy, so it acts like a spring, and you can show that we obtain the same result that the tension at the lowest point is independent of ℓ. So although a long fall might be scarier, it isn't any harder on your body.

5.5. PROBLEM SOLUTIONS

large acceleration, the eyeballs are the thing to worry about, for any orientation of your body.[3]
From Eq. (5.44), the ratio of T to the mg normal force you experience while standing on the platform equals

$$\frac{T}{mg} = 1 + \sqrt{1 + \frac{2\ell k}{mg}}. \tag{5.45}$$

For a given value of ℓ, this ratio increases with k (imagine making the cord out of a stiffer material), as expected. Note that the ratio is greater than or equal to 2 in all cases. For a given value of ℓ, if we make k very small, then the ratio is essentially equal to 2. In this case, you fall a very large distance, and the tension at the bottom is essentially equal to $2mg$, giving you an acceleration of g upward. Basically, in this limit you undergo a large oscillation up and down, with your acceleration ranging from $+g$ at the bottom to $-g$ at the top.

5.14. Mass on a spring

(a) The work done by friction in moving to position x is

$$W_f = \int_0^x F_f\, dx = -\int_0^x \mu N\, dx = -\int_0^x (Ax) mg\, dx = -\frac{mgAx^2}{2}. \tag{5.46}$$

The heat lost to friction equals the absolute value of this. We want to find the value of x where the block stops. Since the kinetic energy is zero at the start and the end, conservation of energy tells us that (using the fact that at position x, the spring is stretched by $|\ell - x|$)

$$U^i_{spring} = U^f_{spring} + \text{Heat} \implies \frac{k\ell^2}{2} = \frac{k(\ell - x)^2}{2} + \frac{mgAx^2}{2}$$

$$\implies \frac{k}{2}\left(2\ell x - x^2\right) = \frac{mgAx^2}{2} \implies 2\ell - x = \frac{mgAx}{k}$$

$$\implies x = \frac{2\ell}{1 + \dfrac{mgA}{k}}. \tag{5.47}$$

This stopping value of x is larger than ℓ if

$$\frac{mgA}{k} < 1 \implies A < \frac{k}{mg}. \tag{5.48}$$

Alternatively, we can think in terms of the work-energy theorem. In the second equation on the first line in Eq. (5.47), if we put all three terms on the left-hand side, the result is just the statement that the total work done on the block by the spring and friction equals zero (which is the change in kinetic energy).

UNITS: A has dimensions of inverse length (to make $\mu = Ax$ dimensionless), and k has dimensions of force per length, so mgA/k is dimensionless, which means that the x in Eq. (5.47) correctly has units of length.

LIMITS: Large m, g, or A implies small x; friction dominates the spring. Large k implies $x = 2\ell$; friction is irrelevant, so we have a simple spring bouncing back and forth symmetrically around the anchor.

(b) If the block is instantaneously at rest at position x, where $x > \ell$, then the spring force is directed to the left with magnitude $k(x - \ell)$. And the static friction force is directed to the right with a maximum magnitude of $\mu N = (Ax)mg$. (You should

---

[3] The record for the largest acceleration experienced (voluntarily) by a human is held by Colonel John Paul Stapp (1910-1999). In the 1950's, he performed numerous potentially fatal experiments involving accelerations as large as $45g$. These left him with permanent vision problems. Due to his first-hand experience with extreme accelerations, he was a strong proponent of the use of seatbelts in cars. Modern-day seatbelt laws are owed largely to his efforts. If there was ever anyone to trust on the subject of safety mechanisms for large accelerations, it was John Stapp!

convince yourself that in order to have any chance of moving again, the block must be to the right of the anchor.) The block starts moving again if

$$k(x - \ell) > (Ax)mg \implies x > \frac{\ell}{1 - \frac{mgA}{k}}. \tag{5.49}$$

Using the x value we found in Eq. (5.47) (which is the stopping distance in all cases), this inequality becomes

$$\frac{2\ell}{1 + \frac{mgA}{k}} > \frac{\ell}{1 - \frac{mgA}{k}} \implies 1 > \frac{3mgA}{k} \implies A < \frac{k}{3mg}, \tag{5.50}$$

which is correctly a stricter bound than the one we found in part (a). Plugging $A = k/3mg$ into the expression for x in Eq. (5.47) yields a stopping position of $x = 3\ell/2$. This result is independent of k, m, and g. So for any values of these parameters, the block will start moving again, provided that it goes beyond the $x = 3\ell/2$ mark. Said in a different way, if $x > 3\ell/2$ then Eq. (5.47) tells us that $A < k/3mg$, in which case Eq. (5.50) tells us that the block will start moving again.

Limits: The A in the $A < k/3mg$ result is correctly on the "less than" side of the inequality, because the block will certainly start moving again in the frictionless case with $A = 0$. Likewise, it will start moving again if k is large or mg is small.

5.15. Falling with a spring

(a) From the work-energy theorem, the masses stop when the total work done on them is zero, because their speed is zero at both the start and the end. Gravity does positive work, while the spring and friction do negative work. If x is the stretching distance, then setting the total work equal to zero gives

$$mgx - \frac{1}{2}kx^2 - \mu_k mgx = 0. \tag{5.51}$$

Equivalently, we can write this as

$$mgx = \frac{1}{2}kx^2 + \mu_k mgx, \tag{5.52}$$

which is the statement that the loss in gravitational potential energy shows up as spring potential energy plus heat from friction. Solving either of the above equations for x gives the stopping distance x_0 as

$$x_0 = \frac{2mg}{k}(1 - \mu_k) = \frac{2mg}{k}(1 - 1/4) = \frac{3mg}{2k}. \tag{5.53}$$

Another solution is $x = 0$, of course, because that corresponds to the initial location.

(b) If the masses are at rest at $x_0 = 3mg/2k$, then the forces (which must sum to zero) on the left mass are shown in Fig. 5.22. The tension in the string is mg since it is supporting the right mass. We have drawn the static friction force pointing rightward, because we are concerned with the case where the mass is about to slip and head leftward. Setting the total force equal to zero, and using the fact that $F_f \leq \mu_s N$, gives

$$kx_0 = mg + F_f \implies k\left(\frac{3mg}{2k}\right) \leq mg + \mu_s mg$$

$$\implies \frac{3}{2} \leq 1 + \mu_s \implies \mu_s \geq \frac{1}{2}. \tag{5.54}$$

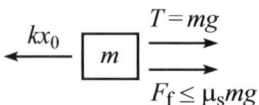

Figure 5.22

5.5. PROBLEM SOLUTIONS

If this condition isn't satisfied, that is, if μ_s is smaller than $1/2$, then it will be impossible to make the net force zero, so the block will head leftward.

REMARK: As in Problem 5.14, the block's stopping point is the point where the total *work* done equals zero, whereas the cutoff case to start moving again involves the point where the total *force* equals zero.

(c) We now have only the left mass. It ends up stopping a second time when the total new work done on it is zero. Let d be the ending compression *distance* (so d is a positive quantity). The total work done on the mass by the spring and friction is

$$W = \left(\frac{1}{2}kx_0^2 - \frac{1}{2}kd^2\right) - \mu_k mg(x_0 + d). \tag{5.55}$$

Equivalently, the loss in potential energy of the spring ends up as heat from friction (plus kinetic energy in the general case where W isn't zero). Setting the work W equal to zero gives (using the x_0 in Eq. (5.53))

$$0 = \frac{1}{2}k(x_0 + d)(x_0 - d) - \mu_k mg(x_0 + d)$$

$$\implies 0 = \frac{1}{2}k(x_0 - d) - \mu_k mg$$

$$\implies d = x_0 - \frac{2\mu_k mg}{k} = \frac{2mg}{k}(1 - \mu_k) - \frac{2\mu_k mg}{k}$$

$$= \frac{2mg}{k}(1 - 2\mu_k) = \frac{mg}{k}, \tag{5.56}$$

where we have used $\mu_k = 1/4$. If we instead had $\mu_k = 1/2$, then the mass would make it exactly back to the starting position at $d = 0$. Note that parts (a) and (c) of this problem involved energy (or work), whereas part (b) involved force.

REMARK: $d = -x_0$ is also a solution to the first equation in Eq. (5.56), because we divided by $x_0 + d$. But this solution simply corresponds to the left mass's location when the string is cut. If we hadn't realized that $x_0 + d$ is a common factor in the above equation, we would have had to use the quadratic formula to solve the quadratic equation in Eq. (5.55).

5.16. Bead, spring, and rail

(a) The spring, normal, and friction forces are shown in Fig. 5.23. (When the bead is moving back toward the vertex, the friction force points in the other direction.) With x being the distance traveled along the rail, the spring is stretched by a distance $x \sin \theta$, so the spring force is

$$F_s = k(x \sin \theta). \tag{5.57}$$

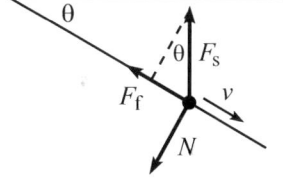

Figure 5.23

The component of this force that is perpendicular to the rail is $F_s \cos \theta = kx \sin \theta \cos \theta$. Since the bead isn't accelerating perpendicular to the rail, the normal force must be

$$N = F_s \cos \theta = kx \sin \theta \cos \theta. \tag{5.58}$$

The friction force is then

$$F_f = \mu N = \mu kx \sin \theta \cos \theta. \tag{5.59}$$

(b) The component of the spring force along the rail points back toward the vertex with magnitude $F_s \sin \theta = kx \sin^2 \theta$. When the bead reaches its farthest point along the rail, it is instantaneously at rest. So the work-energy theorem gives

$$W_{\text{spring}} + W_{\text{friction}} = K_f - K_i$$

$$\implies \int_0^x (-kx \sin^2 \theta)\, dx + \int_0^x (-\mu kx \sin \theta \cos \theta)\, dx = 0 - \frac{1}{2}mv_0^2$$

$$\implies -\frac{1}{2}kx^2(\sin^2 \theta + \mu \sin \theta \cos \theta) = -\frac{1}{2}mv_0^2. \tag{5.60}$$

The distance traveled by the bead when it comes to rest is therefore

$$x_{\max} = \sqrt{\frac{mv_0^2}{k(\sin^2\theta + \mu\sin\theta\cos\theta)}}. \tag{5.61}$$

Alternatively, we can find $x_{\max}$ by saying that the initial $mv_0^2/2$ kinetic energy ends up in the form of $(1/2)k(x\sin\theta)^2$ potential energy in the spring (because the spring is stretched by $x\sin\theta$), plus the $|W_f| = (1/2)kx^2\sin\theta\cos\theta$ heat due to friction. This gives Eq. (5.60) without the minus signs.

LIMITS: $x_{\max}$ is large if m or v_0 is large, or if k or θ is small. And $x_{\max}$ decreases with μ. These limits make sense.

(c) The bead accelerates from rest back toward the vertex if the spring force component along the rail is larger than the maximum opposing static friction force. So it accelerates if

$$kx_{\max}\sin^2\theta > \mu k x_{\max}\sin\theta\cos\theta \implies \tan\theta > \mu. \tag{5.62}$$

Note that this result is independent of x. So if the bead accelerates in one place, then it accelerates in any other. Therefore, if $\tan\theta > \mu$ (so that the bead starts moving again), then the bead will make it all the way back to the vertex.

REMARK: This $\tan\theta > \mu$ result is the same as the standard result for a block resting on an inclined plane. The block will accelerate from rest and slide down the plane if $\tan\theta > \mu$. The result in this problem is the same because the spring force in this problem takes the place of the gravitational force in the inclined-plane problem (imagine flipping Fig. 5.23 upside down). Eq. (5.62) can be written alternatively as $F_s\sin\theta > \mu F_s\cos\theta$, which takes the place of the standard $mg\sin\theta > \mu mg\cos\theta$ equation for the inclined plane. The fact that F_s varies with x is irrelevant; it cancels just as the mg cancels.

LIMITS: In Eq. (5.62), large μ implies that θ needs to be essentially equal to $90°$, if the bead is to start moving again. And small μ means that even if θ is fairly small, the bead will start moving again. These results make sense.

(d) We'll use the work-energy theorem again, as we did in part (b), but now for the trip back to the vertex. $W = \Delta K$ gives (note that the dx in the integrals is now negative and that the friction force points in the positive x direction)

$$W_{\text{spring}} + W_{\text{friction}} = K_f - K_i$$

$$\implies \int_{x_{\max}}^{0}(-kx\sin^2\theta)\,dx + \int_{x_{\max}}^{0}(\mu kx\sin\theta\cos\theta)\,dx = \frac{1}{2}mv_f^2 - 0$$

$$\implies \frac{1}{2}kx_{\max}^2(\sin^2\theta - \mu\sin\theta\cos\theta) = \frac{1}{2}mv_f^2. \tag{5.63}$$

Again, you could alternatively say that the initial $(1/2)k(x\sin\theta)^2$ potential energy in the spring ends up as $mv_f^2/2$ kinetic energy plus the $|W_f| = (1/2)kx^2\sin\theta\cos\theta$ heat due to friction.

Plugging the value of $x_{\max}$ from Eq. (5.61) into Eq. (5.63) gives the final speed as

$$v_f^2 = \frac{k}{m}\left(\frac{mv_0^2}{k(\sin^2\theta + \mu\sin\theta\cos\theta)}\right)(\sin^2\theta - \mu\sin\theta\cos\theta)$$

$$\implies v_f = v_0\sqrt{\frac{\sin\theta - \mu\cos\theta}{\sin\theta + \mu\cos\theta}}. \tag{5.64}$$

LIMITS: If $\mu = 0$, then $v_f = v_0$; the friction force is zero, so no energy is lost to heat. And if $\theta = 90°$, then $v_f = v_0$, because the friction force is again zero since the normal force is zero. If $\tan\theta \to \mu$, then $v_f \to 0$, because the bead's acceleration is very small on the way back to the vertex, so it hardly ever gets going, even though it does indeed make it back to the vertex.

5.5. PROBLEM SOLUTIONS

5.17. Ring on a pole

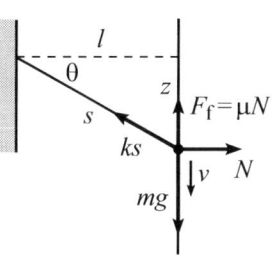

Figure 5.24

(a) If s is the length of the spring at a general later time, then the spring force is ks (because the relaxed length is zero). So the free-body diagram is shown in Fig. 5.24. The four forces are the spring, normal, friction, and gravity. The friction force points opposite to the velocity; we've drawn the figure for a downward velocity.

There is no acceleration in the horizontal direction, so the normal force must balance the horizontal component of the spring force:

$$N = (ks)\cos\theta = k(s\cos\theta) = k\ell. \tag{5.65}$$

We see that the normal force is independent of the ring's height on the pole. This is a consequence of the spring's relaxed length being zero.

(b) Let z be the *distance* fallen (so z is defined to be a positive quantity). The velocity at the start and at the bottom of the motion is zero, so conservation of energy tells us that the loss in gravitational potential energy goes into the gain in spring potential energy plus heat from friction, which is the magnitude of the work done by the (constant) friction force, namely $(\mu N)z$. So we have (using $s^2 - \ell^2 = z^2$, and $N = k\ell$)

$$mgz = \left(\frac{1}{2}ks^2 - \frac{1}{2}k\ell^2\right) + (\mu N)z$$

$$\implies mgz = \frac{1}{2}kz^2 + \mu k\ell z \implies z = \frac{2mg}{k} - 2\mu\ell. \tag{5.66}$$

(Another solution is $z = 0$, of course, because that is the initial position.) If you want think in terms of work instead of conservation of energy, note that the first equation on the second line of Eq. (5.66) can be written as

$$mgz + (-\mu N)z + \int_0^z (-kz')\,dz' = 0. \tag{5.67}$$

(We have put a prime on the integration variable, so that we don't confuse it with the limit of integration.) This is the statement that the net work done is zero, as it should be, because the kinetic energy is zero at the start and finish. Gravity does positive work, while friction and the spring do negative work. Only the vertical component of the spring force matters, and it equals $(ks)\sin\theta = k(s\sin\theta) = kz$ upward. We are taking z to be positive downward, hence the minus signs in the friction and spring forces in Eq. (5.67).

LIMITS: In Eq. (5.66), large m or g implies large z. And large k, μ, or ℓ implies small z. But if the values of these parameters exceed certain values, then z simply stays at zero.

(c) i. The ring will initially fall if the downward gravitational force is larger than the maximum upward static friction force. So we need

$$mg > \mu N \implies mg > \mu(k\ell) \implies \mu < \frac{mg}{k\ell}. \tag{5.68}$$

So $mg/k\ell$ is the desired cutoff value of μ. This is consistent with Eq. (5.66); if $\mu = mg/k\ell$, then $z = 0$.

LIMITS: If m or g is large, then μ can be large. If k or ℓ is large, then μ must be small.

ii. The ring will rise back up at the bottom of its motion if the upward spring force is larger than the sum of the downward gravitational force and the maximum static friction force (friction now points downward). So we need (using the value of z we found in Eq. (5.66))

$$kz > mg + \mu N \implies k\left(\frac{2mg}{k} - 2\mu\ell\right) > mg + \mu k\ell$$

$$\implies mg > 3\mu k\ell \implies \mu < \frac{mg}{3k\ell}. \tag{5.69}$$

So $mg/3k\ell$ is the desired cutoff value of μ. It makes sense that this is smaller than the above $mg/k\ell$ cutoff for initially falling, because if μ is only slightly smaller than $mg/k\ell$, then the ring will fall only a tiny bit. So it certainly won't rise back up, because the upward spring force is negligible in this case.

5.18. Block on a plane

(a) If the spring ends up uncompressed, then there is no spring potential energy at the end. And since there is no kinetic energy at the start or end, conservation of energy tells us that the initial spring potential energy $k\ell^2/2$ gets converted into gravitational potential energy (which equals $mgh = mg\ell\sin\theta$) plus heat, which is the magnitude of the work done by kinetic friction (which equals $F_f \ell = (\mu N)\ell = (mg\cos\theta)\ell$, where we have used $\mu = 1$). So we have

$$\frac{1}{2}k\ell^2 = mg\ell\sin\theta + mg\ell\cos\theta \implies \ell = \frac{2mg}{k}(\sin\theta + \cos\theta). \tag{5.70}$$

(Another solution is $\ell = 0$, of course.) Note that if we write the above conservation-of-energy equation as $k\ell^2/2 - mg\ell\sin\theta - mg\ell\cos\theta = 0$, this is simply the $W = \Delta K$ statement, because it says that $W_{\text{spring}} + W_{\text{gravity}} + W_{\text{friction}} = 0$.

LIMITS: If $k \to 0$ then $\ell \to \infty$; the spring needs to be compressed a huge distance to overcome the gravitational force and start rising back up. If $\theta = 90°$ then $\ell = 2mg/k$; the block just bounces vertically (and frictionlessly) up and down around the equilibrium point where the spring is compressed by mg/k. If $\theta = 0$ then again we have $\ell = 2mg/k$. This limit isn't quite as intuitive, although we do know that the answer must be the same as in the $\theta = 90°$ case, because in both cases the spring is working against a force of mg (the friction force equals mg in the $\theta = 0$ case since $\mu = 1$).

(b) At the moment when the spring is uncompressed, the only forces (along the plane) on the block are friction and gravity. The block will slide back down if the gravitational force downward is larger than the maximal static friction force upward (which is μN), So the block will slide down if

$$mg\sin\theta > \mu mg\cos\theta \implies \tan\theta > \mu \implies \tan\theta > 1 \implies \theta > 45°. \tag{5.71}$$

(c) This is similar to part (a). There is no spring potential energy at the start, and no kinetic energy at the start or end. So conservation of energy tells us that the loss in gravitational potential energy goes into spring potential energy plus heat. Therefore, if x is the resulting compression distance (so x is defined to be a positive quantity), we have

$$mgx\sin\theta = \frac{1}{2}kx^2 + mgx\cos\theta \implies x = \frac{2mg}{k}(\sin\theta - \cos\theta). \tag{5.72}$$

(Again, another solution is $x = 0$.) Note that since we are assuming $\theta > 45°$, this value of x is positive, as it should be. As in part (a), if we write the above conservation-of-energy equation with all the terms on the left-hand side, the result is the statement that the total work done on the block is zero.

LIMITS: If $\theta = 45°$ then $x = 0$, consistent with Eq. (5.71). And if $\theta = 90°$ then $x = 2mg/k$, consistent with the remark in part (a).

(d) We need the upward spring force to be larger than the sum of the downward gravitational force and the maximum static friction force (friction now points downward).

Using the value of x from Eq. (5.72), this translates to (with $\mu = 1$)

$$kx > mg\sin\theta + \mu mg\cos\theta$$

$$\implies k\left(\frac{2mg}{k}(\sin\theta - \cos\theta)\right) > mg(\sin\theta + \cos\theta)$$

$$\implies 2\sin\theta - 2\cos\theta > \sin\theta + \cos\theta$$

$$\implies \sin\theta > 3\cos\theta$$

$$\implies \tan\theta > 3, \tag{5.73}$$

which gives $\theta > 71.6°$.

5.19. Tangential acceleration

(a) At an angle θ with respect to the vertical, the height above the center of the hoop is $R\cos\theta$. So the height fallen from the top is $h = R - R\cos\theta$. Conservation of energy therefore gives

$$\frac{1}{2}mv^2 = mgh \implies \frac{1}{2}mv^2 = mgR(1-\cos\theta)$$

$$\implies v = \sqrt{2gR(1-\cos\theta)}. \tag{5.74}$$

(b) Copious use of the chain rule gives (remembering to go all the way to the final $d\theta/dt \equiv \dot\theta$ derivative)

$$\frac{dv}{dt} = \sqrt{2gR} \cdot \frac{1}{2} \cdot \frac{1}{\sqrt{1-\cos\theta}} \cdot \sin\theta \cdot \dot\theta. \tag{5.75}$$

But the tangential speed v can be written as $v = R\dot\theta$, so

$$\dot\theta = \frac{v}{R} = \frac{\sqrt{2gR(1-\cos\theta)}}{R}. \tag{5.76}$$

Substituting this value of $\dot\theta$ into Eq. (5.75) gives

$$\frac{dv}{dt} = \sqrt{2gR} \cdot \frac{1}{2} \cdot \frac{1}{\sqrt{1-\cos\theta}} \cdot \sin\theta \cdot \frac{\sqrt{2gR(1-\cos\theta)}}{R} = g\sin\theta, \tag{5.77}$$

as desired.

5.20. Comparing the tensions

(a) At the angle θ shown in the left diagram in Fig. 5.13, the mass is a distance $\ell\cos\theta$ below the top of the pendulum motion. So conservation of energy gives

$$mg(\ell\cos\theta) = \frac{1}{2}mv^2 \implies v^2 = 2g\ell\cos\theta. \tag{5.78}$$

The radial forces are the tension inward and the $g\cos\theta$ component of gravity outward. So the radial $F = ma$ equation at angle θ is

$$T_1 - mg\cos\theta = \frac{mv^2}{\ell}. \tag{5.79}$$

Using the above value of v^2, this becomes

$$T_1 - mg\cos\theta = \frac{m(2g\ell\cos\theta)}{\ell} \implies T_1(\theta) = 3mg\cos\theta. \tag{5.80}$$

Note that this is independent of the radius ℓ.

After the pendulum is stopped, the net force is zero, so the vertical component of T_2 must be $T_{2,y} = mg$. Therefore, since the string is tilted at angle θ, we have

$$T_2(\theta) = \frac{T_{2,y}}{\cos\theta} = \frac{mg}{\cos\theta}. \qquad (5.81)$$

Equating $T_1(\theta)$ and $T_2(\theta)$ yields

$$3mg\cos\theta = \frac{mg}{\cos\theta} \implies \cos\theta = \frac{1}{\sqrt{3}} \implies \theta \approx 54.7°. \qquad (5.82)$$

LIMITS: If $\theta = 0$ (the bottom of the pendulum motion), then $T_1 = 3mg$, which is plausible; the tension certainly needs to be larger than mg, to cancel out gravity and provide a net upward centripetal force. If $\theta = \pi/2$ (the top of the pendulum motion), then $T_1 = 0$, which is correct; the mass is momentarily at rest (which means that the centripetal acceleration is zero), and the string is horizontal (which means that it isn't supporting any of the weight).

If $\theta = 0$, then $T_2 = mg$, which is correct; the mass simply hangs straight down. If $\theta \to \pi/2$, then $T_2 \to \infty$, which is also correct; the vertical component of the tension of the nearly horizontal string needs to be mg, so the tension must be huge.

REMARK: The above $T_1 = 3mg$ result nearly cost five people their lives when they attempted a pendulum swing off the Skyway Bridge in Tampa in 1997. Their combined weight was about 900 lbs, and the steel cable they used was rated at 1400 lbs. This was plenty strong to support their mg weight if they were simply hanging at rest, but not strong enough to provide the necessary $3mg$ tension at the bottom point in the circular motion. The cable broke, and they hit the water at nearly 70 mph. Fortunately no one died, but it was close; two people had broken necks. In planning the jump, they apparently thought that the circular motion's effect on the tension was either nonexistent, or at most minor. But it isn't minor; it's a factor of 3 (or equivalently an additive difference of $2mg$). Knowledge of physics is critical in certain endeavors!

(b) At $\theta = 0$, T_1 is larger than T_2, because an extra upward force is needed in the first case to provide the necessary centripetal acceleration. But at $\theta = \pi/2$, T_1 is smaller than T_2, because as we saw in the above limits, $T_1 = 0$ and $T_2 \to \infty$. So somewhere in between $\theta = 0$ and $\theta = \pi/2$, T_1 must be equal to T_2.

5.21. **Semicircular tube**

The net height fallen by the time the mass leaves the tube is $2R\cos\theta$. So conservation of energy gives the initial speed of the projectile motion as

$$\frac{1}{2}mv_0^2 = mg(2R\cos\theta) \implies v_0^2 = 4gR\cos\theta. \qquad (5.83)$$

The given angle θ is also the launch angle of the projectile motion, as you can verify. So from Problem 3.1, the range of the projectile motion is $d = (2v_0^2/g)\sin\theta\cos\theta$. Using the above value of v_0^2, this becomes

$$d = \frac{2(4gR\cos\theta)\sin\theta\cos\theta}{g} = 8R\cos^2\theta\sin\theta. \qquad (5.84)$$

Maximizing this be setting the derivative equal to zero gives

$$-2\cos\theta\sin\theta \cdot \sin\theta + \cos^2\theta \cdot \cos\theta = 0 \implies \cos\theta(-2\sin^2\theta + \cos^2\theta) = 0$$
$$\implies \tan\theta = \frac{1}{\sqrt{2}}, \qquad (5.85)$$

which gives $\theta \approx 35.3°$. (The other solution, $\cos\theta = 0 \implies \theta = 90°$, yields zero distance; the projectile motion is vertical.) Substituting this value of θ into Eq. (5.84) gives the maximum distance d as $(16/3\sqrt{3})R$, which is slightly larger than $3R$.

5.5. PROBLEM SOLUTIONS

5.22. Horizontal force

We'll first need to find the speed when the bead is at an angle θ below the horizontal. The height fallen from the top is $R + R\sin\theta$, so conservation of energy gives

$$\frac{1}{2}mv^2 = mg(R + R\sin\theta) \implies v = \sqrt{2gR(1 + \sin\theta)}. \tag{5.86}$$

The radial $F = ma$ equation (with inward taken to be positive) at an angle θ below the horizontal is then

$$N - mg\sin\theta = \frac{mv^2}{R} \implies N - mg\sin\theta = \frac{m(2gR(1 + \sin\theta))}{R}$$
$$\implies N = mg(2 + 3\sin\theta). \tag{5.87}$$

LIMITS: As a check on this, when $\theta = -90°$ (which corresponds to the top of the hoop) we have $N = -mg$. This is correct, because the normal force initially points upward (outward) to cancel the weight mg, and we have taken positive N to point inward.

The total force on the bead is horizontal when the vertical component of the normal force (which is $N\sin\theta$) cancels the downward weight mg. So we want

$$N\sin\theta = mg \implies mg(2 + 3\sin\theta)\sin\theta = mg$$
$$\implies 3\sin^2\theta + 2\sin\theta - 1 = 0 \implies (3\sin\theta - 1)(\sin\theta + 1) = 0$$
$$\implies \sin\theta = 1/3 \implies \theta \approx 19.5°. \tag{5.88}$$

The mirror-image point on the other side of the hoop, with $\theta = 160.5°$, is also a solution because it likewise has $\sin\theta = 1/3$.

The other root of the quadratic equation in Eq. (5.88), namely $\sin\theta = -1 \implies \theta = -90°$, corresponds to the top of the hoop, where the total force is zero. It's semantics as to whether the zero vector is horizontal. (It has zero vertical component, if you take that as your definition of horizontal.)

REMARK: Since a net horizontal force implies a horizontal acceleration, this problem could alternatively have been phrased as, "For what θ is the acceleration of the bead horizontal?" This was Problem 3.23, and the answer to that problem is the same, as it should be.

5.23. Maximum vertical normal force

(a) We'll use conservation of energy to find the speed of the bead when it is at an angle θ with respect to the vertical. We'll then use this speed to find the normal force.

If we choose $y = 0$ to be located at the center of the hoop, then the bead's height as as function of θ is $R\cos\theta$. So conservation of energy from the start to a general later position gives

$$\frac{1}{2}mv_0^2 + mgR = \frac{1}{2}mv^2 + mg(R\cos\theta)$$
$$\implies v^2 = v_0^2 + 2gR(1 - \cos\theta). \tag{5.89}$$

Let N be the normal force from the *hoop* on the *bead* (the opposite of what the problem asks for), with inward defined to be positive. The radially inward component of the gravitational force is $mg\cos\theta$, so the radial $F = ma$ equation for the bead is (using the above result for v^2)

$$N + mg\cos\theta = \frac{mv^2}{R}$$
$$\implies N + mg\cos\theta = \frac{mv_0^2}{R} + 2mg(1 - \cos\theta)$$
$$\implies N = \frac{mv_0^2}{R} + 2mg - 3mg\cos\theta. \tag{5.90}$$

By Newton's third law, this is also the normal force from the *bead* on the *hoop* (which is the force we are asked about), with *outward* defined to be positive. The desired vertical component is then

$$N_y = N\cos\theta = \left(\frac{mv_0^2}{R} + 2mg - 3mg\cos\theta\right)\cos\theta. \tag{5.91}$$

(b) Taking the derivative of N_y to maximize it yields

$$0 = \frac{dN_y}{d\theta} = -\frac{mv_0^2}{R}\sin\theta - 2mg\sin\theta + 6mg\sin\theta\cos\theta$$

$$\implies \cos\theta = \frac{1}{3} + \frac{v_0^2}{6gR}. \tag{5.92}$$

This is valid only if $v_0^2 \leq 4gR$ (see part (c)); this is what we meant in the statement of the problem by saying that v_0 was "relatively small." Assuming $v_0^2 \leq 4gR$, you can show that the second derivative of N_y is negative at the value of θ given by Eq. (5.92), which means that we have a local maximum of N_y. So this is the angle at which the bead is pushing upward on the hoop with the maximum force. If you plug this value of $\cos\theta$ back into Eq. (5.91), you will find that the maximum value of N_y is $(mg/3)(1 + v_0^2/2gR)^2$. If the hoop is resting on a table, and if this maximum value of N_y exceeds the weight of the hoop, then the hoop will briefly rise up off the table. Another solution to Eq. (5.92) is $\sin\theta = 0 \implies \theta = 0$. But if $v_0^2 \leq 4gR$, you can show that the second derivative of N_y is positive at $\theta = 0$, which means that we have a local minimum. This is consistent with the plots shown below in part (c).

(c) If $v_0^2/6gR = 2/3$, or equivalently if $v_0^2 = 4gR$, then Eq. (5.92) gives $\cos\theta = 1 \implies \theta = 0$, which corresponds to the top of the hoop. If $v_0^2 > 4gR$, then the derivative of N_y is zero for a value of $\cos\theta$ that is larger than 1. This isn't possible, so the maximum value of N_y must occur at the boundary of the legal region, that is, at $\cos\theta = 1 \implies \theta = 0$. (This coincides with the other solution to Eq. (5.92), namely $\sin\theta = 0 \implies \theta = 0$, which is now a local maximum.) So the desired special value of v_0 is $\sqrt{4gR} = 2\sqrt{gR}$.

LIMITS: The plots in Fig. 5.25 show the N_y in Eq. (5.91) (in units of mg) as a function of θ for $-\pi \leq \theta \leq \pi$, that is, for all points on the hoop. We can consider some special cases:

- If $v_0 = 0$, then the bead is at rest at $\theta = 0$, so it pushes down on the hoop with a force with magnitude mg; that is, $N_y = -mg$, in agreement with Eq. (5.91). From Eq. (5.92) the maximum N_y occurs at $\cos\theta = 1/3 \implies \theta = \pm 70.5° = \pm 1.23$ radians, which is consistent with the locations of the peaks in the first plot in Fig. 5.25. You can show from scratch that $N_y = -5mg$ at $\theta = \pm\pi$ (that is, at the bottom of the hoop), which is also consistent with the plot.

- If $v_0 = \sqrt{gR}$, then from Problem 4.19 we know that $N_y = 0$ at $\theta = 0$; the second plot in Fig. 5.25 confirms this. From Eq. (5.92) the maximum N_y occurs at $\cos\theta = 1/2 \implies \theta = \pm 60° = \pm\pi/3 = \pm 1.05$ radians, which is consistent with the plot.

- If $v_0 = 2\sqrt{gR}$, then as we saw above, Eq. (5.92) tells us that the maximum N_y occurs at $\theta = 0$, which is consistent with the third plot in Fig. 5.25. Additionally, you can show from scratch that the $N_y = 3mg$ value at $\theta = 0$ in the plot is correct. This plot is the transition between the curve being concave up or concave down at $\theta = 0$.

- For larger values of v_0, the result for $\cos\theta$ in Eq. (5.92) isn't valid. The maximum is always at $\theta = 0$. The bump in the plot of N_y becomes somewhat more pointed at the top, as shown in the fourth plot in Fig. 5.25. For very large values of v_0, the plot of N_y approaches a simple cosine curve, because the v_0^2 term dominates in Eq. (5.91).

5.5. PROBLEM SOLUTIONS

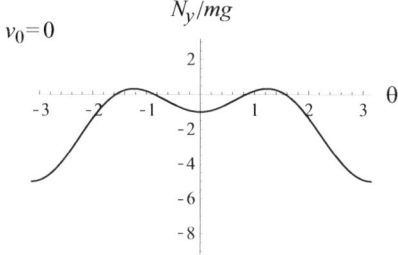
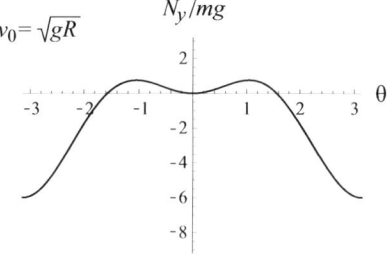
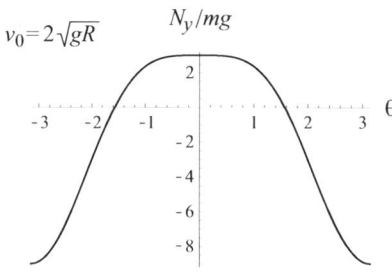
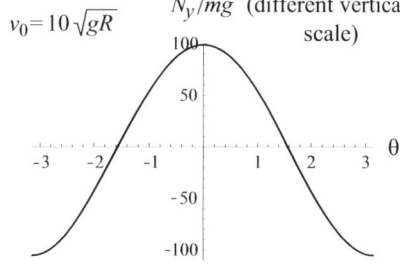

Figure 5.25

5.24. Falling stick on a table

When the stick has fallen through an angle θ, the height of the mass is $\ell \cos \theta$. So from conservation of energy, the mass's velocity v as a function of θ is

$$\frac{1}{2}mv^2 = mg(\ell - \ell \cos \theta) \implies v^2 = 2g\ell(1 - \cos \theta). \tag{5.93}$$

The radial acceleration of the mass is then

$$a_r = \frac{v^2}{\ell} = 2g(1 - \cos \theta). \tag{5.94}$$

We will also need to know the tangential acceleration (assuming that no slipping has occurred). Breaking up the gravitational force into its radial and tangential components gives the tangential acceleration of the mass as

$$a_t = g \sin \theta. \tag{5.95}$$

Our task is to find the friction force F_f and the normal force N from the table (assuming no slipping), and then impose the $F_f \leq \mu N$ condition. F_f and N can be found by writing down the horizontal and vertical $F = ma$ equations for the mass. (The radial and tangential equations would also work, but they would be messier since F_f and N would each appear in both equations.) Using the acceleration components shown in Fig. 5.26, the two $F = ma$ equations are

$$\begin{aligned} F_x &= ma_x \\ \implies F_f &= ma_t \cos \theta - ma_r \sin \theta \\ &= mg[\sin \theta \cos \theta - 2(1 - \cos \theta) \sin \theta] \\ &= mg \sin \theta (3 \cos \theta - 2), \end{aligned} \tag{5.96}$$

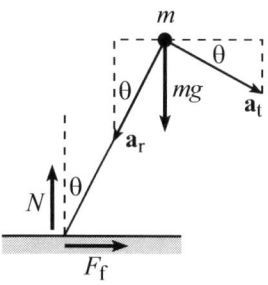

Figure 5.26

and

$$F_y = ma_y$$
$$\implies mg - N = ma_t \sin\theta + ma_r \cos\theta$$
$$\implies N = mg[1 - \sin^2\theta - 2(1 - \cos\theta)\cos\theta]$$
$$= mg[1 - (1 - \cos^2\theta) - 2(1 - \cos\theta)\cos\theta]$$
$$= mg\cos\theta(3\cos\theta - 2). \tag{5.97}$$

The condition $F_f \leq \mu N$ therefore becomes

$$mg\sin\theta(3\cos\theta - 2) \leq \mu mg\cos\theta(3\cos\theta - 2). \tag{5.98}$$

Assuming that $\cos\theta > 2/3$ (that is, $\theta < 48.2°$), we can divide both sides of Eq. (5.98) by $3\cos\theta - 2$, which yields

$$\tan\theta \leq \mu. \tag{5.99}$$

This is a necessary condition for no slipping. However, it isn't sufficient, because if $\cos\theta < 2/3$ (that is, $\theta > 48.2°$) then Eq. (5.97) gives a negative normal force N. But the normal force can't be negative; what happens is that the stick comes up off the table. So the $F_f \leq \mu N$ condition has no chance of being true. The stick will therefore slip if $\theta > 48.2°$, no matter how large μ is. But it will slip sooner if μ is small enough, at the angle given by $\tan\theta = \mu$. Since $\cos^{-1}(2/3) = \tan^{-1}(\sqrt{5}/2)$, we can summarize our results by saying that the stick slips at the angle θ_s, where θ_s is given by

$$\theta_s = \tan^{-1}(\mu) \qquad \text{if } \mu \leq \sqrt{5}/2,$$
$$\theta_s = \tan^{-1}(\sqrt{5}/2) = 48.2° \qquad \text{if } \mu \geq \sqrt{5}/2. \tag{5.100}$$

REMARK: A quicker way to obtain the above $\tan\theta \leq \mu$ result (although you still need to recognize the $\cos\theta > 2/3$ condition) is to apply $\tau = I\alpha$ around the point mass. (Torque is the subject of Chapter 7.) Since the moment of inertia I is zero for a point mass relative to itself, and since the stick is massless, the torque τ on the stick-plus-mass system must be zero. This means that the total force from the table must point *along* the stick. Hence $F_f/N = \tan\theta$. So the $F_f \leq \mu N$ condition becomes $N\tan\theta \leq \mu N \implies \tan\theta \leq \mu$. But again, you must still realize that the stick will necessarily slip at any angle larger than the specific angle (which is $48.2°$) where the normal force becomes zero.

5.25. Bead, spring, and hoop

(a) When the bead has traveled a quarter of the way around the circle, the length of the spring has decreased from $2R$ to $\sqrt{2}R$. So conservation of energy gives

$$\frac{1}{2}k(2R)^2 = \frac{1}{2}k(\sqrt{2}R)^2 + \frac{1}{2}mv^2 \implies 2kR^2 = mv^2. \tag{5.101}$$

After a quarter circle, the radial component of the spring force F_s is $F_s \cos 45°$. The radial $F = ma$ equation is therefore

$$N + F_s \cos 45° = \frac{mv^2}{R}, \tag{5.102}$$

with inward taken to be positive. The spring force is $F_s = k(\sqrt{2}R)$ at this point, so Eq. (5.102) becomes (using Eq. (5.101))

$$N + (k\sqrt{2}R)\frac{1}{\sqrt{2}} = \frac{2kR^2}{R} \implies N = kR. \tag{5.103}$$

Since N is positive here, the normal force points inward.

(b) At the start, when the bead is barely moving, the normal force points radially outward (to balance the spring force to yield essentially zero acceleration). And we just found in part (a) that after a quarter circle, the normal force points radially inward. Therefore, by continuity, we must have $N = 0$ somewhere in between. So the answer is "yes." (Actually, even without solving part (a), we know that after a semicircle the spring force will be zero, so the normal force will certainly have to point radially inward to provide the necessary centripetal force.)

REMARK: As an exercise, you can show that as a general function of θ (measured relative to the start), the normal force is $N = kR(1 - 3\cos\theta)$. If $\theta = 0$ this yields $N = -k(2R)$, which is correct; N points outward, equal and opposite to the spring force. If $\theta = 90°$ it yields $N = kR$, in agreement with the result in part (a). And if $\theta = 180°$ it yields $N = 4kR$, which you can verify from scratch. We see that the special value of $N = 0$ is achieved when $\cos\theta = 1/3 \Longrightarrow \theta = 70.5°$.

5.26. **Bead, spring, hoop, and pole**

If the bead hangs at rest, the upward kR force from the spring equals the downward mg force from gravity. So k is indeed equal to mg/R, as claimed.

With N defined to be positive outward, the free-body diagram for the bead is shown in Fig. 5.27. The spring is stretched a distance $x = R\sin\theta$, so the spring force is $kx = k(R\sin\theta)$. The radial component of this force is obtained by multiplying by $\sin\theta$, as you can verify. The radial $F = ma$ equation is therefore

$$mg\cos\theta + k(R\sin\theta)\sin\theta - N = \frac{mv^2}{R}$$

$$\Longrightarrow N = mg\cos\theta + kR\sin^2\theta - \frac{mv^2}{R}. \quad (5.104)$$

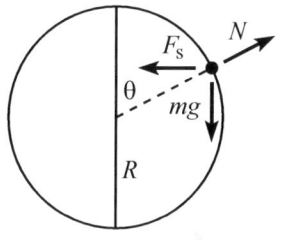

Figure 5.27

We can use conservation of energy to find v. The loss in gravitation potential energy shows up as spring potential energy plus kinetic energy. The height fallen is $R - R\cos\theta$, so we have

$$mg(R - R\cos\theta) = \frac{1}{2}k(R\sin\theta)^2 + \frac{1}{2}mv^2$$

$$\Longrightarrow mv^2 = 2mgR(1 - \cos\theta) - kR^2\sin^2\theta. \quad (5.105)$$

REMARK: If you invoke $kR = mg$, and if you make a plot of this expression for v^2 as a function of θ, you will see that it is always positive (or zero, at $\theta = 0$), as it must be, if the stated motion is possible. However, if kR were larger than mg, then v^2 would be negative for (at least) small values of θ, as you can show by using the Taylor series for $\sin\theta$ and $\cos\theta$. Since v^2 of course can't be negative, this means that the bead simply remains at $\theta = 0$; the backward tangential component of the spring force is larger than the forward tangential component of gravity. You can explicitly check this fact by writing out these two force components.

Plugging the value of mv^2 from Eq. (5.105) into Eq. (5.104), and invoking $kR = mg$, gives

$$N = mg\cos\theta + kR\sin^2\theta - [2mg(1 - \cos\theta) - kR\sin^2\theta]$$

$$= 3mg\cos\theta - 2mg + 2kR\sin^2\theta$$

$$= mg(3\cos\theta - 2 + 2\sin^2\theta). \quad (5.106)$$

To find where N is zero, we can replace $\sin^2\theta$ with $1 - \cos^2\theta$:

$$0 = 3\cos\theta - 2 + 2(1 - \cos^2\theta) \implies 3\cos\theta - 2\cos^2\theta = 0$$

$$\implies \cos\theta(3 - 2\cos\theta) = 0. \quad (5.107)$$

Therefore $\cos\theta = 0$, which means that $\theta = 90°$. (The other solution, $\cos\theta = 3/2$, isn't physical.) So the desired point is halfway down the circle, where the spring is horizontal.

REMARK: As a double check, you can solve the problem from scratch, under the assumption that $\theta = 90°$. The gravitational potential energy lost is mgR, while the potential energy gained in the spring is $kR^2/2 = mgR/2$ (using $kR = mg$). The difference in these is the kinetic energy, so $mv^2/2 = mgR/2$, which gives $v^2 = gR$. The necessary centripetal force is therefore $mv^2/R = mg$. But this is exactly the value of the spring force kR (since $kR = mg$). There is therefore no need for a normal force from the hoop.

5.27. Entering freefall

(a) Since energy is conserved in the collision (because we are told that ball B picks up whatever speed ball A had), we can just use conservation of energy between the starting point at the top of the circle and an arbitrary later point when ball B is at an angle θ with respect to the vertical. The net height fallen between these two points is $R - R\cos\theta$. Since the loss in gravitational potential energy shows up as kinetic energy, the speed of B at an arbitrary later point is given by

$$\frac{1}{2}mv^2 = mg(R - R\cos\theta) \implies v = \sqrt{2gR(1 - \cos\theta)}. \quad (5.108)$$

At a point in the upper half of the circle, the radially inward component of the gravitational force is $mg\cos\theta$. So if T is the tension in the string (while it is still taut), the radial $F = ma$ equation is $T + mg\cos\theta = mv^2/R$.

We are concerned with the point where the ball enters projectile motion, which is where the string goes limp, that is, where the tension becomes zero. So the radial $F = ma$ equation with $T = 0$ becomes (using the v we found above)

$$mg\cos\theta = \frac{mv^2}{R} \implies mg\cos\theta = \frac{m(2gR(1 - \cos\theta))}{R}$$
$$\implies \cos\theta = 2 - 2\cos\theta \implies \cos\theta = \frac{2}{3} \implies \theta \approx 48.2°. \quad (5.109)$$

REMARK: This result is exactly the same as the result for the standard problem: A mass is initially at rest at the top of a frictionless sphere. It is given a tiny kick and slides off. At what point does it lose contact with the sphere? The answer to this new problem is the same because the $F = ma$ equation is $mg\cos\theta - N = mv^2/R$ (the normal force points outward, hence the minus sign), and we are concerned with the point where $N = 0$, so we again end up with the $mg\cos\theta = mv^2/R$ condition. And since the conservation-of-energy statement is the same, we arrive at the same value of θ.

(b) FIRST SOLUTION: From Eq. (5.108) the speed of the ball when $\cos\theta = 2/3$ is $v = \sqrt{2gR(1 - 2/3)} = \sqrt{2gR/3}$. At this point the velocity of the ball makes an angle θ with respect to the horizontal, because the string makes an angle θ with respect to the vertical.

We must now solve the following problem: A projectile is fired at speed $v = \sqrt{2gR/3}$ at an angle of θ with respect to the horizontal, where $\cos\theta = 2/3$. How high does it go?

The height of the projectile motion is determined solely by the v_y component of the velocity at launch. Conservation of energy (or a standard kinematic argument) gives the maximum height as $mgy = mv_y^2/2 \implies y = v_y^2/2g$. But $v_y = v\sin\theta$, where $\sin\theta = \sqrt{1 - \cos^2\theta} = \sqrt{1 - (2/3)^2} = \sqrt{5}/3$. Therefore,

$$v_y = v\sin\theta = \sqrt{\frac{2gR}{3}} \cdot \frac{\sqrt{5}}{3} \implies y = \frac{v_y^2}{2g} = \frac{5R}{27}. \quad (5.110)$$

Adding this to the height (relative to the center of the circle) at the start of the projectile motion, which is $R\cos\theta = 2R/3$, we see that the total maximum height of the projectile motion, relative to the center of the circle, is $(2/3 + 5/27)R = 23R/27$.

5.5. PROBLEM SOLUTIONS

SECOND SOLUTION: At the top of the projectile motion, the vertical component of the velocity is zero. So conservation of energy tells us that the loss in potential energy from the top of the circle down to the top of the projectile motion goes entirely into the kinetic energy of the horizontal motion. The v_x during the projectile motion has the constant value of $v_x = v\cos\theta = \sqrt{2gR/3} \cdot (2/3)$. So if Δh is the net height fallen, we have

$$mg\,\Delta h = \frac{1}{2}mv_x^2 \implies mg\,\Delta h = \frac{1}{2}m\left(\frac{8gR}{27}\right) \implies \Delta h = \frac{4R}{27}. \qquad (5.111)$$

This agrees with the above result, because $4R/27$ below the top implies $23R/27$ above the center.

Chapter 6

Momentum

6.1 Introduction

Momentum

The momentum of a mass m moving with velocity $\mathbf{v}$ is defined to be

$$\mathbf{p} = m\mathbf{v}. \tag{6.1}$$

Because velocity is a vector, momentum is also a vector. The momentum is what appears in Newton's second law, $\mathbf{F} = d\mathbf{p}/dt$. In the common case where m is constant, this law reduces to $\mathbf{F} = m\, d\mathbf{v}/dt \implies \mathbf{F} = m\mathbf{a}$.

Impulse

Consider the time integral of the force, which we shall define as the *impulse* $\mathbf{J}$:

$$\mathbf{J} \equiv \int \mathbf{F}\, dt \quad \text{(impulse)}. \tag{6.2}$$

This is just a definition, so there's no content here. But if we invoke Newton's second law, then we can produce some content. If we multiply both sides of $\mathbf{F} = d\mathbf{p}/dt$ by dt and then integrate, we obtain $\int \mathbf{F}\, dt = \Delta \mathbf{p}$. The left-hand side of this relation is just the impulse. We therefore see that the impulse $\mathbf{J}$ associated with a time interval Δt equals the total change in momentum $\Delta \mathbf{p}$ during that time:

$$\mathbf{J} = \Delta \mathbf{p}. \tag{6.3}$$

The impulse $\mathbf{J} \equiv \int \mathbf{F}\, dt$ is the area under the force vs. time curve (technically three different areas for the three different directions in 3-D). So an extended gradual force and a hard quick strike will impart the same momentum to an object if they have the same area under the force vs. time curve.

The above definition of impulse closely parallels the definition of work: $W = \int F\, dx$, or more generally $W = \int \mathbf{F} \cdot d\mathbf{x}$. The time integral of the force equals the change in momentum (by Newton's second law), while the space integral of the force equals the change in kinetic energy (by the work-energy theorem, which can be traced to Newton's second law).

Conservation of momentum

Consider an isolated system of two particles, labeled 1 and 2. If $\mathbf{F}_{ij}$ represents the force on particle i due to particle j, then Newton's third law tells us that $\mathbf{F}_{12} = -\mathbf{F}_{21}$. But $\mathbf{F} = d\mathbf{p}/dt$, so we have

$$\frac{d\mathbf{p}_1}{dt} = -\frac{d\mathbf{p}_2}{dt} \implies \frac{d(\mathbf{p}_1 + \mathbf{p}_2)}{dt} = 0 \implies \mathbf{p}_1 + \mathbf{p}_2 = \text{Constant}. \tag{6.4}$$

6.1. INTRODUCTION

In other words, the total momentum is conserved. This derivation wasn't much of a derivation, being only one line. We can therefore view Newton's third law as basically postulating conservation of momentum for two particles. If we have an isolated system of many particles, you can show that the forces cancel in pairs (by Newton's third law), so again the total momentum is conserved; see Problem 6.1.

If the system isn't isolated and there are external forces, then we can break up the force on the ith particle into external and internal forces: $d\mathbf{p}_i/dt = \mathbf{F}_i^{\text{ext}} + \mathbf{F}_i^{\text{int}}$. Summing over all the particles gives

$$\frac{d(\sum \mathbf{p}_i)}{dt} = \sum \mathbf{F}_i^{\text{ext}} + \sum \mathbf{F}_i^{\text{int}} \quad \Longrightarrow \quad \frac{d\mathbf{p}_{\text{total}}}{dt} = \mathbf{F}_{\text{total}}^{\text{ext}} + 0, \tag{6.5}$$

where we have used Newton's third law to say that the sum of the *internal* forces is zero since they cancel in pairs. We can therefore omit the "ext" superscript in this result, because the total force on a system of particles is the same as the total external force.

Collisions

Consider a collision between two particles. Assuming that there are no external forces, we know from Eq. (6.4) that the total momentum of the particles is conserved. Since momentum is a vector, this means that each component is conserved separately. So in N dimensions, we can write down N independent statements that must be true. Note that we don't have to worry about the messy specifics of what goes on during the collision. The momentum of an isolated system is conserved, period.

Is the total energy of the particles conserved? The answer to this question is technically yes, because energy is always conserved. However some of the energy may be "lost" to heat, in which case it doesn't show up in the form of $mv^2/2$ terms for the macroscopic particles in the system.[1] As mentioned in the discussion of heat on page 108, it is common in such a situation to say that energy isn't conserved, even though it is of course conserved when heat is taken into account. If no heat is created, then we call the collision *elastic*. If heat is created, so that energy "isn't" conserved, then we call the collision *inelastic*. In the first case, the sum of the $mv^2/2$ terms for the macroscopic particles in the system is the same before and after the collision. In the second case, it isn't. The degree to which it isn't depends on how inelastic the collision is. (The maximally inelastic case occurs when the particles stick together.) So in summary, in an isolated collision,

1. Momentum is always conserved.

2. Mechanical energy (by which we mean the $mv^2/2$ energies of the macroscopic particles involved; that is, excluding heat) is conserved only if the collision is elastic (by definition).

A helpful fact that is valid for 1-D elastic collisions is:

- In a 1-D elastic collision, the final relative velocity between the two particles equals the negative of the initial relative velocity.

That is, if one particle initially sees the other coming toward it with speed v, then after the collision, it sees the other moving away from it with the same speed v; see Problem 6.2 for a proof. This linear relation among the various velocities often simplifies calculations, because it can be used (in tandem with the linear conservation-of-p relation) as a substitute for the more complicated conservation-of-E relation, which is *quadratic* in the velocities.

---

[1] Imagine a ball of clay that is thrown at a wall and sticks to it. The "lost" energy shows up in $mv^2/2$ terms for the individual random motions of the microscopic molecules in the clay, not in the $mv^2/2$ term for the motion of the ball as a whole, because that v is now zero.

Center of mass

The location of the *center of mass*, or CM, of two objects lying along the x axis is defined to be

$$x_{\rm CM} = \frac{m_1 x_1 + m_2 x_2}{m_1 + m_2}. \tag{6.6}$$

This location has the property that the distances to the two masses are inversely proportional to the masses; see Problem 6.4. If one mass is ten times the other, then the CM is ten times closer to the larger mass. The analogous expressions hold for $y_{\rm CM}$ and $z_{\rm CM}$ in the general 3-D case. The definition generalizes to any number of particles; the vector location of the CM of many masses is

$$\mathbf{r}_{\rm CM} = \frac{\sum m_i \mathbf{r}_i}{M}, \tag{6.7}$$

where $M \equiv \sum m_i$ is the total mass of the system. Basically, $\mathbf{r}_{\rm CM}$ is the weighted average of the various positions, with each position being weighted by the associated mass. In the case of a continuous distribution of mass, we have

$$\mathbf{r}_{\rm CM} = \frac{\int \mathbf{r}\, dm}{M}. \tag{6.8}$$

When calculating the location of the CM of an object, a useful fact that simplifies things is that a given subpart of the object can be replaced with a point mass (with the same mass as the subpart) located at the CM of the subpart. See Problem 6.5 for a proof.

Center of mass, v and a

By taking time derivatives of Eq. (6.7), the velocity and acceleration of the CM are

$$\mathbf{v}_{\rm CM} = \frac{\sum m_i \mathbf{v}_i}{M} \quad \text{and} \quad \mathbf{a}_{\rm CM} = \frac{\sum m_i \mathbf{a}_i}{M}. \tag{6.9}$$

The first of these equations can be rewritten as $\sum m_i \mathbf{v}_i = M\mathbf{v}_{\rm CM}$, or equivalently,

$$\mathbf{p}_{\rm total} = M\mathbf{v}_{\rm CM}. \tag{6.10}$$

So the total momentum of a system is the same as if all the mass is lumped together and moves along with the velocity of the CM.

The second of the equations in Eq. (6.9) can be rewritten as $\sum m_i \mathbf{a}_i = M\mathbf{a}_{\rm CM}$. But since $\mathbf{F} = m\mathbf{a}$, the left-hand side is just the sum of the forces on all of the masses. So we have

$$\mathbf{F}_{\rm total} = M\mathbf{a}_{\rm CM}. \tag{6.11}$$

In other words, the CM moves just as if all of the force were applied to a mass M located at the CM. As we mentioned above, all of the internal forces cancel in pairs, so we can write $\mathbf{F}_{\rm total}$ alternatively as $\mathbf{F}_{\rm external}$. Therefore, a corollary is that if there are no external forces, the CM moves with constant velocity, independent of how the various particles in the system may be moving with respect to each other, and independent of any complicated internal forces.

Collisions in the CM frame

When viewed in the CM frame (the frame that travels along with the CM), collisions are particularly simple. In a 1-D elastic collision, the masses head toward the (stationary) CM, and then after the collision they simply reverse direction and head out with the same speeds they originally had. This scenario satisfies conservation of momentum (the total momentum is zero both before and after the collision),[2] and it also satisfies conservation of energy (the speeds don't

---

[2] The total momentum is always zero in the CM frame, because we saw in Eq. (6.10) that $\mathbf{p}_{\rm total} = M\mathbf{v}_{\rm CM}$. And the velocity $\mathbf{v}_{\rm CM}$ of the CM is zero in the CM frame, because the CM isn't moving in that frame, by definition.

change, so neither do the $mv^2/2$ kinetic energies). Since everything that needs to be conserved is in fact conserved, this scenario must be what happens.

In the more general case of a 2-D (or 3-D) elastic collision in the CM frame, the particles must still end up moving in opposite directions, with the same speeds they originally had, otherwise we wouldn't have conservation of both p and E; see Problem 6.6. But there is now freedom to choose the orientation of the line of the final velocities; see Fig. 6.1. This line can make any angle with respect to the original line, and both momentum and energy will still be conserved. To determine the direction (which can be specified by one angle in 2-D or two angles in 3-D), we need to be given more information about how exactly the particles collide.

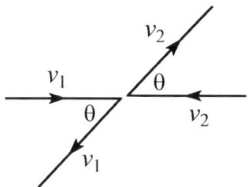

Figure 6.1

In the case of *inelastic* collisions in the CM frame, the only modification to the above results is that both speeds are scaled down by the *same* factor (this will keep the total momentum at zero). The size of this factor depends on how inelastic the collision is. If the collision is completely inelastic (so that the particles stick together), then the factor is zero.

If you want to solve a collision (let's assume it's elastic) by utilizing the CM frame, there are three steps to perform:

1. Assuming that the setup was given in the lab frame, you need to switch to the CM frame. This involves finding the velocity of the CM and then subtracting this velocity from the lab-frame velocities to obtain the CM-frame velocities.

2. Find the final velocities in the CM frame. In a 1-D elastic collision, this step is trivial; the velocities simply reverse. In a 2-D elastic collision, the final line containing the velocities may be different from the initial line. Additional information needs to be given in order to determine the direction of the line.

3. Assuming that the problem asks for the final velocities in the lab frame, you need switch back to the lab frame. This involves adding on the CM velocity to the velocities you found in the CM frame.

Variable mass

Since the momentum p equals mv (we'll work in just one dimension here), Newton's second law can be written as

$$F = \frac{dp}{dt} = \frac{d(mv)}{dt} = m\frac{dv}{dt} + \frac{dm}{dt}v = ma + \frac{dm}{dt}v. \tag{6.12}$$

If the mass m of an object is constant, this reduces to $F = ma$. But if the mass changes, then we need to keep the $(dm/dt)v$ term. In the general case, the momentum can change because the velocity changes, or because the mass changes,[3] or both. For setups involving changing mass, it is important to label clearly the system that you are applying $F = dp/dt$ to, because the F and p here must apply to the *same* system.

6.2 Multiple-choice questions

6.1. A ping-pong ball and a bowling ball have the same momentum. Which one has the larger kinetic energy?

(a) the ping-pong ball

(b) the bowling ball

(c) They have the same kinetic energy.

---

[3]Imagine pushing a bucket at constant speed v, with someone dropping sand into it. You need to apply a force, and this force equals $(dm/dt)v$. This force doesn't speed up the sand that is already in the bucket, but rather it gives momentum to the new sand by suddenly bringing it up to speed v.

6.2. You are stuck in outer space and want to propel yourself as fast as possible in a certain direction by throwing an object in the opposite direction. You note that every object you've ever thrown in your life has had the *same kinetic energy*. Assuming that this trend continues, you should

(a) throw a small object with a large speed

(b) throw a large object with a small speed

(c) It doesn't matter.

6.3. How should you build a car in order to reduce the likelihood of injury in a head-on crash?

(a) Make the front bumper be rigid.

(b) Make the front of the car crumple when large forces are applied.

(c) Make the front of the car crumple easily when small forces are applied.

(d) Make the front of the car be rigid, so that it does not crumple at all.

(e) Install a set of springs in the front of the car, so that it bounces backward after the collision.

6.4. An apple falls from a tree. Which of the following does *not* explain why the apple speeds up as it falls?

(a) The momentum of the earth-apple system is conserved.

(b) There is a downward gravitational force acting on the apple.

(c) Gravity does positive work on the apple as it falls.

(d) The apple loses potential energy as it falls.

6.5. Two people stand on opposite ends of a long sled on frictionless ice. The sled is oriented in the east-west direction, and everything is initially at rest. The western person then throws a ball eastward toward the eastern person, who catches it. The sled

(a) moves eastward, and then ends up at rest

(b) moves eastward, and then ends up moving westward

(c) moves westward, and then ends up at rest

(d) moves westward, and then ends up moving eastward

(e) does not move at all

6.6. On a frictionless table, a mass m moving at speed v collides with another mass m initially at rest. The masses stick together. How much energy is converted to heat?

(a) 0 (b) $\frac{1}{4}mv^2$ (c) $\frac{1}{3}mv^2$ (d) $\frac{1}{2}mv^2$ (e) mv^2

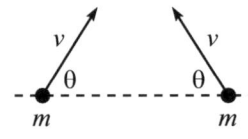

Figure 6.2

6.7. Two masses move toward each other as shown in Fig. 6.2. They collide and stick together. How much energy is converted to heat?

(a) mv^2

(b) $mv^2 \sin\theta$

(c) $mv^2 \sin^2\theta$

(d) $mv^2 \cos\theta$

(e) $mv^2 \cos^2\theta$

6.8. N balls with mass m lie at rest in a line on a frictionless table, with a small separation between adjacent balls. The first ball is given a kick and acquires a speed v. It collides and sticks to the second ball, and the resulting blob collides and sticks to the third ball, and so on. What is the final speed of the resulting blob of mass Nm?

(a) 0 (b) v/N (c) $v/\sqrt{N}$ (d) v (e) Nv

6.2. MULTIPLE-CHOICE QUESTIONS

6.9. A very light ping-pong ball bounces elastically head-on off a very heavy bowling ball that is initially at rest. The fraction of the ping-pong ball's initial kinetic energy that is transferred to the bowling ball is approximately

(a) 0   (b) 1/4   (c) 1/2   (d) 3/4   (e) 1

6.10. A small marble with mass m moves with speed v toward a large block with mass M (assume $M \gg m$), which sits at rest on a frictionless table. In Scenario A the marble sticks to the block. In Scenario B the marble bounces elastically off the block and heads backward. The ratio of the block's resulting kinetic energy in Scenario B to that in Scenario A is approximately

(a) 1/4   (b) 1/2   (c) 1   (d) 2   (e) 4

6.11. A ball of clay is thrown at a wall and sticks to it. Virtually all of its momentum ends up in

(a) the ball
(b) the earth
(c) heat
(d) sound
(e) It doesn't end up anywhere, because momentum isn't conserved.

6.12. A ball of clay is thrown at a wall and sticks to it. Virtually all of its kinetic energy ends up in

(a) kinetic energy of the ball
(b) kinetic energy of the earth
(c) heat
(d) sound
(e) It doesn't end up anywhere, because energy isn't conserved.

6.13. In a one-dimensional elastic collision, Ball 1 collides with Ball 2, which is initially at rest. Can the masses be chosen so that the final speed of Ball 2 is larger than the initial speed of Ball 1?

Yes   No

6.14. A mass m moves with a given speed and collides (not necessarily head-on) elastically with another mass m that is initially at rest, as shown in Fig. 6.3. Which of the figures shows a possible outcome for the two velocities? (The velocities are drawn to scale.)

Figure 6.3

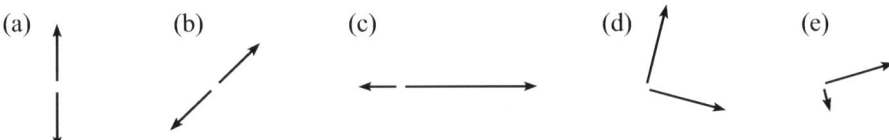

6.15. A paddle hits a ping-pong ball. The paddle moves with speed v to the right, and the ping-pong ball moves with speed $3v$ to the left. The collision is elastic. What is the resulting speed of the ping-pong ball? (Assume that the paddle is much more massive than the ping-pong ball.)

(a) $2v$ (b) $3v$ (c) $4v$ (d) $5v$ (e) $6v$

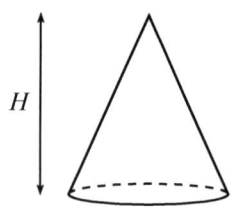

Figure 6.4

6.16. Let y be the height (above the base) of the center of mass of a hollow cone (like an ice cream cone) with total height H, as shown in Fig. 6.4. Then

(a) $y = H$

(b) $H/2 < y < H$

(c) $y = H/2$

(d) $0 < y < H/2$

(e) $y = 0$

6.17. A mass $2m$ moves rightward with speed v toward a mass m that is at rest. What is the speed of the mass $2m$ in the CM frame?

(a) 0 (b) $v/3$ (c) $v/2$ (d) $2v/3$ (e) v

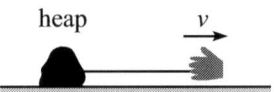

Figure 6.5

6.18. A heap of rope with mass density λ (per unit length) lies on a table. You grab one end and pull horizontally with constant speed v, as shown in Fig. 6.5. (Assume that the rope has no friction with itself in the heap.) The force that you must apply to maintain the constant speed v is

(a) 0

(b) λv

(c) λv^2

(d) $\lambda \ell g$, where ℓ is the length that you have pulled straight

(e) $\lambda v^4 / g\ell$

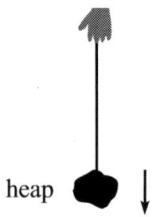

Figure 6.6

6.19. With your hand at a fixed position, you hold onto one end of a heap of rope with mass m and then let the heap fall, as shown in Fig. 6.6. The rope has no friction with itself in the heap. Eventually the heap runs out and you are left holding a vertical piece of rope with mass m. Is there any time during this process when the upward force you exert with your hand exceeds mg?

Yes   No

6.20. A dustpan is accelerated with acceleration a across a frictionless floor, and it gathers up dust as it moves, as shown in Fig. 6.7. The mass of the dustpan itself is M, and the linear mass density of the dust on the floor is λ. If the dustpan starts empty at $x = 0$, then the force that must be applied to it later on when it is moving with speed v at position x is

(a) Ma

(b) $Ma + \lambda v^2$

(c) $(M + \lambda x)a$

(d) λv^2

(e) $(M + \lambda x)a + \lambda v^2$

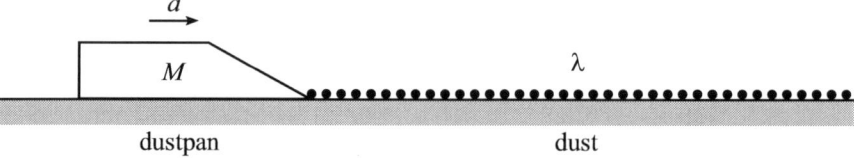

Figure 6.7

6.3 Problems

The first six problems are foundational problems.

6.1. **Conservation of momentum**

In an isolated system of many particles, show that the total momentum is conserved, by showing that the forces cancel in pairs (by Newton's third law).

6.2. **Same relative speed**

Show that in a 1-D elastic collision, the final relative velocity between the two masses equals the negative of the initial relative velocity.

Note: Although this can be shown very easily by working in the CM frame, demonstrate it here by working in the lab frame and using conservation of energy and momentum. One way to do this is to solve for the final two velocities in terms of the initial ones. But that gets very messy. A sneakier way is to put the terms associated with each particle on different sides of the conservation of E and p equations, and then divide the E equation by the p equation.

6.3. **1-D collision**

In a 1-D elastic collision, a mass M moving with velocity V collides with a mass m that is initially at rest. Show that the resulting velocities are

$$V_M = \frac{(M-m)V}{M+m} \quad \text{and} \quad v_m = \frac{2MV}{M+m}. \tag{6.13}$$

6.4. **Distances to the CM**

Show that the location of the center of mass, given in Eq. (6.6), has the property that the distances to the two masses are inversely proportional to the masses.

6.5. **Equivalent subparts**

Show that the location of the CM of the object in Fig. 6.8 can be found by replacing the two subparts S_1 and S_2 with point masses (with the same masses as the subparts) located at the CM's of the subparts.

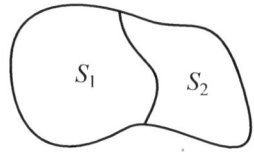

Figure 6.8

6.6. **Collision in the CM frame**

Consider an elastic collision (in any dimension), as viewed in the CM frame. Show that after the collision, the particles must move in opposite directions, with the same speeds they originally had.

6.7. **Hemispherical-shell CM**

Find the location of the CM of a hollow hemispherical shell with radius R and uniform surface mass distribution.

6.8. **Atwood's machine**

In Problem 4.5 we considered the Atwood's machine shown in Fig. 6.9. In the solution to that problem, we found that the accelerations of the masses and the tension in the upper rope are given by

$$a = g\frac{m_2 - m_1}{m_2 + m_1} \quad \text{and} \quad T = g\frac{4m_1 m_2}{m_2 + m_1}.$$

Let's assume $m_2 > m_1$ so that a is positive.

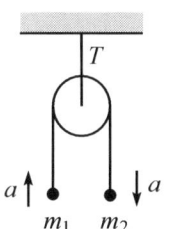

Figure 6.9

(a) After each mass has moved a distance d, find the potential and kinetic energies, and verify that energy is conserved. Assume that the masses start at rest.

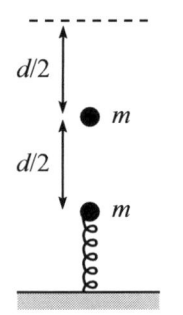

Figure 6.10

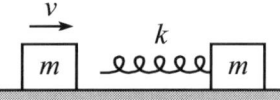

Figure 6.11

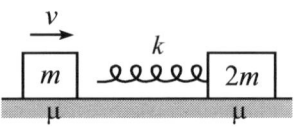

Figure 6.12

(b) After a time t, verify that $P_{\text{total}} = F_{\text{total}}t$ (which is a consequence of integrating $F = dP/dt$ when F is constant). Be careful to include *all* the external forces acting on the system.

6.9. Rising and colliding

A vertical spring is compressed a distance d relative to its relaxed length. A mass m is attached to the top end and held at rest. Another mass m is suspended a distance $d/2$ above it, as shown in Fig. 6.10. The spring is released. The bottom mass rises up, smashes and sticks to the upper mass, and the resulting blob continues to rise up. What must the spring constant k be, so that the blob reaches its maximum height at the dotted line shown (where the spring is at its relaxed length)?

6.10. Maximum compression

A block with mass m slides with speed v along a frictionless table toward a stationary block that also has mass m. A massless spring with spring constant k is attached to the second block, as shown in Fig. 6.11. What is the maximum distance the spring gets compressed? *Hint*: How do the blocks' speeds compare at maximal compression?

6.11. Collision and a spring

A spring with spring constant k is attached to a stationary mass $2m$, as shown in Fig. 6.12. A mass m travels to the right and sticks to the left end of the spring and compresses the spring. The coefficient of friction (both static and kinetic) between the masses and the ground is μ. Let v be the speed of m right before it hits the spring.

(a) What is the largest value of v for which the mass $2m$ never moves? (Don't forget that m feels the friction too.) Assume that the spring is sufficiently long so that m doesn't collide with $2m$.

(b) Assume that v takes on this largest value. The spring will compress and then expand back out and push m to the left. What will m's speed be by the time the spring expands back to its relaxed length? (Again, don't forget that m feels the friction.)

6.12. Colliding balls

A stream of N clay balls with mass m move with speed v in a line across a frictionless table. The spacing between them is ℓ. An additional ball with mass m sits at rest in front of them, as shown in Fig. 6.13. The front moving ball collides with the stationary ball and sticks to it and forms a blob of mass $2m$. Then the second ball collides with the blob and forms a blob of mass $3m$. And so on. How much time elapses between the instant shown below (when all the balls are separated by ℓ) and the last collision? Solve this by working in (a) the lab frame, and then (b) the frame in which the N balls are initially at rest (this is the quicker way). *Note*: As with any problem involving a general number N, it is often helpful to first work things out for small values of N.

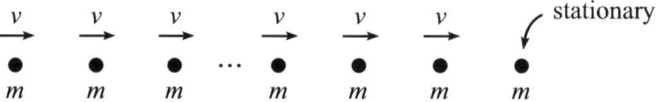

Figure 6.13

6.13. Block and balls

A block with large mass M moves with initial speed V_0 across frictionless ice. It encounters a long row of stationary balls, each with small mass m, and it collides elastically with

6.3. PROBLEMS

each ball, one after the other; see Fig. 6.14.[4]

(a) If at a later time the block is moving with speed v right before it collides with a given ball, what is the block's speed right after the collision? Assume $m \ll M$, and give your answer to leading order in m. *Hint*: To save yourself some algebra, find the speed of the ball after the collision by working in the frame of the block.

(b) If the balls are essentially continuously distributed, with the row having mass density λ (kg/m), what is the decrease in the block's speed (in terms of v) during a small interval of time dt? Give your answer to leading order in dt.

(c) Find $v(t)$, assuming that $t = 0$ corresponds to the time right before the first collision, when the block has speed V_0.

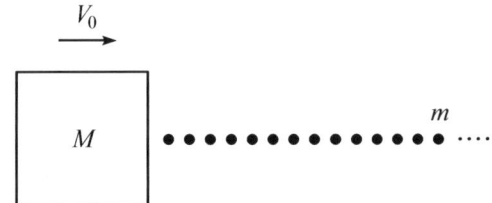

Figure 6.14

6.14. Maximum final speed

The result of Problem 6.3 was that in a 1-D elastic collision, if a mass M moving with velocity V collides with a stationary mass m, the resulting velocities are

$$V_M = \frac{(M-m)V}{M+m} \quad \text{and} \quad v_m = \frac{2MV}{M+m}. \tag{6.14}$$

(a) Given M and m, imagine putting a stationary mass (call it x) between M and m. If your goal is to have m pick up as large a speed as possible, what should x be in terms of M and m? Assume that both collisions are elastic.

(b) If you instead put two stationary masses, x and y, between M and m, what should they be so that m picks up as large a speed as possible? What about a general number of masses, say 10? In answering these questions, there is no need to do any calculations; your result from part (a) is all you need. A proof by contradiction is your best bet. (You are also encouraged to think about where the initial energy and momentum end up, in the $M \gg m$ limit with a very large number of optimally-chosen masses in between.)

6.15. Throwing a block in pieces

(a) You sit on a crate at rest on frictionless ice. The combined mass of you and the crate is m. You hold a block also of mass m and then throw it horizontally so that the relative speed between you and it is v_0. What is your resulting speed?

(b) Now assume that you cut the block in half and throw one half, and then wait a moment and throw the other half. What is your resulting speed? Assume that whenever you throw something, the relative speed between you and it is v_0.

(c) What is your resulting speed if you divide the block in thirds and throw the pieces successively? Or fourths? A pattern should emerge. In general, if you divide the block into n pieces, your resulting speed will take the form of $f(n)v_0$. What is the function $f(n)$?

---

[4]Assume that when each ball bounces off the block, it (the ball) magically passes through the other balls as it moves forward; so you don't have to worry about it bouncing back off the next ball and hitting the block. Equivalently, assume that the balls are displaced slightly from one another in the transverse direction.

(d) What is $f(100)$? If you have *Mathematica* or something similar handy, make a plot of $f(n)$ vs. n, for n from 1 to 100. (Take v_0 to be 1.) Make a guess for the limit of $f(n)$ as $n \to \infty$.

(e) The $n \to \infty$ limit is identical to rocket motion, because it involves a continuous ejection of mass. So we actually already know the final speed in this case, from Problem 6.22 below. What is the speed? How much does it differ from the speed in the $n = 100$ case?

6.16. A collision in two frames

Consider the following one-dimensional collision. A mass $4m$ moves to the right, and a mass m moves to the left, both with speed v. They collide elastically. Find their final lab-frame velocities. Solve this by working in (a) the lab frame, and (b) the CM frame.

6.17. 45-degree deflections

A ball with mass m moves in the positive x direction with speed v, as shown in Fig. 6.15. It collides elastically with two other balls (also with mass m) which are situated so that they both move at equal speeds at equal angles of 45° with respect to the x axis after the collision. What are the final speeds of all three balls?

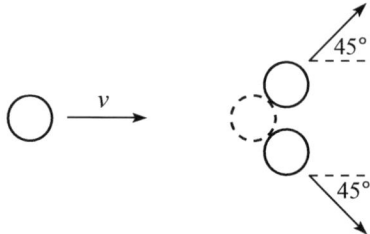

Figure 6.15

6.18. Northward deflection 1

A mass $2m$ moves to the east, and a mass m moves to the west, both with speed v_0. If they collide elastically, but not head-on, and if it is observed that the mass $2m$ ends up moving northward (that is, perpendicular to the original direction of motion), what is its speed?

6.19. Northward deflection 2

A mass m moving to the east with speed v_0 collides elastically, but not head-on, with a mass $2m$ at rest. If it is observed that the mass m ends up moving northward (that is, perpendicular to the original direction of motion), what angle does the resulting velocity of the mass $2m$ make with the east-west direction? Solve this by working in (a) the lab frame, and (b) the CM frame.

6.20. Equal energies

A mass m moving with speed v_0 collides elastically with a stationary mass $2m$, as shown in Fig. 6.16. Assuming that the final energies of the masses turn out to be equal, find the two final speeds, v_1 and v_2. Also, find the two angles of deflection, θ_1 and θ_2. *Hint*: The best way to solve for angles is usually to square equations (in appropriate form) and use the fact that $\sin^2 \theta + \cos^2 \theta = 1$.

6.21. Equal speeds

A mass m moving with speed v_0 collides elastically with a stationary mass $2m$, as shown in Fig. 6.17. Assuming that the final speeds of the masses turn out to be equal, find this speed, and also find the two angles of deflection, θ_1 and θ_2. (Same hint as in the preceding problem.)

6.3. PROBLEMS

Figure 6.16

Figure 6.17

6.22. Rocket motion

A rocket ejects mass backward with a constant speed u relative to the rocket.[5] If the initial mass of the rocket is M, what is the rocket's speed at a later time when the mass is m? *Hint*: Consider a short interval of time, and determine the increase in speed after a small amount of mass has been ejected. This will yield a differential equation relating dv and dm, which you can then integrate.

6.23. Hovering board

(a) A hose shoots a stream of water vertically upward. The water leaves the hose at speed v_0 and at a mass rate R (kg/s). A horizontal board with mass m is placed a very small distance above the hose and then released. What should m be so that the board hovers at this height? Assume that when the water crashes into the board, it bounces off essentially sideways.

(b) If you break the board in half, so that its mass is now $m/2$, how high above the hose should it be located if you want it to hover in place?

(c) In part (a), what should m be if the stream of water is replaced by a stream of marbles that bounce off the board elastically (that is, they bounce off downward with the same speed v_0)?[6]

6.24. Falling heap

A rope with total length L and mass density λ (kg/m) is held in a heap, and you grab an end that protrudes a tiny bit out of the top. The heap is then released, and it falls downward. As a function of time, what is the force that your hand must apply to the top end of the rope, to keep it motionless? Assume that the rope has no friction with itself, so that the remaining part of the heap is always in freefall. The setup at a general later time is shown in Fig. 6.18.

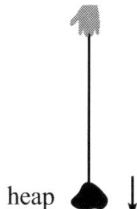

Figure 6.18

---

[5]To emphasize, u is the speed relative to the rocket. It wouldn't make sense to say "relative to the ground," because the rocket's engine shoots out the matter relative to itself, and the engine has no way of knowing how fast the rocket is moving with respect to the ground.

[6]Even though the marbles are discrete objects, assume that they have an essentially continuous mass rate equal to R. Also, assume that the downward-moving marbles that have bounced off the board somehow magically pass through the upward-moving ones without colliding.

6.25. Bucket and chain

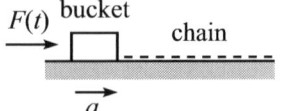

Figure 6.19

A massless bucket is initially at rest next to one end of a chain that lies in a straight line on the floor, as shown in Fig. 6.19. The chain has uniform mass density λ (kg/m). You push on the bucket (so that it gathers up the chain) with the force $F(t)$ that gives the bucket (plus whatever chain is inside) a constant acceleration a at all times. There is no friction between the bucket and the floor.

(a) What is $F(t)$?

(b) How much work do you do up to time t?

(c) How much energy is lost to heat up to time t? That is, by how much does the work you do exceed the kinetic energy of the system?

6.4 Multiple-choice answers

6.1. $\boxed{a}$ The kinetic energy can be written as $mv^2/2 = (mv)v/2 = pv/2$. Since the p's are equal, the object with the larger v has the larger energy. And equal p's imply that the ping-pong ball has the larger v because its mass is smaller. So in general we want to throw a small object with a large speed.

Alternatively, the kinetic energy can be written as $mv^2/2 = (mv)^2/2m = p^2/2m$. So if the p's are equal, the object with the smaller m (the ping-pong ball) has the larger energy.

Alternatively again, just pick some numbers: the combination of $m = 1$ and $v = 10$ yields the same momentum as the combination of $m = 10$ and $v = 1$, and the former has a larger energy $mv^2/2$.

REMARK: Another way of seeing why the ping-pong ball has the larger energy is the following. Imagine both objects being decelerated with the same constant force F. Since $|\Delta p| = Ft$, the objects will take the same *time* to stop if they start with the same momentum. However, since the smaller object is moving faster at any given time, it travels for a larger *distance* x by the time it stops. But the work-energy theorem tells us that $|\Delta K| = Fx$. So the object with the larger x (which is the smaller object) must have started off with a larger kinetic energy. In short, the same impulse $\int F\,dt$ is imparted on both objects, but more work $\int F\,dx$ is done on the smaller object.

6.2. $\boxed{b}$ This question is basically the opposite of the previous one, where the momentum was assumed to be the same. In the present setup, conservation of momentum tells us that we want to give the object as much momentum as possible, so that you will have as much momentum as possible in the opposite direction. The kinetic energy $mv^2/2$ is assumed to be given. Since this energy can be written as $(mv)v/2 = pv/2$, a small v means a large p. And a small v in turn means a large m (since $mv^2/2$ is given). So we want to throw a large object with a small speed.

Alternatively, the energy can be written as $p^2/2m$. So if the energies are the same, the object with the larger m has the larger p.

Alternatively again, just pick some numbers: the combination of $m = 1$ and $v = 10$ yields the same energy as the combination of $m = 100$ and $v = 1$, and the latter has a larger momentum $p = mv$.

REMARK: Is the assumption of every thrown object having the same kinetic energy a reasonable one? Certainly not in all cases. If a baseball pitcher throws a ball with 1/4 the mass of a baseball, he definitely isn't going to be able to throw it twice as fast. The limiting factor in this case is arm speed.

But what about large objects, for which a limiting arm speed isn't an issue? Ignoring myriad real-life complications, it's reasonable to make the approximation that you apply the same force on any large object, assuming that your hand is moving relatively slowly and that you are pushing as hard as you can. You apply this same force for the same distance, assuming that your throwing motion looks roughly the same for any large object. The work (force times distance) that you do is therefore the same for all objects, which means that the resulting kinetic energy is the same. So the assumption of equal kinetic energies stated in the question is reasonable in this case.

Note that this reasoning doesn't work with *momentum*, because you will accelerate a smaller (but still large) object more quickly, which means that you will spend less *time* applying a force to it, which in turn means that the *impulse* you apply will be smaller. Hence a smaller object will have less momentum than a larger object, even though their energies are the same.

6.3. $\boxed{b}$ The work done on you, which is $\int F\,dx$, equals the change in your kinetic energy. Since you end up at rest, this integral (which is the area under the F vs. x curve) takes on the given value $mv_0^2/2$ (in magnitude), independent of the specifics of the collision. To reduce injury, the goal is to have the force F be spread out evenly over the largest possible range of x, instead of being spiked at a small interval of x, because a large force will cause injury. Choice (b) correctly spreads out the force over the largest distance.

The other options are inferior because: Choice (a) involves only the front bumper (which is a small fraction of the total possible crumpling distance), so it is hardly different from a completely rigid car, (c) leads to a spike at the end of the collision, (d) leads to a spike at the beginning, and (e) actually causes the whole process to be repeated, because after you stop, the springs throw you backward. So you're even more likely to get injured. (If the spread-out force was on the threshold of seriously injuring you, making it last twice as long can't be good.) Additionally, the spring force isn't uniform, so it somewhat spikes halfway through the process.

REMARK: You can also answer this question by working with momentum instead of energy. The impulse imparted on you, which is $\int F\,dt$, equals the change in your momentum, which is mv_0 (in magnitude). However, things are a little trickier now because while the collision has a specific maximum possible *distance*, it doesn't have a maximum possible *time*. The time could be long if you suddenly decelerate to a small speed and then continue to move forward for a while with this small speed. But this sudden deceleration will injure you. So the question reduces to: If you want the maximum value of $F(t)$ to be as small as possible, what should the function $F(t)$ look like if an object takes a given distance to go from a given speed v_0 to speed zero? The answer is that you want $F(t)$ to be a constant function. As an exercise, you can think about why.

6.4. $\boxed{a}$ Although momentum is indeed conserved, it doesn't explain why the apple speeds up. There are many different possible motions consistent with momentum conservation. The other three choices do in fact explain the acceleration. In choice (b), the force directly causes the acceleration. And in choices (c) and (d), the work and the loss in potential energy both imply an increase in kinetic energy, which means an increase in speed.

6.5. $\boxed{c}$ By conservation of momentum, the sled moves westward while the ball moves eastward in the air. Also by conservation of momentum, the whole system must be at rest at the end of the process, because nothing is moving with respect to anything else, and the total momentum must be zero, as it was at the start.

6.6. $\boxed{b}$ The initial kinetic energy is $mv^2/2$. Since the masses stick together to form a blob of mass $2m$, conservation of momentum gives the final speed as $mv = (2m)v_f \implies v_f = v/2$. So the final kinetic energy is $(2m)(v/2)^2/2 = mv^2/4$. Therefore, $mv^2/2 - mv^2/4 = mv^2/4$ energy is converted to heat.

Alternatively, in the CM frame the masses head toward each other, each with speed $v/2$. The total kinetic energy of both masses in the CM frame is therefore $2 \cdot m(v/2)^2/2 = mv^2/4$. But all of this energy must end up as heat, because the resulting blob is at rest in the CM frame. And the heat is independent of the frame.

6.7. $\boxed{e}$ The kinetic energy can be written as $mv^2/2 = m(v_x^2 + v_y^2)/2$. In the given setup, the v_x components of the two velocities are brought to zero by the collision, while the v_y components are unchanged (by conservation of momentum). So the loss in kinetic energy is $2 \cdot mv_x^2/2 = 2m(v\cos\theta)^2/2 = mv^2\cos^2\theta$.

LIMITS: The $\theta \to 90°$ limit quickly eliminates choices (a), (b), and (c), because the energy loss is zero in this case, since the masses collide (or "touch" would be a better word) with zero relative velocity.

6.8. $\boxed{b}$ Momentum is conserved throughout the entire process. The initial momentum is mv, and the final momentum is $(Nm)v_f$. Setting these equal gives $v_f = v/N$.

REMARK: It is incorrect to use conservation of energy in this setup. That would give $mv^2/2 = (Nm)v_f^2/2 \implies v_f = v/\sqrt{N}$, which is incorrect (it is too large) because energy is lost to heat in each of the (completely inelastic) collisions.

6.9. $\boxed{a}$ The ping-pong ball has nearly the same speed afterward as before, because the bowling ball is much more massive and essentially acts like a brick wall. So the ping-pong ball's kinetic energy remains essentially the same. Therefore, approximately none of the energy goes into the bowling ball.

REMARKS: If you want to solve this problem quantitatively, you can use the results from Problem 6.3. In the notation of that problem, the small ping-pong ball has mass M, and the large bowling ball has mass m. Using the final speed of m given in Eq. (6.13), the final kinetic energy of the bowling ball is $(1/2)m[2MV/(M+m)]^2$. This goes to zero in the $m \gg M$ limit, because there is an extra power of m in the denominator.

This zero-energy result is a general result that holds whenever a small object collides (elastically or inelastically) with a stationary large object. The large object picks up essentially zero energy, even though it picks up nonzero momentum. In the present case of an elastic collision, if the ping-pong ball's initial momentum is p, then its final momentum is essentially $-p$. The change is $-2p$, so the bowling ball's final momentum is $2p$. Consistent with this, its energy, which can be written as $(2p)^2/2m$, is essentially zero since m is very large. If the collision is instead completely inelastic and the ping-pong ball sticks to the bowling ball, then the bowling ball's final momentum is just p, and the same result of zero energy still holds.

The word "stationary" in the first sentence of the preceding paragraph is important. If a small object collides with a *moving* large object, some energy will be transferred between the two objects. This is true because (among other reasons) if the large object is moving, the speed of the small object will be different after the collision. This is most easily seen by utilizing the CM frame.

6.10. $\boxed{e}$ In the $M \gg m$ limit, the block's final momentum equals mv in scenario A, because the marble's momentum goes from mv to zero. But the block's final momentum equals $2mv$ in scenario B, because the marble's momentum goes from $+mv$ to $-mv$. The block therefore has twice the speed in B as in A. Since the kinetic energy $mv^2/2$ is proportional to v^2, the desired ratio is $2^2 = 4$. Note that the preceding multiple-choice question (where this question was basically answered in the remark given there) tells us that the two energies of the block will be small, whatever they are.

6.11. $\boxed{b}$ Momentum is conserved in the ball-earth system during the collision (or ball-wall-earth system, if you don't want to define the wall as part of the earth). The ball and the earth have the same final speed, so the earth has virtually all of the momentum because its mass is so much larger. Choices (c) and (d) can't be correct, because heat and sound don't have units of momentum.

6.12. $\boxed{c}$ As we saw in Multiple-Choice Question 6.9, the large object (the earth here) picks up essentially no energy. And the clay ball doesn't have any energy in the end, because it is at rest. But energy *is* always conserved, so it has to go somewhere. It goes into heat – the kinetic energy of the random motion of the molecules, mostly in the clay, but some in the wall. There is certainly also some energy in sound, since you will undoubtedly hear the splat on the wall. But this energy is much smaller than the heat energy.

6.13. $\boxed{\text{Yes}}$ For example, if $m_1 \gg m_2$ then the final speed of Ball 2 is twice the initial speed of Ball 1. This is true because the relative speed of the balls is unchanged by the elastic collision; see Problem 6.2. So if v is the initial rightward speed of Ball 1, then after the

6.4. MULTIPLE-CHOICE ANSWERS

collision, Ball 2 must be moving rightward with speed v relative to Ball 1. But Ball 1 is still plowing forward with speed v relative to the ground, because it is essentially unaffected by the collision. So the speed of Ball 2 relative to the ground is $v + v = 2v$.

6.14. $\boxed{e}$ Choices (a) and (b) have zero total p_x, so they don't satisfy conservation of p_x. Choice (c) doesn't satisfy conservation of E, because the final energy is larger than the initial energy, due to the large velocity of the right mass. Choice (d) doesn't satisfy conservation of p_y (it has a nonzero final value); it also doesn't satisfy conservation of E (the final E is twice the initial). The correct answer must be (e), which does indeed satisfy conservation of p_x, p_y, and E (at least to the accuracy of how well the vectors are drawn).

REMARK: The velocities in choice (e) appear to be orthogonal. And indeed, it can be shown that if the masses are equal and if one of them is initially at rest, then if the collision is elastic, the final velocities are always orthogonal, no matter how the masses glance off each other. See the example in Section 5.7.2 of Morin (2008) for a proof of this fact.

6.15. $\boxed{d}$ This problem is most easily solved by working in the reference frame of the paddle. In this frame the (heavy) paddle is at rest and the ball comes in with speed $v + 3v = 4v$ leftward. So the ball must go out with the same speed $4v$ (now rightward), because the paddle essentially doesn't move. This is the final speed of the ball with respect to the paddle. But in the frame of the ground, the paddle is still moving to the right with speed v. So the speed of the ball with respect to the ground is $v + 4v = 5v$.

6.16. $\boxed{d}$ If the cone is sliced into horizontal rings, the lower rings have larger radii and are therefore more massive. Each ring can effectively be replaced by a point mass at its center, so the cone is equivalent to a series of point masses (lying along the axis) that are larger the lower they are. The CM is therefore less than halfway to the top.

6.17. $\boxed{b}$ The speed of the CM is $((2m)v + 0)/3m = 2v/3$, so the speed of the mass $2m$ relative to the CM is $v - 2v/3 = v/3$.

REMARK: More generally, if $2m$ is replaced by Nm, where N is a numerical factor, the answer becomes $v/(N + 1)$. This correctly goes to zero in the $N \to \infty$ limit (the CM coincides with the mass Nm as it moves along). And it correctly goes to v in the $N \to 0$ limit (the CM is located at the stationary mass m).

6.18. $\boxed{c}$ Your force is what adds momentum to the system, according to $\Delta p = F \Delta t$, or $F = dp/dt$ in derivative form. The new momentum shows up in the new mass that gets moving as it leaves the stationary heap and joins the moving straight part of the rope. If a new mass dm gets moving in a small time dt, the new momentum is $dp = (dm)v = (\lambda\, dx)v$. Therefore $F = dp/dt = \lambda v\, dx/dt = \lambda v^2$.

Equivalently, we can use Eq. (6.12). The ma term is zero because the rope isn't accelerating. So we have $F = (dm/dt)v = (\lambda\, dx/dt)v = \lambda v^2$.

REMARKS: Note that you can rule out the other answers: (a) is incorrect because the force must be nonzero since the momentum of the rope is increasing. (b) has the wrong units. And (d) and (e) depend on g, but the problem doesn't involve vertical forces.

We should emphasize that the new momentum comes from the new mass that gets moving, and *not* from any increase in speed of the straight part. Imagine painting a dot on the straight part of the rope, and consider the part of the rope that lies between the dot and your hand. This part always moves with the same speed v, and its length (and hence mass) doesn't change, so its momentum is constant. The total force on this part must therefore be zero. And indeed, the rightward force from your hand is canceled by the leftward tension from the part of the rope just to the left of the dot. If we instead look at the rope to the *left* of the dot (which includes the heap plus a straight part), then the only horizontal force acting on it is the tension from the part of the rope just to the right of the dot. This force is what produces the new momentum of the rope, as sections from the stationary heap join the moving part.

6.19. **Yes** In addition to balancing the weight of the straight part of the rope that hangs at rest, you must also provide the force needed to change the momentum of the pieces of rope as they make the transition from the moving heap to the stationary straight part. (From the reasoning in the preceding question, this force happens to be λv^2, where λ is the mass density. But the exact value isn't critical here. All we need to know is that the force isn't zero.) So at least right before the rope is straightened out (when the weight of the stationary hanging part is nearly mg), your force will need to exceed mg. See Problem 6.24 for a quantitative treatment.

6.20. **e** The mass of the dustpan plus dust inside at a later time is $M + \lambda x$, so the momentum is $p = (M + \lambda x)\dot{x}$. The necessary force is therefore (using the product rule)

$$F = \frac{dp}{dt} = (M + \lambda x)\ddot{x} + \lambda \dot{x}^2 \equiv (M + \lambda x)a + \lambda v^2. \tag{6.15}$$

Physically, the first term in this result is the force needed to accelerate the dustpan plus the dust already inside it (this term depends on a, and not on v), while the second term is the force needed to get the new bits of dust moving (this term depends on v, and not on a). The applied force F makes the momentum of the system increase, and these are the two ways it increases.

6.5 Problem solutions

6.1. Conservation of momentum

The case of two particles was treated in Eq. (6.4). Consider now the case of three particles. If $\mathbf{F}_{ij}$ represents the force on particle i due to particle j, then the rate of change of the total momentum of the system equals

$$\begin{aligned}\frac{d(\mathbf{p}_1 + \mathbf{p}_2 + \mathbf{p}_3)}{dt} &= \frac{d\mathbf{p}_1}{dt} + \frac{d\mathbf{p}_2}{dt} + \frac{d\mathbf{p}_3}{dt} \\ &= \mathbf{F}_1 + \mathbf{F}_2 + \mathbf{F}_3 \\ &= (\cancel{\mathbf{F}_{12}} + \mathbf{F}_{13}) + (\cancel{\mathbf{F}_{21}} + \cancel{\mathbf{F}_{23}}) + (\mathbf{F}_{31} + \cancel{\mathbf{F}_{32}}) \\ &= 0. \end{aligned} \tag{6.16}$$

We have used the fact that Newton's third law, $\mathbf{F}_{ij} = -\mathbf{F}_{ji}$, tells us that the forces cancel in pairs, as indicated. This reasoning clearly extends to a general number N of particles. For every term $\mathbf{F}_{ij}$ in the generalization of the third line above, there will also be a term $\mathbf{F}_{ji}$. And the sum of these two terms is zero, by Newton's third law.

6.2. Same relative speed

Energy is conserved because we are assuming that the collision is elastic. With the notation shown in Fig. 6.20 (where some of the v's may be negative), the conservation of E and p equations are

$$\frac{1}{2}m_1 v_1^2 + \frac{1}{2}m_2 v_2^2 = \frac{1}{2}m_1 v_1'^2 + \frac{1}{2}m_2 v_2'^2,$$
$$m_1 v_1 + m_2 v_2 = m_1 v_1' + m_2 v_2'. \tag{6.17}$$

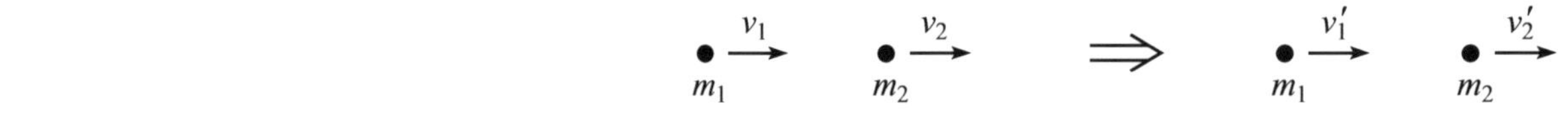

Figure 6.20

6.5. PROBLEM SOLUTIONS

If we put all the "1" terms on the left and all the "2" terms on the right, these two equations become

$$m_1(v_1^2 - v_1'^2) = m_2(v_2'^2 - v_2^2),$$
$$m_1(v_1 - v_1') = m_2(v_2' - v_2). \tag{6.18}$$

Dividing the first equation by the second gives

$$v_1 + v_1' = v_2' + v_2 \implies v_1' - v_2' = -(v_1 - v_2), \tag{6.19}$$

which is the desired result that the final relative velocity between the two masses equals the negative of the initial relative velocity.

REMARKS: In the CM frame, this result is clear. The masses simply head out with the same speeds they came in, because this satisfies conservation of both p (which is zero since we're dealing with the CM frame) and E. Therefore, because both velocities simply change sign, Eq. (6.19) is certainly true. And since this result holds in one frame, it holds in all others, because it involves only *differences* in velocities. Differences are frame independent, because both velocities shift by the same amount when going from one frame to another.

The linear equation in Eq. (6.19) can be combined with the linear conservation-of-p equation in Eq. (6.17) to solve for the final velocities, v_1' and v_2'. There will be only one solution to this system of linear equations. So you might wonder how we ended up with only one solution for v_1' and v_2', given that we started with a quadratic equation in Eq. (6.17). Quadratic equations should have two solutions. The explanation is that we lost the other solution when we divided the equations in Eq. (6.18). Another solution is clearly $v_1' = v_1$ and $v_2' = v_2$, because that makes all the terms in Eq. (6.18) be zero (in which case we divided by zero). This solution is simply the one involving the initial conditions, which are what we would end up with if the masses missed each other. The initial conditions certainly satisfy conservation of E and p with the initial conditions. A fine tautology, indeed.

6.3. 1-D collision

FIRST SOLUTION: This solution involves some brute force; the other two solutions below will be shorter. Let's switch notation and have V' and v' be the final velocities. Since m is initially at rest, the conservation of E and p equations are

$$\frac{1}{2}MV^2 = \frac{1}{2}MV'^2 + \frac{1}{2}mv'^2 \quad \text{and} \quad MV = MV' + mv'. \tag{6.20}$$

Solving for v' in the p equation and substituting the result into the E equation (and then multiplying through by m/M) gives

$$\frac{1}{2}MV^2 = \frac{1}{2}MV'^2 + \frac{1}{2}m\left(\frac{M(V-V')}{m}\right)^2$$
$$\implies 0 = (M+m)V'^2 - 2MVV' + (M-m)V^2. \tag{6.21}$$

This quadratic equation for V' can be solved with the quadratic formula, or you can just note that it factors nicely:

$$0 = \bigl((M+m)V' - (M-m)V\bigr)(V' - V). \tag{6.22}$$

In retrospect, we know that in problems like this, $V' = V$ and $v' = v$ must always be a solution, because the initial conditions certainly satisfy conservation of E and p with the initial conditions. So that tells us that $(V - V')$ must be a factor of the above quadratic equation. However, we are concerned with the other (nontrivial) solution, $V' = (M-m)V/(M+m)$. Plugging this into the conservation-of-p equation then gives $v' = 2MV/(M+m)$, as desired.

LIMITS: If $M \ll m$, the final velocities of M and m are $-V$ and 0; M essentially bounces backward off a brick wall. If $M = m$, the final velocities are 0 and V; m picks up whatever velocity M had, and M ends up at rest (an outcome familiar to pool players). If $M \gg m$, the final velocities are V and $2V$; M plows forward with the same velocity, as expected, and m picks up twice this velocity (not obvious in the lab frame, but consistent with the relative-speed result from Problem 6.2). These limits were the subject of Problem 1.7.

SECOND SOLUTION: Instead of using both of the E and p conservation equations, we can use the p equation along with the relative-velocity relation from Problem 6.2, which says that the final relative velocity equals the negative of the initial relative velocity. That is, $v' - V' = -(0 - V)$. So $v' = V + V'$. Substituting this into the conservation-of-p equation in Eq. (6.20) gives

$$MV = MV' + m(V + V') \implies V' = \frac{(M - m)V}{(M + m)}. \tag{6.23}$$

We then have $v' = V + V' = 2MV/(M + m)$, as desired. This solution was quicker than the first one above, because it involved two linear equations instead of one linear and one quadratic.

THIRD SOLUTION: We can also solve this problem by working in the CM frame, following the strategy outlined on page 145.

1. Switch to the CM frame: The CM moves with velocity $MV/(M + m)$ with respect to the lab frame. So the velocities of M and m with respect to the CM are, respectively, $V - MV/(M + m) = mV/(M + m)$, and $0 - MV/(M + m)$.

2. Do the collision in the CM frame: In the CM frame, the velocities simply reverse themselves during the collision (see the discussion at the start of the "Collisions in the CM frame" section on page 144). So the final velocities of M and m in the CM frame are $-mV/(M + m)$ and $MV/(M + m)$.

3. Switch back to the lab frame: To get back to the original lab frame, we must add on the velocity of the CM, namely $MV/(M + m)$, to each of the velocities in the CM frame. The final velocities of M and m in the lab frame are therefore $V' = (M - m)V/(M + m)$ and $v' = 2MV/(M + m)$, as desired.

6.4. Distances to the CM

Assume without loss of generality that $x_2 > x_1$. Then the distance from the CM to x_1 is

$$d_1 = x_{\text{CM}} - x_1 = \frac{m_1 x_1 + m_2 x_2}{m_1 + m_2} - x_1 = \frac{m_2(x_2 - x_1)}{m_1 + m_2}. \tag{6.24}$$

And the distance from the CM to x_2 is

$$d_2 = x_2 - x_{\text{CM}} = x_2 - \frac{m_1 x_1 + m_2 x_2}{m_1 + m_2} = \frac{m_1(x_2 - x_1)}{m_1 + m_2}. \tag{6.25}$$

Therefore, $d_1/d_2 = m_2/m_1$, as desired. If one mass is ten times the other, then the CM is ten times closer to the larger mass. That is, it is located $1/11$ of the way from the larger mass to the smaller one.

6.5. Equivalent subparts

The location of the CM of the entire object is given by Eq. (6.8), where the integral runs over the whole object. We can break the integral up into two integrals over the subparts S_1 and S_2. To demonstrate the desired result, we will judiciously multiply each integral by 1, in the form of a mass divided by itself. This gives

$$\begin{aligned} \mathbf{r}_{\text{CM}} &= \frac{1}{M} \int \mathbf{r}\, dm = \frac{1}{M} \int_{S_1} \mathbf{r}\, dm + \frac{1}{M} \int_{S_2} \mathbf{r}\, dm \\ &= \frac{1}{M} \cdot m_1 \frac{\int_{S_1} \mathbf{r}\, dm}{m_1} + \frac{1}{M} \cdot m_2 \frac{\int_{S_2} \mathbf{r}\, dm}{m_2} \\ &= \frac{1}{M}(m_1 \mathbf{r}_{\text{CM},1} + m_2 \mathbf{r}_{\text{CM},2}), \end{aligned} \tag{6.26}$$

6.5. PROBLEM SOLUTIONS

where we have obtained the last line by using Eq. (6.8) for each subpart. By comparing this result with the CM for a discrete system (see Eq. (6.7)), we see that the given object can be treated like two point masses m_1 and m_2 located at the CM's of the two subparts. By repeated application of this result, we can subdivide the given object into an arbitrary number of subparts.

6.6. Collision in the CM frame

In the CM frame, the initial momentum (and hence the final momentum) is zero; see Footnote 2. The final momenta of the two particles must therefore be equal and opposite. Hence the particles must move in opposite directions.

The final speeds are the same as the initial speeds, for the following reason. Since the momenta are equal (and opposite), we know that the speeds are inversely proportional to the masses, as they are initially. So the final speeds must be in the same ratio as the initial speeds. This means that the only way they can possibly differ from the initial speeds is if they are both scaled up or scaled down by the same factor. But since kinetic energy is proportional to v^2, any such scaling would have the effect of scaling the total final kinetic energy up or down by the square of this factor, causing energy to not be conserved. The scaling (or lack thereof) factor must therefore simply be 1. That is, the final speeds must be the same as the initial speeds.

REMARKS: All of the above reasoning is valid in any dimension. But 2-D and 3-D, the orientation of the line containing the final velocities requires additional information about how exactly the particles collide.

Note that conservation of p and E are both required in the reasoning. Conservation of p alone would allow for a scaling of the speeds (which is what happens in an inelastic collision). And conservation of E alone would allow for increasing one speed and decreasing the other, in such a way that the total kinetic energy stays the same. Additionally, the particles could head off in arbitrary directions.

6.7. Hemispherical-shell CM

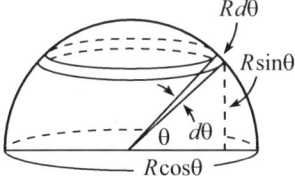

Figure 6.21

First note that the x coordinate of the CM in Fig. 6.21 is zero, by symmetry. Let the surface mass density be σ (kg/m^2). Consider a circular strip located at angle θ above the horizontal, subtending an angle $d\theta$, as shown. The radius of this circle is $R\cos\theta$ and the width of the strip is $R\,d\theta$, so the mass of the strip is

$$dm = \sigma(\text{area}) = \sigma(\text{length})(\text{width})$$
$$= \sigma(2\pi R\cos\theta)(R\,d\theta) = 2\pi R^2 \sigma \cos\theta\,d\theta. \quad (6.27)$$

All points on the strip have a y value of $R\sin\theta$ (at least in the limit where $d\theta$ is infinitesimal), so the CM of the strip is located at the point $(0, R\sin\theta)$. From Problem 6.5, we can replace the circular strip with a point mass at $(0, R\sin\theta)$, with a mass given by the above dm. The hemisphere is therefore equivalent to a string of point masses located on the y axis. The total mass of the hemisphere is $\sigma(\text{area}) = \sigma(2\pi R^2)$, so Eq. (6.8) gives the height of the CM as

$$y_{\text{CM}} = \frac{1}{M}\int y\,dm = \frac{1}{2\pi R^2 \sigma}\int_0^{\pi/2}(R\sin\theta)(2\pi R^2 \sigma \cos\theta\,d\theta)$$
$$= R\int_0^{\pi/2}\sin\theta\cos\theta\,d\theta = R\frac{\sin^2\theta}{2}\bigg|_0^{\pi/2} = \frac{R}{2}. \quad (6.28)$$

REMARKS: The fact that the answer comes out to be so simple implies that there is probably a quicker way to figure it out. And indeed, consider slicing the hemisphere into circular strips with equal *heights* dy (instead of equal angular spans $d\theta$, although technically we never said anything about equal $d\theta$'s when doing the above integral). We claim that all of these strips have the *same mass*, which means that we effectively have equal masses evenly distributed on the y axis, which implies that the CM is located halfway up at $y = R/2$. The equality of the strips' masses follows from the facts that the radius of each strip is $R\cos\theta$, while the width (the slant height along the surface of the hemisphere) equals $dy/\cos\theta$, as you can verify. These factors of $\cos\theta$ cancel when calculating the area, so all strips spanning the same height dy have the same area, and hence the same mass.

As an exercise, you can find the height of the CM for a *solid* hemisphere, which involves slicing the hemisphere into disks instead of circular strips. By modifying the reasoning in the preceding paragraph, you should convince yourself that the height of the CM must be less than $R/2$. See the remark in the solution to Problem 7.12 for further discussion of the difference between 2-D and 3-D objects.

6.8. Atwood's machine

(a) The speed of each mass after it has moved a distance d is $v = \sqrt{2ad}$. (This kinematic result can be derived in many ways. For example, the time it takes an object with acceleration a to move a distance d is given by $d = at^2/2 \implies t = \sqrt{2d/a}$. The speed at this time is then $v = at = \sqrt{2ad}$.) The total kinetic energy of the masses is therefore

$$K = \frac{1}{2}m_1 v^2 + \frac{1}{2}m_2 v^2 = \frac{1}{2}(m_1 + m_2)(2ad)$$
$$= \frac{1}{2}(m_1 + m_2)\left(2 \cdot g\frac{m_2 - m_1}{m_2 + m_1} \cdot d\right) = (m_2 - m_1)gd. \quad (6.29)$$

The potential energy of the masses (relative to their initial positions) is

$$U = m_1 g y_1 + m_2 g y_2 = m_1 g d + m_2 g(-d) = (m_1 - m_2)gd. \quad (6.30)$$

Therefore, $K + U = 0$, so energy is indeed conserved.

(b) After a time t, the total momentum of the masses is (with upward taken to be positive)

$$P_{\text{total}} = m_1 v_1 + m_2 v_2 = m_1 at + m_2(-at) = (m_1 - m_2)at$$
$$= (m_1 - m_2)\left(g\frac{m_2 - m_1}{m_2 + m_1}\right)t = -gt\frac{(m_2 - m_1)^2}{m_2 + m_1}. \quad (6.31)$$

Let the system be defined to be the two masses and the pulley. The external forces acting on the system are the two weights *and* the tension T in the upper string. The tension in the string connecting the masses is an internal force, so we can ignore it. (If you instead define the system to be just the two masses, then the upward tensions $T/2$ acting on each mass are now relevant external forces, while the upper string is irrelevant. So the total force is the same.) The total external force is therefore

$$F_{\text{total}} = -m_1 g - m_2 g + T = -(m_1 + m_2)g + g\frac{4m_1 m_2}{m_1 + m_2}$$
$$= g\frac{-(m_1 + m_2)^2 + 4m_1 m_2}{m_1 + m_2} = -g\frac{(m_2 - m_1)^2}{m_2 + m_1}. \quad (6.32)$$

Comparing this with Eq. (6.31), we see that P_{total} is indeed equal to $F_{\text{total}}t$, as desired.

6.9. Rising and colliding

Conservation of energy gives the kinetic energy of the bottom mass right before the collision as (taking zero height for the gravitational U to be at that start)

$$K^i + U_s^i + U_g^i = K^f + U_s^f + U_g^f$$
$$\implies 0 + \frac{1}{2}kd^2 + 0 = \frac{1}{2}mv^2 + \frac{1}{2}k\left(\frac{d}{2}\right)^2 + mg\frac{d}{2}$$
$$\implies \frac{3}{8}kd^2 - mg\frac{d}{2} = \frac{1}{2}mv^2. \quad (6.33)$$

6.5. PROBLEM SOLUTIONS

During the collision, half of this kinetic energy is lost to heat, because conservation of momentum gives the speed of the resulting mass $2m$ as $v/2$, so the resulting kinetic energy is $(1/2)(2m)(v/2)^2 = mv^2/4$. Therefore, from Eq. (6.33) the kinetic energy of the mass $2m$ right after the collision is $3kd^2/16 - mgd/4$. Applying conservation of energy from this moment (which we will now take to be at zero height for the gravitational U) up to the moment when the mass $2m$ reaches its maximum height at the relaxed length of the spring, we obtain (using the fact that there is no kinetic energy or spring potential energy at the end)

$$K^i + U_s^i + U_g^i = K^f + U_s^f + U_g^f$$

$$\implies \left(\frac{3}{16}kd^2 - mg\frac{d}{4}\right) + \frac{1}{2}k\left(\frac{d}{2}\right)^2 + 0 = 0 + 0 + (2m)g\frac{d}{2}$$

$$\implies \frac{5}{16}kd^2 = \frac{5mgd}{4}$$

$$\implies k = \frac{4mg}{d}. \tag{6.34}$$

REMARK: The following continuity argument shows that there must exist a value of k that makes the masses stop at the relaxed length of the spring. If k is small, then the masses won't make it back up to the dotted line. (If k is smaller than mg/d, then the bottom mass will actually fall when released.) And if k is very large, then we can essentially ignore gravity, so we can imagine that the setup lies on a horizontal table, in which case the mass $2m$ certainly travels beyond the relaxed position of the spring. The spring force must reverse direction if the mass is to slow down and stop.

6.10. **Maximum compression**

FIRST SOLUTION: At maximum compression the speeds of the blocks are equal; that is, the relative speed is zero. This is true because if the relative speed weren't zero, then the distance between the blocks would either be decreasing to a smaller value, or increasing from a smaller value, contradicting the assumption of maximum compression. Let the common speed be u. Then conservation of momentum gives

$$mv = (2m)u \implies u = \frac{v}{2}. \tag{6.35}$$

And if the maximum compression is x, then conservation of energy gives

$$\frac{1}{2}mv^2 = \frac{1}{2}(2m)u^2 + \frac{1}{2}kx^2. \tag{6.36}$$

Substituting $u = v/2$ into this equation gives the maximum compression as

$$\frac{1}{2}mv^2 = \frac{1}{2}(2m)\left(\frac{v}{2}\right)^2 + \frac{1}{2}kx^2 \implies \frac{1}{4}mv^2 = \frac{1}{2}kx^2 \implies x = v\sqrt{\frac{m}{2k}}. \tag{6.37}$$

At maximum compression, we see that half of the initial kinetic energy $mv^2/2$ remains kinetic energy, and half ends up as potential energy of the spring.

LIMITS: x increases with m and decreases with k, as expected.

REMARKS: After the spring stretches back to its relaxed length and loses contact with the left block, what do the motions of the blocks look like? You can show by writing down the initial/final conservation of E and p equations (in which the spring doesn't come into play, because it is uncompressed initially and finally) that the left block ends up at rest and the right block moves at speed v. So the velocities of the blocks are simply interchanged from their initial values. In retrospect, this must be what happens, because this scenario certainly satisfies conservation of E and p with the initial motion.

The above collision involving two blocks and a spring is a good model for a completely elastic collision between two balls. If a rubber ball collides with another one, they will both compress and then uncompress during the collision, just like the spring. The ball that was initially moving will now be at rest (in 1-D).

In a completely *inelastic* collision where the balls stick together, the energy that goes into the spring eventually ends up as heat (which is the internal kinetic energy of the random motion of the molecules in the balls). You can model the collision roughly as having the spring vibrate back and forth, with an amplitude that gradually decreases (eventually reaching zero, with the balls being at rest with respect to each other) due to some kind of friction force. This friction generates the heat.

SECOND SOLUTION: Consider the collision in the CM frame. In this frame the blocks move toward each other with equal speeds of $v/2$. At maximum compression the blocks are both instantaneously at rest, so all of the initial kinetic energy must end up as potential energy of the spring. That is,

$$2 \cdot \frac{1}{2} m \left(\frac{v}{2}\right)^2 = \frac{1}{2} k x^2 \implies x = v \sqrt{\frac{m}{2k}}, \tag{6.38}$$

in agreement with the first solution.

6.11. Collision and a spring

(a) Let x be the compression distance of the spring when m comes to rest. This is the maximum compression (assuming that $2m$ doesn't move), so this is the moment when the spring exerts the maximum force on the $2m$ block. If the $2m$ block is to remain at rest at all times, then we need kx to be less than or equal to the maximum friction force on $2m$, which is $\mu(2mg)$. So $kx \leq 2\mu mg \implies x \leq 2\mu mg/k$. This is the condition on x. We must now convert it to the desired condition on v.

We can relate x and v by conservation of energy. The m block ends up instantaneously at rest, so the initial kinetic energy $mv^2/2$ goes into the spring's potential energy $kx^2/2$ plus heat, which equals the magnitude of the work done by friction, $W_f = F_f x = (\mu mg)x$. So we have

$$\frac{mv^2}{2} = \frac{kx^2}{2} + (\mu mg)x. \tag{6.39}$$

Equivalently, if we multiply this equation through by -1, it is the statement that the change in kinetic energy of the m block (which is negative) equals the total work done by the spring and friction (both of which are negative).

There is no need to solve the quadratic equation in Eq. (6.39). Instead, we can simply take the $x \leq 2\mu mg/k$ condition from above and plug it into Eq. (6.39). This gives the desired condition on v:

$$\frac{mv^2}{2} \leq \frac{k}{2}\left(\frac{2\mu mg}{k}\right)^2 + (\mu mg)\left(\frac{2\mu mg}{k}\right)$$

$$\implies \frac{mv^2}{2} \leq \frac{4\mu^2 m^2 g^2}{k} \implies v \leq \sqrt{\frac{8\mu^2 m g^2}{k}}. \tag{6.40}$$

You can check that this result has the correct units.

LIMITS: If $\mu = 0$ then the maximum v equals 0; that is, the $2m$ mass always slips, which is correct. Large μ, m, or g implies large v. These all make sense, although the m case requires a little thought because there are partially canceling effects; you can see the various effects of m in the first line of Eq. (6.40). And large k implies small v; the mass m stops very quickly, so the spring force would be very large if v weren't very small.

(b) If v takes on the value we just found, then the maximum compression is $x = 2\mu mg/k$. When m bounces back, the spring's potential energy goes into kinetic energy plus

6.5. PROBLEM SOLUTIONS

heat due to friction. So if u is the speed at the point when the spring reaches its relaxed length, we have

$$\frac{kx^2}{2} = \frac{mu^2}{2} + (\mu mg)x$$
$$\implies \frac{k}{2}\left(\frac{2\mu mg}{k}\right)^2 = \frac{mu^2}{2} + (\mu mg)\left(\frac{2\mu mg}{k}\right), \tag{6.41}$$

which gives $u = 0$. So m barely makes it back to the position where it first came in contact with the spring.

REMARK: If the mass of the right block is larger than $2m$, and if m comes in with the speed that yields the maximum compression without the right block moving, then as m bounces back, its speed as it passes through the relaxed length of the spring will be positive. Conversely, if the mass of the right block is smaller than $2m$, then in the analogous scenario, m won't reach the relaxed length of the spring as it bounces back. These results quickly follow by replacing the two $\mu(2m)g$ terms in Eq. (6.41) with the more general $\mu(nm)g$ term, where n is a numerical factor.

6.12. Colliding balls

(a) It takes the first moving ball a time of ℓ/v to reach the stationary ball and produce the blob of mass $2m$. After this collision, conservation of momentum gives the speed of the $2m$ blob as

$$mv + 0 = (2m)v_f \implies v_f = v/2. \tag{6.42}$$

How long does it take the second moving ball (which is moving at speed v) to catch up with the $2m$ blob (which is moving at speed $v/2$)? It must close the gap of ℓ between them at a relative speed of $v - v/2 = v/2$, so the time is $\ell/(v/2) = 2\ell/v$. After it collides, conservation of momentum gives the speed of the resulting $3m$ blob as

$$mv + 2m(v/2) = (3m)v_f \implies v_f = 2v/3. \tag{6.43}$$

How long does it take the third moving ball (which is moving at speed v) to catch up with the $3m$ blob (which is moving at speed $2v/3$)? It must close the gap of ℓ between them at a relative speed of $v - 2v/3 = v/3$, so the time is $\ell/(v/3) = 3\ell/v$. From conservation of momentum, you can show that the speed of the resulting $4m$ blob is $3v/4$.

Continuing in this manner, it takes the fourth moving ball a time of $4\ell/v$ to catch up with the blob in front of it, and the speed of the resulting $5m$ blob is $4v/5$.

In general, the blob of mass nm moves at speed $(n-1)v/n$. (In short, this is because it has mass nm and a total momentum of $(n-1)mv$ from the $n-1$ balls that were originally moving.) And it takes the nth moving ball a time of $n\ell/v$ to catch up with the nm blob, because the relative speed is $v - (n-1)v/n = v/n$. The total time for the entire process is therefore

$$\frac{\ell}{v}\Big(1 + 2 + 3 + \cdots + (N-1) + N\Big) = \frac{\ell}{v}\left(\frac{N(N+1)}{2}\right), \tag{6.44}$$

where we have used the formula for the sum of the first N integers.

LIMITS: If $N = 1$, the time equals ℓ/v, as expected. For large N, the time goes like N^2, which isn't intuitively obvious.

REMARK: As an exercise, you can calculate the total distance the blob moves, between the first and last collisions. The quick way to do this is to subtract $N\ell$ from the total distance the leftmost ball moves (which is just v times the total time we found above). But you should verify that you obtain the same result by adding up the distances the blob moves between successive collisions.

(b) In the frame in which the N balls are initially at rest, one ball moves toward N stationary balls that lie in a line. From the instant indicated in Fig. 6.13, it takes a time of ℓ/v for the moving ball to reach the first stationary ball. After this collision, conservation of momentum gives the speed of the $2m$ blob as $v/2$. This blob therefore takes a time of $\ell/(v/2) = 2\ell/v$ to reach the second stationary ball.

After this collision, conservation of momentum gives the speed of the $3m$ blob as $v/3$. (This follows from $(2m)(v/2) + 0 = (3m)v_f$, or from simply noting that the total momentum is still just the mv from the initially moving ball, but the mass is now $3m$.) This blob therefore takes a time of $\ell/(v/3) = 3\ell/v$ to reach the third stationary ball.

Continuing in this manner, we see that after the collision with the $(n-1)$th stationary ball, the speed of the nm blob is v/n, and the time to reach the nth stationary ball is $n\ell/v$. The result holds for all n from 1 to N, so the total time is

$$\frac{\ell}{v}\Big(1 + 2 + 3 + \cdots + (N-1) + N\Big) = \frac{\ell}{v}\left(\frac{N(N+1)}{2}\right), \tag{6.45}$$

in agreement with the result in part (a).

6.13. Block and balls

(a) In the frame of the heavy block, the light ball comes in at speed v and bounces out at (essentially) speed v. The final relative speed is therefore v. But the relative speed doesn't depend on the frame, so it is also v in the original lab frame. And since the block keeps moving forward at (essentially) speed v, the final speed of the ball after the collision must be $2v$ in the lab frame. Conservation of momentum in the lab frame then gives the final speed v' of the block as

$$Mv + 0 = Mv' + m(2v) \implies v' = v - \frac{2mv}{M}. \tag{6.46}$$

The block's speed therefore decreases by $2mv/M$. You can also find v' by making a Taylor-series approximation to the V_M in Eq. (6.13).

(b) In a small time dt, the block effectively hits a "ball" (which is actually more like a little tube) with mass $dm = \lambda\, dx$, where $dx = v\, dt$ is the distance the block travels in time dt. So $dm = \lambda v\, dt$. Using this as the m in part (a), we see that the change in the block's speed is

$$dv = -\frac{2(dm)v}{M} = -\frac{2(\lambda v\, dt)v}{M} = -\frac{2\lambda v^2\, dt}{M}. \tag{6.47}$$

REMARK: Note the v^2 dependence here. One power of v comes from the fact that if we make v larger, then a given ball bounces off with a larger speed, which means that it has a larger increase in momentum, which in turn means that the block has a larger decrease in momentum, and hence speed. The other power of v comes from the fact that the faster the block moves, the more balls it hits in a given time. This v^2 dependence is a general feature of drag forces where the actual motion of mass (shoving things out of the way) is the dominant effect. Even if the balls stuck to the block, we would still obtain the v^2 factor; we just wouldn't have the factor of 2 in Eq. (6.47).

(c) Equation (6.47) is a differential equation involving v and t. Separating variables and integrating yields (we won't bother putting primes on the integration variables)

$$\int_{V_0}^{v} \frac{dv}{v^2} = -\int_0^t \frac{2\lambda\, dt}{M} \implies -\frac{1}{v}\bigg|_{V_0}^{v} = -\frac{2\lambda t}{M}$$

$$\implies -\frac{1}{v} + \frac{1}{V_0} = -\frac{2\lambda t}{M} \implies v(t) = \frac{1}{\dfrac{1}{V_0} + \dfrac{2\lambda t}{M}}. \tag{6.48}$$

6.5. PROBLEM SOLUTIONS

LIMITS: If $\lambda \to 0$ or $M \to \infty$ then $v(t) \to V_0$, as expected. If $t \to 0$ then $v \to V_0$, which is correct. And if $t \to \infty$ then $v \to M/2\lambda t$, which correctly goes to zero and which interestingly is independent of V_0. The total distance traveled by the block equals $\int v\, dt$, which diverges like $\ln(t)$ as $t \to \infty$.

6.14. Maximum final speed

(a) Using the second of the equations given in Eq. (6.14), the collision between M and x gives x a speed of $v_x = 2MV/(M+x)$. We then need to use this speed as the initial speed for the collision between x and m. This yields a speed of m equal to

$$v_m = \frac{2xv_x}{x+m} = \left(\frac{2x}{x+m}\right)\left(\frac{2MV}{M+x}\right). \tag{6.49}$$

Our goal is to maximize this, which means that we want to maximize the function $f(x) = x/(x+m)(x+M)$. Setting the derivative equal to zero (and ignoring the denominator of the result) gives

$$0 = (x+m)(x+M)(1) - x(2x+m+M)$$
$$\implies 0 = mM - x^2 \implies x = \sqrt{Mm}, \tag{6.50}$$

which correctly has units of mass. We see that the optimal value of x is the geometric mean of the original masses, which is about as nice a result as we could hope for. The three masses therefore form a geometric progression. From Eq. (6.49) the maximum final speed of m equals $4MV/(\sqrt{m}+\sqrt{M})^2$.

LIMITS: If $m = M$ then $x = m = M$. This makes sense, because with three equal masses, M and x end up at rest. So m has all of the energy. And we can't do any better than that, by conservation of energy.

(b) With two masses inserted between M and m, we claim that the largest speed of m is obtained if the four masses form a geometric progression, that is, if the ratio of any two successive masses is the same. In other words, the masses should take the form of M, $M^{2/3}m^{1/3}$, $M^{1/3}m^{2/3}$, m, with the common ratio here being $(m/M)^{1/3}$. This claim can be proved by contradiction, as follows.

Assume that in the optimal case, the first three masses, M, x, and y, do *not* form a geometric progression. If this is the case, then from part (a), we can increase the speed of y (which would then increase the speed of m) by making x be the geometric mean of M and y. This contradicts our initial assumption of optimal-ness, so it must be the case that M, x, and y form a geometric progression. Likewise, x, y, and m must form a geometric progression, because otherwise we could increase the speed of m by making y be the geometric mean of x and m. Since these two progressions have the x/y ratio as overlap, all four masses must form a geometric progression, as we wanted to show. The values of x and y are therefore $M^{2/3}m^{1/3}$ and $M^{1/3}m^{2/3}$.

If we have a general number of masses, say 10, then they must all form a geometric progression. This follows from the fact that if this *weren't* the case, then we could take a group of three of the masses that aren't in geometric progression and increase the speed of the third mass by making the middle one be the geometric mean of the other two. This would then increase the speeds of all the masses to the right of this group, in particular the last one, m. This means that we didn't actually have the optimal scenario in the first place. In short, any sequence that isn't in geometric progression can be improved, so the only possibility for optimal-ness is the sequence where the masses are all in geometric progression.

REMARK: It turns out that if $M \gg m$, and if there is a very large number of masses between M and m (all in geometric progression), then essentially *all* of the *kinetic energy* ends up in m (so the other masses have essentially none), but essentially *none* of the *momentum* ends up in m (so the other masses have essentially all of it). In the more general case where M isn't much greater than m, the former of these statements is still true, but the latter isn't.

The basic reason for these divisions of the energy and momentum is that the final speeds of all the masses except the last one are small, but not *too* small. More precisely, on one hand the speeds are small enough so that when their *squares* are multiplied by the masses to obtain the kinetic energy, the smallness of the squares wins out over the fact that there is a large number of masses. So the energy ends up being essentially zero, which means that m must have essentially all of the energy. But on the other hand the speeds are large enough so that when their *first powers* are multiplied by the masses to obtain the momentum, the result isn't zero. And in fact it equals the initial momentum.

If you want to verify the above claims rigorously, the calculations get a bit messy. But the basic strategy is the following. Let the ratio of the masses be r, where r is very close to 1. Then up to a factor of M, the masses are 1, r, r^2, r^3, etc. You can use the expressions for the velocities given in Eq. (6.14) to show that the final speeds of all the masses except the last one are (up to a factor of V)

$$\frac{1-r}{1+r}, \quad \left(\frac{2}{1+r}\right)\frac{1-r}{1+r}, \quad \left(\frac{2}{1+r}\right)^2\frac{1-r}{1+r}, \quad \left(\frac{2}{1+r}\right)^3\frac{1-r}{1+r}, \quad \ldots. \quad (6.51)$$

And the final speed of the last mass is $(2/(1+r))^{N-1}$ if there are N masses. You can then explicitly calculate all the energies and momenta and verify the above claims. In some of the calculations it will be advantageous to set $r \equiv 1 - \epsilon$, where ϵ is very small. If you (quite reasonably) don't want to go through all the algebra, you are encouraged to at least check that things work out numerically. For example, if $m/M = 10^{-4}$ and $N = 10^4$ (which implies that $r = (10^{-4})^{1/9999} = 0.99908$), then you will find that the last mass has essentially all (99.8%, don't forget to square the v) of the kinetic energy and essentially none (1%) of the momentum.

One thing that is quick enough to do here is to verify that the above claims are at least consistent. That is, we can show that if $M \gg m$ and *if* all of the initial energy goes into m, then m has (essentially) no momentum. This can be done as follows. Let the ratio of the masses be $M/m = R \gg 1$. Then given the assumption of equal energies, the final speed of m must be $\sqrt{R}$ times the initial speed of M (because the kinetic energy is proportional to mv^2). The final momentum of m is then $mv = (M/R)(\sqrt{R}\,V) = MV/\sqrt{R}$, which goes to zero for large R.

Figure 6.22

6.15. Throwing a block in pieces

Let's first solve the general case where you throw a mass m_1, and where the mass of you plus the cart plus any other mass you are holding is m_2. Let the final speeds *with respect to the ground* be v_1 and v_2, as shown in Fig. 6.22.

The initial momentum is zero, so conservation of momentum gives $m_2 v_2 - m_1 v_1 = 0$. And we are told that the final relative speed is $v_1 + v_2 = v_0$. Solving this system of two equations and two unknowns (v_1 and v_2) quickly gives your final speed as $v_2 = m_1 v_0/(m_1 + m_2)$. Additionally, the speed of m_1 is $v_1 = m_2 v_0/(m_1 + m_2)$, although we won't need this. These two speeds correctly add up to v_0 and are inversely proportional to the masses.

In words: your final speed v_2 equals v_0 times the mass of the piece you threw, divided by the *total* mass of you plus what you threw. We can write this as

$$v_{\text{you}} \equiv v_2 = v_0 \frac{m_{\text{throw}}}{M_{\text{total}}}. \quad (6.52)$$

This correctly equals zero when $m_{\text{throw}} = 0$ and correctly equals v_0 when $m_{\text{throw}} = M_{\text{total}}$; if an ant "throws" a rock with speed v_0, the ant is really just propelling itself backward with speed v_0. Eq. (6.52) allows us to quickly solve the various parts of this problem.

(a) In this case we have $m_{\text{throw}} = m$, and $M_{\text{total}} = 2m$, so your final speed is $v_0/2$.

(b) The first stage involves $m_{\text{throw}} = m/2$ and $M_{\text{total}} = 2m$. So you gain a speed of $v_0/4$. The second stage involves $m_{\text{throw}} = m/2$ and $M_{\text{total}} = 3m/2$. So you gain an additional speed of $v_0/3$, relative to how fast you were going after the first throw. (The speeds do indeed simply add, because we can apply conservation of momentum in a new inertial frame moving at your speed of $v_0/4$ after the first throw.) Your final speed is therefore $v_0(1/4 + 1/3) = 7v_0/12 \approx (0.583)v_0$.

6.5. PROBLEM SOLUTIONS

(c) Let $r \equiv m_{\text{throw}}/M_{\text{total}}$. If the block is divided in thirds, the first stage has $r = 1/6$, the second has $r = 1/5$, and the third has $1/4$. So your final speed is $v_0(1/6 + 1/5 + 1/4) = 37v_0/60 \approx (0.617)v_0$. Similarly, for fourths we obtain a final speed of $v_0(1/8 + 1/7 + 1/6 + 1/5) \approx (0.635)v_0$. In general, for n pieces we end up with the sum from $1/2n$ to $1/(n+1)$, so

$$f(n) = \sum_{n+1}^{2n} \frac{1}{k}. \tag{6.53}$$

(d) The plot of $f(n)$ for n from 1 to 100 is shown in Fig. 6.23. $f(100)$ equals 0.69065, and $f(n)$ appears to be approaching a number just a hair above that.

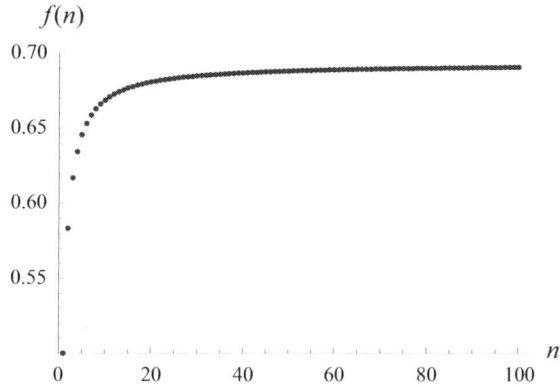

Figure 6.23

(e) For a continuous stream of particles, we have rocket motion. The initial mass is $2m$ and the final mass is m, so from Problem 6.22 the final speed is

$$v_{\text{final}} = v_0 \ln\left(\frac{m_{\text{initial}}}{m_{\text{final}}}\right) = (\ln 2)v_0 \approx (0.693)v_0. \tag{6.54}$$

This is very close to the $n = 100$ result. The difference is only $(0.69315)v_0 - (0.69065)v_0 = (0.0025)v_0$.

6.16. A collision in two frames

(a) FIRST SOLUTION: Let the final velocities of $4m$ and m be v_4 and v_1, respectively, with rightward taken to be positive. Conservation of momentum gives

$$(4m)v + m(-v) = (4m)v_4 + mv_1 \implies 3v = 4v_4 + v_1. \tag{6.55}$$

And conservation of energy gives

$$\frac{1}{2}(4m)v^2 + \frac{1}{2}mv^2 = \frac{1}{2}(4m)v_4^2 + \frac{1}{2}mv_1^2 \implies 5v^2 = 4v_4^2 + v_1^2. \tag{6.56}$$

Solving for v_1 in Eq. (6.55) and plugging the result into Eq. (6.56) gives

$$5v^2 = 4v_4^2 + (3v - 4v_4)^2 \implies 0 = 20v_4^2 - 24vv_4 + 4v^2$$
$$\implies 0 = 4(5v_4 - v)(v_4 - v) \implies v_4 = \frac{v}{5}. \tag{6.57}$$

(Another solution is $v_4 = v$, of course, because that is the initial velocity of $4m$.) Equation (6.55) then gives $v_1 = 3v - 4v_4 = 11v/5$. Both velocities are positive, so both masses move to the right.

SECOND SOLUTION: We can solve this problem by combining Eq. (6.55) with the relative-velocity statement from Problem 6.2, instead of with Eq. (6.56). This way, we won't need to deal with a quadratic equation. Eq. (6.19) gives

$$v_1 - v_4 = -((-v) - v) \implies v_1 - v_4 = 2v. \tag{6.58}$$

Plugging the v_1 from this equation into Eq. (6.55) gives

$$3v = 4v_4 + (2v + v_4) \implies v_4 = \frac{v}{5}. \tag{6.59}$$

And then $v_1 = 2v + v_4 = 11v/5$.

(b) The CM frame's velocity with respect to the lab frame is given by $v_{CM} = [(4m)v + m(-v)]/5m = 3v/5$. Our strategy will be to switch to the CM frame, then do the collision in the CM frame (which will be trivial), and then switch back to the lab frame.

- Switch to the CM frame: The initial velocities of the two masses in the CM frame are $v - v_{CM} = 2v/5$ and $-v - v_{CM} = -8v/5$, as shown:

$$\begin{array}{cc} 4m & m \\ \bullet \rightarrow & \leftarrow \bullet \\ 2v/5 & 8v/5 \end{array} \quad \text{(before)}$$

- Do the collision in the CM frame: The masses simply reverse their velocities (because this satisfies conservation of p and E). So after the collision, the velocities are $-2v/5$ and $8v/5$, as shown:

$$\begin{array}{cc} 4m & m \\ \leftarrow \bullet & \bullet \rightarrow \\ 2v/5 & 8v/5 \end{array} \quad \text{(after)}$$

- Switch back to the lab frame: To get back to the lab frame, we must add v_{CM} to the CM-frame velocities. So the final velocities in the lab frame are

$$v_4 = -\frac{2v}{5} + v_{CM} = -\frac{2v}{5} + \frac{3v}{5} = \frac{v}{5},$$
$$v_1 = \frac{8v}{5} + v_{CM} = \frac{8v}{5} + \frac{3v}{5} = \frac{11v}{5}. \tag{6.60}$$

6.17. 45-degree deflections

Let the final velocities be labeled as in Fig. 6.24. In drawing $\mathbf{v}_1$ horizontal, we have already used conservation of p_y (the initial p_y was zero). Conservation of p_x gives

$$mv = mv_1 + 2 \cdot mv_2 \cos 45° \implies v = v_1 + \sqrt{2}v_2. \tag{6.61}$$

Conservation of E gives

$$\frac{1}{2}mv^2 = \frac{1}{2}mv_1^2 + 2 \cdot \frac{1}{2}mv_2^2 \implies v^2 = v_1^2 + 2v_2^2. \tag{6.62}$$

Plugging the v_1 from the p_x equation into the E equation gives

$$v^2 = (v - \sqrt{2}v_2)^2 + 2v_2^2 \implies 0 = -2\sqrt{2}vv_2 + 4v_2^2 \implies v_2 = \frac{v}{\sqrt{2}}. \tag{6.63}$$

(Technically, another solution is $v_2 = 0$, but this corresponds to the balls missing each other. This is simply the initial value of v_2, which of course is guaranteed to be a solution to the conservation equations.) We then find v_1 to be $v_1 = v - \sqrt{2}v_2 = 0$. So the ball that was initially moving ends up at rest.

REMARK: We've noted on various occasions that a ball will end up at rest if it collides elastically head-on with an identical ball; the second ball absorbs all of the energy and momentum of the first

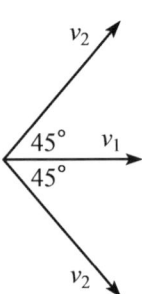

Figure 6.24

6.5. PROBLEM SOLUTIONS

ball. The result of this problem therefore tells us that the right two balls look effectively like a single ball with mass m, as far as the left ball is concerned. That is, the right two balls absorb all of the energy and momentum of the left ball. As an exercise, you can show that if the right two balls scatter at equal angles of θ instead of $45°$, and if they each have a mass of $m/(2\cos^2\theta)$, then the left ball will end up at rest. This mass correctly equals $m/2$ when $\theta = 0$, and ∞ when $\theta \to 90°$.

6.18. Northward deflection 1

Let the northward direction point along the y axis, as shown in Fig. 6.25. Let u be the desired final velocity of the mass $2m$ in the y direction. Then conservation of p_y (which is initially zero) quickly gives the final y velocity of the mass m as $-2u$. Also, if v_x is the final x velocity of m, then conservation of p_x gives

$$2mv_0 + m(-v_0) = 2m(0) + mv_x \implies v_x = v_0, \quad (6.64)$$

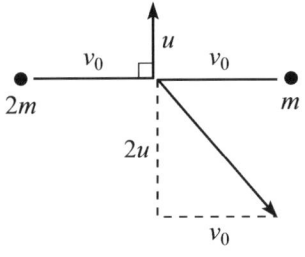

Figure 6.25

as shown in the figure. We now have only one unknown, u, so we can use conservation of E to solve for u:

$$\frac{1}{2}(2m)v_0^2 + \frac{1}{2}mv_0^2 = \frac{1}{2}(2m)u^2 + \frac{1}{2}m[(2u)^2 + v_0^2]$$
$$\implies 2v_0^2 + v_0^2 = 2u^2 + 4u^2 + v_0^2 \implies u = \frac{v_0}{\sqrt{3}}. \quad (6.65)$$

This is the final northward speed of the mass $2m$.

REMARK: We can also solve this problem by working in the CM frame. We'll just sketch the solution here, by stating some facts you can justify. The velocity of the CM is $v_0/3$ eastward. So the velocities of the $2m$ and m masses in the CM frame are $2v_0/3$ eastward and $4v_0/3$ westward. The speeds are the same after the collision, although the directions change (but they are still antiparallel). The $2m$ mass's velocity in the CM frame must have a westward component of $v_0/3$, so that it has no east-west component when we transform back to the original frame by adding on the eastward $v_0/3$ velocity of the CM. So the velocity of $2m$ in the CM frame must look like the vector shown in Fig. 6.26, with magnitude $2v_0/3$ and westward component $v_0/3$. The vertical component, which is the northward velocity in both the CM frame and the original frame, is therefore $v_0/\sqrt{3}$. Additionally, since the final velocity of m in the CM frame is antiparallel to the vector in Fig. 6.26 and has twice the length, you can quickly show that the final eastward component of the velocity of m in the original frame equals v_0, consistent with Eq. (6.64).

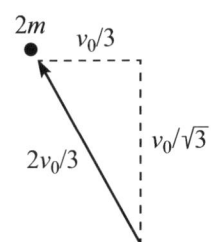

Figure 6.26

6.19. Northward deflection 2

(a) LAB FRAME: Let the final northward speed of m be u, as shown in Fig. 6.27. Conservation of p_y (which is initially zero) quickly tells us that the final southward speed of $2m$ is $u/2$. And conservation of p_x (which is initially mv_0) quickly tells us that the final eastward speed of $2m$ is $v_0/2$, as shown in the figure.

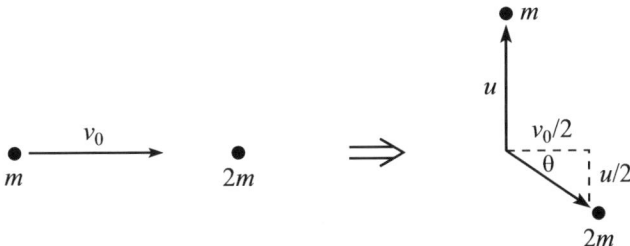

Figure 6.27

We now have only one unknown, u, so we can use conservation of E to solve for u:

$$\frac{1}{2}mv_0^2 = \frac{1}{2}mu^2 + \frac{1}{2}(2m)\left(\left(\frac{v_0}{2}\right)^2 + \left(\frac{u}{2}\right)^2\right)$$

$$\implies v_0^2 = u^2 + \frac{v_0^2}{2} + \frac{u^2}{2} \implies u = \frac{v_0}{\sqrt{3}}. \quad (6.66)$$

The desired angle is therefore

$$\tan\theta = \frac{u/2}{v_0/2} = \frac{1}{\sqrt{3}} \implies \theta = 30°. \quad (6.67)$$

Additionally, the 30-60-90 triangle tells us that the final speed of $2m$ is $v_0/\sqrt{3}$. So the two masses end up with the same speeds.

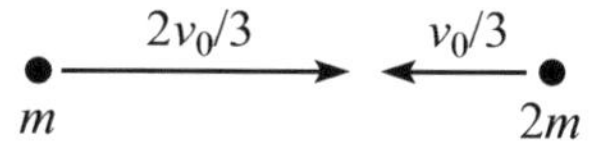

(b) CM FRAME: The CM moves with velocity $v_{CM} = mv_0/(m + 2m) = v_0/3$. So the velocities of m and $2m$ in the CM frame are, respectively, $v_0 - v_{CM} = 2v_0/3$ and $0 - v_{CM} = -v_0/3$. The situation is shown in Fig. 6.28.

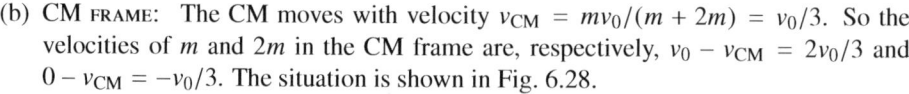

Figure 6.28

In the CM frame, the speeds of the masses are unchanged by the elastic collision, and the velocities come off in antiparallel directions. The direction of m's final velocity must be such that when we add the vector $\mathbf{v}_{CM}$ back on, to get back to the lab frame, the result is a vertical (northward) velocity for m. So the final velocity of m in the CM frame must have a westward component of $v_0/3$, as shown in Fig. 6.29. From the right triangle shown, we see that the angle ϕ is given by $\sin\phi = 1/2 \implies \phi = 30°$.

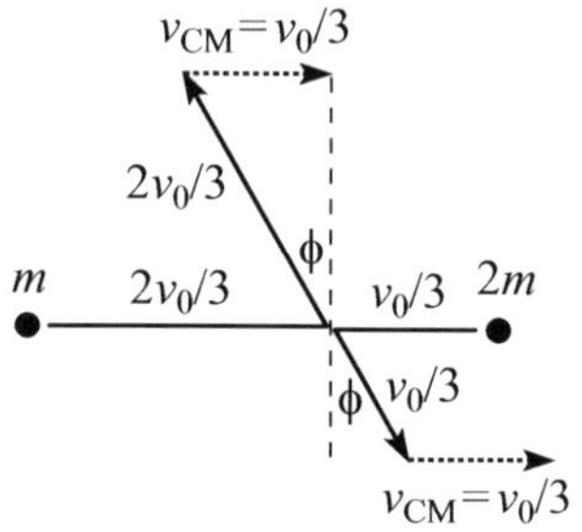

The final velocity of $2m$ in the lab frame is obtained by adding $\mathbf{v}_{CM}$ to the diagonally downward vector with length $v_0/3$ in Fig. 6.29. The resulting lab-frame velocity of $2m$ is shown in Fig. 6.30. This shape is a rhombus, so the final lab-frame velocity of $2m$ bisects the 60° angle between the initial and final velocities of $2m$ in the CM frame. The desired angle with respect to the horizontal is therefore 30°.

Additionally, the 30° angles in the rhombus tell us that the length of the long diagonal is $\sqrt{3}$ times the length of a side. So the final speed of $2m$ in the lab frame is $v_0/\sqrt{3}$.

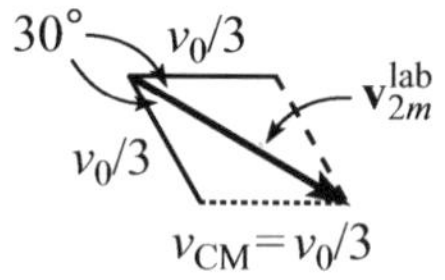

Figure 6.29

Figure 6.30

6.20. **Equal energies**

Since the final energies are equal, they must each be half of the initial energy. So we quickly obtain

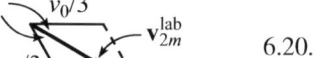

$$\frac{1}{2}mv_1^2 = \frac{1}{2}\left(\frac{1}{2}mv_0^2\right) \implies v_1 = \frac{v_0}{\sqrt{2}},$$

$$\frac{1}{2}(2m)v_2^2 = \frac{1}{2}\left(\frac{1}{2}mv_0^2\right) \implies v_2 = \frac{v_0}{2}. \quad (6.68)$$

Conservation of p_x and p_y will allow us to solve for the angles θ_1 and θ_2. Conservation of p_x gives

$$mv_0 = m \cdot \frac{v_0}{\sqrt{2}} \cdot \cos\theta_1 + 2m \cdot \frac{v_0}{2} \cdot \cos\theta_2 \implies 1 = \frac{1}{\sqrt{2}}\cos\theta_1 + \cos\theta_2. \quad (6.69)$$

And conservation of p_y gives

$$0 = m \cdot \frac{v_0}{\sqrt{2}} \cdot \sin\theta_1 - 2m \cdot \frac{v_0}{2} \cdot \sin\theta_2 \implies 0 = \frac{1}{\sqrt{2}}\sin\theta_1 - \sin\theta_2. \quad (6.70)$$

Putting the θ_1 terms on the left-hand sides of the previous two equations, squaring and adding, and using $\sin^2\theta + \cos^2\theta = 1$, gives

$$1 - \sqrt{2}\cos\theta_1 + \frac{1}{2} = 1 \implies \cos\theta_1 = \frac{1}{2\sqrt{2}} \implies \theta_1 \approx 69.3°. \quad (6.71)$$

6.5. PROBLEM SOLUTIONS

Plugging $\cos\theta_1 = 1/(2\sqrt{2})$ into Eq. (6.69) then gives $\cos\theta_2 = 3/4 \implies \theta_2 \approx 41.4°$. Note that the sum of the two angles isn't 90°, as it would be if the masses where equal.

REMARK: If a marble collides with a stationary bowling ball, then no matter what the angle of deflection is, the marble will end up with essentially all of the energy, which means that the energies certainly can't be equal. Similarly, if a bowling ball collides with a stationary marble, then the bowling ball will end up with essentially all of the energy. So if we replace the $2m$ in this problem with a general mass Nm, then N must lie within a certain range in order for it to be possible for the two masses to end up with equal energies. As an exercise, you can show that this range is $3 - 2\sqrt{2} \leq N \leq 3 + 2\sqrt{2}$, that is, $0.17 \leq N \leq 5.8$. In the $N = 0.17$ case, the collision is head-on, and the mass m continues moving forward. In the $N = 5.8$ case, the collision is head-on, and the mass m bounces directly backward.

6.21. **Equal speeds**

Conservation of energy gives the common final speed as

$$\frac{1}{2}mv_0^2 = \frac{1}{2}mv^2 + \frac{1}{2}(2m)v^2 \implies v = \frac{v_0}{\sqrt{3}}. \tag{6.72}$$

Conservation of p_x and p_y will allow us to solve for the angles θ_1 and θ_2. Conservation of p_x gives

$$mv_0 = m \cdot \frac{v_0}{\sqrt{3}} \cdot \cos\theta_1 + 2m \cdot \frac{v_0}{\sqrt{3}} \cdot \cos\theta_2 \implies \sqrt{3} = \cos\theta_1 + 2\cos\theta_2. \tag{6.73}$$

And conservation of p_y gives

$$0 = m \cdot \frac{v_0}{\sqrt{3}} \cdot \sin\theta_1 - 2m \cdot \frac{v_0}{\sqrt{3}} \cdot \sin\theta_2 \implies 0 = \sin\theta_1 - 2\sin\theta_2. \tag{6.74}$$

Putting the θ_1 terms on the left-hand sides of the previous two equations, squaring and adding, and using $\sin^2\theta + \cos^2\theta = 1$, gives

$$3 - 2\sqrt{3}\cos\theta_1 + 1 = 4 \implies \cos\theta_1 = 0 \implies \theta_1 = 90°. \tag{6.75}$$

Plugging $\cos\theta_1 = 0$ into Eq. (6.73) then gives $\cos\theta_2 = \sqrt{3}/2 \implies \theta_2 = 30°$. Note that the sum of the two angles isn't 90°, as it would be if the masses where equal.

REMARK: As in Problem 6.20, if we replace the $2m$ in this problem with a general mass Nm, then N must lie within a certain range in order for it to be possible for the two masses to end up with equal speeds. As an exercise, you can show that this range is $0 < N \leq 3$ (the 0 means that the stationary ball can be made arbitrarily small). In the $N = 3$ case, the collision is head-on, and the mass m bounces directly backward. The $N \approx 0$ case is a little tricker. A head-on collision will give the tiny stationary mass Nm a speed of $2v_0$ (as you can show). And a completely glancing collision, where m doesn't quite touch the tiny mass, will give it no speed at all. So by continuity there must be a particular glancing collision that causes the tiny mass Nm to come off at a specific intermediate angle with speed v_0 (which is the final speed of m, which just plows through with essentially the same speed). As an exercise, you can show that the $N = 0$ modification of the above solution leads to θ_2 equaling the nice angle of 60°. Alternatively, there is a quick way to derive this 60° result by working in the CM frame.

6.22. **Rocket motion**

Since the given u is a speed, it is defined to be positive. This means that the velocity of the particles ejected at a given instant is obtained by subtracting u from the velocity of the rocket at that instant. The rocket's initial mass is M, and m is the (decreasing) mass at a general later time. The rate of change of the rocket's mass is dm/dt, which is negative. So mass is ejected at a rate $|dm/dt| = -dm/dt$, which is positive. In other words, during a

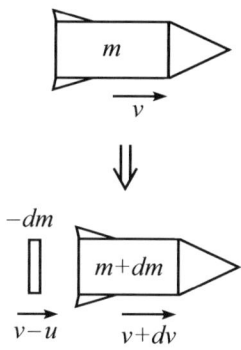

Figure 6.31

small time dt, a negative mass dm gets added to the rocket (so the rocket's mass decreases), and a positive mass $-dm$ gets shot out the back.[7]

Consider a moment when the rocket has mass m and velocity v (with respect to the ground). Then at a time dt later (see Fig. 6.31), the rocket has mass $m + dm$ and velocity $v + dv$, while the exhaust has mass $-dm$ and velocity $v - u$ (which may be positive or negative, depending on the relative size of v and u).[8] There are no external forces, so the total momenta at these two times must be equal. Therefore,

$$mv = (m + dm)(v + dv) + (-dm)(v - u). \tag{6.76}$$

Ignoring the second-order term $dm\,dv$, this simplifies to $m\,dv = -u\,dm$.

REMARK: This relation is actually much easier to see in the inertial reference frame that coincides with the rocket's frame at the start of the dt time interval. In this frame the rocket is initially at rest, and then the ejected mass picks up a momentum of $u\,dm$ (which is negative) while the rocket picks up a momentum of $m\,dv$. (Technically these two quantities should be $(u - dv)\,dm$ and $(m + dm)\,dv$, but the corrections are of second order, and they cancel anyway). These momenta must be equal and opposite, hence $m\,dv = -u\,dm$.

Dividing the $m\,dv = -u\,dm$ equation by m and integrating from t_1 to t_2 gives

$$\int_{v_1}^{v_2} dv = -\int_{m_1}^{m_2} u\frac{dm}{m} \implies v_2 - v_1 = u\ln\frac{m_1}{m_2}. \tag{6.77}$$

For the case where the initial mass m_1 is M and the initial speed v_1 is 0, we obtain

$$v = u\ln\left(\frac{M}{m}\right). \tag{6.78}$$

REMARKS: Note that we didn't assume anything about the ejection rate dm/dt in this derivation. There is no need for it to be constant; it can change in any way it wants. The only thing that matters (assuming that M and u are given) is the final mass m. In the special case where dm/dt is constant (call it $-\eta$, where η is positive), we have $m(t) = M - \eta t$, so $v(t) = u\ln[M/(M - \eta t)]$.

The log in the result in Eq. (6.78) is not very encouraging. If the mass of the metal in the rocket is m, and if the mass of the fuel is $9m$, then the final speed after all the fuel has been used up is only $u\ln 10 \approx (2.3)u$. If the mass of the fuel is increased by a factor of 11 up to $99m$ while holding the mass m of the metal constant (which is probably not even structurally possible),[9] then the final speed only doubles to $u\ln 100 = 2(u\ln 10) \approx (4.6)u$. So the factor of 11 in the fuel gives only a factor of 2 in the speed. How, then, do you make a rocket go significantly faster?

For concreteness, let's assume that it is impossible to build a structurally sound container that can hold fuel of more than, say, 19 times its mass. It would then seem like the limit for the speed of a rocket is $u\ln 20$. The strategy for beating this limit is to simply put a little rocket on top of another one. The final speed of the little rocket is the sum of the $u\ln 20$ limits for each rocket, which gives $2u\ln 20$. This is the same as having one rocket with a fuel-to-container mass ratio of $20^2 - 1 = 399$, which is huge. The point of using these "stages" is that if you jettison the container of the big rocket after its fuel is used up, then the little rocket doesn't have to keep accelerating it.

---

[7] If you wanted, you could define dm to be positive, and then *subtract* it from the rocket's mass, and have dm get shot out the back. However, you would then have to be careful to switch the order of the m_i limits of integration in Eq. (6.77) below.

[8] Technically the exhaust's velocity is $(v + dv) - u$, because we are told that the resulting speed relative to the rocket is u. But the distinction is irrelevant in the $dt \to 0$ limit, because the dv would lead to a second-order term $dm\,dv$ in Eq. (6.76) below. This term would be proportional to dt^2 and hence negligible in comparison with the first-order terms proportional to dt.

[9] The space shuttle's external fuel tank, just by itself, has a fuel-to-container mass ratio of only about 20.

6.5. PROBLEM SOLUTIONS

6.23. Hovering board

(a) The magnitude of the upward force on the board due to the water equals (by Newton's third law) the magnitude of the downward force on the water due to the board. And Newton's second law, $F = dp/dt$, tells us that this force equals the rate of change of momentum of the water. The water is initially moving upward at speed v_0 and is then brought to rest (at least vertically). So the magnitude of the change in momentum of a little drop of water with mass dm is $|dp| = (dm)v_0$. The magnitude of the rate of change of momentum is therefore (with dt being the small time that it takes the small dm to crash into the board)

$$\frac{dp}{dt} = \frac{(dm)v_0}{dt} = \frac{dm}{dt} v_0 = Rv_0. \tag{6.79}$$

This is the downward force on the water, and hence also the upward force on the board. The board will hover in place if this force equals mg, so we want $m = Rv_0/g$. The units of R are kg/s, so the units of m are correct. And m correctly grows with R and v_0, and decreases with g.

(b) Since the mass (and hence weight) of the board is half of what it was in part (a), we want the force from the water to also be half. So we want the dp/dt of the water to be half of what it was. The water still hits the board at the same mass rate R (because if it didn't, mass would be piling up or magically appearing at some intermediate height). But it is going slower because it loses speed as it rises. (What happens is that although the water is moving slower higher up, its stream is wider, so the same amount of water crosses a given plane as at a lower height.) From the expression for dp/dt in Eq. (6.79), we want the speed to be $v_0/2$. At what height is the speed reduced to $v_0/2$? This can quickly be answered by using conservation of energy:

$$\frac{1}{2}mv_0^2 = mgh + \frac{1}{2}m\left(\frac{v_0}{2}\right)^2 \implies h = \frac{3v_0^2}{8g}. \tag{6.80}$$

This height is 3/4 of the maximum height of $v_0^2/2g$ that the water would reach if the board weren't present. The height h correctly grows with v_0 and decreases with g.

Alternatively, you can also obtain h by using standard constant-acceleration kinematics. The time is given by $v_0 - gt = v_0/2 \implies t = v_0/2g$. Plugging this into $h = v_0 t - gt^2/2$ gives $h = 3v_0^2/8g$.

(c) The rate of change of momentum is now doubled, because each marble of mass dm goes from having momentum $(dm)v_0$ upward to $(dm)v_0$ downward. Therefore, the magnitude of dp is $2(dm)v_0$. The solution proceeds in exactly the same way as in part (a), except with an extra factor of 2, so we obtain a mass of $m = 2Rv_0/g$.

6.24. Falling heap

FIRST SOLUTION: At time t, the distance the heap has fallen is $gt^2/2$, because we are assuming that it is always in freefall. Therefore, the length left in the heap is $L - gt^2/2$. The heap is moving with speed gt, so its momentum is $p = \lambda(L - gt^2/2)(-gt)$, with upward taken to be positive. This is the momentum of the entire rope, because only the heap is moving.

The net force on the entire rope is $F_{\text{hand}} - \lambda Lg$. (The entire weight is $(\lambda L)g$, and gravity doesn't care that part of the rope is moving and part of it isn't.) So $F = dp/dt$ gives

$$F_{\text{hand}} - \lambda Lg = \frac{d}{dt}\left(-\lambda Lgt + \frac{1}{2}\lambda g^2 t^3\right) = -\lambda Lg + \frac{3}{2}\lambda g^2 t^2$$

$$\implies F_{\text{hand}} = \frac{3}{2}\lambda g^2 t^2. \tag{6.81}$$

This result holds until the rope straightens out when $gt^2/2 = L \implies t = \sqrt{2L/g}$. Just before this time, Eq. (6.81) gives $F_{\text{hand}} = 3\lambda Lg$. And just after, F_{hand} simply equals the weight λLg of the stationary hanging rope. So F_{hand} drops abruptly from $3\lambda Lg$ to λLg.

SECOND SOLUTION: F_{hand} is responsible for holding up the straight part of the rope, which weighs $\lambda(gt^2/2)g$, and also for stopping the atoms that join the straight part. In a small time dt, a mass of $dm = \lambda\, dx = \lambda(v\, dt)$ joins the straight part. This mass initially has momentum with magnitude $(\lambda v\, dt)v$ downward, and then it comes to rest. So the change in momentum is $dp = +\lambda v^2\, dt$ (it increases from a negative quantity to zero). Hence, $dp/dt = \lambda v^2 = \lambda(gt)^2$. This much additional force must be supplied by your hand, so the total force you apply is

$$F_{\text{hand}} = \lambda\left(\frac{gt^2}{2}\right)g + \lambda(gt)^2 = \frac{3}{2}\lambda g^2 t^2. \tag{6.82}$$

6.25. Bucket and chain

(a) Since the acceleration a is constant, the position and speed of the bucket are $x = at^2/2$ and $v = at$. The mass of the chain that has been gathered up is $m = \lambda x$, so the momentum of the bucket (which is massless) plus whatever chain is inside is

$$p = mv = (\lambda x)v = \lambda\left(\frac{at^2}{2}\right)(at) = \frac{1}{2}\lambda a^2 t^3. \tag{6.83}$$

The force that you apply is what causes the change in this momentum, so

$$F = \frac{dp}{dt} = \frac{d}{dt}\left(\frac{1}{2}\lambda a^2 t^3\right) = \frac{3}{2}\lambda a^2 t^2. \tag{6.84}$$

(You should verify that this answer is consistent with the answer to Multiple-Choice Question 6.20.) This force can also be written as $F = 3\lambda v^2/2$. Since λ has units of kg/m, F correctly has units of kg m/s^2.

REMARK: It would be incorrect to use $F = ma$ to say that $F = (\lambda \cdot at^2/2)a = \lambda a^2 t^2/2$, which equals $\lambda v^2/2$. This answer is too small by a factor of 3. The error is that your force is responsible for doing *two* things: It accelerates the mass that is already in the bucket (this yields ma), and it also gives momentum to the new bits of chain that are suddenly brought from speed 0 to speed v. Mathematically,

$$F = \frac{dp}{dt} = \frac{d(mv)}{dt} = m\frac{dv}{dt} + \frac{dm}{dt}v. \tag{6.85}$$

The first term here is simply ma, which we just showed equals $\lambda v^2/2$. The second term is $(d(\lambda x)/dt)v = \lambda(dx/dt)v = \lambda v^2$. The sum of these two terms gives the correct result of $3\lambda v^2/2$. Note that the second term is twice as large as the first term; the force needed to get the new bits of mass moving is twice as large as the force needed to accelerate the mass that is already in the bucket.

(b) The work that you do up to time t is (we won't bother putting a prime on the integration variable)

$$W = \int F\, dx = \int_0^t Fv\, dt = \int_0^t \left(\frac{3}{2}\lambda a^2 t^2\right)(at)\, dt$$
$$= \frac{3}{2}\lambda a^3 \int_0^t t^3\, dt = \frac{3}{8}\lambda a^3 t^4. \tag{6.86}$$

Since λ has units of kg/m, you can quickly verify that F correctly has units of $\text{kg m}^2/\text{s}^2$.

(c) The kinetic energy of the chain inside the bucket at time t is

$$K = \frac{1}{2}mv^2 = \frac{1}{2}\left(\lambda\frac{at^2}{2}\right)(at)^2 = \frac{1}{4}\lambda a^3 t^4. \tag{6.87}$$

This is $\lambda a^3 t^4/8$ less than the $3\lambda a^3 t^4/8$ work you do (which is the energy that you put into the system). So this difference of $\lambda a^3 t^4/8$, which is 1/3 of the work you do, is what is lost to heat.

Chapter 7

Torque

7.1 Introduction

Basics of rotations

Consider a rigid planar object rotating around the origin, as shown in Fig. 7.1. More generally, we can have a rigid 3-D object rotating around a fixed axis perpendicular to the page. All points in the object move in circles centered on the axis. A cross-sectional slice made with a plane parallel to the page will simply yield a planar object like the one shown in Fig. 7.1. Since the object is rigid, all points have the same the angular velocity ω at any given instant.

If θ is the angle through which the object has rotated, then the angular velocity is $\omega = d\theta/dt$. And the angular acceleration is $\alpha = d\omega/dt = d^2\theta/dt^2$. In the special case where α is constant, we can integrate it twice to obtain $\theta(t)$:

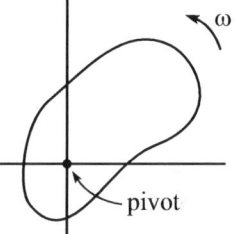

Figure 7.1

$$\theta(t) = \theta_0 + \omega_0 t + \frac{\alpha t^2}{2}, \tag{7.1}$$

where the constants of integration θ_0 and ω_0 are the initial angle and initial angular velocity, respectively. If we multiply Eq. (7.1) by the radius r of the circular motion of a given point, and if we use the relations $s = r\theta$ (where s is the arclength), $v = r\omega$ (where v is the velocity), and $a = r\alpha$ (where a is the tangential acceleration), we obtain

$$s(t) = s_0 + v_0 t + \frac{at^2}{2}, \tag{7.2}$$

which is the standard "linear" relation for the case of constant acceleration.

Moment of inertia

If an object consists of many masses m_i, whose distances from the fixed axis of rotation are r_i, then the *moment of inertia*, I, of the object is defined as

$$I = \sum m_i r_i^2. \tag{7.3}$$

I depends on the location of the axis of rotation (or the point of rotation, for planar objects), and also on the distribution of mass. The question "What is I for an object with mass M?" is meaningless. We must specify where the axis of rotation is chosen and how the mass is distributed. The I of an object does *not* depend on the angular velocity ω. Even if an object is sitting at rest, it still has an I relative to any particular chosen axis.

In the case of an object with a continuous mass distribution, the sum in Eq. (7.3) is replaced by an integral:

$$I = \int r^2 \, dm, \tag{7.4}$$

where r is the distance from each little piece dm to the axis of rotation. The moments of inertia of some common objects (with uniform mass distribution) are:

1. Stick (around center): $ML^2/12$
2. Stick (around end): $ML^2/3$
3. Ring (around center): MR^2
4. Ring (around diameter): $MR^2/2$
5. Disk (around center): $MR^2/2$
6. Disk (around diameter): $MR^2/4$
7. Spherical shell (around diameter): $2MR^2/3$
8. Solid sphere (around diameter): $2MR^2/5$

The phrase "around center" (or end) is shorthand for "around an axis that passes through the center and is perpendicular to the object." Note that all of the above I's take the general form of a numerical factor times a mass times a length squared. So when calculating any of these I's, it's just a question of what the numerical factor is. See, for example, Problems 7.11 and 7.12.

The moment of inertia appears in two fundamental relations in rotational dynamics. One relation involves the kinetic energy, the other involves the torque. We will discuss these below.

Two theorems

A very useful theorem involving moments of inertia is the *parallel-axis theorem*. Consider the moment of inertia, I_P, of an object (not necessarily planar) around a given axis through the point P in Fig. 7.2. Consider also the moment of inertia, I_{CM}, around a parallel axis through the center of mass. Then the parallel-axis theorem states that I_P is related to I_{CM} by (see Problem 7.1 for a proof)

$$I_P = I_{CM} + Md^2 \quad \text{(parallel-axis theorem)}, \tag{7.5}$$

where M is the mass of the object and d is the distance between the two axes. (This distance is equal to the distance between P and the CM only if the two axes are perpendicular to the line joining P and the CM.) This result is valid only if the CM is used on the right-hand side. If you want to compare the I's around two parallel axes, neither of which passes through the CM, then they do *not* in general differ by Md^2, where d is the separation. To compare them, you must compare them each to I_{CM} via Eq. (7.5). Note that Eq. (7.5) implies that I_{CM} is smaller than any other I_P, since Md^2 is always a positive quantity.

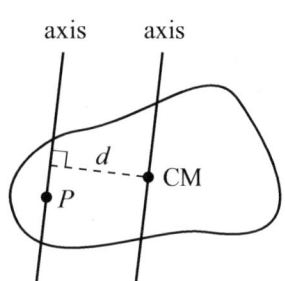

Figure 7.2

Another theorem is the *perpendicular-axis theorem*. This theorem isn't as widely applicable as the parallel-axis theorem, because it holds only for *planar objects*. But for the pancake object in the x-y plane shown in Fig. 7.3, the perpendicular-axis theorem states that the moments of inertia around the three coordinate axes are related by (see Problem 7.2 for a proof)

$$I_z = I_x + I_y \quad \text{(perpendicular-axis theorem)}. \tag{7.6}$$

Remember that this applies only to planar objects.

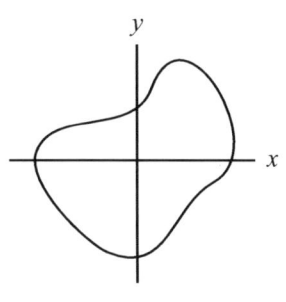

Figure 7.3

Kinetic energy

For an object that is rotating with angular velocity ω around a fixed axis, the kinetic energy K is given by (see Problem 7.3 for a derivation)

$$K = \frac{1}{2}I\omega^2. \tag{7.7}$$

This relation is the analog of the $K = mv^2/2$ relation in linear dynamics; I takes the place of m, and ω takes the place of v. For an object with mass M that is both translating and rotating, as

7.1. INTRODUCTION

shown in Fig. 7.4, the kinetic energy is given by (see Problem 7.4 for a derivation)

$$K = \frac{1}{2}MV_{\text{CM}}^2 + \frac{1}{2}I_{\text{CM}}\omega^2, \qquad (7.8)$$

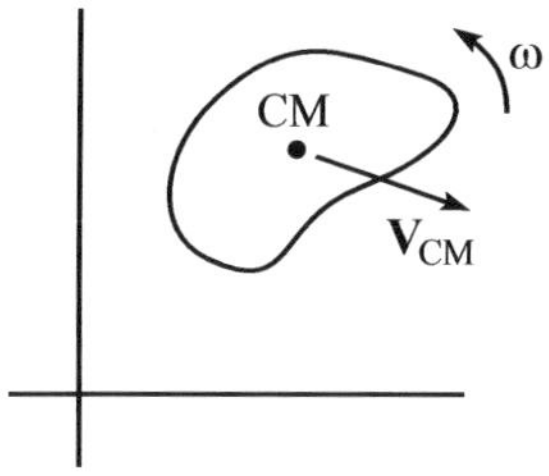

where V_{CM} is the speed of the center of mass, and I_{CM} is the moment of inertia around the center of mass. In words: K equals the sum of the energy of the entire object treated like a point mass M traveling at V_{CM}, plus the rotational energy relative to the CM (imagine that you are riding along with the CM as the object is spinning around you). The way that the translational and rotational energies combine in Eq. (7.8) is about as nice a result as we could hope for. It certainly reduces properly in the special cases where the object is only translating or only rotating. But note well that Eq. (7.8) isn't valid if the CM is replaced by any other point.

Figure 7.4

Torque

If a force **F** acts on an object, then the *torque* (denoted by τ) on the object, relative to a given origin, is defined via the "cross product" (see Section 13.1.7 in Appendix A for the definition of the cross product) as

$$\boldsymbol{\tau} \equiv \mathbf{r} \times \mathbf{F}, \qquad (7.9)$$

where **r** is the vector from the origin to the point where the force **F** is applied. If different forces are applied at different locations, then the total torque is the sum of all of the individual torques: $\boldsymbol{\tau} = \sum \mathbf{r}_i \times \mathbf{F}_i$. The same origin must be chosen for all of the individual torques.

When drawing a free-body diagram for the purpose of calculating torques for the $\tau = I\alpha$ relation that we will discuss below, it is critical (due to the appearance of **r** in the above expression for τ) that you draw the forces at the correct locations where they act on an object. In contrast, it doesn't matter where the forces act if you are using only $\mathbf{F} = m\mathbf{a}$.

Torque is a vector, being the cross product of two other vectors. However, we will invariably deal only with situations where both the position **r** and the force **F** lie in the plane of the page, in which case the torque τ points perpendicular to the page and has magnitude $rF\sin\theta$, where θ is the angle between **r** and **F**. Since τ is always perpendicular to the page in this case, we can ignore the fact that it is actually a vector and deal simply with its magnitude, $rF\sin\theta$. You can think of this magnitude in either of two ways, depending on which quantity you want to group with the $\sin\theta$ (see Fig. 7.5):

$$\tau = r(F\sin\theta) = (\text{radius})(\text{tangential force}), \text{ or}$$
$$\tau = F(r\sin\theta) = (\text{force})(\text{lever arm}). \qquad (7.10)$$

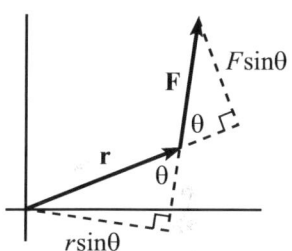

Figure 7.5

That is, the torque equals the entire distance r times the tangential component of the force (the radial component of the force won't cause an object to rotate). And it also equals the entire force times the "lever arm." The lever arm is the component of the **r** vector that is perpendicular to **F**. To geometrically construct it, draw a line pointing along **F** (extending in both directions) and look at the closest approach to the origin.

Many setups involve the torque due to the gravitational force. A very useful fact is:

- When finding the gravitational torque on an object with mass m, the object can be treated like a point mass m located at the CM.

See Problem 7.5 for a proof. Note that this result holds only for the torque. As far as the moment of inertia goes, the object *cannot* be treated like a point mass at the CM; it matters how the mass is distributed.

The $\tau = I\alpha$ relation

The most important thing about the torque τ on an object is that it is proportional to the angular acceleration α of the object, with the constant of proportionality being equal to the moment of inertia, I:

$$\tau = I\alpha. \qquad (7.11)$$

See Problem 7.6 for a proof. This relation is the bread-and-butter relation for rotational dynamics. It is the rotational analog of $F = ma$; τ takes the place of F, I takes the place of m, and α takes the place of a. If you apply a force to an object, you give it a linear acceleration that is inversely proportional to the mass. Likewise, if you apply a torque to an object, you give it an angular acceleration that is inversely proportional to the moment of inertia.

Equation (7.11) is valid if the choice of origin is (1) a fixed point (or more generally a point moving with constant velocity), or (2) the CM of the object, or (3) a third possibility which rarely comes up. (See Section 8.4.3 in Morin (2008) for a discussion of the third condition.) If you forget about this restriction, it probably won't end up mattering, because it's unlikely that you would pick a point as your origin that isn't a fixed point or the CM. Those are the natural choices. But keep the restriction in mind. The same origin must be chosen for both τ and I, of course.

7.2 Multiple-choice questions

7.1. A solid cylinder has mass M, length L, and radius R, as shown in Fig. 7.6. What is the moment of inertia around an axis that passes through the CM and is perpendicular to the symmetry axis? (Don't solve this from scratch; just check limiting cases of the shape.)

(a) $\dfrac{ML^2}{12} + \dfrac{MR^2}{2}$

(b) $\dfrac{ML^2}{12} + \dfrac{MR^2}{4}$

(c) $\dfrac{ML^2}{3} + \dfrac{MR^2}{2}$

(d) $\dfrac{ML^2}{3} + \dfrac{MR^2}{4}$

Figure 7.6

7.2. All three objects shown below have the same total mass m, and they all have the same vertical span. The dots in the first object are point masses. The square and the circle are made from wires (that is, they are *not* solid planar objects). Which object(s) has/have the largest moment of inertia around an axis that passes through the CM and is perpendicular to the page?

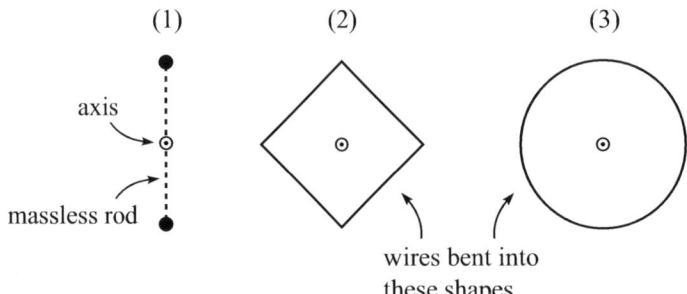

7.3. What is the kinetic energy of a stick rotating around an axis located at one of the ends and perpendicular to the stick?

(a) $\frac{1}{2} I_{\text{end}} \omega^2$

(b) $\frac{1}{2} I_{\text{CM}} \omega^2$

(c) $\frac{1}{2} I_{\text{CM}} \omega^2 + \frac{1}{2} m v_{\text{CM}}^2$

(d) Both (a) and (c)

(e) None of the above

7.4. A ball rolls down into a valley, as shown in Fig. 7.7. The left side of the valley has friction, and the ball rolls without slipping there. But the right side is frictionless. The ball is released from rest at height h on the left side. Which of the following is correct?

(a) Because of friction, energy is not conserved, so the ball reaches a height less than h on the right side.

(b) By conservation of energy the ball reaches a height less than h on the right side.

(c) By conservation of energy the ball reaches the same height h on the right side.

(d) The rolling of the ball causes it to reach a height greater than h on the right side.

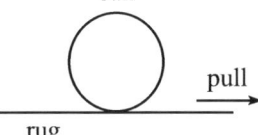

Figure 7.7

7.5. You give a quick push on three identical sticks at the locations shown below. Your force is constant, and it acts over the same small distance Δx in each case. Which stick has the largest resulting CM velocity?

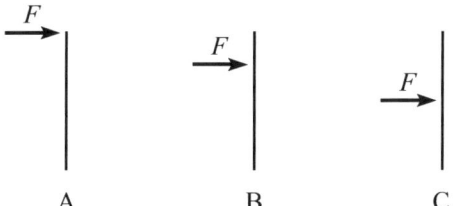

(a) A (b) B (c) C (d) They all have the same CM velocity.

7.6. A ball lies on a rug, and the rug is accelerated to the right, as shown in Fig. 7.8. Assume that there is at least some friction between the ball and the rug. Let the ball's acceleration a be defined positive rightward, and let its angular acceleration α be defined positive clockwise. Then

(a) a is positive, and α is positive

(b) a is positive, and α is negative

(c) a is negative, and α is positive

(d) a is negative, and α is negative

(e) a is zero, and α is negative

Figure 7.8

7.7. If all of the following wheels are simultaneously released from rest at the same height on an inclined plane, and if none of the wheels slip, which one will reach the bottom first? Assume that darker regions indicate higher densities.

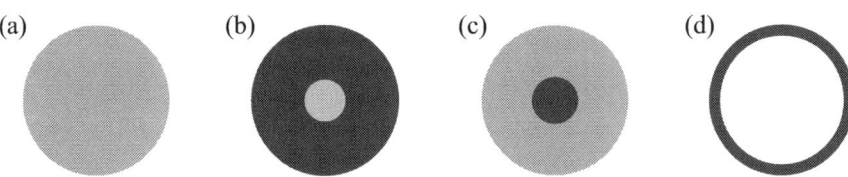

7.8. A ball rolls without slipping down a plane inclined at angle θ. The friction force from the plane on the ball

(a) makes ω increase, and makes $a_{\rm CM}$ be larger than $g \sin \theta$

(b) makes ω increase, and makes $a_{\rm CM}$ be smaller than $g \sin \theta$

(c) makes ω decrease, and makes $a_{\rm CM}$ be larger than $g \sin \theta$

(d) makes ω decrease, and makes $a_{\rm CM}$ be smaller than $g \sin \theta$

7.9. Consider a ball that rolls down a plane (which has friction) without slipping. And consider a block that slides down a different plane, which is frictionless. Both planes are inclined at the same angle. The acceleration of the CM of the ball is smaller than the acceleration of the block because (circle all that apply)

(a) the component of **g** pointing down the plane is smaller in the case of the ball

(b) the ball has one contact point with its plane, whereas the block has a whole surface of contact points with its plane

(c) there is a friction force on the ball pointing up the plane

(d) energy is contained in the rotational motion of the ball

7.10. Imagine that you are holding a 20-pound weight in a "curling" position, with your upper arm vertical and your forearm horizontal. The main muscle holding your forearm in place is your biceps muscle. By looking at your arm and considering torques around your elbow joint, which of the following is your best guess for the tension in your biceps?

(a) 2 lbs   (b) 10 lbs   (c) 20 lbs   (d) 40 lbs   (e) 200 lbs

7.11. If you apply a force to the pedals of your bike and accelerate, the wheels have an angular acceleration. Consider the rim of the back wheel. The things that apply a torque to this rim are

(a) the ground

(b) the spokes

(c) both (a) and (b)

(d) neither (a) nor (b)

7.12. A stick has its bottom end attached to a wall by a pivot and is held up by a massless string attached to its other end. Which of the following scenarios has the smallest tension in the string?

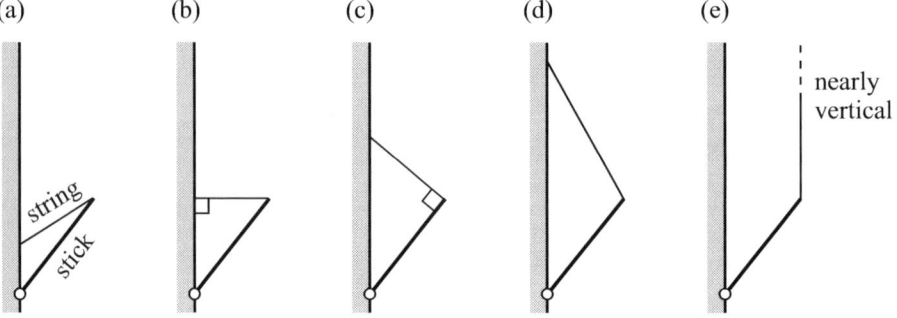

7.13. The objects shown below are released from rest, all from the same initial angle. They are all pivoted on the ground, and the CM's are all the same distance d from the pivot. The total mass of each object is the same (although this assumption isn't necessary). Dotted lines denote massless sticks. Which object falls the fastest?

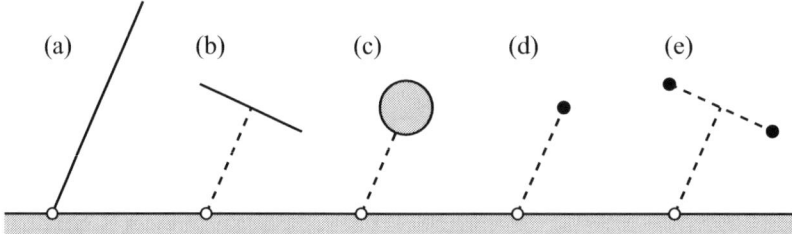

7.3. PROBLEMS

7.14. The left end of a *massless* stick with length ℓ is placed on the corner of a table, as shown in Fig. 7.9. A point mass m is attached to the center of the stick, which is initially held horizontal. It is then released. Immediately afterward, what normal force does the table exert on the stick?

(a) 0 (b) $mg/6$ (c) $mg/2$ (d) mg (e) $2mg$

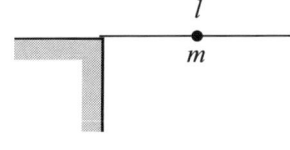

Figure 7.9

7.15. You pull on a roll of toilet paper with a given force F and observe the angular acceleration α. A few days later when the radius is half of what it was, you pull with the same force F. (For simplicity, ignore the hollow tube; assume that the toilet paper goes all the way down to zero radius.) The ratio of the new α to the old α is

(a) 1 (b) 2 (c) 4 (d) 8 (e) 16

7.3 Problems

The first eight problems are foundational problems.

7.1. Parallel-axis theorem

Prove the parallel-axis theorem, stated in Eq. (7.5). Feel free to deal only with the special case where the object is planar and the axes are perpendicular to this plane. The reasoning in this case has all the ingredients of the general proof. *Hint*: Write the coordinates of a general point in the object as the sum of the coordinates of the CM plus the coordinates relative to the CM.

7.2. Perpendicular-axis theorem

Prove the perpendicular-axis theorem, stated in Eq. (7.6).

7.3. Rotational kinetic energy

For an object that is rotating with angular velocity ω around a fixed axis, show that the kinetic energy is given by Eq. (7.7). *Hint*: Divide the object into a large number of tiny masses m_i, and find the kinetic energy of each mass in terms of ω.

7.4. Translation plus rotation

For an object that is both translating and rotating, show that the kinetic energy is given by Eq. (7.8). *Hint*: Write the velocity of a general point in the object as the sum of the velocity of the CM plus the velocity relative to the CM. Also, it will be helpful to use the dot product (see Section 13.1.6 in Appendix A) to write v^2 as $\mathbf{v} \cdot \mathbf{v}$.

7.5. Gravitational torque

Show that when finding the gravitational torque on an object with mass m, the object can be treated like a point mass m located at the CM.

7.6. The $\tau = I\alpha$ relation

Derive the $\tau = I\alpha$ relation. There are various ways to do this, one of which is the following. Imagine applying a force to an object pivoted at the origin; see Fig. 7.10. If the force is applied over a tangential displacement $d\mathbf{s}$, you can calculate the work done and then equate it with the change in the kinetic energy, $dK = d(I\omega^2/2) = I\omega\,d\omega$. This should give you $\tau = I\alpha$.

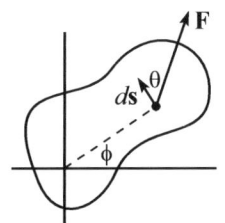

Figure 7.10

7.7. Force along a massless stick

A massless stick is attached by pivots to objects at its two ends. Explain why the force that the stick applies to each object is always directed along the stick.

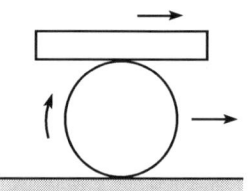

Figure 7.11

7.8. Non-slipping condition

(a) A wheel with radius R rolls without slipping on the ground. Let the distance it moves be d, and let the angle through which it rolls be θ. Show that the non-slipping condition is $d = R\theta$ (or equivalently $v = R\omega$, or equivalently $a = R\alpha$).

(b) Consider now the case where a horizontal board rolls on top of a wheel, which rolls on the ground, as shown in Fig. 7.11. If there is no slipping anywhere, show that the board moves twice as far as the wheel.

7.9. *I* for a square

Find the moment of inertia of a solid square (mass M, side L) around the axis through the center and perpendicular to the plane of the square. Do this by making use of a scaling argument, along with the parallel-axis theorem. You will need to find the quantities A, B, and C in the following equations, where an object represents the moment of inertia around the given dot. These are three equations and three unknowns, so the unknowns can be determined.

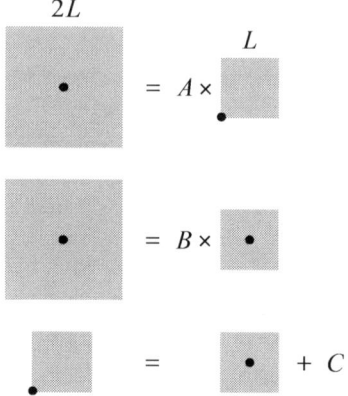

7.10. Another *I* for a square

(a) Find the moment of inertia of a solid square (mass M, side L) around an axis connecting the midpoints of two opposite sides. Do this by slicing the square into thin strips and integrating over the strips.

(b) Same task, but now let the axis connect two opposite corners.

(c) Explain why your answers to parts (a) and (b) must agree.

7.11. *I* for a disk

Find the moment of inertia of a solid disk with mass m and radius R, around the axis that passes through the center and is perpendicular to the plane of the disk.

7.12. *I* for a spherical shell

Find the moment of inertia of a hollow spherical shell with mass m and radius R, around a diameter.

7.13. Bending

The wooden beams beneath the floors in houses are often "2 by 8's" (inches). These are oriented with the "8" direction vertical. This orientation makes sense because intuitively the beams would sag much more if the "2" direction were vertical. Why exactly would they sag more? More precisely, how does the sagging depend on the height of the beam?

To answer this, consider the model of a beam shown in Fig. 7.12. We have drawn two beams, one with twice the height of the other. Imagine breaking each beam into two halves

7.3. PROBLEMS

that are pivoted on a short rod. The two halves are connected by many short springs (the thin lines shown) at their relaxed length. When each beam is bent (due to the application of an external torque), the springs in the top half compress, and the springs in the bottom half stretch, as shown in the right half of the figure. This is basically what happens on a cellular level; the cells in the wood act like little springs.

Here is the question you need to answer: Consider the torque (relative to the pivot) on one half of the beam with height $2h$ due to the springs in the beam. For the same angle of deflection, how much larger is this torque than the analogous torque on one half of the beam with height h? *Hint*: Consider a given small section of springs in the taller beam, and compare the torque due to these springs with the torque due to the corresponding section (which is half as large) in the shorter beam. No calculations needed! Just determine how the relevant quantities scale with length.

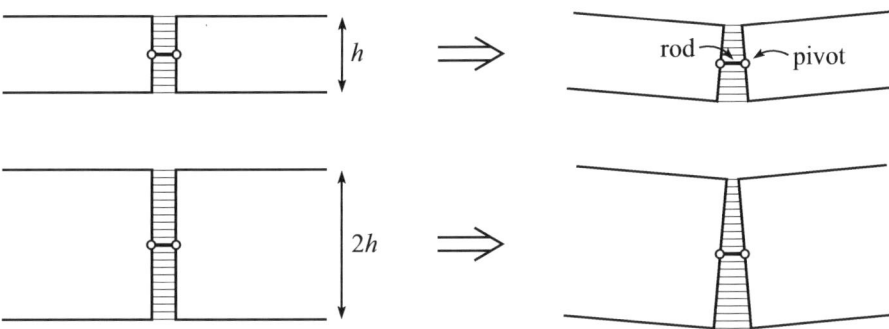

Figure 7.12

7.14. Initial angular acceleration

A uniform disk with mass m and radius R lies in a vertical plane and is pivoted at its center. A stick with length ℓ and uniform mass density λ (kg/m) is glued tangentially at its top end to the disk, as shown in Fig. 7.13, so that it forms a rigid object with the disk. If the system is held with the stick vertical and then released, what is the *initial* angular acceleration of the system? Given m, R, and λ, for what ℓ is this angular acceleration maximum?

7.15. Atwood's with a massive pulley

Two masses, m and $2m$, hang over a pulley with mass m and radius R (and $I = mR^2/2$), as shown in Fig. 7.14. Assuming that the (massless) string doesn't slip on the pulley, find the accelerations of the masses.

7.16. Equivalent mass

In Fig. 7.15(a) two masses m_1 and m_2 hang over a pulley with mass m, radius R, and moment of inertia $I = \beta m R^2$, where β is a numerical factor. The (massless) string doesn't slip on the pulley. The two masses will accelerate at exactly the same rate as they will in the system in Fig. 7.15(b) (with two massless pulleys and a frictionless table), provided that the value of m' is chosen correctly. What is this value?

7.17. Braking on a bike

Let the distance between the centers of the front and back wheels of a bicycle be $2L$, and assume for simplicity that the CM of the bike plus rider is located halfway between the wheels, at a height L above the ground. The coefficient of kinetic friction between the wheels and the ground is μ. If the brakes are applied hard enough so that skidding occurs in all of the following situations (with both wheels always remaining in contact with the ground), what is the deceleration of the bike if

(a) the brakes are applied to both wheels?

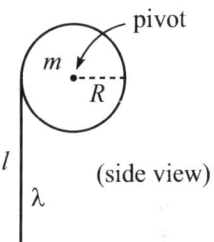

Figure 7.13

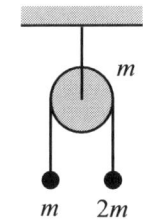

Figure 7.14

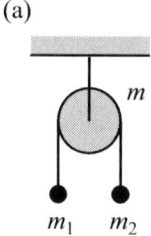
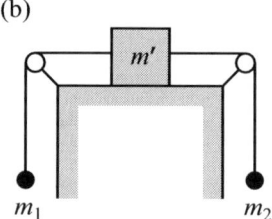

Figure 7.15

(b) the brake is applied only to the back wheel?

(c) the brake is applied only to the front wheel?

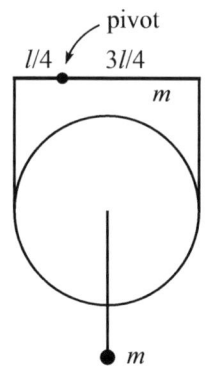

Figure 7.16

7.18. **Pulley below a stick**

A stick with mass m and length ℓ is pivoted at a point $\ell/4$ from an end, as shown in Fig. 7.16. A massless string hangs from each end and wraps around the underside of a massless pulley with diameter ℓ. A mass m hangs from the pulley. If the stick is held horizontal and then released, what is the initial acceleration of the mass m?

7.19. **Falling stick 1**

A stick with mass m, length ℓ, and uniform mass density is initially held horizontal with its left end attached to a pivot. It is then released. Immediately after, what is the acceleration of the right end?

7.20. **Falling stick 2**

A stick with mass m, length ℓ, and uniform mass density initially stands vertically with its bottom end attached to a pivot. It is given an infinitesimal push, and it swings down around the pivot. At the instant the stick is horizontal (after a quarter turn), what is the force (specify the horizontal and vertical components) that the pivot applies to the stick?

7.21. **Sticks on a table**

(a) A uniform stick with mass m and length ℓ has its bottom end pivoted on a table, as shown in Fig. 7.17. It is held at rest at an angle θ with respect to the horizontal and then released. What is its angular acceleration right after it is released?

(b) Answer the same question, but now let the bottom end *not* be pivoted on the table. Instead, let the table be frictionless, with the bottom end of the stick free to slide on it.

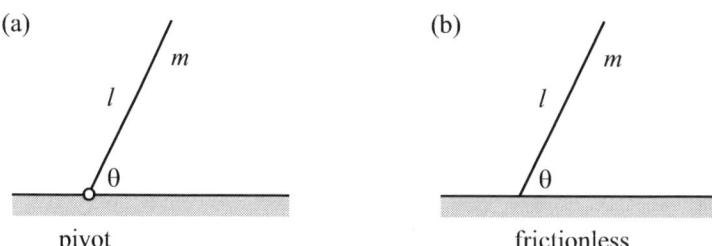

Figure 7.17

7.3. PROBLEMS

7.22. Hoop on a plane

A hoop (that is, a wheel with all of its mass on the rim) is given an initial speed up a plane inclined at angle θ. The initial linear speed is v_0, and the initial angular speed is zero. The coefficient of kinetic friction between the hoop and the plane is $\mu = 1/2$. How much time passes before the hoop starts to roll without slipping? For what θ is this time minimum?

7.23. Cylinder and board

A uniform solid cylinder with mass m and radius R lies on top of a long board that also has mass m, as shown in Fig. 7.18. There is sufficient friction between them to prevent slipping. The coefficient of kinetic friction between the table and the board is μ_k. The initial velocity of both objects is v_0, which means that the cylinder is initially at rest with respect to the board.

(a) What is the acceleration of the board?

(b) What is the speed of the cylinder when the board finally comes to rest?

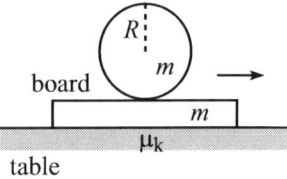

Figure 7.18

7.24. Another cylinder and board

A uniform solid cylinder with mass m and radius R lies on top of a long board which also has mass m, as shown in Fig. 7.19. The board is free to slide frictionlessly on a table, but there is kinetic friction between the cylinder and the board, with the coefficient of kinetic friction equal to μ. If the board is initially at rest, and if the cylinder is given an initial speed v_0 to the right, but *without* any initial rotation, what is the speed of the cylinder (with respect to the ground) when it finally rolls without slipping on the board?

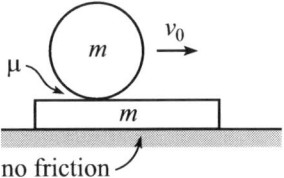

Figure 7.19

7.25. Rolling down a plane

A wheel with mass m, radius R, and moment of inertia $\beta m R^2$ (where β is a numerical factor) rolls without slipping down a plane inclined at angle θ. What is the linear acceleration of the wheel?

7.26. Pulling a cylinder on a board

A board with mass m lies on a frictionless plane inclined at angle θ. A uniform solid cylinder with mass m and radius R lies on top of the board. A string that is attached to the cylinder and wrapped around it is pulled with a tension T parallel to the plane, as shown in Fig. 7.20. There is sufficient friction between the cylinder (or string) and the board so that the cylinder rolls on the board without slipping.

(a) Draw the free-body diagram for the board. Likewise for the cylinder. Be sure to include the forces in all directions.

(b) Given T, write down all of the equations necessary to solve for the various accelerations. (Don't worry about actually solving the equations.) How many equations and unknowns are there?

(c) What should T be so that the board has zero acceleration? What is the acceleration of the cylinder in this case? You do need to solve some equations here, but they're not so bad.

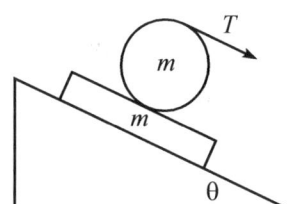

Figure 7.20

7.27. Cylinder on board on plane

A uniform solid cylinder with mass m and radius R lies on a board also with mass m, which lies on a plane inclined at angle θ, as shown in Fig. 7.21. The coefficient of kinetic friction between the board and the plane is $\mu = (1/2)\tan\theta$. Assume that there is sufficient friction between the cylinder and the board so that the cylinder doesn't slip with respect to the board. Draw the free-body diagrams for the board and the cylinder. What is the linear acceleration of the cylinder?

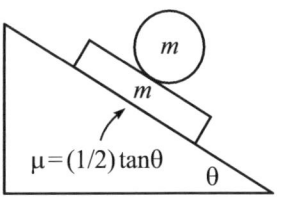

Figure 7.21

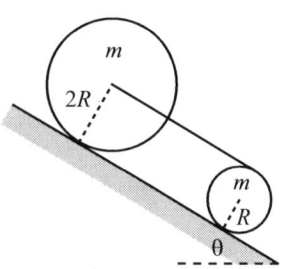

Figure 7.22

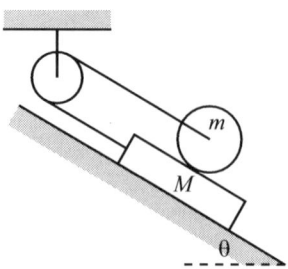

Figure 7.23

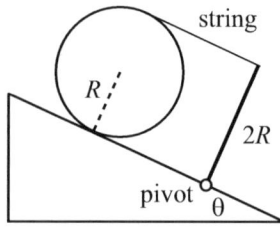

Figure 7.24

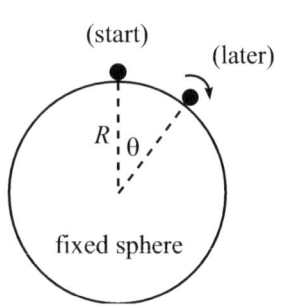

Figure 7.26

7.28. Two cylinders on a plane

Two uniform solid cylinders are placed on a plane inclined at angle θ. Both cylinders have mass m, but one radius is twice the other. A massless string connects the axle of the large cylinder to the rim of the small one, as shown in Fig. 7.22. The cylinders are released from rest.

(a) Assuming that the cylinders roll without slipping down the plane, what are their accelerations?

(b) If the coefficient of static friction between both cylinders and the plane is $\mu = 1$, what is the largest angle θ for which no slipping occurs between either cylinder and the plane?

7.29. Wheel and board

A uniform wheel with mass m and radius R lies on top of a board with mass M (you can assume $M > m$), which lies on a frictionless plane inclined at angle θ, as shown in Fig. 7.23. A massless string wraps around a massless pulley and has its ends tied to the board and the wheel's axle. Assume that the wheel rolls without slipping on the board, and assume that the two string segments shown are parallel to the plane. What is the acceleration of the board?

7.30. Cylinder and stick

A uniform solid cylinder with mass m and radius R and a uniform stick with mass m and length $2R$ are situated on a plane inclined at angle θ, with the stick perpendicular to the plane, as shown in Fig. 7.24. The cylinder rolls without slipping on the plane, and the stick is connected to the plane by a pivot. A massless string connects the top of the stick to the "top" of the cylinder, as shown. If the objects are held in this position and then released, what is the acceleration of the center of the cylinder immediately afterward?

7.31. Infinite set of disks

An infinite number of disks are cut out from a flat sheet of metal (so they all have the same mass density per unit area), and they are situated in a horizontal line with their centers at the same height, as shown in Fig. 7.25 (side view). They are free to spin around their (fixed) centers. There is no friction between any of the disks; equivalently, assume that they are separated by infinitesimal distances. The largest disk has radius R, and the radius of each successive disk is 1/3 of the radius of the disk to its left.

A stick, whose mass m is the same as the mass of the largest disk, is placed on top of the disks (touching them all) and is released. It is possible to show (you can just accept this) that the above factor of 1/3 implies that the stick makes an angle of 30° with the horizontal. Assuming that the stick doesn't slip with respect to any of the disks, what is its diagonally downward acceleration, during the time when it is still in contact with all the disks? You will need to use the sum of an infinite geometric series, which is $1 + b + b^2 + \cdots = 1/(1-b)$.

7.32. Rolling off a sphere

(a) A small ball with mass m and radius r is initially at rest on the top of a large fixed sphere with radius R. The ball's moment of inertia around its center is $I = \beta m r^2$, where β is a numerical factor. The ball is given an infinitesimal kick and rolls down the sphere, as shown in Fig. 7.26. Assume that the coefficient of static friction, μ, is essentially infinite, which implies that the ball doesn't slip with respect to the sphere, for essentially all of the time that it is in contact with the sphere. At what angle θ with respect to the vertical does the ball lose contact with the sphere? (You can assume that r is much smaller than R, to avoid a few complications, although the answer actually comes out the same in any case.)

7.4. MULTIPLE-CHOICE ANSWERS

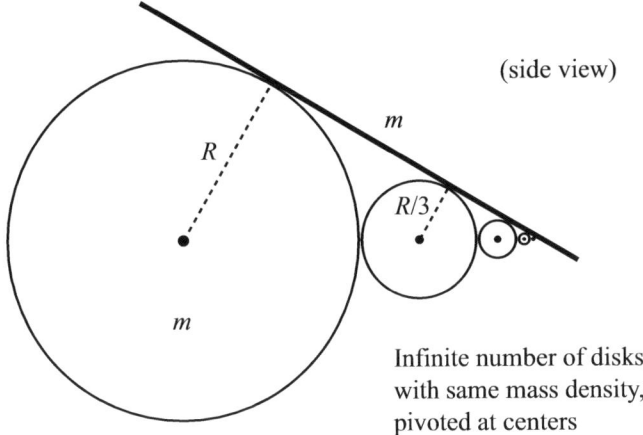

Figure 7.25

(b) Now assume that μ takes on a normal finite value. At what angle θ does the ball start to slip? It suffices to generate an equation that θ must satisfy. You need not solve for θ, but you should check the $\mu \to 0$ and $\mu \to \infty$ limits.

7.33. Yo-yo

If you drop a yo-yo from rest, while holding on to the end of the string, it will accelerate downward, reach a maximum speed, and then slow down. Eventually it will reach its lowest point when the string is completely unwound. (A yo-yo consists of two disks connected by an axle, with a string wrapped many times around the axle.)

(a) Explain qualitatively why the speed reaches its maximum value at an intermediate point, instead of at the bottom, as it would for a dropped ball. *Hint*: As the string unwinds, the radius of the wound-up spiral of string in the yo-yo becomes smaller.

(b) Now work things out quantitatively. Assume for simplicity that the (massless) string is wound in a single-layer spiral (like a movie reel) and that the spiral starts with the radius R of the yo-yo. What is the radius of the spiral when the yo-yo achieves its maximum speed? *Hints:* Let the string have a tiny thickness ϵ when viewed from the side. At any time, the area (in a cross-sectional side view) of the string that is "missing" from the spiral must equal the area of the string that is hanging down to the yo-yo. You can assume that the spiral is essentially circular at all times. Let the yo-yo have mass m, and assume that its moment of inertia is $mR^2/2$.

7.4 Multiple-choice answers

7.1. **b** In the $R = 0$ limit, we have a stick of length L with the axis at the center, so $I = ML^2/12$. In the $L = 0$ limit, we have a flat disk with the axis along a diameter, so $I = MR^2/4$ (from the list of I's on page 178). The answer must therefore be (b).

REMARK: If you want to solve the problem from scratch, you can slice the cylinder into disks and then use the parallel-axis theorem to find the I of each disk relative to the given axis (in terms of the position y of a disk). You can then integrate over all the disks.

7.2. **1 and 3** If r is half of the vertical span, then the first and third objects have $I = mr^2$, because all parts of the objects are a distance r from the axis. But the second object has $I < mr^2$, because all points (except the corners) are less than r from the axis.

7.3. $\boxed{\text{d}}$ Choice (a) is valid by Eq. (7.7). And choice (c) is valid by Eq. (7.8), which tells us how the kinetic energy breaks up for an object that is rotating around its CM while the CM is translating. The fact that one of the stick's ends happens to be at rest in the given setup is irrelevant in the application of Eq. (7.8).

REMARK: Both of Eqs. (7.7) and (7.8) are relevant here, but which one you use depends on how you want to view the object. The equality of choices (a) and (c) is basically the statement that the parallel-axis theorem is valid. To see why, let d be half the length of the stick. Then the speed of the CM is ωd, because it is rotating around the (fixed) end with angular speed ω. Choice (c) can therefore be written as $I_{CM}\omega^2/2 + m(\omega d)^2/2$. (And yes, the same ω appears in both terms here. The stick makes a given number of revolutions per second, independent of which point you consider it to be rotating around.) Equating this with choice (a) and canceling a factor of $\omega^2/2$ yields $I_{end} = I_{CM} + md^2$, which is the parallel-axis theorem, as desired. If you consider a general object (rotating around a given axis) instead of a stick, this reasoning actually provides a much quicker proof of the parallel-axis theorem than the one given in Problem 7.1, as you can verify.

7.4. $\boxed{\text{b}}$ The static friction force does no work in the left side of the valley, because the thing applying the force (the ground) isn't moving. So energy is conserved. At the highest point on the right side, the ball is still rotating. This is true because it was rolling at the bottom, and there is no friction on the right side, and hence no torque to slow down the rotation. So at the highest point on the right side, there is energy contained in the rotation. The potential energy must therefore be less than at the start. In other words, the ball doesn't go as high.

7.5. $\boxed{\text{c}}$ All of the sticks have the same total energy E, because you do the same work $F\Delta x$. From Eq. (7.8) the energy is $E = mv_{CM}^2/2 + I_{CM}\omega^2/2$. For the first two sticks, ω is nonzero, so some of the energy is "wasted" in rotational motion. Since $\omega = 0$ for the third stick, v_{CM} acquires its largest possible value.

REMARK: We can ask a similar question, but now with the force acting for the same small *time* Δt for all three sticks, instead of the same small *distance* Δx. In this new setup, you apply the same *impulse*, $F\Delta t$. So all three sticks have the same final *momentum*. But the momentum of an object is determined solely by the motion of the CM and not by whatever motion may be taking place relative to the CM; see Eq. (6.10). So all three sticks end up with the same v_{CM}.

However, we now have a puzzle in the same-Δt case. The first two sticks have additional energy associated with rotation, so from Eq. (7.8) they must have *more* total energy than the third stick, since all the v_{CM}'s are the same. How is this possible if you push on all three sticks for the same time? As we saw above in the solution to the original question, the energy equals the *work* you do, $F\Delta x$. So in the present scenario it must be the case that although you push on the sticks for the same *time* Δt, you apparently push on the first two sticks for a longer *distance* Δx (with Δx being the longest for the first stick, because it will rotate the fastest since it experiences the largest torque). And indeed, as you push on the first stick, the end "recoils" as it rotates away from you. So your hand moves a longer distance Δx, even though the centers of all three sticks move the same amount.

7.6. $\boxed{\text{b}}$ The friction force on the ball at its bottom point is directed rightward. This causes a rightward (positive) acceleration and a counterclockwise (negative) angular acceleration.

7.7. $\boxed{\text{c}}$ The mass is more concentrated in the center in choice (c). So when I is written in the form of βmr^2, where β is a numerical coefficient, the β for choice (c) is the smallest. This means that when potential energy is converted into kinetic energy as the wheel rolls down the plane, less energy goes into rotational energy (the second term in Eq. (7.8), which involves I_{CM}) and more energy goes into translational energy (the first term in Eq. (7.8), which involves v_{CM}). The CM speed for (c) is therefore largest, so it reaches the bottom first.

REMARK: The complete ordering, from fastest to slowest, is (c), (a), (b), (d). Equivalently, this is the ordering of the I's, from smallest to largest. This is true because the mass in (a) is generally farther away from the center than in (c). And likewise farther in (b) than in (a). And farther in (d) than in (b).

Note that the actual mass of each wheel doesn't matter; only the distribution (which determines β) matters. This is true for basically the same reason that the speed of a falling object is independent of its mass (ignoring air drag). The mass m appears in every term in the conservation-of-energy statement (always to the first power), so it cancels out. Equivalently, the speed can't depend on m, due to dimensional analysis; there is no other quantity involving units of kg in the setup, to cancel the kg's. (I isn't an independent quantity, since it can be written as $I = \beta m r^2$.)

If you want to quantitatively solve for the speed of a particular wheel (given β) after it rolls a distance d down the plane (inclined at angle θ), you can quickly find it by applying conservation of energy along with the non-slipping condition, $v_{CM} = R\omega$ (see Problem 7.8). You can check your answer by noting that it must reduce to the free-fall result of $\sqrt{2gd}$ when $\theta = 90°$ and $\beta = 0$. See Problem 7.25.

7.8. $\boxed{b}$ The angular speed ω increases as the ball rolls down the plane and picks up speed. (The linear speed increases too, of course. It is related to ω by $v = R\omega$.) The friction force produces the torque (around the CM) that makes ω increase, so the friction force must point up the plane. It therefore causes a_{CM} to be smaller than $g \sin \theta$ (the component of g pointing down the plane).

Note that the friction force is the only force that can produce a torque around the CM. Gravity can't do the job, because it effectively acts at the CM and therefore has zero lever arm relative to the CM. The normal force from the plane also has zero lever arm.

REMARK: The friction force F_f is responsible for doing *two* things. It is the force that produces the torque that increases ω, and it is also the force that makes the ball accelerate slower than $g \sin \theta$. The *entire* friction force F_f appears in *each* of these effects. It is *not* the case that the force gets divided into two pieces, where one piece produces the torque and the other piece works against gravity. The entire friction force F_f appears on both the torque and force equations that you would write down if you wanted to solve things quantitatively. You might think that you are "double counting" the force this way, but you aren't.

7.9. $\boxed{c,d}$ The friction force in choice (c) works against the $mg \sin \theta$ gravitational force pulling the ball down the plane, making the acceleration of the ball be smaller than the $g \sin \theta$ acceleration of the block. In choice (d), the energy contained in the rotational motion of the ball implies (by conservation of energy and Eq. (7.8)) that there is less energy contained in the translational motion, which means that v_{CM} is smaller than it would be if the ball were sliding down a frictionless plane (in which case it would have the same $g \sin \theta$ acceleration as the block). Choice (a) is incorrect, because the planes are inclined at the same angle. The statement in choice (b) is true but isn't relevant to the acceleration of the objects.

7.10. $\boxed{e}$ Your forearm is pivoted at your elbow, and since your forearm is at rest, the torques on it relative to your elbow must cancel. The torques come from the 20-lb weight which has your whole forearm as its lever arm, and from your biceps muscle which has a much smaller lever arm since your biceps attaches to your forearm fairly close to your elbow. As a rough guess, this lever arm might be 1/10 of your whole forearm. (There is also the torque from the weight of your arm, but this is small in comparison.) Since the torques must balance, the products of the forces and lever arms must be the same. This means that the biceps force must be 10 times the 20-lb weight. That is, it must be 200 lbs. The next largest choice, 40 lbs, would imply a lever arm that is half of your forearm, but it certainly isn't that large.

REMARK: The lever arms of the muscles in our bodies are generally much smaller than the lengths of the bones they attach to, which means (due to the torque reasoning above) that the internal forces in our bodies are generally much larger than the external forces we apply or feel. For example, when standing on your toes, the tension in your calf muscles is much larger than your weight. And when standing up from a squat, the tension in your quad muscles is much larger than your weight.

You might wonder why humans didn't evolve in a way that gave us larger lever arms, so that we could apply larger forces. If we did evolve that way, we wouldn't have thin limbs, but instead would be rather "blobby" (imagine your biceps muscle attaching to the middle of your forearm). But looks aside, the reason for small lever arms is undoubtedly one of speed. With small lever arms, a slight movement in a muscle causes a large movement in the end of the bone that the muscle is attached to. So small lever arms mean that we can move our feet and hands quickly. (But they can't be *too* small, otherwise the required forces would be prohibitive.) This is good for running and throwing, which in turn is good for eating (and avoiding being eaten).

7.11. $\boxed{c}$ There is certainly a friction force from the ground. This force points forward and is what accelerates the bike via $F = ma$. However, this force produces a torque in the direction that corresponds to the wheel moving *backward*. So there must be another torque on the rim, and the only other things that touch the rim are the spokes. So the spokes must apply a (slightly larger) torque that corresponds to the wheel moving forward. This torque wins, and the bike moves forward.

REMARK: A radial force can't apply a torque, because the lever arm around the center of the wheel is zero. So if you look at the back wheel of a bicycle, you will see that the spokes aren't radial; they have a slight tangential component. This might not be obvious from looking at how they attach to the rim, but it is clear when looking at how they attach to the hub. The non-radial connection to the hub allows the hub (which is connected to the chain, pedals, and you) to apply a torque to the spokes. In contrast, the front-wheel spokes *can* be radial, because that wheel just goes along for the ride. The force that provides the necessary "forward" torque on the front wheel is a small *backward* friction force from the ground.

7.12. $\boxed{c}$ The torque from the string (relative to the pivot) is the same in all five cases, because it must balance the gravitational torque on the stick (which is the same in all five cases), so that the stick remains at rest. The torque from the string equals the product of the tension and the lever arm. So the smallest tension is associated with the largest lever arm. The lever arm of a force is defined to be the distance from the pivot to the line drawn through the force (so we just need to extend the string in the present setup). The lever arm is largest in choice (c), because it equals the length of the stick in this case, and there is no way for it to be any larger than that. So the tension is smallest in (c).

7.13. $\boxed{d}$ Gravity effectively acts at the CM, as far as the torque is concerned. So all of the objects experience the same torque around the pivot. Therefore, $\tau = I\alpha$ tells us that the object with the smallest I around the pivot will have the largest α. The parallel-axis theorem says that $I_{\text{pivot}} = I_{\text{CM}} + md^2$. Since m and d are the same for all of the objects, we want the object with the smallest I_{CM}. This is choice (d) because $I_{\text{CM}} \approx 0$ for a (nearly) point mass.

REMARK: Assuming that the cross sticks in (b) and (e) have length d, and that the disk in (c) has radius $d/4$, you can show as an exercise that the ordered list of the I's, from smallest to largest, is (d), (c), (b), (e), (a).

7.14. $\boxed{a}$ Relative to the center of the massless stick where the point mass is, the moment of inertia I of the stick (plus point mass) is zero. So $\tau = I\alpha$ (around the center) implies that $\tau = 0$. (A nonzero τ with a zero I would lead to an infinite α.) A zero torque then implies that the force from the table is zero. (It then follows that the mass is in freefall; the stick effectively doesn't exist).

7.15. $\boxed{d}$ In the $\tau = I\alpha$ equation, τ_{new} is half of τ_{old} because the force is the same but the lever arm (the radius) has been cut in half. How has I changed? The I of a cylinder is $mR^2/2$. And $m_{\text{new}} = m_{\text{old}}/4$ because the cross-sectional area (and hence volume, and hence mass) of the cylinder has been decreased by a factor of $(R_{\text{new}}/R_{\text{old}})^2 = 1/4$, because the area is proportional to R^2. So mR^2 has decreased by a factor of $(1/4)(1/2)^2 = 1/16$. Therefore,

$$\alpha_{\text{new}} = \frac{\tau_{\text{new}}}{I_{\text{new}}} = \frac{\tau_{\text{old}}/2}{I_{\text{old}}/16} = 8\frac{\tau_{\text{old}}}{I_{\text{old}}} = 8\alpha_{\text{old}}. \tag{7.12}$$

REMARK: We can also answer this question by using dimensional analysis. What does α (with units of $1/s^2$) depend on? The various parameters that describe the setup are the force F (units kg m/s^2), the radius R (units m), the mass density ρ (units kg/m^3), and the length ℓ (units m) of the cylinder. As mentioned in, for example, Problem 1.5, the four parameters here imply that we need an additional fact. This fact is that α must be proportional to $1/\ell$, because making the cylinder twice as long will double the moment of inertia, all other things being equal. You can then quickly show that the only way to obtain the desired units of α is to have $\alpha \propto F/\rho\ell R^3$. The R^3 in the denominator yields the factor of $2^3 = 8$ in choice (d).

7.5 Problem solutions

7.1. Parallel-axis theorem

As suggested, let's consider the special case where the object is planar and the axes are perpendicular to this plane. In Fig. 7.27, P is the origin of the coordinate system, and the position of the CM of the object is $\mathbf{R}_{CM} = (x_{CM}, y_{CM})$. Let $\mathbf{r}' = (x', y')$ be the position of a general point in the object relative to the CM. Then the position $\mathbf{r} = (x, y)$ of this point relative to the origin is

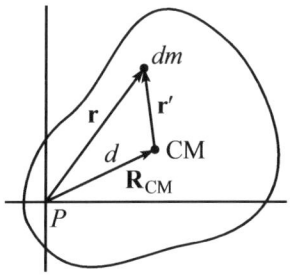

Figure 7.27

$$\mathbf{r} = \mathbf{r}' + \mathbf{R}_{CM} \implies (x, y) = (x' + x_{CM}, y' + y_{CM}). \tag{7.13}$$

The moment of inertia around the axis passing through P and perpendicular to the page is then

$$\begin{aligned} I_P &= \int r^2\, dm = \int (x^2 + y^2)\, dm = \int \left[(x' + x_{CM})^2 + (y' + y_{CM})^2 \right] dm \\ &= \int (x'^2 + y'^2)\, dm + 2 x_{CM} \int x'\, dm + 2 y_{CM} \int y'\, dm + (x_{CM}^2 + y_{CM}^2) \int dm \\ &= \int r'^2\, dm + 0 + 0 + R_{CM}^2 M \\ &= I_{CM} + M d^2, \end{aligned} \tag{7.14}$$

as desired. The two zeros in the third line appear due to the definition of the CM. In short, Eq. (6.8) tells us that $(\int x'\, dm)/M$ is the x coordinate of the position of the CM, calculated with respect to the CM (because x' is measured with respect to the CM), which is zero by definition.

REMARKS: The above derivation makes it clear why the parallel-axis theorem holds only if the CM is used on the right-hand side of Eq. (7.5). If we had picked a general point instead of the CM, then the two zeros in Eq. (7.14) wouldn't appear. Instead we would have integrals associated with the position of the CM relative to the point we had chosen, which would mean that at least one of the two integrals would be nonzero.

In the more general case of a non-planar object, the proof is nearly the same. Let us again orient our coordinate system so that the rotation axes are perpendicular to the page. Then Fig. 7.27 is still relevant, with the only exception being that the CM might not lie in the plane of the page. But this doesn't matter. If we define $\mathbf{R}_{CM}$ to be the cylindrical-coordinate position of the CM relative to the *axis* through P, and if we likewise define $\mathbf{r}'$ to be the position of a general point in the object relative to the *axis* through the CM, then the position $\mathbf{r}$ of this point relative to the *axis* through P is $\mathbf{r} = \mathbf{r}' + \mathbf{R}_{CM}$. So all of the steps in the above proof are still valid; the z coordinates of the points never come into play.

7.2. Perpendicular-axis theorem

In Fig. 7.3 the z axis points out of the page. The moment of inertia around the z axis is

$$I_z = \int r^2\, dm = \int (x^2 + y^2)\, dm = \int x^2\, dm + \int y^2\, dm. \tag{7.15}$$

Since x is the distance from the y axis (because the object lies in the x-y plane; this is where the planar assumption comes in), the $\int x^2\, dm$ integral is just I_y (not I_x!). Likewise, the $\int y^2\, dm$ integral is I_x. So we have $I_z = I_y + I_x$, as desired.

7.3. Rotational kinetic energy

If we divide the object into many little masses m_i, the kinetic energy of each mass is $m_i v_i^2/2$. But since the object is undergoing pure rotation, the speed of each mass is given by $v_i = r_i \omega$, where r_i is the distance from the axis. The total kinetic energy of all the masses (in other words, the kinetic energy of the entire object) is therefore

$$K = \sum \frac{1}{2} m_i v_i^2 = \sum \frac{1}{2} m_i (r_i \omega)^2 = \frac{1}{2} \left(\sum m_i r_i^2 \right) \omega^2 = \frac{1}{2} I \omega^2, \tag{7.16}$$

as desired. The critical part of this derivation is the fact that although the masses in general have different speeds v_i, they all have the *same* angular speed ω. This is why we could take ω outside the above sum.

In the continuum limit, the sum $\sum m_i r_i^2$ is replaced by $\int r^2\, dm$. Equivalently, you can just use integrals throughout the above derivation, starting with $K = (1/2) \int v^2\, dm$.

7.4. Translation plus rotation

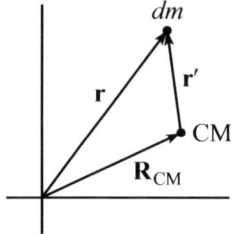

Figure 7.28

In Fig. 7.28 let the position of the CM of the object (we haven't drawn the object) relative to an arbitrary fixed origin be $\mathbf{R}_{\mathrm{CM}}$, and let the position of a general point in the object relative to the CM be $\mathbf{r}'$. (The figure is planar since we're drawing it on the page, but this restriction isn't necessary. This derivation holds for a general 3-D object.) Then the position of a general point relative to the origin is $\mathbf{r} = \mathbf{R}_{\mathrm{CM}} + \mathbf{r}'$. So the velocity of the point is $\mathbf{v} = \mathbf{V}_{\mathrm{CM}} + \mathbf{v}'$. The total kinetic energy of the object is therefore

$$\begin{aligned} K &= \int \frac{1}{2} v^2\, dm = \frac{1}{2} \int \mathbf{v} \cdot \mathbf{v}\, dm = \frac{1}{2} \int (\mathbf{V}_{\mathrm{CM}} + \mathbf{v}') \cdot (\mathbf{V}_{\mathrm{CM}} + \mathbf{v}')\, dm \\ &= \int \frac{1}{2} V_{\mathrm{CM}}^2\, dm + \int \mathbf{V}_{\mathrm{CM}} \cdot \mathbf{v}'\, dm + \int \frac{1}{2} v'^2\, dm, \end{aligned} \tag{7.17}$$

where we have used the fact that the square of the length of a vector $\mathbf{a}$ can be written as $a^2 = \mathbf{a} \cdot \mathbf{a}$. Let's look at each of the three terms in the above result. The first term equals

$$\frac{1}{2} V_{\mathrm{CM}}^2 \int dm = \frac{1}{2} M V_{\mathrm{CM}}^2, \tag{7.18}$$

which is the first term in Eq. (7.8).

The second term equals $\mathbf{V}_{\mathrm{CM}} \cdot \int \mathbf{v}'\, dm$. But $\int \mathbf{v}'\, dm$ is the total momentum of the object in the CM frame, which is zero. Equivalently, $\int \mathbf{v}'\, dm = (d/dt) \int \mathbf{r}'\, dm = 0$, because $\int \mathbf{r}'\, dm$ is the position of the CM of the object as measured with respect to the CM, which is zero (and hence constant).

The third term equals

$$\int \frac{1}{2} (r' \omega)^2\, dm = \frac{1}{2} \left(\int r'^2\, dm \right) \omega^2 = \frac{1}{2} I_{\mathrm{CM}} \omega^2, \tag{7.19}$$

which is the second term in Eq. (7.8). We have used the fact that $v' = r' \omega$, because if you are riding along on the CM, you see the object simply rotating with angular velocity ω around the CM.

The proof here was very similar to the proof in Problem 7.1. In both cases the cross term(s) were zero due to the definition of the CM.

7.5. PROBLEM SOLUTIONS

7.5. Gravitational torque

We'll divide the object into many little masses dm and then integrate the gravitational torque over all the masses. The gravitational force on a mass dm is $d\mathbf{F} = dm\,\mathbf{g}$, where the vector $\mathbf{g}$ points downward with magnitude g. The total gravitational torque on the object is therefore

$$\tau = \int d\tau = \int \mathbf{r} \times d\mathbf{F} = \int \mathbf{r} \times (dm\,\mathbf{g})$$
$$= \left(\int \mathbf{r}\,dm\right) \times \mathbf{g} = (m\mathbf{R}_{\text{CM}}) \times \mathbf{g}$$
$$= \mathbf{R}_{\text{CM}} \times (m\mathbf{g}), \tag{7.20}$$

which is the gravitational torque on a point mass m located at the CM, as desired. The fifth of the above equalities follows from the fact that the location of the CM is defined to be $\mathbf{R}_{\text{CM}} = (\int \mathbf{r}\,dm)/m$; see Eq. (6.8).

REMARK: Although we just showed that an object can be treated like a point mass at the CM when calculating the gravitational torque, it *cannot* be treated like a point mass when calculating the moment of inertia I. The actual distribution of mass matters in I.

7.6. The $\tau = I\alpha$ relation

For the setup given in the problem, the small amount of work done on the object by the applied force over the small tangential displacement $d\mathbf{s}$ is $dW = F\,ds\cos\theta$. But $ds = r\,d\phi$, where ϕ is the angle of rotation. So the work done is

$$dW = F(r\,d\phi)\cos\theta = (F\cos\theta)(r)\,d\phi$$
$$= (\text{tangential force})(\text{radius})\,d\phi$$
$$= \tau\,d\phi. \tag{7.21}$$

Basically, work is force times distance, but if we take a factor of length out of the distance and put it with the force, then we see that work also equals torque times angle.

The change in the kinetic energy of the object is

$$dK = d\left(\frac{1}{2}I\omega^2\right) = I\omega\,d\omega. \tag{7.22}$$

Equating this with the work done gives

$$dW = dK \implies \tau\,d\phi = I\omega\,d\omega \implies \tau\,d\phi = I\frac{d\phi}{dt}\,d\omega$$
$$\implies \tau = I\frac{d\omega}{dt} \implies \tau = I\alpha, \tag{7.23}$$

as desired. It was indeed legal to cancel the $d\phi$'s here and put the $d\omega$ over the dt. If you want, you can imagine working with finite quantities, $\Delta\phi$, $\Delta\omega$, Δt, for which these manipulations are certainly legal, and then taking the infinitesimal limits of these quantities. See Section 8.4 in Morin (2008) for another proof of $\tau = I\alpha$.

7.7. Force along a massless stick

Consider the torque on the stick around one of its ends; call this end A. If the stick is massless, then the torque on it (around any point) must be zero, because otherwise there would be infinite angular acceleration. The only forces on the massless stick are the forces at the two pivots, because there is no gravitational force. The pivot at end A produces zero torque around A. The pivot at the other end B must therefore also produce zero torque around A. This implies that the force at B must point radially long the stick, as we wanted to show. The same reasoning works with A and B interchanged. Note that this reasoning holds whether the system is static or moving.

REMARKS: If the stick is instead *massive*, then there is now (1) in general a gravitational torque and (2) a nonzero moment of inertia, which means that the total torque need not be zero. Each of these facts allows for there to be a nonzero torque due to the force at end B. So the force doesn't necessarily point along the stick.

Note that the above conclusion for a massless stick relies on the stick being connected to the objects at its ends by *pivots*. If we instead have a pivot at one end and a rigid clamp at the other, then the conclusion doesn't hold. The force at the ends *can* have a component that is transverse to the stick. You can see this by imagining trying to twist the left object in Fig. 7.29. There will be an upward transverse force on the right object at the pivot, and also a downward transverse force on the left object at the clamp.

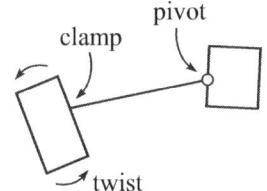

Figure 7.29

In addition to massless sticks, a completely flexible stationary string (massive or massless) is another type of object for which the force at an end is always directed along the object. There can be no transverse force at an end, because if there is, the string will simply bend until the force *does* point along it. You can demonstrate this by painting a dot on the string near the end and looking at the torque on the little piece between the dot and the end, relative to the dot. However, there is actually no need for this reasoning, because the definition of a "completely flexible" string is one that can't apply a transverse force anywhere.

7.8. Non-slipping condition

(a) In Fig. 7.30 consider a point A on the rim of the wheel that is an angle θ from the present contact point C on the ground. Point A will eventually hit the ground at the point B shown. The distance along the rim of the wheel from C to A is $R\theta$ (that's how an angle θ measured in radians is defined). And the distance along the ground from C to B is the distance d the wheel travels. But the non-slipping condition is the statement that the rim distance from C to A equals the ground distance from C to B. Hence $R\theta = d$, as desired. Taking the first time derivative of $d = R\theta$ gives $v = R\omega$, and then taking another derivative gives $a = R\alpha$.[1]

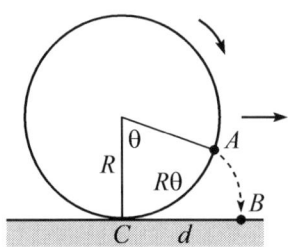

Figure 7.30

REMARK: If you want to be more rigorous about the statement that "the rim distance from C to A equals the ground distance from C to B," you can imagine the wheel spinning in place, with the ground moving by to the left. If there is no slipping, then in a short interval of time, the ground must move to the left by exactly the same distance that the bottom points on the rim (in the region near the contact point) move to the left. This is true because locally we simply have two flat surfaces, and the non-slipping condition implies that the two surfaces must move the same distance. When all the little distances are added up, the end result is that the rim distance from C to A equals the ground distance from C to B.

(b) Perhaps the easiest way to show this is to imagine (as in the above remark) the wheel spinning in place. In this reference frame, the ground moves to the left and the board moves to the right. Since there is no slipping, they both move a distance $R\theta$ if the wheel rotates by an angle θ. If we now shift back to the reference frame of the ground, the wheel moves to the right by $R\theta$, and the board (which moves to the right by $R\theta$ more than the wheel) moves to the right by $R\theta + R\theta = 2R\theta$. So the board does indeed move twice as far as the wheel. The various relations between the linear quantities of the board and the angular quantities of the wheel are $d_b = 2R\theta$, $v_b = 2R\omega$, and $a_b = 2R\alpha$.

REMARK: Another way of understanding this result is to note that at a given instant, the wheel can be considered to be rotating around the contact point on the ground, because this point is instantaneously at rest. In particular, both the top point on the wheel and the center instantaneously move along arcs of circles around the contact point. Therefore, since the top point

---

[1] The $a = R\alpha$ relation is actually only a necessary, but not sufficient, condition for not slipping. This can be seen by noting that $v = R\omega + C$ yields $a = R\alpha$ but involves slipping. However, this technicality won't concern us, because the true statement, "If there is no slipping, then a must equal $R\alpha$," is the one we will make use of.

7.5. PROBLEM SOLUTIONS

on the wheel is twice as far from the contact point as the center is, the top point (and hence also the board, since there is no slipping) moves twice as far as the center, for a given angle of rotation.

7.9. *I* for a square

In the first equation, we have $A = 4$, because the big square can be constructed by putting four of the little squares together (with their dots at a common point). And moments of inertia simply add.

To find B, we'll use a scaling argument. Consider a small piece of the little square, and look at the corresponding piece of the big square. The latter piece has four times the mass (because it has $2^2 = 4$ times the area), and it is twice as far from the center. The expression for the moment of inertia, $I = \int r^2\, dm$, therefore picks up a factor of $2^2 \cdot 4 = 16$ when comparing I_{2L} to I_L. So $B = 16$.

To find C, we can use the parallel-axis theorem. The axes (the dots) are $L/\sqrt{2}$ away from each other, so $C = M(L/\sqrt{2})^2 = ML^2/2$.

We now have three equations and three unknowns (represented by the figures). The first two equations give $I_{L,\text{corner}} = 4 I_{L,\text{center}}$. Plugging this into the third equation then yields

$$I_{L,\text{center}} = \frac{ML^2}{6}. \tag{7.24}$$

REMARKS: You should convince yourself why the perpendicular-axis theorem says that this result is correctly twice the $ML^2/12$ result for a uniform stick around its center.

Note that at least one of the three given equations (the third one) needs to have an additive constant in it. If all of the equations involved only multiplicative factors (like the first two), then we would be able to determine the unknowns only up to an overall scaling factor.

7.10. Another *I* for a square

(a) As suggested, we'll slice the square into thin strips, as shown in Fig. 7.31(a). Let the mass of each strip be dm. The moment of inertia of each strip around its center is $(dm)L^2/12$, so the moment of inertia of the whole square is

$$I = \int \frac{1}{12}(dm)L^2 = \frac{L^2}{12}\int dm = \frac{1}{12}ML^2. \tag{7.25}$$

The square has the same factor of $1/12$ in I that a strip has, because the extension of the square in the vertical direction is irrelevant. I depends only on the distance from the axis of rotation, not on the position along it.

(b) Let's now slice the square into the strips shown in Fig. 7.31(b). Let y be the height above the bottom corner. We'll find the moment of inertia of just the bottom half, and then we'll double the result. In terms of y, the length of a strip in the bottom half is $2y$, so the area of the strip is $(2y)\,dy$. The mass is then $dm = (2y\,dy)\sigma$, where $\sigma = M/L^2$ is the mass density (per unit area). The moment of inertia of a strip is therefore

$$dI = \frac{1}{12}(dm)(2y)^2 = \frac{1}{12}(2y\sigma\, dy)(2y)^2 = \frac{2}{3}\sigma y^3\, dy. \tag{7.26}$$

Integrating this from $y = 0$ to $y = L/\sqrt{2}$ to find the moment of inertia of the bottom half of the square gives

$$I_{\text{half}} = \int_0^{L/\sqrt{2}} \frac{2}{3}\sigma y^3\, dy = \frac{\sigma}{6} y^4 \Big|_0^{L/\sqrt{2}} = \frac{\sigma L^4}{24} = \frac{(M/L^2)L^4}{24} = \frac{ML^2}{24}. \tag{7.27}$$

Doubling this to get the I for the whole square gives $I = ML^2/12$.

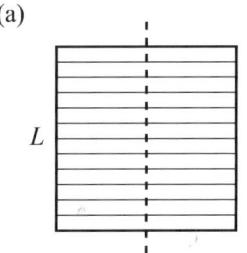

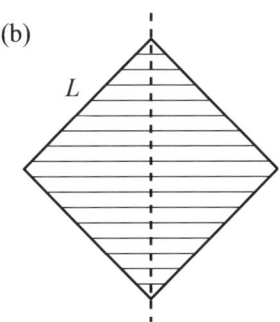

Figure 7.31

(c) The answers to parts (a) and (b) are the same, and the following reasoning involving the perpendicular-axis theorem explains why they must be equal. In Fig. 7.32(a), the I's for the two axes shown are equal, by symmetry. Since the sum of these I's equals the I for the axis pointing perpendicular to the page (by the perpendicular-axis theorem), these two I's must each be half of that I. The same reasoning holds for the case shown in Fig. 7.32(b). Therefore, the results in parts (a) and (b) must both be equal to half of the I for the axis pointing perpendicular to the page (which apparently equals $ML^2/6$, in agreement with the result in Problem 7.9).

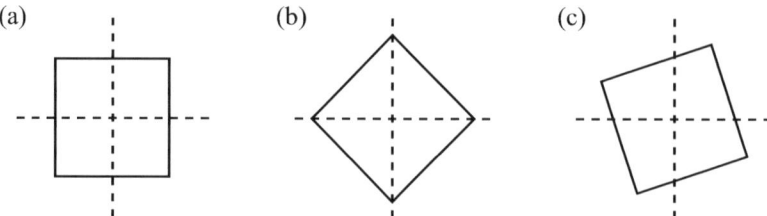

Figure 7.32

REMARK: Note that the above perpendicular-axis-theorem reasoning, applied to the axes shown in Fig. 7.32(c), tells us that *any* axis through the center of the square (and lying in the plane of the square) has an I equal to $ML^2/12$.

7.11. *I* for a disk

A little patch of the disk spanning dr in the radial direction and $d\theta$ in the tangential direction is essentially a little rectangle with sides dr and $r\,d\theta$. So the area is $dr(r\,d\theta)$, which means that the mass is $dm = \sigma r\,dr\,d\theta$, where $\sigma = m/(\pi R^2)$ is the mass density (per unit area). The moment of inertia of the whole disk is therefore

$$I = \int r^2\,dm = \int_0^{2\pi}\int_0^R r^2 \sigma r\,dr\,d\theta = \left(\sigma\int_0^{2\pi}d\theta\right)\left(\int_0^R r^3\,dr\right)$$
$$= 2\pi\sigma\frac{R^4}{4} = \frac{(\sigma\pi R^2)R^2}{2} = \frac{mR^2}{2}. \tag{7.28}$$

REMARK: You can save the (trivial) step of integrating over θ by considering the disk to be made up of many concentric rings. Every point in a ring is the same distance from the center, so the moment of inertia of a ring with mass dm equals $(dm)r^2$. The mass of a thin ring with thickness dr is $dm = \sigma(2\pi r\,dr)$, so the moment of inertia is $\sigma(2\pi r\,dr)r^2$. Integrating over all the rings gives $I = \int_0^R \sigma(2\pi r\,dr)r^2 = \sigma\pi R^4/2 = mR^2/2$, as above. Although slicing up the disk into rings doesn't save much (if any) time in this setup, it is often helpful to slice up an object into sub-pieces whose I is already known.

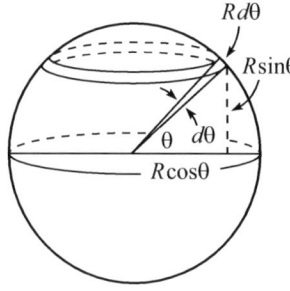

Figure 7.33

7.12. *I* for a spherical shell

Let's slice the shell into horizontal ring-like strips, as shown in Fig. 7.33. In terms of the angle θ shown, a strip has radius $r = R\cos\theta$ and width $R\,d\theta$, as indicated. The area is therefore $2\pi(R\cos\theta)R\,d\theta$, so the mass of the strip is $(2\pi R^2\cos\theta\,d\theta)\sigma$, where $\sigma = m/(4\pi R^2)$ is the mass density (per unit area). Using

$$\int\cos^3\theta = \int\cos\theta(1-\sin^2\theta) = \sin\theta - \frac{\sin^3\theta}{3} \tag{7.29}$$

(or you can just look up the integral in a table or plug it into a computer), and integrating over the rings from $-\pi/2$ to $\pi/2$, we find the moment of inertia of the shell to be

$$I = \int r^2\,dm = \int_0^\pi (R\cos\theta)^2(2\pi\sigma R^2\cos\theta\,d\theta) = 2\pi\sigma R^4\int_{-\pi/2}^{\pi/2}\cos^3\theta\,d\theta$$
$$= 2\pi\sigma R^4 \cdot \frac{4}{3} = \frac{2}{3}(4\pi R^2\sigma)R^2 = \frac{2}{3}mR^2. \tag{7.30}$$

7.5. PROBLEM SOLUTIONS

As an exercise, you can calculate the I for a solid sphere by building up the solid sphere from spherical shells and integrating over the shells. Alternatively, you can build up the solid sphere from flat disks. In ether case, the result is $I = (2/5)mR^2$. This is smaller than the $(2/3)mR^2$ result for the hollow shell, because the mass in the solid sphere is generally closer to the axis.

REMARK: Note that in the above calculation, we (correctly) used the "slant height" $R\,d\theta$ as the width of a strip. You might instead be tempted to use the vertical height dy of the strip as the width. After all, if you performed the suggested exercise of finding the I for a solid sphere by integrating over flat disks, then you *would* use the height dy as the thickness of a disk. So why did we use the slant height $R\,d\theta$ for the width of a strip in the case of the hollow shell? Imagine that you are a tiny bug on the strip. Then locally it looks like a flat strip with width $R\,d\theta$. This is how far you would need to walk to get from one edge of the strip to the other.

As an analogy, consider the curve shown in Fig. 7.34. If we want to find the *length* of the curve, then we need to add up all the *slanted* lengths along the curve. (Simply adding up all the dx components of the lengths would always just give the same total horizontal span of the curve, so it can't be correct.) On the other hand, if we want to find the *area* under the curve, then we need to add up the areas of many infinitesimally thin rectangles. And to calculate these areas, we *do* want to use the simple horizontal span dx (and *not* the slanted length) as the width of the rectangle.

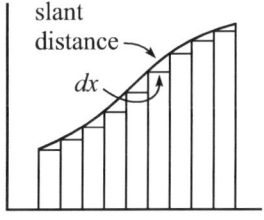

Figure 7.34

In short, the slant of the curve leads to two effects. First, the area under a section of the curve isn't exactly equal to the area of the thin rectangle, due to the existence of the tiny triangle on top of the rectangle. However, the area of this triangle is negligible compared with the area of the rectangle (in the limit where dx becomes small). Second, the slanted length of a section of the curve isn't equal to the horizontal span dx of the section. This difference *is* important. If, for example, a piece of the curve is slanted at, say, a 45° angle, then the piece is $\sqrt{2}$ times as long as the horizontal span dx. This factor cannot be ignored. Similar reasoning applies in the above case of the hollow and solid spheres, when dealing with the *area* of a strip and the *volume* of a disk; things are just kicked up by one dimension compared with the present example involving the *length* and *area* of the curve in Fig. 7.34.

7.13. Bending

As suggested, consider a given small section of the taller beam, along with the corresponding section of the shorter beam; see Fig. 7.35. The taller section is twice as tall, so it has *twice* as many springs applying a force. Additionally, each spring in the taller section is stretched (or compressed) *twice* as far as the corresponding spring in the shorter section, because it is twice as far from the pivot.[2] Furthermore, the springs in the taller section are twice as far from the pivot, so each spring provides *twice* as much torque relative to the pivot, for a given amount of stretching.

Therefore, the corresponding section of the taller beam has twice as many springs, each of which provides twice as much force with twice the lever arm. The corresponding section of the taller beam therefore provides $2^3 = 8$ times as much torque as the section of the shorter beam. Since this result holds for all corresponding sections, and since the beams are built up from all of the subsections, the same factor of 8 holds for the complete beams.

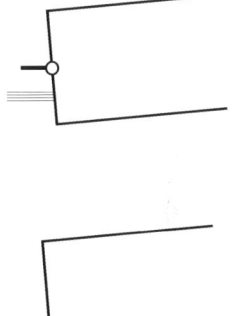

Figure 7.35

We conclude that for a *given angle* of bending, the external torque on the taller beam (due to, for example, upward forces on the ends) needs to be eight times as large as the external torque on the shorter beam. Equivalently, for a *given external torque*, the bending angle in the taller beam is 1/8 the bending angle in the shorter beam. This is true because with 1/8 the angle, each spring is stretched 1/8 of what it was in the above case (the stretching distance is proportional to the bending angle). So in the present case with 1/8 the angle, the corresponding section of the taller beam has (still) twice as many springs, each of which now provides $(1/8) \cdot 2 = 1/4$ as much force with (still) twice the lever arm. So the torque from the taller section is $2 \cdot (1/4) \cdot 2 = 1$ times the torque from the corresponding

---
[2]Since we are assuming that the bending angle is the same in the two beams, you can draw some similar triangles to demonstrate the factor of 2 in the stretching. Each spring itself in the taller beam isn't twice as long as the corresponding spring in the shorter beam, but the stretching distance is.

shorter section. These identical torques from the springs will therefore be balanced by identical external torques.

REMARK: With regard to the amount of sagging in a real beam, you should convince yourself that 1/8 the bending angle implies 1/8 the sag in the middle, at least for small bending angles. This requires a little thought, because a real beam is slightly bent along its entire length, not just in the middle as we assumed in our model. But we can imagine any location along the beam to consist of our little springs and pivot. So the 1/8 factor is relevant at all corresponding locations along the two beams.

7.14. Initial angular acceleration

We'll apply $\tau = I\alpha$ around the center of the disk. The torque relative to the center of the disk is due only to the stick, so the initial torque is $\tau = (mg)R = (\lambda\ell)gR$. The center of the stick is a distance $\sqrt{(\ell/2)^2 + R^2}$ from the center of the disk, so the parallel-axis theorem gives the total moment of inertia of the system around the center of the disk as

$$I = I_{\text{disk}} + I_{\text{stick}}$$
$$= \frac{1}{2}mR^2 + \left[\frac{1}{12}(\lambda\ell)\ell^2 + (\lambda\ell)\left(\left(\frac{\ell}{2}\right)^2 + R^2\right)\right]$$
$$= \frac{1}{2}mR^2 + \frac{1}{3}\lambda\ell^3 + \lambda\ell R^2.$$
(7.31)

The initial angular acceleration is then found from $\tau = I\alpha$ to be

$$\alpha = \frac{\tau}{I} = \frac{\lambda\ell gR}{mR^2/2 + \lambda\ell^3/3 + \lambda\ell R^2}.$$
(7.32)

This equals zero if $\ell = 0$ (there is no stick), and it also equals zero if $\ell \to \infty$ (the huge I of the stick dominates). So it must reach a maximum for some intermediate value of ℓ. Taking the derivative with respect to ℓ and setting the result equal to zero (and ignoring the overall factor of λgR, and also ignoring the denominator of the result because we are setting the derivative equal to zero) gives

$$0 = \left(\frac{1}{2}mR^2 + \frac{1}{3}\lambda\ell^3 + \lambda\ell R^2\right)(1) - \ell(\lambda\ell^2 + \lambda R^2)$$
$$= \frac{1}{2}mR^2 - \frac{2}{3}\lambda\ell^3 \implies \ell = \left(\frac{3mR^2}{4\lambda}\right)^{1/3}.$$
(7.33)

Since the units of λ are kg/m, this result correctly has units of meters.

LIMITS: You can verify that both the α in Eq. (7.32) and the ℓ in Eq. (7.33) have the correct behavior when the various parameters become very large or very small. For example, if $R \to \infty$ then $\alpha \to 0$, because the moments of inertia of both the disk and the stick are so large. And if $R \to 0$ then we again have $\alpha \to 0$, because the lever arm of the stick's torque is so small.

7.15. Atwood's with a massive pulley

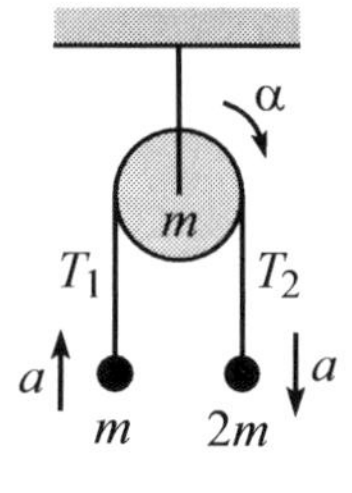

The main point here is that the tension in the string is different on either side of the pulley. Different tensions are required if there is to be a nonzero net torque on the pulley, which there must be because the (massive) pulley accelerates angularly as the masses accelerate linearly.[3] Let the two tensions be T_1 and T_2, as shown in Fig. 7.36. (The tension varies continuously as the string wraps around the pulley, but we care only about the tensions where the string leaves the pulley. These are the external forces causing the torques.) With the sign conventions shown, the various force and torque equations are (using the non-slipping condition, $\alpha = a/R$)

Figure 7.36

---

[3]The pulleys in the Atwood's problems in Chapter 4 were massless. Or more precisely, they had zero moment of inertia. This implied that the tensions were the same on either side (otherwise there would be infinite angular acceleration).

7.5. PROBLEM SOLUTIONS

- $F = ma$ on m: $T_1 - mg = ma$.
- $F = ma$ on $2m$: $(2m)g - T_2 = (2m)a$.
- $\tau = I\alpha$ on pulley: $T_2 R - T_1 R = (mR^2/2)(a/R) \implies T_2 - T_1 = ma/2$.

Our unknowns are T_1, T_2, and a. Solving for T_1 and T_2 in the first two equations and plugging the results into the third equation gives

$$(2mg - 2ma) - (mg + ma) = \frac{1}{2}ma \implies mg = \frac{7}{2}ma \implies a = \frac{2g}{7}. \qquad (7.34)$$

This is the acceleration of both masses; m goes up and $2m$ goes down.

REMARK: If the pulley were massless, then a would be larger, of course. The only modification is that we would need to erase the $ma/2$ term in Eq. (7.34), because that term can be traced to the moment of inertia. So we would end up with $a = g/3$. This acceleration makes sense, because a net force of $2mg - mg$ pulls down on the right side, and this force accelerates the total mass of $m + 2m$.

7.16. Equivalent mass

As in Problem 7.15, the tension in the string is different on either side of the pulley in the first setup. Let the tensions be T_1 and T_2, as shown in Fig. 7.37. Using the non-slipping condition, $\alpha = a/R$, the $\tau = I\alpha$ equation for the pulley is

$$T_2 R - T_1 R = (\beta m R^2)\alpha \implies T_2 R - T_1 R = (\beta m R^2)(a/R)$$
$$\implies T_2 - T_1 = (\beta m)a. \qquad (7.35)$$

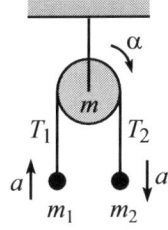

Figure 7.37

Now consider the second of the given setups. If the accelerations of the masses m_1 and m_2 are to be the same as they are in the first setup, then we must have the same tensions T_1 and T_2 as in the first setup, because the gravitational forces are the same in both setups. The $F = ma$ equation for the block with mass m' is then

$$T_2 - T_1 = m'a. \qquad (7.36)$$

Comparing this with Eq. (7.35), we see that $m' = \beta m$. Equivalently, since $I = \beta m R^2$, we have $m' = I/R^2$.

LIMITS: If $\beta = 0$ (that is, if all of the pulley's mass is at its center), then the equivalent mass m' equals zero. This makes sense, because the pulley is effectively massless (more precisely, it is moment-of-inertia-less), so it provides no resistance to the acceleration of the hanging masses.

If $\beta = 1$ (that is, if all of the pulley's mass is on its rim), then $m' = m$. This makes sense, because we can equivalently collapse all of the mass to one point on the rim, in which case the $T_2 - T_1$ net tension needs to accelerate a single mass, just as it does in the second setup.

If $\beta = 1/2$ (that is, if we have a uniform pulley), then $m' = m/2$. If we let $m_1 = m$ and $m_2 = 2m$ as in Problem 7.15, then in the second setup with the m' mass, we have a total mass of $m_{\text{tot}} = m + 2m + m/2 = 7m/2$. A net gravitation force of $F_{\text{net}} = 2mg - mg = mg$ pulls down on the right side, so the acceleration of the system is $a = F_{\text{net}}/m_{\text{tot}} = mg/(7m/2) = 2g/7$, in agreement with the result in Eq. (7.34).

7.17. Braking on a bike

(a) Let the normal forces on the back and front wheels be N_1 and N_2, respectively. Then if both wheels are skidding, the friction forces on them are μN_1 and μN_2. If the bike is moving to the right, the various forces on it are shown in the free-body diagram in Fig. 7.38(a). The total friction force is $\mu(N_1 + N_2)$ leftward. But $N_1 + N_2 = mg$ because the net force in the vertical direction must be zero (since we're assuming that both wheels stay on the ground). So the friction force has magnitude $F_f = \mu mg$. From $F = ma$, the deceleration of the bike is therefore

$$a = \frac{F_f}{m} = \frac{\mu mg}{m} = \mu g. \qquad (7.37)$$

(a)

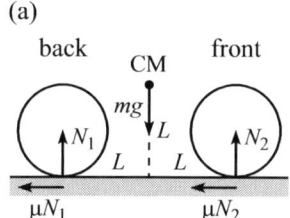

(b)

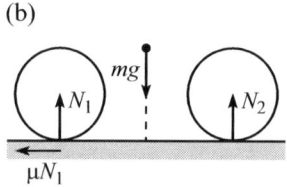

(c)

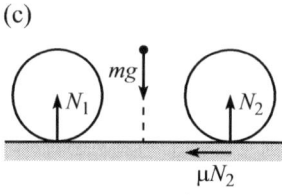

(bike is moving rightward)

Figure 7.38

REMARK: We didn't need to determine the individual normal forces N_1 and N_2 to solve this problem. But if you want to figure out what they are, you can consider the total torque around the CM. Since there is no rotation around the CM (because both wheels stay on the ground), the total torque around the CM must be zero. The mg force produces zero torque around the CM. Of the other four forces in Fig. 7.38(a), three produce clockwise torques, and one produces a counterclockwise torque. Balancing these torques gives

$$N_1 L + \mu N_1 L + \mu N_2 L = N_2 L \implies N_1(1 + \mu) = N_2(1 - \mu). \tag{7.38}$$

We also know that $N_1 + N_2 = mg$. So we have a system of two equations and two unknowns. Solving for N_1 and N_2 gives

$$N_1 = \frac{1-\mu}{2} mg \quad \text{and} \quad N_2 = \frac{1+\mu}{2} mg. \tag{7.39}$$

But again, we didn't need to know these values to answer the given question.

(b) If only the back wheel is skidding and the front wheel is rotating freely, the free-body diagram is shown in Fig. 7.38(b). There is no friction force on the front wheel (assuming it is massless; see the last remark below), so the total friction force is just μN_1. We therefore do need to determine N_1 in this case. Demanding that the total torque around the CM is zero gives

$$N_1 L + \mu N_1 L = N_2 L \implies N_1(1 + \mu) = N_2. \tag{7.40}$$

Combining this with $N_1 + N_2 = mg$ gives $N_1 = mg/(2 + \mu)$. The deceleration is therefore

$$a = \frac{F_f}{m} = \frac{\mu N_1}{m} = \frac{\mu g}{2 + \mu}. \tag{7.41}$$

(c) If only the front wheel is skidding and the back wheel is rotating freely, the free-body diagram is shown in Fig. 7.38(c). There is no friction force on the back wheel, so the total friction force is μN_2. Demanding that the total torque around the CM is zero gives

$$N_1 L + \mu N_2 L = N_2 L \implies N_1 = N_2(1 - \mu). \tag{7.42}$$

Combining this with $N_1 + N_2 = mg$ gives $N_2 = mg/(2 - \mu)$. The deceleration is therefore

$$a = \frac{F_f}{m} = \frac{\mu N_2}{m} = \frac{\mu g}{2 - \mu}. \tag{7.43}$$

LIMITS: In parts (a) and (c), we must have $\mu \leq 1$, otherwise the N_1's in Eqs. (7.39) and (7.42) would need to be negative (which they can't be) if the back wheel is to remain in contact with the ground. In the limiting case where $\mu = 1$, the deceleration in Eq. (7.43) in part (c) is g, which is the same as in Eq. (7.37) in part (a) when $\mu = 1$. This equality is due to the fact that if $\mu = 1$ then $N_1 = 0$ in parts (a) and (c), which means that the back wheel is just barely touching the ground, so it doesn't matter if the back brake is applied in addition to the front brake.

In part (b), nothing special happens when $\mu = 1$; the normal forces are $N_1 = mg/3$ and $N_2 = 2mg/3$, and the deceleration in Eq. (7.41) is $g/3$. There is no upper bound on μ as there is in parts (a) and (c); the back wheel will always stay on the ground in part (b). In the $\mu \to \infty$ limit, Eq. (7.41) gives $a \approx g$, which isn't so obvious.

If $0 < \mu < 1$, the deceleration in part (a) is larger than in (c), which is larger than in (b). If $\mu = 0$, the deceleration in all three cases is zero, of course.

REMARKS: In all three scenarios, the normal force N_2 on the front wheel is larger than the normal force N_1 on the back wheel (or equal if $\mu = 0$). This is due to the fact that the torques from the friction forces always get added to the torque from N_1, and the sum of these must be canceled by the torque from N_2. It makes intuitive sense that the bike should want to tip forward, necessitating a larger normal force on the front wheel.

Note that the bottom point on a skidding wheel isn't a legal point to use as the origin for calculating torques, because this point is accelerating (see the discussion on page 180). This

7.5. PROBLEM SOLUTIONS

is why we chose the CM as our origin. Technically we could have chosen a fixed point on the ground, but that would make things more complicated because the rate of change of the bike's angular momentum around such a point is *not* zero, due to the deceleration of the CM. After covering angular momentum in Chapter 8, you are encouraged to work things out by using such an origin. The easiest point on the ground to pick as the origin is the point directly below the instantaneous location of the CM.

The reasoning in this problem applies to cars too, of course. But for cars, the height of the CM is less than half the distance between the wheels. And the CM is also closer to the front, due to the (heavy) engine in the front. The parameters are therefore modified, but the general reasoning is the same.

If you want to make as long a skid mark as possible on a bike, which brake should you use, and how should you shift your weight? The answer is that you should shift your weight forward as much as possible and use only the back brake. (In theory, shifting your weight backward and using only the front brake will also work, but then you can't reach the brake with your hand!) If you lean forward enough, you can make the normal force on the back wheel be very small, which means that the friction force will be small. So your acceleration a will be small, and you will skid a large distance (although the skid mark will be very light).

Here's a picky point: Since the wheels of a bike have nonzero mass, the above results in parts (b) and (c) aren't quite correct. The non-skidding wheel rolls without slipping on the ground, so it has a nonzero angular deceleration as the bike decelerates. There must therefore be a torque on it, and the only force that can cause this torque is the (static) friction force from the ground. You can quickly show that this friction force must point *forward*. This partially cancels the (larger) backward friction force on the skidding wheel (which itself will be slightly modified), making the deceleration be slightly smaller than what we calculated above. Basically, a rolling massive wheel wants to keep rolling. But in the limit of massless wheels, the above results are correct.

7.18. Pulley below a stick

We have three unknowns: the acceleration a of the mass, the angular acceleration α of the stick, and the tension T in the long string. (The tension in the string hanging down to the mass m is $2T$, because the net force on the massless pulley must be zero.) Let us define positive a to be downward and positive α to be clockwise. We need three equations to solve for our three unknowns.

- The first equation is the $\tau = I\alpha$ equation for the stick, relative to the pivot. We will need the moment of inertia of the stick around the pivot. The parallel-axis theorem gives the I of a stick around a point $\ell/4$ from the CM as $m\ell^2/12 + m(\ell/4)^2 = 7m\ell^2/48$. The torque on the stick comes from the tensions acting at the ends and from gravity effectively acting at the CM. So the $\tau = I\alpha$ equation for the stick is

$$T \cdot \frac{3\ell}{4} - T \cdot \frac{\ell}{4} + mg \cdot \frac{\ell}{4} = \left(\frac{7m\ell^2}{48}\right)\alpha \implies T = \frac{7m\ell\alpha}{24} - \frac{mg}{2}. \quad (7.44)$$

- The $F = ma$ equation for the mass m is

$$mg - 2T = ma. \quad (7.45)$$

- Our third equation is the conservation-of-string relation. We claim that this takes the form of $a = \ell\alpha/4$. The reasoning is as follows. Imagine temporarily holding the pulley in place and rotating the stick through a small clockwise angle θ. Then the right end of the stick goes down by $(3\ell/4)\theta$, and the left end goes up by $(\ell/4)\theta$. This means that an effective length of $(3\ell/4)\theta - (\ell/4)\theta = \ell\theta/2$ string disappears from above the pulley. This length has to go somewhere, so it ends up below the pulley. It gets divided evenly on either side of the pulley, so if we now allow the pulley to move, the pulley (and hence also the mass m) goes down by a distance $d = (\ell\theta/2)/2$. Taking two time derivatives of this relation yields $a = \ell\alpha/4$, as claimed.

Plugging $\ell\alpha = 4a$ into Eq. (7.44) gives $T = 7ma/6 - mg/2$. Substituting this T into Eq. (7.45) then gives

$$mg - \left(\frac{7ma}{3} - mg\right) = ma \implies a = \frac{3g}{5}. \tag{7.46}$$

7.19. Falling stick 1

FIRST SOLUTION: Let's apply $\tau = I\alpha$ around the pivot (the left end). The torque is due to gravity acting at the CM. Since $I = m\ell^2/3$ for a stick around an end, we have

$$\tau = I\alpha \implies mg\left(\frac{\ell}{2}\right) = \left(\frac{m\ell^2}{3}\right)\alpha \implies \alpha = \frac{3g}{2\ell}. \tag{7.47}$$

The right end moves along the arc of a circle of radius ℓ centered at the left end, so the acceleration of the right end is $a_{\text{end}} = \ell\alpha = 3g/2$.

REMARK: It might seem odd that the acceleration is larger than g. But it is certainly possible for the acceleration of a point in a falling extended object to exceed g. For example, in the limit where all of the mass is concentrated at the center of a pivoted horizontal stick, the center simply freefalls (briefly) with acceleration g, which means that the far end accelerates downward at $2g$. Since the acceleration of the end of our original uniform stick exceeds g, there must be a tangential force acting within the stick. If we paint a dot on the stick near the right end, then the part of the stick to the left of the dot must apply a downward force on the part of the stick to the right of the dot. Otherwise the downward acceleration of the right part would be at most g.

SECOND SOLUTION: Let's now apply $\tau = I\alpha$ around the center of the stick. Right after the stick is released, there is an upward force F_p from the pivot. This force must exist, because we know that the stick angularly accelerates, and F_p is the only force that can produce a torque around the center (because gravity effectively acts at the center). Since $I = m\ell^2/12$ for a stick around its center, we have

$$\tau = I\alpha \implies F_p\left(\frac{\ell}{2}\right) = \left(\frac{m\ell^2}{12}\right)\alpha \implies F_p = \frac{m\ell\alpha}{6}. \tag{7.48}$$

We have two unknowns here (F and α), so we need another equation. This is the vertical $F = ma$ equation. With downward taken to be positive, we have

$$F = ma \implies mg - F_p = ma_{\text{CM}}. \tag{7.49}$$

However, with the introduction of a_{CM} we now have three unknowns, so we still need another equation. This is the equation that relates a_{CM} and α. Since the CM moves along the arc of a circle of radius $\ell/2$ centered at the left end, we have

$$a_{\text{CM}} = \frac{\ell}{2}\alpha \implies \alpha = \frac{2a_{\text{CM}}}{\ell}. \tag{7.50}$$

Plugging this expression for α into Eq. (7.48) gives $F_p = ma_{\text{CM}}/3$. And then plugging this expression for F_p into Eq. (7.49) gives $mg = (4/3)ma_{\text{CM}} \implies a_{\text{CM}} = 3g/4$. Hence $a_{\text{end}} = 2a_{\text{CM}} = 3g/2$, in agreement with the first solution. And we also find that F_p equals $mg/4$.

REMARK: The first solution was simpler because choosing our origin to be the left end meant that we never needed to deal with the F_p force from the pivot. So there was never any need for the vertical $F = ma$ equation. The single $\tau = I\alpha$ equation sufficed.

7.5. PROBLEM SOLUTIONS

7.20. Falling stick 2

During the first quarter turn, the CM of the stick falls a distance $\ell/2$. The loss in potential energy shows up as kinetic energy of the stick's rotation around the pivot. So conservation of energy gives the resulting angular velocity as (using $I = m\ell^2/3$)

$$mg\frac{\ell}{2} = \frac{1}{2}\left(\frac{m\ell^2}{3}\right)\omega^2 \implies \omega = \sqrt{\frac{3g}{\ell}}. \tag{7.51}$$

The CM moves along a circle with radius $\ell/2$, so its speed is $v = \omega(\ell/2) = \sqrt{3g\ell}/2$.

The horizontal component F_x of the force from the pivot accounts for the centripetal acceleration of the CM as it moves in its circle of radius $\ell/2$. Therefore,

$$F_x = \frac{mv^2}{r} = \frac{m(3g\ell/4)}{\ell/2} = \frac{3mg}{2}. \tag{7.52}$$

This force points in the direction from the CM to the pivot.

The find the vertical component F_y of the force from the pivot, we can use a combination of two torque equations – around the pivot and around the center; these will involve two different I's. In the former case, only gravity provides a torque; and in the latter case, only F_y provides a torque. $\tau = I\alpha$ around the pivot gives

$$mg\frac{\ell}{2} = \left(\frac{m\ell^2}{3}\right)\alpha, \tag{7.53}$$

because gravity effectively acts at the CM. And $\tau = I\alpha$ around the CM gives

$$F_y\frac{\ell}{2} = \left(\frac{m\ell^2}{12}\right)\alpha. \tag{7.54}$$

Dividing the second of these equations by the first yields

$$\frac{F_y}{mg} = \frac{1}{4} \implies F_y = \frac{mg}{4}. \tag{7.55}$$

The direction of this force is upward. It has 1/6 the magnitude of F_x.

REMARK: Another way to find F_y is the following. Equation (7.53) gives $\alpha = 3g/2\ell$, so the acceleration of the CM is $a_y = \alpha(\ell/2) = 3g/4$. The vertical $F_y = ma_y$ equation for the stick (with downward taken to be positive) is then

$$mg - F_y = m\left(\frac{3g}{4}\right) \implies F_y = \frac{mg}{4}. \tag{7.56}$$

Note that these $a_y = 3g/4$ and $F_y = mg/4$ values are the same as the ones we found in the second solution to Problem 7.19. The reason for this is that in both cases the sticks are horizontal, and a_y and F_y aren't affected by whatever downward speed the stick may have (whereas F_x is).

7.21. Sticks on a table

(a) We'll apply $\tau = I\alpha$ around the pivot, so that we won't have to worry about the messy force acting at with pivot (which has both vertical and horizontal components). The moment of inertia of a stick around an end is $m\ell^2/3$. And the torque in Fig. 7.39(a) comes from gravity with a lever arm of $(\ell/2)\cos\theta$, so $\tau = I\alpha$ gives

$$mg\frac{\ell}{2}\cos\theta = \left(\frac{m\ell^2}{3}\right)\alpha \implies \alpha = \frac{3g\cos\theta}{2\ell}. \tag{7.57}$$

If $\theta = 90°$ then $\alpha = 0$, which is correct. If $\theta = 0$ then $\alpha = 3g/2\ell$, in agreement with Eq. (7.47) in the solution to Problem 7.19.

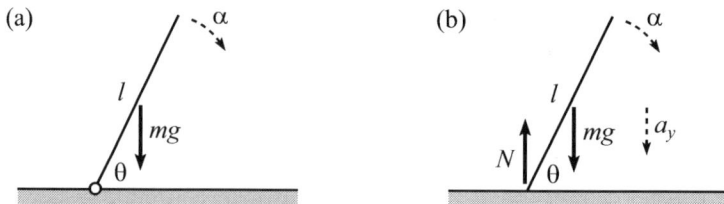

Figure 7.39

(b) We now can't apply $\tau = I\alpha$ around the bottom end of the stick, because it is accelerating. So let's use the center of the stick as our origin. (The CM is a legal point to use as the origin for $\tau = I\alpha$, even if it is accelerating.) The moment of inertia of a stick around its center is $m\ell^2/12$. And the torque in Fig. 7.39(b) comes only from the normal force from the table (because there is no friction force). So $\tau = I\alpha$ gives

$$N\frac{\ell}{2}\cos\theta = \left(\frac{m\ell^2}{12}\right)\alpha \implies N = \frac{m\ell\alpha}{6\cos\theta}. \quad (7.58)$$

We have one equation and two unknowns (N and α), so need another equation. This is the vertical $F = ma$ equation. The CM of the stick moves vertically because there is no friction. So if a_y is the acceleration of the CM, then the vertical $F = ma$ equation for the stick (with downward taken to be positive) is

$$mg - N = ma_y. \quad (7.59)$$

However, we have introduced a new unknown, a_y, so we now have two equations and three unknowns. We therefore still need one more equation. This is the relation between α and a_y, which can be found as follows.

If the stick were pivoted at the bottom end, and if it rotated through a small angle $d\theta$, then the CM would travel in a tiny arc with length $(\ell/2)\,d\theta$. The vertical component of this displacement would be $dy = (\ell/2)\,d\theta \cdot \cos\theta$ (with downward positive). The actual motion with the stick *not* pivoted is a combination of the preceding motion plus a shift to the left, to keep the CM moving in a vertical line. But this leftward shift doesn't affect the vertical displacement, so the CM still falls a distance $dy = (\ell/2)\,d\theta \cdot \cos\theta$. Dividing both sides of this relation by dt gives $v_y = (\ell/2)\omega\cos\theta$. Taking the time derivative and using the fact that $\dot\theta = 0$ at the start (so that we don't need to worry about the derivative of the $\cos\theta$ term) gives the desired relation between α and a_y,

$$a_y = \frac{\ell}{2}\alpha\cos\theta. \quad (7.60)$$

Plugging this a_y into Eq. (7.59), and also using the N from Eq. (7.58), we obtain

$$mg - \frac{m\ell\alpha}{6\cos\theta} = \frac{m\ell\alpha}{2}\cos\theta \implies \alpha = \frac{g}{\ell}\cdot\frac{1}{\dfrac{\cos\theta}{2}+\dfrac{1}{6\cos\theta}}. \quad (7.61)$$

For comparison with the result in part (a), we can rewrite this as

$$\alpha = \frac{\dfrac{3g\cos\theta}{2\ell}}{\dfrac{1}{4}+\dfrac{3}{4}\cos^2\theta}. \quad (7.62)$$

REMARK: The denominator of this result is always less than or equal to 1, so this α is always greater than or equal to the α in part (a). For $\theta \to 0$ (horizontal sticks) both α's are equal to

7.5. PROBLEM SOLUTIONS

$3g/2\ell$; the bottom end of the stick in part (b) hardly wants to move, so it's essentially the same system as in part (a). For $\theta \to 90°$ (vertical sticks) both α's approach zero, but the α in part (b) is four times the α in part (a). It is believable that the α in part (b) is larger, because if we have a *nonuniform* stick in the limit where all of its mass is located at a point mass at the CM, then the CM in part (b) simply freefalls (while the bottom end of the stick gets kicked out to the left), whereas the CM in part (a) very slowly moves away from the (nearly) vertical initial position with a tangential acceleration of $g\cos\theta \approx 0$.

7.22. Hoop on a plane

The situation at a general later time, when the hoop is still slipping, is shown in Fig. 7.40. The kinetic friction force is $F_f = \mu N = \mu mg\cos\theta$. (We'll work with a general μ and then set $\mu = 1/2$ at the end.) We have chosen the positive direction for v and a to point up the plane, which means that a will be negative. The component of the gravitational force down the plane is $mg\sin\theta$. So $F = ma$ along the plane gives

$$-mg\sin\theta - \mu mg\cos\theta = ma \implies a = -g\sin\theta - \mu g\cos\theta. \quad (7.63)$$

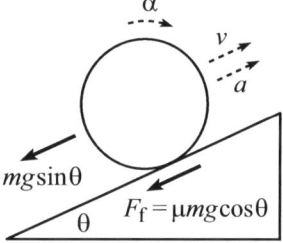

Figure 7.40

The torque on the hoop relative to the center, with clockwise taken to be positive, is $\tau = F_f R$. So $\tau = I\alpha$ gives (with $I = mR^2$, since all the mass is on the rim)

$$F_f R = I\alpha \implies (\mu mg\cos\theta)R = (mR^2)\alpha \implies \alpha = \frac{\mu g\cos\theta}{R}. \quad (7.64)$$

When the hoop finally starts to roll without slipping, we have $v_f = R\omega_f$. But $v_f = v_0 + at$ and $\omega_f = 0 + \alpha t$. So $v_f = R\omega_f$ becomes

$$v_0 - (g\sin\theta + \mu g\cos\theta)t = R\left(\frac{\mu g\cos\theta}{R}\right)t$$

$$\implies v_0 = g(\sin\theta + 2\mu\cos\theta)t. \quad (7.65)$$

With $\mu = 1/2$, we see that the hoop starts to roll without slipping when

$$t = \frac{v_0}{g(\sin\theta + \cos\theta)}. \quad (7.66)$$

This is minimum when $\sin\theta + \cos\theta$ is maximum. Setting the derivative equal to zero gives $\cos\theta - \sin\theta = 0 \implies \tan\theta = 1 \implies \theta = 45°$. The minimum t is $t_{\min} = v_0/(\sqrt{2}g)$.

LIMITS: If $\theta = 90°$ then Eq. (7.66) gives $t = v_0/g$. This makes sense because there is no friction force from the vertical plane, so ω is always zero. The $v_f = R\omega_f$ condition is therefore met only at the top of the vertical motion where $v_f = 0$, which happens when $t = v_0/g$. If $\theta = 0$ then we again have $t = v_0/g$ (or more generally $v_0/(2\mu g)$). This answer isn't obvious, but at least it is correctly larger than $t_{\min}$.

For a general value of μ, you can show that the minimum t occurs at $\tan\theta = 1/(2\mu)$, with the associated $t_{\min}$ being $v_0/(g\sqrt{1+4\mu^2})$. It makes sense that μ is in the denominator of $\tan\theta$, because if μ is very large, we want the plane to be essentially horizontal so that the friction force is large; the hoop will then quickly roll without slipping. And if μ is very small, then the hoop will take a long time to get rolling appreciably, so our best hope is to make v become small as quickly as possible. This is obtained by throwing the hoop essentially straight up.

7.23. Cylinder and board

(a) The horizontal forces in the free-body diagrams for the cylinder and the board are shown in Fig. 7.41. Let the static friction force from the cylinder on the board be F_s (it is indeed static friction, because there is no slipping), with rightward taken to be positive (which will turn out to be the direction it does point). Then by Newton's third law the static friction force from the board on the cylinder is F_s pointing leftward. The kinetic friction force from the table on the board is $F_k = \mu_k N = \mu_k(2mg)$. Since the board is moving rightward, we know that this force points leftward. With the sign

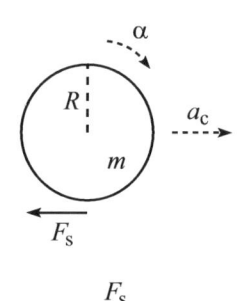

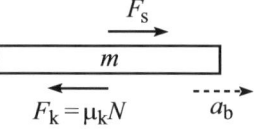

Figure 7.41

conventions for the accelerations indicated in the diagram (the a's will be negative), the various statements we can make are:

$$F = ma \text{ on board}: \quad F_s - F_k = ma_b \implies F_s - 2\mu_k mg = ma_b,$$
$$F = ma \text{ on cylinder}: \quad -F_s = ma_c,$$
$$\tau = I\alpha \text{ on cylinder}: \quad F_s R = \left(\frac{mR^2}{2}\right)\alpha,$$
$$\text{no slipping}: \quad a_c = a_b + R\alpha. \tag{7.67}$$

The last of these equations comes from the fact that the acceleration of the cylinder (with respect to the table) equals its acceleration with respect to the board plus the acceleration of the board with respect to the table. The former equals $R\alpha$ (due to the non-slipping condition), while the latter is by definition a_b.

We have four equations and four unknowns: F_s, a_b, a_c, α. (Remember that although we know that $F_k = \mu_k N$, we don't know what F_s is without doing some calculations. The $F_s \leq \mu_s N$ relation gives only an upper bound on F_s.) There are various ways to solve for the unknowns. One way is: Plugging the α from the 4th equation into the 3rd yields $F_s = m(a_c - a_b)/2$. Adding this to the 2nd equation gives $a_b = 3a_c$. The sum of the first two equations yields $-2\mu_k g = a_b + a_c$. Combining this with $a_b = 3a_c$ gives

$$a_c = -\frac{\mu_k g}{2} \quad \text{and} \quad a_b = -\frac{3\mu_k g}{2}. \tag{7.68}$$

So $-3\mu_k g/2$ is the desired acceleration of the board. Both of the accelerations negative, as expected. Note that the speeds of the two objects are irrelevant in finding the accelerations (assuming there is no slipping between the cylinder and the board).

REMARK: From the second equation in Eq. (7.67), the static friction force equals $F_s = -ma_c = \mu_k mg/2$. If there is no slipping, the $F_s \leq \mu_s N$ condition tells us that $\mu_k mg/2 \leq \mu_s (mg) \implies \mu_s \geq \mu_k/2$. This condition on μ_s is therefore what we meant by the "sufficient friction" phrase in the statement of the problem.

(b) We found above that the acceleration of the cylinder is $1/3$ of the acceleration of the board. So by the time the board slows down by a speed of v_0 and stops, the cylinder will have slowed down by only $v_0/3$. Its final speed is therefore $v_0 - v_0/3 = 2v_0/3$. Ignoring any dissipative effects, the cylinder will roll at this speed indefinitely on the stationary board. Note that this $2v_0/3$ result is independent of the coefficient of kinetic friction μ_k between the table and the board. If μ is small, the process will take a long time, but in this case all of the accelerations are small, and the final speed of the cylinder will still be $2v_0/3$.

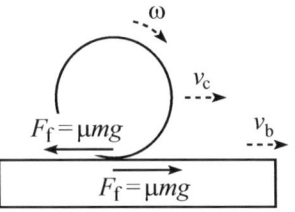

Figure 7.42

7.24. Another cylinder and board

The situation at a general later time, when the cylinder is still slipping, is shown in Fig. 7.42. The kinetic friction force is $F_f = \mu N = \mu mg$ (forward on the board, backward on the cylinder), so the accelerations of the two objects are $a_c = -\mu g$ and $a_b = \mu g$. The speeds are therefore given by $v_c = v_0 - \mu g t$ and $v_b = \mu g t$.

The torque on the cylinder, with clockwise taken to be positive, is $\tau = F_f R$. So $\tau = I\alpha$ gives

$$F_f r = I\alpha \implies (\mu mg)R = \left(\frac{mR^2}{2}\right)\alpha \implies \alpha = \frac{2\mu g}{R}. \tag{7.69}$$

Since the initial angular velocity of the cylinder is zero, the angular velocity at a later time t is therefore $\omega = \alpha t = 2\mu g t/R$.

When the cylinder finally rolls without slipping on the board, we have $v_c = v_b + R\omega$. This is true because the velocity of the (non-slipping) cylinder with respect to the board is $R\omega$, due to the forward rotation. Adding this to the velocity v_b of the board with respect to the ground gives the velocity of the cylinder with respect to the ground as $v_c = v_b + R\omega$. Using the various quantities we found above, this condition becomes

$$v_0 - \mu g t = \mu g t + R\left(\frac{2\mu g t}{R}\right) \implies v_0 = 4\mu g t \implies t = \frac{v_0}{4\mu g}. \quad (7.70)$$

The velocity of the cylinder with respect to the ground at this time is

$$v_c = v_0 - \mu g t = v_0 - \mu g \left(\frac{v_0}{4\mu g}\right) = \frac{3v_0}{4}. \quad (7.71)$$

Additionally, the velocity of the board is $v_b = \mu g t = v_0/4$, so the relative velocity is $v_0/2$. You can quickly verify that this correctly equals $R\omega$. These velocities will be maintained indefinitely, because there is no friction anywhere once the slipping ends (assuming an idealized setup).

As in Problem 7.23, the final speeds are independent of the coefficient of kinetic friction μ. If μ is small, the process will take a long time, but in this case all of the accelerations are small, and the final speed of the cylinder will still be $3v_0/4$.

7.25. Rolling down a plane

FIRST SOLUTION: Let the (static) friction force F_f be defined with the positive direction pointing up the plane (which is in fact the direction it points), as shown in Fig. 7.43. Then the $F = ma$ equation along the plane, the $\tau = I\alpha$ equation around the center of the wheel, and the non-slipping condition are

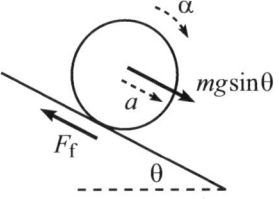

Figure 7.43

$$\begin{aligned} F = ma &\implies mg\sin\theta - F_f = ma, \\ \tau = I\alpha &\implies F_f R = (\beta m R^2)\alpha, \\ \text{No slipping} &\implies a = R\alpha. \end{aligned} \quad (7.72)$$

The last two equations give $F_f = \beta m a$. Plugging this into the first equation gives

$$mg\sin\theta - \beta m a = ma \implies a = \frac{g\sin\theta}{1+\beta}. \quad (7.73)$$

Note that this is independent of both R and m.

LIMITS: If $\beta = 0$ (all the mass is at the center) then we simply have $a = g\sin\theta$. So the wheel behaves just like a frictionless sliding block. This makes sense, because in this case the friction force must be zero (as with a frictionless sliding block), because otherwise the nonzero torque would produce an infinite angular acceleration since $I = 0$. If $\beta = 1/2$ (a uniform cylinder) then $a = (2/3)g\sin\theta$. If $\beta = 1$ (all the mass is on the rim) then $a = (1/2)g\sin\theta$. If $\beta \to \infty$,[4] then $a \to 0$. This makes sense; I is very large, so α (and hence a) must be very small.

REMARK: Since $\beta \geq 0$, the acceleration $a = g\sin\theta/(1+\beta)$ is always less than or equal to the $g\sin\theta$ result for a block sliding down a frictionless plane. The physical reason for this is that there is a nonzero friction force pointing up the plane (unless $\beta = 0$), which decreases the net force down the plane. This friction force *must* be present, because otherwise there would be nothing to provide the torque to get the wheel rolling. Alternatively, as you will see in the second solution below, a is smaller than $g\sin\theta$ because kinetic energy is "wasted" in the rotational motion. This makes the kinetic energy associated with the CM motion be smaller than it would be with no rotation (that is, if the wheel just slid down a frictionless plane).

SECOND SOLUTION: We'll use conservation of energy to determine the speed v of the center of the wheel after it has moved a distance d down the plane. And then we'll find a by using the standard constant-acceleration kinematic relation, $v = \sqrt{2ad}$. Even though our setup

---

[4]A very large β can be obtained by adding on long massive spokes protruding from the wheel and having them pass through a deep groove in the plane. Alternatively, we can have a spool with a very small inner radius roll down a thin plane with only its inner "axle" rolling on the plane.

210   CHAPTER 7. TORQUE

involves rotation, the center of the wheel undergoes simple linear motion with constant acceleration, so the $v = \sqrt{2ad}$ relation is perfectly valid.

If the wheel moves a distance d down the plane, the loss in potential energy is $mgd\sin\theta$. This shows up as kinetic energy, which equals $mv^2/2 + I\omega^2/2$ from Eq. (7.8), because the wheel is both translating and rotating. The non-slipping condition is $v = R\omega \Longrightarrow \omega = v/R$. Conservation of energy therefore gives

$$\begin{aligned} mgd\sin\theta &= \frac{1}{2}mv^2 + \frac{1}{2}I\omega^2 \\ &= \frac{1}{2}mv^2 + \frac{1}{2}\left(\beta mR^2\right)\left(\frac{v}{R}\right)^2 \\ &= \frac{1+\beta}{2}mv^2. \end{aligned} \quad (7.74)$$

The speed as a function of distance is therefore

$$v = \sqrt{\frac{2gd\sin\theta}{(1+\beta)}} = \sqrt{2 \cdot \frac{g\sin\theta}{1+\beta} \cdot d}. \quad (7.75)$$

We have written v this way because the $v = \sqrt{2ad}$ relation allows us to quickly read off the acceleration as $a = g\sin\theta/(1+\beta)$, in agreement with the first solution.

7.26. Pulling a cylinder on a board

(a) The free-body diagrams for the board and the cylinder are shown in Fig. 7.44. We have arbitrarily picked diagonally upward to be the positive direction for the friction force from the cylinder on the board. If it comes out negative, that just means it actually points downward. (We'll eventually be demanding that it equals $mg\sin\theta$ in part (c).) By Newton's third law, the friction force from the board on the cylinder points in the opposite direction, diagonally downward. Since there is no net force on either object perpendicular to the plane, the second diagram quickly gives $N_2 = mg\cos\theta$. And then the first diagram gives $N_1 = 2mg\sin\theta$. But these normal forces are irrelevant for the present purposes.

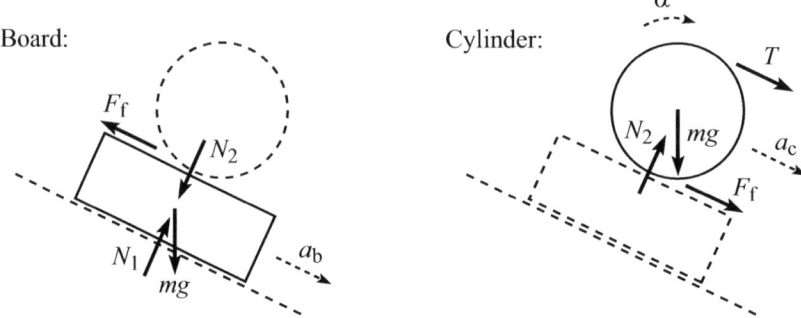

Figure 7.44

(b) Let's take the linear accelerations to be positive down the plane, and the angular acceleration to be positive clockwise. Then the various equations are:

- $F = ma$ for the board, along the plane:

$$mg\sin\theta - F_f = ma_b. \quad (7.76)$$

- $F = ma$ for the cylinder, along the plane:

$$T + mg\sin\theta + F_f = ma_c. \quad (7.77)$$

- $\tau = I\alpha$ for the cylinder, around the center:

$$(T - F_\text{f})R = \left(\frac{mR^2}{2}\right)\alpha. \qquad (7.78)$$

- Non-slipping condition between the cylinder and the board:

$$a_\text{c} = a_\text{b} + R\alpha, \qquad (7.79)$$

because $R\alpha$ is the acceleration of the cylinder with respect to the board, which itself is accelerating at a_b.

Given T, we have four equation and four unknowns: a_b, a_c, α, and F_f. (Or six and six, if you count the two normal forces and the two $F = ma$ equations perpendicular to the plane.)

(c) If $a_\text{b} = 0$, the first of the above equations above tells us that $F_\text{f} = mg\sin\theta$. And the fourth equation says that $a_\text{c} = R\alpha \Longrightarrow \alpha = a_\text{c}/R$. The second and third equations then become

$$T + mg\sin\theta + mg\sin\theta = ma_\text{c} \implies T + 2mg\sin\theta = ma_\text{c},$$
$$(T - mg\sin\theta)R = \left(\frac{mR^2}{2}\right)\left(\frac{a_\text{c}}{R}\right) \implies T - mg\sin\theta = \frac{1}{2}ma_\text{c}. \qquad (7.80)$$

Subtracting these equations quickly gives $a_\text{c} = 6g\sin\theta$. And then either of the equations gives $T = 4mg\sin\theta$. If T is smaller (or larger) than this, then the board accelerates down (or up) the plane.

7.27. Cylinder on board on plane

Let F_1 be the kinetic friction force between the board and the plane, and let F_2 be the static friction force between the board and the cylinder. Define the normal forces N_1 and N_2 similarly. Then the free-body diagrams for the board and the cylinder are shown in Fig. 7.45. The normal forces are $N_1 = 2mg\cos\theta$ and $N_2 = mg\cos\theta$, because neither object has any acceleration perpendicular to the plane. The kinetic friction force F_1 is therefore

$$F_1 = \mu N_1 = \left(\frac{1}{2}\tan\theta\right)(2mg\cos\theta) = mg\sin\theta. \qquad (7.81)$$

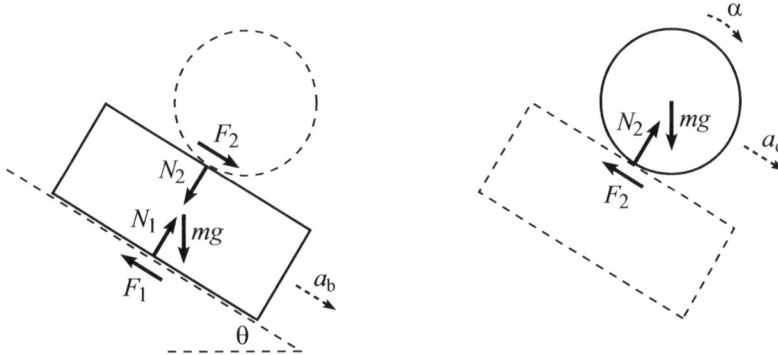

Figure 7.45

We have four unknowns: a_b, a_c, α, and F_2. (F_2 is a static friction force, so we can't say that it equals μN_2.) We can write down four equations:

- $F = ma$ for the board, along the plane:

$$mg\sin\theta - F_1 + F_2 = ma_b$$
$$\implies mg\sin\theta - mg\sin\theta + F_2 = ma_b \implies F_2 = ma_b. \tag{7.82}$$

- $F = ma$ for the cylinder, along the plane:

$$mg\sin\theta - F_2 = ma_c. \tag{7.83}$$

- $\tau = I\alpha$ for the cylinder, around the center:

$$F_2 R = \left(\frac{mR^2}{2}\right)\alpha \implies F_2 = \frac{1}{2}mR\alpha. \tag{7.84}$$

- Non-slipping condition between the cylinder and the board:

$$a_c = a_b + R\alpha. \tag{7.85}$$

This is just the statement that the cylinder's motion comes from the motion of the board plus the rolling motion of the cylinder with respect to the board.

The first and third of the above equations give $R\alpha = 2a_b$. Plugging this into the fourth equation gives $a_c = 3a_b$. Using this in the 2nd equation, along with the F_2 from the 1st, gives

$$mg\sin\theta - ma_b = m(3a_b) \implies a_b = \frac{g\sin\theta}{4}. \tag{7.86}$$

The acceleration of the cylinder is therefore $a_c = 3a_b = (3/4)g\sin\theta$.

REMARK: If the board isn't present (so that we just have a cylinder rolling without slipping down a plane), then Eq. (7.73) in the solution to Problem 7.25 tells us (with $\beta = 1/2$) that the cylinder has acceleration $a = (2/3)g\sin\theta$. It makes sense that the above result of $(3/4)g\sin\theta$ is larger, because in our original setup the board itself is accelerating down the plane, and this allows the cylinder to accelerate faster linearly (but not angularly, as you can show). This is consistent with the limit where there is no friction between the board and the plane, in which case the linear acceleration of both the board and the cylinder is $g\sin\theta$, while the angular acceleration of the cylinder is zero.

7.28. **Two cylinders on a plane**

(a) Looking at forces along the plane, each cylinder feels the tension in the string, a friction force from the plane, and the $mg\sin\theta$ component of gravity, as shown in Fig. 7.46.

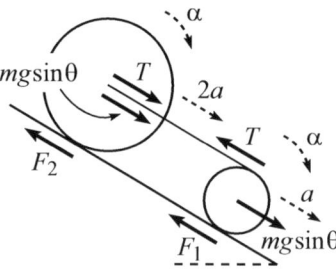

Figure 7.46

Let's first find the relation between the accelerations of the cylinders. Because of the string, the center of the large cylinder has the same linear acceleration as the "top" point on the rim of the small cylinder. But this top point moves with twice the acceleration of the center of the small cylinder, because for the small cylinder

7.5. PROBLEM SOLUTIONS 213

we have $a_{\text{top}} = a_{\text{center}} + \alpha R$, but $a_{\text{center}} = \alpha R$ since there is no slipping, therefore $a_{\text{top}} = 2a_{\text{center}}$. (This was basically just a repeat of the reasoning in Problem 7.8(b).) So the accelerations of the cylinders are $2a$ and a, as shown in the figure. Note that the angular acceleration α of the large cylinder is $2a/2R = a/R$, which is the same as for the small cylinder.

We can write down four $F = ma$ and $\tau = I\alpha$ equations.

- $F = ma$ for the small cylinder, along the plane:

$$mg \sin\theta - F_1 - T = ma. \tag{7.87}$$

- $F = ma$ for the large cylinder, along the plane:

$$T + mg \sin\theta - F_2 = m(2a). \tag{7.88}$$

- $\tau = I\alpha$ for the small cylinder, around the center (using $\alpha = a/R$):

$$F_1 R - TR = \left(\frac{mR^2}{2}\right)\left(\frac{a}{R}\right) \implies F_1 - T = \frac{ma}{2}. \tag{7.89}$$

- $\tau = I\alpha$ for the large cylinder, around the center (using $\alpha = 2a/2R$):

$$F_2(2R) = \left(\frac{m(2R)^2}{2}\right)\left(\frac{a}{R}\right) \implies F_2 = ma. \tag{7.90}$$

We have four equations and four unknowns: F_1, F_2, T, a. The sum of the first and third equations is $mg \sin\theta - 2T = 3ma/2$. And the sum of the second and fourth equations is $T + mg \sin\theta = 3ma$. We now have two equations and two unknowns. These can be quickly solved to yield $T = (1/5)mg \sin\theta$ and $a = (2/5)g \sin\theta$. So the accelerations of the small and large cylinders are, respectively,

$$a = \frac{2g \sin\theta}{5} \quad \text{and} \quad 2a = \frac{4g \sin\theta}{5}. \tag{7.91}$$

REMARK: From Problem 7.25, the acceleration of a single uniform cylinder (with $I = mR^2/2$) rolling without slipping down a plane is $(2/3)g \sin\theta$. The first of the accelerations we just found is correctly smaller than this, and the second is correctly larger. The string slows down the smaller cylinder and speeds up the larger cylinder, relative to the common $(2/3)g \sin\theta$ acceleration they would have if the string weren't present.

(b) Using the values of a and T we found above, Eqs. (7.89) and (7.90) tell us that the friction forces are both equal to $F_1 = F_2 = (2/5)mg \sin\theta$. Since the necessary static friction forces, the coefficients of static friction, and the normal forces are the same for both cylinders, the cylinders will slip at the same angle. In order *not* to slip, we need the friction force F to be less than or equal to μN. That is,

$$F \leq \mu N \implies \frac{2}{5}mg \sin\theta \leq (1)(mg \cos\theta) \implies \tan\theta \leq \frac{5}{2}. \tag{7.92}$$

This cutoff angle is about $68.2°$.

7.29. Wheel and board

FIRST SOLUTION: Since we are assuming $M > m$, the board will move down the plane, and the wheel will move up the plane. So the wheel will rotate counterclockwise. The free-body diagrams for the board and the wheel are shown in Fig. 7.47 (we have drawn only the forces parallel to the plane). Each object feels the tension in the string, the friction force between the objects, and the component of gravity down the plane. With the positive directions for the friction forces defined as shown, we will find that F_f is positive.

Let a be the acceleration of the board down the plane. Then by conservation of string, a is also the acceleration of the wheel up the plane. Let α be the angular acceleration of the

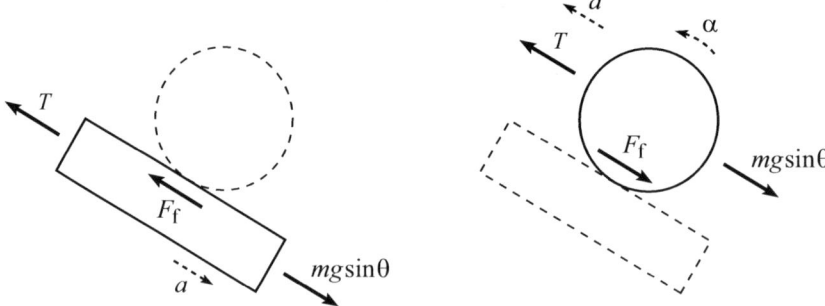

Figure 7.47

wheel, with counterclockwise positive. Then $\alpha = 2a/R$. The factor of 2 comes from the fact that if the board moves down the plane by a distance d and the wheel moves up by d, then the wheel moves a distance of $2d$ with respect to the board. The angle through which the wheel rolls is therefore given by $R\theta = 2d$. Taking two time derivatives of this relation gives $R\alpha = 2a$.

The various force and torque equations are

- $F = ma$ for the board, along the plane:

$$Mg\sin\theta - T - F_f = Ma. \qquad (7.93)$$

- $F = ma$ for the wheel, along the plane:

$$T - F_f - mg\sin\theta = ma. \qquad (7.94)$$

- $\tau = I\alpha$ for the wheel (using $\alpha = 2a/R$):

$$F_f R = I\alpha \implies F_f R = \left(\frac{mR^2}{2}\right)\left(\frac{2a}{R}\right) \implies F_f = ma. \qquad (7.95)$$

Our unknowns are T, F_f, and a. Adding the first two of the above equations eliminates T. Using $F_f = ma$ in the result gives

$$Mg\sin\theta - mg\sin\theta - 2(ma) = Ma + ma$$

$$\implies a = \left(\frac{M - m}{M + 3m}\right)g\sin\theta. \qquad (7.96)$$

LIMITS: If $M \gg m$ then $a \approx g\sin\theta$. This is correct, because the board slides essentially unhindered down the plane. If $m \gg M$, then even though we assumed $M > m$ in the above solution, all of the reasoning is actually still valid (it's just that some of the parameters are negative, given our sign conventions). So we obtain $a \approx -(1/3)g\sin\theta$. The negative sign means that the board accelerates up the plane, and hence the wheel accelerates down the plane, at $(1/3)g\sin\theta$. This result isn't so obvious. A single solid cylinder rolling down a plane without slipping accelerates at $(2/3)g\sin\theta$; see Eq. (7.73) with $\beta = 1/2$. The present result is smaller, because the movement of the board up the plane means that the wheel has to rotate more quickly, for a given linear acceleration. The friction force (which points up the plane, and which produces the torque) is therefore larger, and hence the acceleration down the plane is smaller, compared with the setup involving only a cylinder and no board.

SECOND SOLUTION: We can also solve this problem by applying conservation of energy. (This is true for many of the problems in this chapter, even though we haven't always included that solution.) If the board moves down the plane by a distance d and the wheel moves up by d, then the potential energy decreases by $(M-m)gd\sin\theta$. Let v be the common speed of the board and the wheel, and let ω be the angular speed of the wheel, which is given by $\omega = 2v/R$ from the above reasoning concerning α. Then conservation of energy gives

$$(M-m)gd\sin\theta = \frac{1}{2}Mv^2 + \frac{1}{2}mv^2 + \frac{1}{2}I\omega^2$$

$$= \frac{1}{2}Mv^2 + \frac{1}{2}mv^2 + \frac{1}{2}\left(\frac{mR^2}{2}\right)\left(\frac{2v}{R}\right)^2 = \frac{1}{2}(M+3m)v^2$$

$$\implies v = \sqrt{\frac{2(M-m)gd\sin\theta}{M+3m}}. \tag{7.97}$$

But the general kinematic result for constant acceleration is $v = \sqrt{2ad}$, which immediately yields the a given in Eq. (7.96).

7.30. Cylinder and stick

In Fig. 7.48, there are five unknowns: $F_\mathrm{f}, T, \alpha_\mathrm{c}, \alpha_\mathrm{s}$, and a_c. So we need five equations:

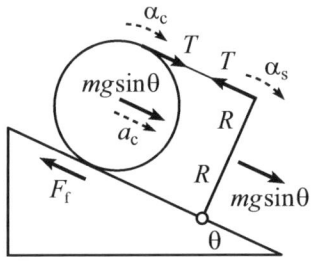

Figure 7.48

- $\tau = I\alpha$ for the cylinder, relative to the center:

$$F_\mathrm{f}R + TR = \left(\frac{mR^2}{2}\right)\alpha_\mathrm{c}. \tag{7.98}$$

- $\tau = I\alpha$ for the stick, relative to the pivot:

$$(mg\sin\theta)R - T(2R) = \left(\frac{m(2R)^2}{3}\right)\alpha_\mathrm{s}. \tag{7.99}$$

- $F = ma$ for the cylinder, along the plane:

$$mg\sin\theta + T - F_\mathrm{f} = ma_\mathrm{c}. \tag{7.100}$$

- Non-slipping condition for the cylinder:

$$a_\mathrm{c} = R\alpha_\mathrm{c}. \tag{7.101}$$

- Conservation-of-string equation:

$$\alpha_\mathrm{c} = \alpha_\mathrm{s}. \tag{7.102}$$

This can be seen from the following reasoning. Because of the string, the linear accelerations of the top of the stick and the "top" of the cylinder must be equal. The first of these is $(2R)\alpha_\mathrm{s}$. The second is $a_\mathrm{c} + R\alpha_\mathrm{c}$, because this is the acceleration of the center of the cylinder plus the acceleration of the rim relative to the center. But $a_\mathrm{c} = R\alpha_\mathrm{c}$ from Eq. (7.101), so we end up with $2R\alpha_\mathrm{c}$ for the acceleration of the "top" of the cylinder. Hence $(2R)\alpha_\mathrm{s} = 2R\alpha_\mathrm{c} \implies \alpha_\mathrm{s} = \alpha_\mathrm{c}$. The intuitive reason for this is that the cylinder rotates around its bottom point (which is instantaneously at rest), just as the stick does.

From Eqs. (7.101) and (7.102), both α's are equal to a_c/R. So we are now down to three unknowns ($F_\mathrm{f}, T, a_\mathrm{c}$) and three equations (the torque and force equations). These equations become, respectively,

$$F_\mathrm{f} + T = \frac{1}{2}ma_\mathrm{c},$$

$$mg\sin\theta - 2T = \frac{4}{3}ma_\mathrm{c},$$

$$mg\sin\theta + T - F_\mathrm{f} = ma_\mathrm{c}. \tag{7.103}$$

Adding these three equations conveniently eliminates both T and F_f, and we obtain

$$2mg\sin\theta = \frac{17}{6}ma_c \implies a_c = \frac{12g\sin\theta}{17}. \tag{7.104}$$

Note that we never needed to use the $F = ma$ equation for the stick. That would serve only to yield the force at the pivot.

REMARKS: We assumed in the above solution that the sting stays taut. This is indeed the case, because if you solve for T you will find that it is positive; it equals $(1/34)mg\sin\theta$. The physical reason why T is positive is that if the string weren't present, then the top of the stick would have a larger linear acceleration than the "top" of the cylinder ($(3/2)g\sin\theta$ vs. $(4/3)g\sin\theta$, as you can show). The tension in the string slows down the former and speeds up the latter, until they are equal. If the stick were instead placed above the cylinder on the plane, then the string would go limp.

For a slight variation on the above solution, we can make use of the $\tau = I\alpha$ equation for the cylinder relative to the contact point on the plane. This gives $T(2R) + (mg\sin\theta)R = (3mR^2/2)\alpha_c$, where we have used the parallel-axis theorem to obtain this moment of inertia. (We do indeed want to use the same α_c here, because the angle through which an object rotates is independent of the chosen origin.) Hence $2T + mg\sin\theta = 3ma_c/2$. Adding this equation to the middle equation in Eq. (7.103) (which was the $\tau = I\alpha$ equation for the stick) quickly yields $a_c = (12/17)g\sin\theta$. Note that the friction force F_f never appeared in this reasoning.

7.31. Infinite set of disks

Let F_i be the friction force between the stick and the ith disk, with $i = 1$ corresponding to the leftmost disk. Let a be the acceleration of the stick. Then the non-slipping condition says that the angular acceleration of the ith disk is $\alpha_i = a/r_i$. So $\tau = I\alpha$ applied to the ith disk gives

$$\tau_i = I_i\alpha_i \implies F_i r_i = \left(\frac{1}{2}m_i r_i^2\right)\left(\frac{a}{r_i}\right) \implies F_i = \frac{1}{2}m_i a. \tag{7.105}$$

Since the mass of each disk is proportional to its area, which in turn is proportional to its radius squared, the m_i are given by $m_1 = m$, $m_2 = m/9$, $m_3 = m/81$, etc.

The $F = ma$ equation for the stick, along its direction of motion, is (using $F_i = m_i a/2$ in the second line below)

$$mg\sin 30° - (F_1 + F_2 + \cdots) = ma$$
$$\implies \frac{1}{2}mg = ma\left(1 + \frac{1}{2}\left[1 + \frac{1}{9} + \frac{1}{81} + \cdots\right]\right)$$
$$\implies \frac{g}{2} = a\left(1 + \frac{1}{2}\left[\frac{1}{1 - 1/9}\right]\right)$$
$$\implies \frac{g}{2} = a\left(\frac{25}{16}\right) \implies a = \frac{8g}{25}. \tag{7.106}$$

REMARK: This acceleration is independent of R (and also m); this follows from dimensional analysis, as you can verify, because the given parameters are g, R, and m. So as long as the mass of the stick equals the mass of the leftmost disk, and as long as the radii of the disks decrease by successive factors of $1/3$, the system can be built to any size and the acceleration of the stick will always be $8g/25$.

7.32. Rolling off a sphere

(a) We first need to find the speed v of the ball as a function of θ. The kinetic energy of the ball comes from both the CM motion and the rotational motion. When the ball is at an angle θ, its height has decreased by $R - R\cos\theta$. So conservation of energy gives

$$mgR(1 - \cos\theta) = \frac{1}{2}mv^2 + \frac{1}{2}(\beta mr^2)\omega^2. \tag{7.107}$$

7.5. PROBLEM SOLUTIONS

The non-slipping condition is $v = r\omega$, so we obtain

$$mgR(1 - \cos\theta) = \frac{1}{2}mv^2 + \frac{1}{2}\beta mv^2 \implies v = \sqrt{\frac{2gR}{1+\beta}(1 - \cos\theta)}. \qquad (7.108)$$

The ball leaves the sphere when the normal force becomes zero. The normal force is determined from the radial $F = ma$ equation. The radially inward component of the gravitational force is $mg\cos\theta$, so with inward taken to be positive, the radial $F = ma$ equation is

$$mg\cos\theta - N = \frac{mv^2}{R} \implies mg\cos\theta - N = \frac{2mg}{1+\beta}(1 - \cos\theta)$$
$$\implies N = \frac{mg}{1+\beta}\big((3+\beta)\cos\theta - 2\big). \qquad (7.109)$$

The ball leaves the sphere when $N = 0$, that is, when

$$\cos\theta = \frac{2}{3+\beta}. \qquad (7.110)$$

LIMITS: If $\beta = 0$ (which is equivalent to a sliding mass, because there is no rotational energy in this case), then $\cos\theta = 2/3 \implies \theta \approx 48.2°$, which is a result that may look familiar to you. If $\beta = 2/5$ (a uniform sphere), then $\cos\theta = 10/17 \implies \theta \approx 54.0°$. The larger the β, the larger the angle, because the speed v of the CM is smaller for a given θ (because the rotational energy is larger), which in turn means that we need a larger θ if we want $mg\cos\theta = mv^2/R$ (which is the condition in Eq. (7.109) that makes $N = 0$). If $\beta \to \infty$ (this would require the ball to have massive extensions that could somehow pass freely through the fixed sphere), then $\theta \to 90°$. The CM speed v (and hence also mv^2/R) is always very small in this case, because essentially all of the energy is contained in the rotational motion. The coefficient of friction needs to be huge in this case, to prevent slipping when θ gets close to $90°$.

(b) The ball slips when the required friction force F_f exceeds the upper limit μN. We already found N above, so we must now find F_f. The tangential $F = ma$ equation and the $\tau = I\alpha$ equation around the center of the ball are

$$F = ma: \quad mg\sin\theta - F_f = ma \qquad (7.111)$$
$$\tau = I\alpha: \quad F_f r = I\alpha \implies F_f r = (\beta mr^2)(a/r) \implies F_f = \beta ma.$$

Plugging this F_f into the $F = ma$ equation gives

$$mg\sin\theta - \beta ma = ma \implies a = \frac{g\sin\theta}{1+\beta} \implies F_f = mg\frac{\beta\sin\theta}{1+\beta}. \qquad (7.112)$$

Recalling the N in Eq. (7.109), the $F_f \leq \mu N$ condition becomes

$$mg\frac{\beta\sin\theta}{1+\beta} \leq \mu\frac{mg}{1+\beta}\big((3+\beta)\cos\theta - 2\big)$$
$$\implies \beta\sin\theta \leq \mu\big((3+\beta)\cos\theta - 2\big). \qquad (7.113)$$

This is the desired equation that θ must satisfy if the ball is to remain rolling without slipping.

LIMITS: If $\mu = 0$, then we have $\sin\theta \leq 0$ (assuming $\beta \neq 0$), so the ball starts to slip right away at $\theta = 0$, which makes sense. If $\mu \to \infty$, then the inequality in Eq. (7.113) will be satisfied as long as $(3+\beta)\cos\theta - 2$ is positive, that is, as long as $\cos\theta > 2/(3+\beta)$. So θ must be smaller than this cutoff angle if there is to be no slipping, in agreement with the result in part (a).

REMARK: It turns out that the $r \ll R$ condition actually isn't necessary; the above results for θ are valid for any r and R. Without the $r \ll R$ assumption, the center of the ball now moves in a circle of radius $R + r$, so the R in the conservation-of-energy statement in Eq. (7.107) must be replaced with $R + r$. But the same is true for the R in the radial $F = ma$ equation in Eq. (7.109). So the $(R + r)$'s cancel, just as the R's did, and we end up with the same result for N as a function of θ in Eq. (7.109).

Note that the $v = r\omega$ and $a = r\alpha$ relations are still true for any r and R. At any instant, the ball may be considered to be rotating around the instantaneous point of contact on the sphere, so the curvature of the sphere doesn't matter; the surface might as well be flat.

7.33. **Yo-yo**

(a) As the yo-yo falls, the loss in potential energy shows up as the gain in kinetic energy. The kinetic energy comes partly from translational motion and partly from rotational motion. As the radius of the string's spiral becomes smaller, the yo-yo must rotate a larger number of times to generate a given vertical displacement. (You can consider the spiral, which is essentially a circle at any given time, to be rolling on the vertical flat "surface" formed by the straight part of the string. So the standard $v = r\omega$ relation applies.) Therefore, as time goes on, a larger fraction of the kinetic energy is contained in the rotational motion. Equivalently, a smaller fraction is contained in the translational motion. After a certain point, this "smaller fraction" effect wins out over the fact that the total kinetic energy is increasing, and the translational kinetic energy starts to decrease. That is, the linear speed starts to decrease.

It is easy to see why this slowdown must happen sooner or later, by looking at a limit. In the limit where the spiral is infinitesimally small (although an actual yo-yo has a nonzero lower bound on the spiral's radius, namely the radius of the axle), the yo-yo basically just spins in place. A non-infinitesimal translation speed would imply a huge rotational speed, corresponding to a far larger kinetic energy than what is available from the given loss in potential energy. In this limit, all of the energy is contained in rotational motion, and none is contained in translational motion.

(b) Let r be the radius of the string's spiral at a general later time. Then the cross-sectional area of the string that has been unwound is $\pi R^2 - \pi r^2$. If the length of the hanging string is ℓ, and if the thickness of the string is ϵ, then the cross-sectional area of the hanging string is $\ell\epsilon$. Since the unwound string is the same as the hanging string, their areas must be equal. So the relation between ℓ and r is $\ell = \pi(R^2 - r^2)/\epsilon$. Conservation of energy tells us that

$$mg\ell = \frac{1}{2}mv^2 + \frac{1}{2}I\omega^2. \qquad (7.114)$$

But ω is related to v by $\omega = v/r$, because as mentioned in part (a), the spiral's radius is the radius of the "axle" on which the yo-yo spins at any given moment. Letting $I = \beta m R^2$ for generality ($\beta = 1/2$ for our yo-yo), and using the ℓ we found above, the conservation-of-energy statement becomes

$$mg\frac{\pi(R^2 - r^2)}{\epsilon} = \frac{1}{2}mv^2 + \frac{1}{2}(\beta m R^2)\left(\frac{v}{r}\right)^2$$

$$\implies v^2 = \frac{2\pi g}{\epsilon} \frac{R^2 - r^2}{1 + \beta R^2/r^2}. \qquad (7.115)$$

This expression gives v in terms of r. At the start when $r = R$, we have $v = 0$, as expected. And in the $r \to 0$ limit, we also have $v \to 0$, consistent with the reasoning in part (a). Therefore, v must achieve a maximum somewhere between $r = R$ and $r = 0$. Taking the derivative of v^2 with respect to r and setting the result equal to zero gives

$$0 = \left(1 + \frac{\beta R^2}{r^2}\right)(-2r) - (R^2 - r^2)\beta R^2 \frac{(-2)}{r^3}$$

$$\implies 0 = r^4 + 2\beta R^2 r^2 - \beta R^4. \qquad (7.116)$$

7.5. PROBLEM SOLUTIONS

This is a quadratic equation in r^2. Using the quadratic formula (and choosing the positive root since r^2 must be positive), and then taking the square root to obtain r, yields

$$r = R\sqrt{-\beta + \sqrt{\beta^2 + \beta}}. \tag{7.117}$$

This is the radius for which v is maximum. For the given case where $\beta = 1/2$, we find $r \approx (0.605)R$. If the axle of the yo-yo is very thin (so that the string's spiral essentially winds all the way down to zero radius), then when $r \approx (0.605)R$, the remaining string in the yo-yo constitutes a fraction $\pi r^2/\pi R^2 = (0.605)^2 = 0.366$ of the total length of string. So the yo-yo is about 63% unwound.

REMARKS: In the limit where β is very small, the r in Eq. (7.117) is also very small; it behaves like $r \approx \beta^{1/4}R$, because the last of the three β's in Eq. (7.117) dominates. This makes sense; the yo-yo has hardly any rotational inertia, so it behaves basically like a dropped ball, and the maximum speed is achieved when the string is essentially completely unwound. In the limit where β is very large (see below for what this physically corresponds to), we can use the $\sqrt{1+\epsilon} \approx 1 + \epsilon/2$ Taylor series to approximate r as

$$r = R\sqrt{-\beta + \beta\sqrt{1 + 1/\beta}} \approx R\sqrt{-\beta + \beta(1 + 1/(2\beta))} = \frac{R}{\sqrt{2}}. \tag{7.118}$$

When r takes on this value, the area of the remaining spiral of string is $\pi(R/\sqrt{2})^2 = \pi R^2/2$. This is half of the area of the initial spiral, so we see that the maximum speed of the yo-yo is achieved when the string is halfway unwound. In the event that the yo-yo has an axle of nonzero size, the above $r = R/\sqrt{2}$ result still holds (because nowhere did we use the value of the inner radius of the spiral), but in this case this value of r corresponds to more than half of the total length of string being unwound.

If the winding of the string has some depth in the direction along the axis of the axle (so that the winding isn't a single "layer" like a movie reel), then the only modification to the above reasoning is that we effectively have a smaller value of the thickness ϵ. But this doesn't affect the functional dependence of v on r in Eq. (7.115). So the result for r in Eq. (7.117) still holds, along with the subsequent remarks.

If the initial spiral doesn't extend all the way out to the rim of the yo-yo, we can still make use of the above results. If we keep the letter R as representing the initial radius of the spiral, then the radius of the yo-yo is something larger; call it nR, where n is a number greater than 1. The moment of inertia of the yo-yo is now $I = \beta m(nR)^2 = (n^2\beta)mR^2$. So the only change in the above calculations is that we need to replace β with $\beta' \equiv n^2\beta$. If n is large (that is, if the initial spiral of string is small compared with the yo-yo), then we are in the "large β'" regime discussed above, so the maximum speed is achieved when the spiral's radius has gone down by a factor of $1/\sqrt{2}$.

Chapter 8

Angular momentum

8.1 Introduction

Angular momentum

If a point mass has momentum $\mathbf{p} = m\mathbf{v}$, then the *angular momentum* (denoted by $\mathbf{L}$) of the mass, relative to a given origin, is defined via the cross product (see Section 13.1.7 in Appendix A for the definition of the cross product) as

$$\mathbf{L} = \mathbf{r} \times \mathbf{p}, \tag{8.1}$$

where $\mathbf{r}$ is the vector from the origin to the mass. If a system consists of a number of masses with various momenta, then the total angular momentum of the system is the sum of all of the individual angular momenta: $\mathbf{L} = \sum \mathbf{r}_i \times \mathbf{p}_i$. The same origin must be chosen for all of the individual angular momenta. In the case of a continuous object, we have $\mathbf{L} = \int \mathbf{r} \times d\mathbf{p} = \int \mathbf{r} \times \mathbf{v}\, dm$.

Angular momentum is a vector, being the cross product of two other vectors. However, as with the torque in Chapter 7, we will invariably deal only with situations where both the position $\mathbf{r}$ and the momentum $\mathbf{p}$ lie in the plane of the page, in which case the angular momentum $\mathbf{L}$ points perpendicular to the page and has magnitude $rp\sin\theta$, where θ is the angle between $\mathbf{r}$ and $\mathbf{p}$. Since $\mathbf{L}$ is always perpendicular to the page in this case, we can ignore the fact that it is actually a vector and deal simply with its magnitude, $rp\sin\theta$. You can think of this magnitude in either of two ways, depending on which quantity you want to group with the $\sin\theta$ (see Fig. 8.1):

$$L = r(p\sin\theta) = \text{(radius)(tangential momentum)}, \text{ or}$$
$$L = p(r\sin\theta) = \text{(momentum)(impact parameter)}. \tag{8.2}$$

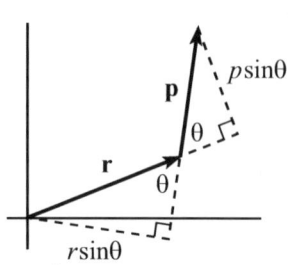

Figure 8.1

That is, the angular momentum equals the entire distance r times the tangential component of the momentum (the radial component of the momentum doesn't have anything to do with angular motion). And it also equals the entire momentum times the "impact parameter." The impact parameter is the component of the $\mathbf{r}$ vector that is perpendicular to $\mathbf{p}$. To geometrically construct it, draw a line pointing along $\mathbf{p}$ (extending in both directions) and look at the closest approach to the origin.

In the case of a rigid pancake object rotating around a fixed axis perpendicular to the page, as in Fig. 8.2, the speed of any point in the object is given by $v = r\omega$. So the total angular momentum points perpendicular to the page and has magnitude (using the fact that $\mathbf{v}$ is perpendicular to $\mathbf{r}$)

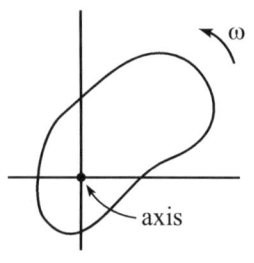

Figure 8.2

$$L = \left| \int \mathbf{r} \times d\mathbf{p} \right| = \left| \int \mathbf{r} \times \mathbf{v}\, dm \right| = \int rv\, dm$$
$$= \int r(r\omega)\, dm = \omega \int r^2\, dm. \tag{8.3}$$

220

8.1. INTRODUCTION

But the integral $\int r^2 dm$ is the moment of inertia I, so we have

$$L = I\omega. \tag{8.4}$$

This $L = I\omega$ relation is the rotational analog of the $p = mv$ relation in linear dynamics; L takes the place of p, I takes the place of m, and ω takes the place of v. The relation holds even if the object extends into and out of the page, provided that we are concerned only with the component of **L** perpendicular to the page.

Translation plus rotation

Consider an object with mass M that is both translating and rotating, as shown in Fig. 8.3. The angular velocity is ω, and the velocity of the CM is $\mathbf{V}_{CM}$. The position of the CM at the instant shown, with respect to the origin, is $\mathbf{R}_{CM}$. Then the total angular momentum of this object, relative to the origin, is given by (see Problem 8.1 for a proof)

$$\mathbf{L} = M\mathbf{R}_{CM} \times \mathbf{V}_{CM} + \mathbf{L}_{\text{around CM}}. \tag{8.5}$$

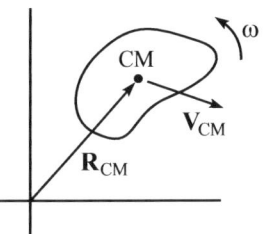

Figure 8.3

In words: the total angular momentum is the sum of the angular momentum of the entire object treated like a point mass M located at the CM and traveling at $\mathbf{V}_{CM}$, plus the angular momentum relative to the CM (imagine that you are riding along with the CM as the object spins around you). As with the energy in Eq. (7.8), the way that the two pieces of the angular momentum combine in Eq. (8.5) is about as nice a result as we could hope for. But note well that this result isn't valid if the CM is replaced by any other point.

For the object shown in Fig. 8.3, the two terms on the right-hand side of Eq. (8.5) point in opposite directions (the first points into the page, the second points out of the page), so we need to subtract their magnitudes when finding the magnitude of the total **L**.

The $\tau = d\mathbf{L}/dt$ relation

Angular momentum is a very useful concept in physics because (among other reasons) it is related to torque. In particular, the net external torque on an object (defined in Eq. (7.9) as $\tau \equiv \mathbf{r} \times \mathbf{F}$) equals the rate of change of the angular momentum:

$$\tau = \frac{d\mathbf{L}}{dt}. \tag{8.6}$$

See Problem 8.2 for a proof. This relation is the rotational analog of $\mathbf{F} = d\mathbf{p}/dt$; τ takes the place of $\mathbf{F}$, and $\mathbf{L}$ takes the place of $\mathbf{p}$. If you apply a force to an object, you change its momentum; likewise, if you apply a torque to an object, you change its angular momentum. As with the $\tau = I\alpha$ relation in Eq. (7.11), Eq. (8.6) is valid if the choice of origin is (1) a fixed point (or more generally a point moving with constant velocity), or (2) the CM of the object, or (3) a third possibility which rarely comes up. The same origin must be chosen for both τ and **L**, of course.

If the special case where a rigid object is rotating around a fixed axis, we have $L = I\omega$, so $\tau = d\mathbf{L}/dt$ reduces to

$$\tau = I\frac{d\omega}{dt} \implies \tau = I\alpha. \tag{8.7}$$

More generally (but still with a fixed rotation axis), if the object deforms so that its I changes, we have $\tau = d(I\omega)/dt$; both I and ω can change here. In cases where the *direction* of **L** changes, things get more complicated (gyroscopic motion, precession, nutation, etc.). See Chapter 9 in Morin (2008) if you are interested in these effects. In this chapter we'll be concerned only with fixed axes of rotation, so the full vector nature of **L** won't be important.

Conservation of angular momentum

If there is zero net external torque on a system, then Eq. (8.6) tells us that $d\mathbf{L}/dt = 0$. That is, the angular momentum **L** doesn't change with time; **L** is conserved. (This exactly mirrors the

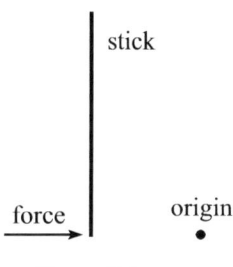

Figure 8.4

fact that $\mathbf{F} = d\mathbf{p}/dt$ tells us that if there is zero net external force on a system, then the linear momentum $\mathbf{p}$ is conserved.) Note that it is possible to have zero net external torque, relative to a given origin, even if there is nonzero net external force. Such is the case in Fig. 8.4, because the applied force has no lever arm relative to the given origin. In this setup, the stick will rotate counterclockwise, and the CM will move in a clockwise sense around the origin. The two terms on the right-hand side of Eq. (8.5) are therefore oppositely pointing vectors (into and out of the page, respectively). This is consistent with the fact that the total L relative to the given origin must be zero (assuming that the stick starts with zero L), because zero torque is applied.

In an isolated system, such as a collision between two objects, there are no external forces, and hence no external torques. So both $\mathbf{p}$ and $\mathbf{L}$ are conserved. Additionally, the energy E is conserved if the collision is elastic. Most of the problems in this chapter are collision problems.

Angular impulse

Consider the time integral of the torque, which we shall define as the *angular impulse* $\mathbf{J}_\theta$:

$$\mathbf{J}_\theta \equiv \int \boldsymbol{\tau}\, dt \quad \text{(angular impulse)}. \tag{8.8}$$

As with the (linear) impulse in Eq. (6.2), this is just a definition. But if we invoke the $\boldsymbol{\tau} = d\mathbf{L}/dt$ relation, then we can produce some content. If we multiply both sides by dt and then integrate, we obtain $\int \boldsymbol{\tau}\, dt = \Delta \mathbf{L}$. The left-hand side of this relation is just the angular impulse. So we see that the angular impulse $\mathbf{J}_\theta$ associated with a time interval Δt equals the total change in angular momentum $\Delta \mathbf{L}$ during that time:

$$\mathbf{J}_\theta = \Delta \mathbf{L}. \tag{8.9}$$

Since we'll be concerned only with planar setups where the vector nature of $\mathbf{L}$ isn't important, it will suffice to write $J_\theta = \Delta L$.

A special case that arises often is where the torque τ is applied at a constant lever arm r. An example of this is a quick strike, one that is quick enough so that the object doesn't have time to move appreciably during the time the force is applied (see Problem 8.22 for another example). In this case we have[1]

$$\Delta L = \int \tau\, dt = \int Fr\, dt = r \int F\, dt = r\Delta p \quad \Longrightarrow \quad \Delta L = r\Delta p, \tag{8.10}$$

where the constant nature of the lever arm r allowed us to pull it outside the integral. This $\Delta L = r\Delta p$ relation is very useful, because even if the force $F(t)$ is a complicated function of time and we don't know what ΔL and Δp are, we still know that they are related by a factor of r (if r is constant).

Since the $\Delta L = r\Delta p$ relation can be traced back to $\tau = dL/dt$, the choice of origin must be a fixed point or the CM of the object (ignoring the rare third possibility). The same origin must be chosen for both the lever arm r and the angular momentum L, of course.

8.2 Multiple-choice questions

Figure 8.5

8.1. A disk spins as shown in Fig. 8.5, with its CM at rest. The angular momentum of this disk, relative to the point P shown, is (circle all that apply)

(a) zero, because the CM is at rest

(b) zero, because for every point with velocity $\mathbf{v}$, there is a point with velocity $-\mathbf{v}$

(c) nonzero, because the first term in Eq. (8.5) is zero, and the second term is nonzero

(d) nonzero, because points moving downward are farther from P than points moving upward

---

[1] We aren't worrying about signs here. So you can take Eq. (8.10) to be a statement about the magnitudes of ΔL and Δp. You can then put in the signs by hand, based on your sign conventions.

8.2. MULTIPLE-CHOICE QUESTIONS

8.2. Which of the following objects has the largest clockwise angular momentum relative to the point P shown? The arrows signify the CM velocity and angular velocity, and the corresponding arrows have the same magnitudes in all of the figures.

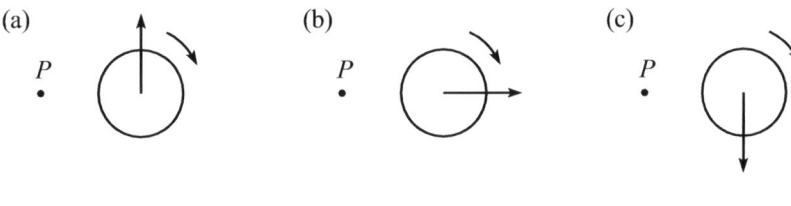

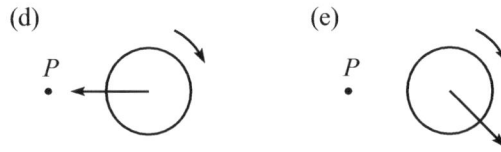

8.3. The CM of a uniform stick moves to the right, while the stick rotates clockwise with a nonzero positive angular velocity, as shown in Fig. 8.6. Assuming that the relative size of v and ω has been chosen properly, which of the three points shown can have zero total angular momentum around it?

(a) A (b) B (c) C (d) It is impossible for any of the points.

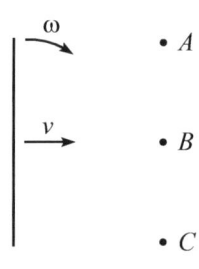

Figure 8.6

8.4. A stick with length ℓ has both translational and rotational motion, as shown in Fig. 8.7. Both v and ω are positive in the directions shown. Taking clockwise angular momentum L to be positive, which one of the following statements is true?

(a) The L around point B is *positive*, but the L around point A can be positive or negative, depending on the (positive) values of v and ω.

(b) The L around point B is *negative*, but the L around point A can be positive or negative, depending on the (positive) values of v and ω.

(c) The L around point A is *positive*, but the L around point B can be positive or negative, depending on the (positive) values of v and ω.

(d) The L around point A is *negative*, but the L around point B can be positive or negative, depending on the (positive) values of v and ω.

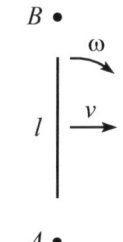

Figure 8.7

8.5. A wheel with $I = mR^2/2$ rolls without slipping on a table, as shown in Fig. 8.8. Its angular speed is ω. What is the angular momentum of the wheel relative to a dot on the table that coincides with the contact point at a certain instant?

(a) 0 (b) $\dfrac{1}{2}mR^2\omega$ (c) $mR^2\omega$ (d) $\dfrac{3}{2}mR^2\omega$ (e) $2mR^2\omega$

8.6. A springboard diver undergoes a forward rotation with her body and extended arms in a straight line, forming what we will model crudely as a uniform stick. She then "tucks" by touching her fingers to her toes, the effect of which is to fold her body-plus-arms in half. The ratio of her new angular speed to her old angular speed is

(a) 1/2 (b) 1 (c) 2 (d) 4 (e) 8

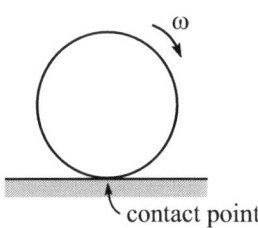

Figure 8.8

8.7. A motorcyclist makes a jump over a long row of cars. Right after he leaves the takeoff ramp, he notes that his motorcycle is angled slightly upward and has zero angular velocity. If this tilt is maintained, it will cause a problem on the landing ramp, because the motorcycle should be pointed slightly downward there. The best way for the rider to correct this problem in the air is to

(a) lean forward (b) lean backward (c) hit the gas (d) hit the brakes

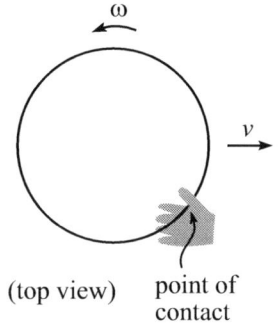

Figure 8.9

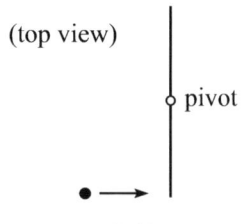

Figure 8.10

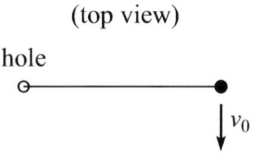

Figure 8.11

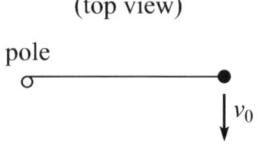

Figure 8.12

8.8. A frisbee™ is moving rightward and spinning counterclockwise. You catch it between your thumb and forefinger, as shown in Fig. 8.9. Assume that after the catch, the frisbee "sticks" to your hand, so that the only possible final motion is a rotation around the point of contact with your hand. Depending on the values of v and ω, which of the following rotations might be possible? (Circle all that apply.)

(a) clockwise

(b) counterclockwise

(c) no rotation

8.9. On a frictionless horizontal table, a uniform stick is pivoted at its middle, and a ball collides elastically with one end, as shown in Fig. 8.10. During the collision, what are all the quantities that are conserved in the stick-plus-ball system?

(a) L around the pivot

(b) L around the pivot, E

(c) L around the pivot, p, E

(d) L around the point of collision, E

(e) L around the point of collision, p, E

8.10. (*Note*: You should answer Multiple-Choice Question 8.11 along with this one before checking your answers.) A mass slides on a frictionless horizontal table. A string connected to the mass passes through a small hole in the table, and someone below the table holds the other end. The mass circles around the hole. Assume that it was given an initial speed v_0; see Fig. 8.11. The person below the table then gradually pulls the string downward, causing the mass to gradually spiral inward. Let E be the kinetic energy of the mass, and let L be the angular momentum of the mass relative to the center of the hole. During this process,

(a) E is conserved, so the speed remains constant

(b) L is conserved, so the speed increases

(c) neither E nor L is conserved

8.11. A mass slides on a frictionless horizontal surface. A string connects it to a pole, and it circles around the pole. Assume that it was given an initial speed v_0; see Fig. 8.12. The radius of the pole is small but nonzero, so as the string wraps around the pole, the mass gradually spirals inward. Let E be the kinetic energy of the mass, and let L be the angular momentum of the mass relative to the center of the pole. During this process,

(a) E is conserved, so the speed remains constant

(b) L is conserved, so the speed increases

(c) neither E nor L is conserved

8.3 Problems

The first two problems are foundational problems.

8.1. Translation plus rotation

For the object shown in Fig. 8.3, show that the angular momentum, relative to the origin, is given by Eq. (8.5).

8.3. PROBLEMS

8.2. **The $\tau = d\mathbf{L}/dt$ relation**

Take the time derivative of $\mathbf{L} \equiv \mathbf{r} \times \mathbf{p}$ and show that the result equals $\mathbf{r} \times \mathbf{F}$, which is by definition the torque τ.

8.3. **Spinning with dumbbells**

A person stands on a stool that is free to rotate. He holds his arms outstretched horizontally, with a 5 kg dumbbell in each hand. He is made to spin with initial angular speed ω_i. If he then draws his hands inward and holds the dumbbells close to his body, what is his new angular speed? Assume that his body can be modeled as a uniform cylinder with mass 70 kg and radius 0.15 m, and assume that each arm is about 1 m long; ignore the mass of the arms. A rough answer will suffice, since we've already made all sorts of crude approximations.

8.4. **Spiraling in**

(a) (*Note*: You should answer Multiple-Choice Questions 8.10 and 8.11 before solving this problem.) Consider the setup in Multiple-Choice Question 8.10. If during a small time dt the radius of the mass's "circular" motion changes by dr (which is negative), show that conservation of angular momentum (that is, a constant value of rv) implies that the work done by the tension equals the increase in kinetic energy. (The spiraling is very gradual, so you can assume that the tension is given by the mv^2/r expression for circular motion.)

(b) Consider the setup in Multiple-Choice Question 8.11. If at a given instant the radius of the mass's "circular" motion is decreasing at a rate $\dot{r}$ (which is negative), show that conservation of energy (that is, a constant value of v) implies that the torque equals the rate of change of angular momentum. (Again assume that the tension is given by mv^2/r.)

8.5. **Spinning coins**

Two identical uniform coins each have clockwise angular velocity ω. They are initially separated by an infinitesimal distance, as shown in Fig. 8.13. They are then bought together and immediately stick to each other. What is the resulting angular velocity of the system?

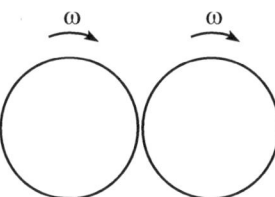

Figure 8.13

Most of the following problems involve sticks, because sticks are easy to visualize. But the principles involved are quite general; if we had chosen other shapes, the only modifications would be the moment of inertia and possibly an impact parameter.

8.6. **Equal velocities**

A ball with mass m travels with speed v perpendicular to a uniform stick with mass m and length ℓ, which is initially lying at rest on a frictionless table. Where along the stick should the ball collide elastically with it, so that the ball and the center of the stick move with equal velocities after the collision?

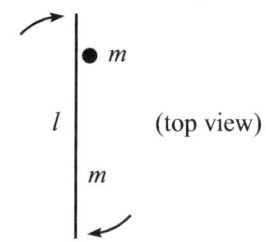

Figure 8.14

8.7. **No final rotation**

On a frictionless table, a uniform stick with mass m and length ℓ rotates around its center, which is initially at rest but *not* attached to a pivot. The stick then collides elastically with a mass m, as shown in Fig. 8.14. At what point along the stick should the collision occur so that the stick has only translational (that is, no rotational) motion afterward?

8.8. **Stick and pivot**

A uniform stick with mass m and length ℓ has one end attached to a pivot, and it swings around on a frictionless table with angular speed ω_0. A ball also with mass m is placed on the table a distance d from the pivot, and the stick collides elastically with it, as shown in Fig. 8.15.

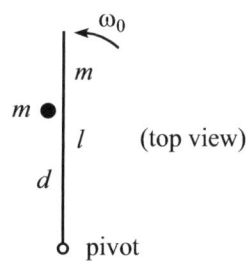

Figure 8.15

(a) What is the resulting speed of the ball?

(b) What value of *d* maximizes this speed? (*Note*: You can take a derivative if you want, or you can think about what the stick must end up doing in this case.)

8.9. Stick hitting a ball

A uniform stick with mass m and length ℓ slides with speed v_0 across a frictionless table, in the direction perpendicular to its length. One end of the stick collides elastically with a ball that also has mass m and that is initially at rest. What is the resulting speed of the ball?

8.10. Collision with a fixed object

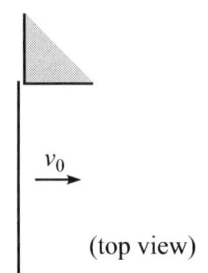

Figure 8.16

A uniform stick with mass m and length ℓ slides with speed v_0 across a frictionless table, in the direction perpendicular to its length. One end of the stick collides elastically with a fixed object, as shown in Fig. 8.16. What is the speed of the other end right after the collision?

8.11. Sticking to a stick

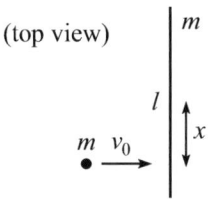

Figure 8.17

A uniform stick with mass m and length ℓ lies at rest on a frictionless table. A ball with the same mass m moves with speed v_0 perpendicular to the stick. It collides completely inelastically with the stick and sticks to it, at a distance x from the center, as shown in Fig. 8.17.

(a) What is the resulting angular velocity of the system?

(b) For what value of x is the angular velocity maximum?

(c) For what value of x does the far end of the stick (the top end in the figure) not move immediately after the collision?

8.12. Collision with a tilted stick

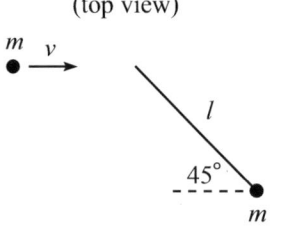

Figure 8.18

On a frictionless table, a massless stick with length ℓ lies at rest at a 45° angle with respect to the x axis. A mass m is attached to its lower-right end, as shown in Fig. 8.18. Another mass m moves with speed v in the x direction and collides with the stick and sticks to it at its upper-left end, forming a dumbbell.

(a) What is the resulting angular velocity of the dumbbell?

(b) How much energy, if any, is lost to heat in this process?

(c) What is the velocity (give both components) of the lower-right mass immediately after the collision?

8.13. Right-angled collision

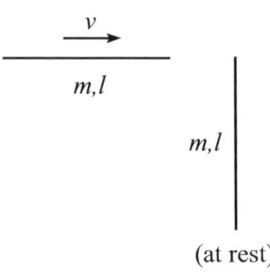

Figure 8.19

A uniform stick with mass m and length ℓ is initially at rest along the y axis on a frictionless table, as shown in Fig. 8.19. An identical stick, oriented in the x direction and traveling with speed v in the x direction, sticks to one of the ends of the first stick, forming a rigid right-angled object. What is the angular speed of the resulting motion? How much energy is lost to heat in the collision?

8.14. Stick on a table

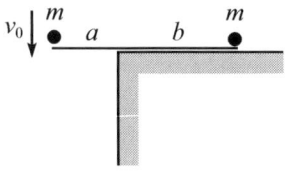

Figure 8.20

A massless stick lies on a table, with a length a hanging over the edge and a length b on the table, as shown in Fig. 8.20. A ball with mass m lies on the stick at its right end. Another ball with mass m is dropped above the left end and hits the end with speed v_0. Assuming that all interactions in the setup are elastic, what are the velocities of the two balls right after the collision?

8.3. PROBLEMS

8.15. Bouncing dumbbell

A dumbbell (consisting of two point masses $m/2$ at the ends of a massless stick with length ℓ) bounces elastically off a frictionless floor. Immediately before the bounce, the dumbbell makes an angle θ with the horizontal, it is not rotating, and the center is moving directly downward with speed v_0, as shown in Fig. 8.21.

What is the velocity of the center immediately after the bounce? Verify that your answer is reasonable in the $\theta = 90°$ and $\theta \to 0$ limits. *Note*: You can ignore gravity, because the bounce takes essentially zero time. Also, the dumbbell will keep bouncing off the floor indefinitely as time goes on, but we're concerned only with the first bounce.

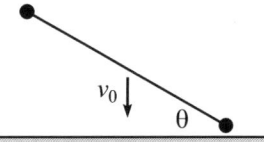

Figure 8.21

8.16. Colliding with a wall

On a frictionless floor, a uniform stick with mass m and length ℓ moves with speed v_0 (without rotating) in the direction perpendicular to a wall. It makes an angle θ with the direction of the wall, as shown in Fig. 8.22. If the stick collides elastically with the wall, what should θ be so that the speed of the CM after the collision is zero?

8.17. Stick hitting a pole

A uniform stick with mass m and length ℓ moves (without rotating) across a frictionless table with velocity v_0 perpendicular to its length, as shown in Fig. 8.23. It collides elastically with a fixed vertical pole. The point of impact is a distance x from the center of the stick.

(a) What should x be so that after the collision, the speed of the stick's CM is zero?

(b) Assume that x takes on the value you found in part (a). After the stick makes a half-rotation, it will collide elastically with the pole again, this time on the left side of the pole. Assuming that the pole has negligible thickness, what does the resulting motion of the stick look like? (It is possible to answer this without doing any calculations, but explain your reasoning clearly. It is also possible to answer this without solving part (a).)

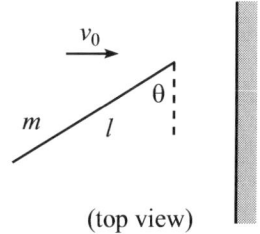

Figure 8.22

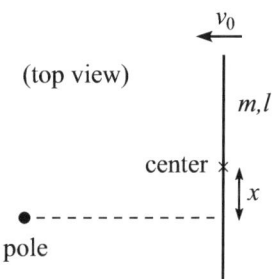

Figure 8.23

8.18. Striking objects

(a) A dumbbell consists of two identical masses at the ends of a massless stick. You give one of the masses a quick strike in the direction perpendicular to the stick. Show that right after the strike, the other mass is instantaneously at rest. *Hint*: Use Eq. (8.10).

(b) You now strike a uniform massive stick (like a pencil) at one of its ends, with the strike again being perpendicular to the stick. If the CM picks up a speed v_0, what is the velocity of the end you don't hit (right after the strike)? Does it move forward (in the same direction as the CM) or backward?

8.19. Center of percussion

You loosely hold one end of a uniform stick of length ℓ, which is then struck with a hammer. Where should this strike occur so that the end you are holding doesn't move (immediately after the strike)? In other words, where should the strike occur so that you don't feel a "sting" in your hand? This point is called the *center of percussion*.

8.20. Dumbbell and pole

A dumbbell consists of two masses m at the ends of a massless stick with length $\ell \equiv 2r$. It lies on a frictionless table, next to a fixed vertical pole that is a distance x from the center of the dumbbell, as shown in Fig. 8.24. The pole and the dumbbell are initially separated by an infinitesimal distance.

The mass closer to the pole is given a swift strike (perpendicular to the dumbbell) and immediately picks up a speed v_0. (The other mass remains instantaneously at rest; see Problem 8.18(a).) The dumbbell then immediately collides elastically with the pole and

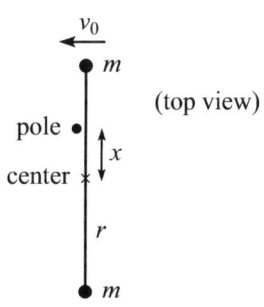

Figure 8.24

bounces off to the right. What should x be so that after the collision the dumbbell has no rotational motion?

8.21. One full revolution

A uniform stick with mass m and length ℓ lies on a frictionless table. It is struck with a quick blow (directed perpendicular to the stick) at a distance x from the center. A dot is painted on the table a distance d from the initial center of the stick, as shown in Fig. 8.25. What should x be (in terms of ℓ and d) so that the stick makes one complete revolution by the time the center reaches the dot? What is the minimum value of d (in terms of ℓ) for which such an x exists?

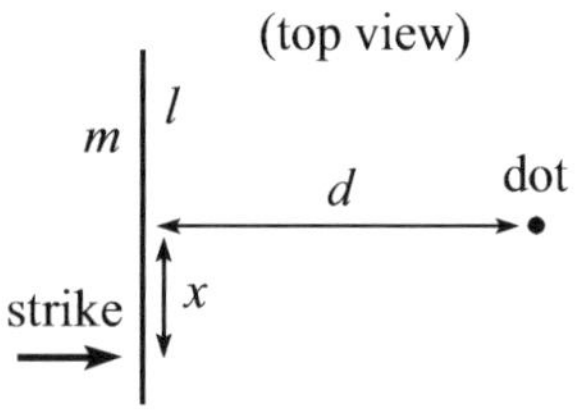

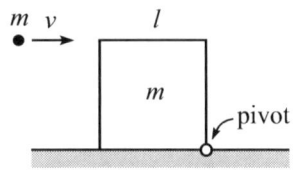

Figure 8.25

8.22. Sliding to rolling

A uniform ball initially slides, without rotating, on the ground. Friction with the ground eventually causes the ball to roll without slipping. If the initial linear speed is v_0, what is the final linear speed?

8.23. Tipping a block

A square block with mass m and side length ℓ sits at rest on a table, with its bottom right corner attached to the table by a pivot, as shown in Fig. 8.26. A ball, also with mass m, moves horizontally to the right with speed v and collides with the block and sticks to it at its upper left corner. The moment of inertia of the block around its *center* is $m\ell^2/6$.

(a) Immediately after the collision, what is the angular speed of the resulting block-plus-ball system around the pivot?

(b) What is the cutoff value of v, above which the block tips over, that is, above which the initial right-hand side of the block lands on the table?

Figure 8.26

8.4 Multiple-choice answers

8.1. $\boxed{c,d}$ Choice (a) is incorrect because it implies only that the *linear* momentum is zero (via Eq. (6.10)). Equivalently, it implies only that the first term in Eq. (8.5) is zero. Choice (b) is incorrect because it likewise implies only that the linear momentum $\mathbf{p}$ is zero. Choice (c) is correct. And choice (d) is correct because $\mathbf{L} = \mathbf{r} \times \mathbf{p}$ has a factor of r in it. So the farther a mass is from the origin, the larger its L will be (all other aspects of the cross product being equal). This implies that the clockwise L from the downward-moving points on the right side of the disk is larger in magnitude than the counterclockwise L from the upward-moving points on the left side. (The upward-moving points do indeed move in a counterclockwise sense relative to P.)

8.2. $\boxed{c}$ The total L is the sum of the L of the object treated like a point mass at the CM, plus the L around the CM; see Eq. (8.5). The latter of these has the same clockwise magnitude in all five choices. The former is largest in choice (c), because the object has the largest "impact parameter" in a clockwise sense. (The arrows seem to "add up" in choice (e) because they point in the same direction. But this alignment is meaningless, because we could have drawn the rotational arrow anywhere along the circumference of the circle.)

8.3. $\boxed{a}$ The total L is the sum of the L around the CM, plus the L of the object treated like a point mass at the CM. The former of these has a clockwise sense relative to A (and any other point, for that matter). The latter has a counterclockwise sense relative to A (but zero relative to B, and clockwise relative to C). So it is possible for these two contributions to cancel around point A if v and ω are related properly ($\omega = 6v/\ell$, as you can show).

8.4. $\boxed{c}$ The total L is the sum of the L around the CM, plus the L of the object treated like a point mass at the CM. Both of these contributions to L are always positive for point A (assuming v and ω are positive), so the answer must be (c).

REMARK: The extra information about point B in choice (c) is also true. In the limit where v is small and ω is large, the L around point B is positive (the second term in Eq. (8.5) dominates with a clockwise sense). But in the limit where v is large and ω is small, the L around point B is negative (the first term in Eq. (8.5) dominates with a counterclockwise sense).

8.5. $\boxed{d}$ The total angular momentum relative to the dot is given by Eq. (8.5). The non-slipping condition tells us that the speed of the CM is $V_{CM} = R\omega$, so Eq. (8.5) gives the magnitude of **L** as

$$L = mRV_{CM} + I\omega = mR(R\omega) + \left(\frac{mR^2}{2}\right)\omega = \frac{3}{2}mR^2\omega. \tag{8.11}$$

Alternatively, the parallel-axis theorem quickly gives the moment of inertia of the wheel around a point on the rim as $I_{rim} = I_{center} + mR^2 = 3mR^2/2$. And since the point on the wheel that is in contact with the ground is instantaneously at rest, we can consider the wheel to be (instantaneously) undergoing simple rotation around a fixed pivot. The simple $L = I\omega$ relation then suffices, and we obtain $L = I_{rim}\omega = 3mR^2\omega/2$, as above.

REMARK: The ω in this second solution is indeed the same as in the first solution where ω was defined to be the angular speed around the CM, because the angular speed ω is independent of the choice of origin. A line painted on the wheel will rotate through the same angle $d\theta$ during a time dt, independent of what point you consider the wheel to be rotating around. You could be riding along on the CM, or you could be standing at the contact point. In contrast with this, ω *is* different if you view things in a rotating reference frame. For example, in the frame rotating along with the wheel, ω is zero. But we'll never use rotating frames when talking about angular momentum. At most, our reference frames will be translating.

8.6. $\boxed{d}$ The moment of inertia of a stick around its center is $I = m\ell^2/12$. The ℓ^2 dependence implies that if we cut ℓ in half (while keeping m the same), I decreases by a factor of 4. But since angular momentum, which is given by $L = I\omega$, is conserved during the dive, the product $I\omega$ remains constant. So if I decreases by a factor of 4, then ω must increase by a factor of 4.

8.7. $\boxed{d}$ The wheels are spinning quickly in the forward direction at takeoff, so they have substantial angular momentum. If the brakes are applied, then some of this angular momentum is transferred to the main body of the bike, because the total angular momentum of the system is conserved. The bike will therefore rotate forward somewhat, as desired.

REMARK: Conversely, if the rider hits the gas, then the back wheel will rotate faster in the forward direction, which means that the body of the bike will rotate backward somewhat. This is relevant if for some reason the bike is tilting too far forward during the jump.

The leaning backward option in choice (b) will also make the bike tilt forward slightly. However, this choice is inferior for two reasons. First, it will put the rider in an untenable position for the landing. And second, the effect is very small due to the fact that the forward tilting angle θ of the bike is proportional (a small fraction of) the backward tilting angle of the rider. In contrast, hitting the brakes changes the angular *velocity* ω of the body of the bike, which means that the forward rotation occurs for as long as the bike is in the air. This effect therefore *extends over time* (and hence can be made arbitrarily large, in the limit of a hypothetically arbitrarily large time in the air), as opposed to being a one-time bounded effect, as is the case with leaning backward.

Of course, if the rotation of the tires is decreased too much by hitting the brakes, then there will be a problem during the landing, because the $v = \omega r$ non-slipping condition won't be satisfied (even approximately). Skidding will therefore occur, making it difficult to keep the bike upright. However, this effect can be mitigated by hitting the gas right before the landing.

8.8. $\boxed{a,b,c}$ All of the choices are possible. This question is basically asking what the total initial angular momentum around the point of contact can be – clockwise, counterclockwise, or zero. (The point of contact can't apply any torque around itself, so L is conserved around this point.) The total L of a translating and rotating object is given by Eq. (8.5).

In the present setup, the two terms in Eq. (8.5) are associated with, respectively, clockwise and counterclockwise angular momentum. So depending on which term has a larger magnitude (or they may be equal), we can obtain any of the three choices.

REMARK: A related question is: Given a rightward v and a counterclockwise ω, where on the rim should you catch the frisbee, if you want it to be motionless afterward? That is, around which point is the total L equal to zero? You need to grab the frisbee at a specific point in the lower-right quadrant, because the CM needs to move in a clockwise sense around your hand if there is to be any chance of L being zero. (If you're allowed to move your hand and grab the trailing edge, then a point in the upper-left quadrant can also yield $L = 0$.) If we demand that the two terms in Eq. (8.5) have equal magnitudes (and opposite directions), we find that the impact parameter b of the center of the frisbee is given by $mvb = I_{CM}\omega \Longrightarrow b = I_{CM}\omega/mv$. However, if ω is too large or v is too small, then this b is larger than the radius R of the frisbee, in which case the solution isn't physical, which means that there is no actual solution for b. This can be traced to the fact that you can control the size of the first term in Eq. (8.5) by adjusting b, whereas the second term takes on a given fixed value.

Assuming $b < R$, catching a frisbee (or a football, or anything spinning) at or near this special point makes it much less likely that the frisbee will roll off your hand. If circumstances dictate that you must instead catch the frisbee in the upper right-hand quadrant (where the two parts of L are both counterclockwise, so they add), the probability of dropping the frisbee is much higher. There is much less room for error in exactly when you clamp down your fingers and firmly grab on to the frisbee.

8.9. $\boxed{b}$ There is an external force at the pivot on the stick-plus-ball system, so p isn't conserved (the earth will gain momentum). But this force doesn't ruin conservation of L around the pivot, because the lever arm of the force is zero, so there is no torque. E is conserved by definition, since we're assuming the collision is elastic. Furthermore, choice (d) isn't correct, because the force from the pivot does provide a torque around the point of collision, since the lever arm is now nonzero.

8.10. $\boxed{b}$ The angular momentum relative to the center of the hole is conserved because the force from the string is always radial; it always points directly toward the center of the hole. So the torque is zero, and choice (b) is correct.

Since the angular momentum mvr is constant, and since the radius r decreases as the mass spirals in, the speed v must increase. The energy $mv^2/2$ must therefore also increase. So choice (a) is incorrect.

REMARK: Since the energy increases, work must be done on the mass. And indeed, because of the slight motion of the mass in the radial direction as it spirals in, there is a component of the velocity that points in the same direction as the radial tension in the string. So a nonzero amount of work is done on the mass. Equivalently, the person below the table does work by applying a force and moving her hand downward along the direction of the string. See Problem 8.4 for a quantitative treatment of this setup.

8.11. $\boxed{a}$ Energy is conserved because the force from the string is always orthogonal to the mass's velocity, so no work is done on the mass. This orthogonality follows from the fact that in contrast with the setup in the previous question, the mass is always instantaneously moving in a circle for which the string points along the radius; the center of the circle is the string's instantaneous point of contact with the pole. Alternatively, you can see that no work is done by noting that the pole doesn't move, so it can't do any work. (And there's nothing else in this setup, like the person in the previous question, that can do any work.) So choice (a) is correct.

Since the energy $mv^2/2$ is constant, the speed v is also constant. This means that the angular momentum mvr (relative to the center of the pole) must decrease, because the radius r decreases as the mass spirals in. So choice (b) is incorrect.

REMARK: Since the angular momentum decreases, there must be a nonzero torque. And indeed, because of the nonzero radius of the pole, the tension in the string isn't radial; it has a tiny lever arm

8.5 Problem solutions

8.1. Translation plus rotation

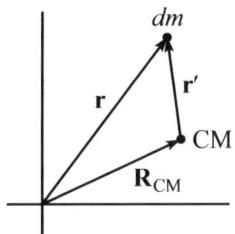

Figure 8.27

In Fig. 8.27 let the position of the CM of the object (we haven't drawn the object) relative to an arbitrary fixed origin be $\mathbf{R}_{CM}$, and let the position of a general point in the object relative to the CM be $\mathbf{r}'$. (The figure is planar since we're drawing it on the page, but this restriction isn't necessary. This derivation holds for a general 3-D object.) Then the position of a general point relative to the origin is $\mathbf{r} = \mathbf{R}_{CM} + \mathbf{r}'$. So the velocity of the point is $\mathbf{v} = \mathbf{V}_{CM} + \mathbf{v}'$. The total angular momentum of the object, relative to the origin, is therefore (see below for the explanations of the zeros in the fifth line)

$$\begin{aligned}
\mathbf{L} &= \int \mathbf{r} \times d\mathbf{p} \\
&= \int \mathbf{r} \times \mathbf{v}\, dm \\
&= \int (\mathbf{R}_{CM} + \mathbf{r}') \times (\mathbf{V}_{CM} + \mathbf{v}')\, dm \\
&= \int \mathbf{R}_{CM} \times \mathbf{V}_{CM}\, dm + \int \mathbf{R}_{CM} \times \mathbf{v}'\, dm + \int \mathbf{r}' \times \mathbf{V}_{CM}\, dm + \int \mathbf{r}' \times \mathbf{v}'\, dm \\
&= M\mathbf{R}_{CM} \times \mathbf{V}_{CM} + 0 + 0 + \mathbf{L}',
\end{aligned} \tag{8.12}$$

where $\mathbf{L}'$ is the angular momentum relative to the CM, which we denoted by $\mathbf{L}_{\text{around CM}}$ in Eq. (8.5).

The first zero in the fifth line above occurs because $\int \mathbf{R}_{CM} \times \mathbf{v}'\, dm = \mathbf{R}_{CM} \times \int \mathbf{v}'\, dm$, and $\int \mathbf{v}'\, dm$ is the total momentum as measured in the CM frame. But the total momentum in any frame is the total mass times the velocity of the CM with respect to that frame (see Eq. (6.10)), and the velocity of the CM is zero in the CM frame, by definition. The second zero occurs because $\int \mathbf{r}' \times \mathbf{V}_{CM}\, dm = (\int \mathbf{r}'\, dm) \times \mathbf{V}_{CM}$, and $\int \mathbf{r}'\, dm$ is the total mass times the position of the CM in the CM frame, which is zero by definition.

REMARK: The proof here was very similar to the proofs in Problems 7.1 and 7.4. As in those problems, the present result holds only if the CM is used on the right-hand side, because otherwise the cross terms in Eq. (8.12) won't be zero.

8.2. The $\tau = d\mathbf{L}/dt$ relation

The time derivative of $\mathbf{L}$ is

$$\begin{aligned}
\frac{d\mathbf{L}}{dt} &= \frac{d(\mathbf{r} \times \mathbf{p})}{dt} \\
&= \left(\frac{d\mathbf{r}}{dt} \times \mathbf{p}\right) + \left(\mathbf{r} \times \frac{d\mathbf{p}}{dt}\right) \\
&= (\mathbf{v} \times m\mathbf{v}) + (\mathbf{r} \times \mathbf{F}) \\
&= 0 + \tau,
\end{aligned} \tag{8.13}$$

where $\mathbf{v} \times \mathbf{v} = 0$ because the cross product of a vector with itself is zero. Note that we used the standard product rule when taking the derivative of the cross product. This is indeed legal, as you should check, because the cross product is built up from standard products; see Eq. (13.14).

8.3. Spinning with dumbbells

Since there are no external torques on the system, the angular momentum is conserved:

$$I_f \omega_f = I_i \omega_i \implies \frac{\omega_f}{\omega_i} = \frac{I_i}{I_f}. \tag{8.14}$$

We must therefore find I_i and I_f. The moment of inertia of a solid cylinder is $mR^2/2$, so the total initial moment of inertia is

$$\begin{aligned} I_i &= \frac{1}{2}(70\,\text{kg})(0.15\,\text{m})^2 + 2 \cdot (5\,\text{kg})(1\,\text{m})^2 \\ &= (0.79 + 10)\,\text{kg}\,\text{m}^2 \approx 11\,\text{kg}\,\text{m}^2. \end{aligned} \tag{8.15}$$

Assuming that the dumbbells end up at a radius of 0.15 m, the total final moment of inertia is

$$\begin{aligned} I_f &= \frac{1}{2}(70\,\text{kg})(0.15\,\text{m})^2 + 2 \cdot (5\,\text{kg})(0.15\,\text{m})^2 \\ &= (0.79 + 0.23)\,\text{kg}\,\text{m}^2 \approx 1\,\text{kg}\,\text{m}^2. \end{aligned} \tag{8.16}$$

Since I_i is about 11 times I_f, Eq. (8.14) tells us that ω_f is about 11 times ω_i.

REMARKS: This $\omega_f/\omega_i \approx 10$ result is probably larger than you would have guessed. If you actually try to perform the experiment, your large final ω will most likely cause you to fall over, or at least look rather silly trying not to (unless you made your initial ω extremely small).

Note that the dumbbells' contribution to I_i in Eq. (8.15) completely dominates the cylinder's (the body's) contribution. This demonstrates how important the r^2 factor is in the moment of inertia. Even though the combined dumbbells have only 1/7 the mass of the body, the initial I of the dumbbells is about 13 times the I of the body.

If you want to solve the problem symbolically, it is perhaps easiest to work with ratios. If r_1 equals the ratio of the combined dumbbell mass to the cylinder mass (so $r_1 = 1/7$ here), and if r_2 equals the ratio of the arm length to cylinder radius (so $r_2 = 1/0.15 = 6.67$ here), then you can show that $\omega_f/\omega_i = (1/2 + r_1 r_2^2)/(1/2 + r_1) \approx 10.7$.

If you simply drop the dumbbells from your outstretched arms instead of drawing them in, will you speed up? Since your final I will be roughly the same size as the I_f in Eq. (8.16) (because the dumbbells' contribution to I_f is reasonably small), you might think that the answer is "yes." However, the answer is "no." You will continue to rotate at the same angular speed (assuming you don't throw the dumbbells backward or forward relative to your hands). The difference between this setup and the original one is that if you just drop the dumbbells, they will keep their angular momentum (until it is transferred to the earth via the ground, a wall, or whatever else you end up smashing), whereas if you draw them inward to your body, their angular momentum will be (mostly) transferred to your body; the dumbbells apply a forward torque on your hands. This torque isn't terrible obvious intuitively. It certainly follows from conservation of momentum via the above reasoning, but you can also think about it in terms of the Coriolis force in the rotating reference frame (see, for example, Problem 12.6).

8.4. Spiraling in

(a) Since $L = mvr$ is constant, the differential of rv equals zero. Using the product rule, this gives

$$r\,dv + v\,dr = 0 \implies dv = -\frac{v\,dr}{r}. \tag{8.17}$$

(dv is positive and dr is negative here.) The increase in kinetic energy is then

$$d\left(\frac{mv^2}{2}\right) = mv\,dv = mv\left(-\frac{v\,dr}{r}\right) = -\frac{mv^2}{r}dr = -T\,dr, \tag{8.18}$$

where $T = mv^2/r$ is the magnitude of the tension force. But $-T\,dr$ is the (positive) work done by the radially inward tension, as the mass moves radially inward by dr (which is negative). So the change in kinetic energy correctly equals the work done by the tension.

8.5. PROBLEM SOLUTIONS

(b) The angular momentum is $L = mvr$, with clockwise taken to be positive in Fig. 8.12. Since v is constant, the rate of change of L is simply $\dot{L} = mv\dot{r}$; this is negative because $\dot{r}$ is negative. What is $\dot{r}$? If we let the small radius of the pole be a, then as the string wraps around the pole, the rate of increase of the length of string touching the pole is $d(a\theta)/dt = a\omega$. But ω is the angular frequency of the (essentially) circular motion, which is given by $\omega = v/r$. So the rate of change of the length of the *straight* part of the string is $-a\omega$ (which is negative). That is, $\dot{r} = -a\omega$. We therefore have

$$\dot{L} = mv\dot{r} = mv(-a\omega) = -mva\frac{v}{r} = -\frac{mv^2}{r}a = -Ta, \tag{8.19}$$

where $T = mv^2/r$ is the magnitude of the tension force. But $-Ta$ is the (negative) torque produced by the tension acting at a lever arm of a (see the solution to Multiple-Choice Question 8.11). So the (negative) rate of change of angular momentum correctly equals the (negative) torque.

8.5. Spinning coins

The center of mass of the system is initially at rest, so it is always at rest. The resulting motion of the system is therefore solely a rotation around the contact point; call this point P. We can find the resulting angular velocity by using conservation of angular momentum. (Energy isn't conserved, because the sticking process is inelastic.) The initial angular momentum of each coin around P is given by Eq. (8.5). The first term in this equation is zero, so the initial L of each coin around P is simply $I_{\text{CM}}\omega = (mR^2/2)\omega$. There are two coins, so the total initial L around P is $mR^2\omega$.

After the coins stick to each other, each coin is rotating around the point P on its rim. The parallel-axis theorem gives the moment of inertia of each coin around P as $I_{\text{CM}} + mR^2 = mR^2/2 + mR^2 = 3mR^2/2$. There are two coins, so the total final moment of inertia around P is $3mR^2$. Conservation of L around point P therefore gives

$$(mR^2)\omega = (3mR^2)\omega_{\text{f}} \implies \omega_{\text{f}} = \frac{\omega}{3}. \tag{8.20}$$

The direction of the spinning is clockwise, because both coins were initially spinning clockwise. If the coins were initially spinning in *opposite* directions with the same angular speed ω, then the total initial L would be zero, so the final ω_{f} would be zero.

REMARK: Note that ω_{f} doesn't depend on the mass m or the radius R of the coins. (This follows from dimensional analysis.) However, ω_{f} does depend on the moment of inertia. We assumed that the coins are uniform, so $I = mR^2/2$. If we instead have rings of mass, with $I = mR^2$, then you can show that the final angular velocity is $\omega/2$. More generally, we can let the moment of inertia be βmR^2, where β is a numerical factor. You can show in this case that $\omega_{\text{f}} = \beta\omega/(1 + \beta)$. You should check that this makes intuitive sense in the $\beta \to 0$ and $\beta \to \infty$ limits. (The $\beta \to \infty$ case would require the coins having massive radial extensions that somehow don't run into each other as the coins spin.)

8.6. Equal velocities

We'll need to apply conservation of all three of p, E, and L. These are all conserved because the collision is elastic and there are no external forces. Let d be the distance from the collision point to the center of the stick. Let the ball's initial speed be v_0; it will cancel out of the equations. The unknowns in Fig. 8.28 are v, ω, and d. Conservation of p gives

$$mv_0 = mv + mv \implies v = \frac{v_0}{2}. \tag{8.21}$$

Conservation of E gives (remembering that the stick's energy comes from a combination

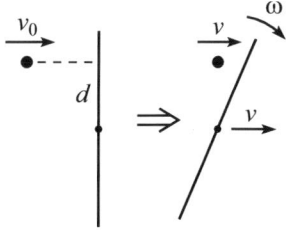

Figure 8.28

of the CM motion and the rotational motion around the CM; see Eq. (7.8))

$$\frac{1}{2}mv_0^2 = \frac{1}{2}mv^2 + \left[\frac{1}{2}mv^2 + \frac{1}{2}I_{CM}\omega^2\right]$$

$$\implies \frac{1}{2}mv_0^2 = \frac{1}{2}m\left(\frac{v_0}{2}\right)^2 + \left[\frac{1}{2}m\left(\frac{v_0}{2}\right)^2 + \frac{1}{2}\left(\frac{1}{12}m\ell^2\right)\omega^2\right]$$

$$\implies \frac{1}{4}mv_0^2 = \frac{1}{24}m\ell^2\omega^2 \implies \omega = \sqrt{6}\frac{v_0}{\ell}. \tag{8.22}$$

Conservation of L, around a fixed dot on the table coinciding with the initial center of the stick, gives (remembering that the stick's angular momentum comes from a combination of the CM motion and the rotational motion around the CM; see Eq. (8.5))

$$mv_0 d = mvd + [0 + I_{CM}\omega]$$

$$\implies mv_0 d = m\left(\frac{v_0}{2}\right)d + \left(\frac{1}{12}m\ell^2\right)\left(\sqrt{6}\frac{v_0}{\ell}\right)$$

$$\implies \frac{1}{2}mv_0 d = \frac{\sqrt{6}}{12}mv_0\ell \implies d = \frac{\ell}{\sqrt{6}} \approx (0.41)\ell. \tag{8.23}$$

The zero in the first line above comes from the fact that the CM of the stick is moving directly away from our chosen origin. This result of $(0.41)\ell$ is about 80% of the $\ell/2$ distance from the center of the stick to the end.

REMARK: In applying conservation of L, any other fixed point will work too. For example, we can pick a fixed dot on the table that coincides with the point of collision. The conservation-of-L statement is then

$$0 = 0 + [-mvd + I_{CM}\omega], \tag{8.24}$$

where the minus sign comes from the fact that the CM of the stick moves in a counterclockwise sense around our chosen origin. Since $v = v_0/2$, Eq. (8.24) is equivalent to the equation in the first line of Eq. (8.23).

8.7. **No final rotation**

In Fig. 8.29, we have three unknowns: v_1, v_2, and d (ω will cancel out of the equations). And we have three equations: conservation of p, E, and L. Conservation of p gives

$$0 = mv_2 - mv_1 \implies v_1 = v_2 \equiv v. \tag{8.25}$$

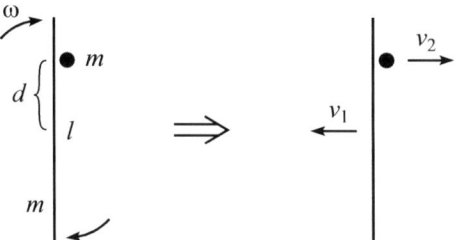

Figure 8.29

Conservation of E gives (using $v_1 = v_2 \equiv v$)

$$\frac{1}{2}I\omega^2 = \frac{1}{2}mv^2 + \left(\frac{1}{2}mv^2 + 0\right) \implies \frac{1}{2}\left(\frac{m\ell^2}{12}\right)\omega^2 = mv^2 \implies v = \frac{\ell\omega}{2\sqrt{6}}. \tag{8.26}$$

8.5. PROBLEM SOLUTIONS

The zero in this equation comes from the fact that the stick isn't rotating, so the second term in Eq. (7.8) is zero. Finally, conservation of L, relative to a dot on the table at the initial location of the center of the stick, gives

$$I\omega = (0+0) + mvd \implies \left(\frac{m\ell^2}{12}\right)\omega = mvd \implies v = \frac{\ell^2\omega}{12d}. \tag{8.27}$$

The zeros in this equation come from the facts that (1) the stick isn't rotating and (2) its CM is moving directly away from our chosen origin, so both terms in Eq. (8.5) are zero. Equating the v in Eq. (8.26) with the v in Eq. (8.27) yields

$$\frac{\ell\omega}{2\sqrt{6}} = \frac{\ell^2\omega}{12d} \implies d = \frac{\ell}{\sqrt{6}} \approx (0.41)\ell. \tag{8.28}$$

This result of $(0.41)\ell$ is about 80% of the $\ell/2$ distance from the center of the stick to the end. You should convince yourself why the answer to this problem is the same as the answer to Problem 8.6.

8.8. Stick and pivot

(a) In Fig. 8.30, we have two unknowns: the final v and ω. And we have two equations: conservation of E and conservation of L (around the pivot). Energy is conserved because the collision is elastic. Linear momentum is *not* conserved, because there is an external force at the pivot. Angular momentum relative to the pivot is conserved because the external force at the pivot produces no torque around it.

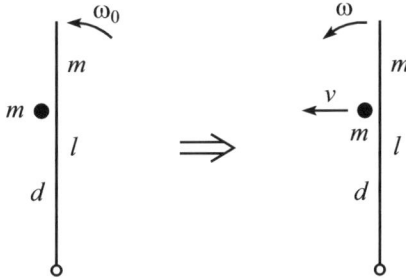

Figure 8.30

Conservation of energy gives

$$\frac{1}{2}I\omega_0^2 = \frac{1}{2}I\omega^2 + \frac{1}{2}mv^2 \implies I(\omega_0^2 - \omega^2) = mv^2. \tag{8.29}$$

Conservation of angular momentum relative to the pivot gives

$$I\omega_0 = I\omega + mvd \implies I(\omega_0 - \omega) = mvd. \tag{8.30}$$

We could solve for ω in Eq. (8.30) and plug the result into Eq. (8.29), and then solve for v. But an easier method, which avoids a quadratic equation, is to divide Eq. (8.29) by Eq. (8.30). This yields $\omega_0 + \omega = v/d$, and so $\omega = v/d - \omega_0$. Plugging this into Eq. (8.30) gives (using $I = m\ell^2/3$ around the pivot at the end of the stick)

$$I[\omega_0 - (v/d - \omega_0)] = mvd \implies 2I\omega_0 = v(md + I/d)$$
$$\implies v = \frac{2I\omega_0}{md + I/d} = \frac{2(m\ell^2/3)\omega_0}{md + (m\ell^2/3)/d} = \frac{2\ell^2\omega_0}{3d + \ell^2/d}. \tag{8.31}$$

LIMITS: As expected, v goes to zero for $d \to 0$ and also for $d \to \infty$, which would require attaching a massless extension to the stick. Also, if $\ell \to \infty$ then $v \to 2\omega_0 d$. You should think about how this relates to the $M \gg m$ case in Problem 1.7.

REMARKS: A second solution to Eqs. (8.29) and (8.30) is $\omega = \omega_0$ and $v = 0$. We dropped this solution when we divided one equation by the other. However, this solution corresponds to the initial conditions, which certainly satisfy conservation of E and L with the initial conditions. If you chose to take the route of solving a quadratic equation, the realization that $\omega = \omega_0$ must be a solution would allow you to easily factor the equation.

The above $\omega_0 + \omega = v/d$ result can be written as $\omega_0 d = v - \omega d$. In words, this equation says that the relative speed of the ball and the contact point on the stick right before the collision (when the gap is decreasing) equals the relative speed right after the collision (when the gap is increasing). This is a general result for elastic collisions. We know from the introduction to Chapter 6 that in an elastic collision (in any dimension) between point particles, the magnitude of the relative velocity is the same before and after the collision. But more generally, in an elastic collision between extended objects that may be rotating, the magnitude of the relative velocity of the *contact points* on the objects is the same before and after, provided that we are talking only about the relative velocity along the line of the (equal and opposite) force that the objects exert on each other. Additionally, the motion needs to be effectively planar. You can think about how this more general result follows from conservation of energy and angular momentum.

(b) FIRST SOLUTION: Maximizing v is equivalent to minimizing the denominator in the result for v in Eq. (8.31). Setting the derivative of $3d + \ell^2/d$ with respect to d equal to zero gives

$$0 = 3 - \frac{\ell^2}{d^2} \implies d = \frac{\ell}{\sqrt{3}} \approx (0.58)\ell. \tag{8.32}$$

SECOND SOLUTION: The maximum v (which corresponds to the maximum kinetic energy of the ball) occurs when the final kinetic energy of the stick is minimum. In other words, when $\omega = 0$. In this case, the conservation of E and L equations become

$$E: \quad \frac{1}{2}I\omega_0^2 = \frac{1}{2}mv^2 \implies v = \sqrt{\frac{I}{m}}\omega_0,$$
$$L: \quad I\omega_0 = mvd. \tag{8.33}$$

Plugging the v from the first of these equations into the second gives

$$I\omega_0 = m\left(\sqrt{\frac{I}{m}}\omega_0\right)d \implies d = \sqrt{\frac{I}{m}} = \sqrt{\frac{m\ell^2/3}{m}} = \frac{\ell}{\sqrt{3}}. \tag{8.34}$$

You are encouraged to let $I = \beta m\ell^2$ and then consider the various limits and special cases for β.

8.9. Stick hitting a ball

Let the various final speeds be ω, v, and u, with their positive directions indicated in Fig. 8.31. Momentum, energy, and angular momentum are all conserved. Conservation of momentum gives

$$mv_0 = mv + mu \implies v = v_0 - u. \tag{8.35}$$

Conservation of energy gives (using Eq. (7.8) for the final energy of the stick)

$$\frac{1}{2}mv_0^2 = \left(\frac{1}{2}mv^2 + \frac{1}{2}\frac{m\ell^2}{12}\omega^2\right) + \frac{1}{2}mu^2 \implies v_0^2 = v^2 + \frac{\ell^2\omega^2}{12} + u^2. \tag{8.36}$$

Conservation of angular momentum, relative to a dot on the table anywhere along the line of the stick's CM motion, gives (using Eq. (8.5) for the stick)

$$0 = \left(0 + \frac{m\ell^2}{12}\omega\right) - m\frac{\ell}{2}u \implies \omega = \frac{6u}{\ell}. \tag{8.37}$$

8.5. PROBLEM SOLUTIONS   237

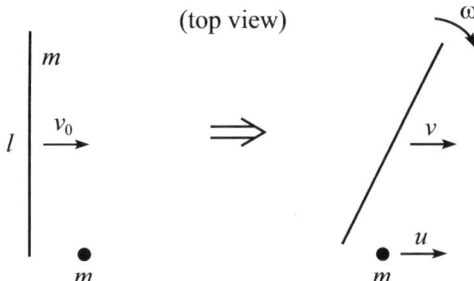

Figure 8.31

Plugging the v and ω from the p and L equations into the E equation gives

$$v_0^2 = (v_0 - u)^2 + \frac{\ell^2}{12}\left(\frac{6u}{\ell}\right)^2 + u^2 \implies 2v_0 u = 5u^2 \implies u = \frac{2v_0}{5}. \quad (8.38)$$

This is the desired final speed of the ball. Technically, $u = 0$ is also a solution. But that is simply the initial speed, which we already know satisfies the conservation equations, by definition.

REMARK: The final speed of the stick's CM is $v = v_0 - u = 3v_0/5$. So the stick moves faster than the ball. You can show that if the distribution of the stick's mass were such that its moment of inertia were $m\ell^2/8$ (which could be obtained by taking an appropriate amount of mass near the center of a uniform stick and moving it out toward the ends), then the ball and the stick's CM would move with the same speed.

8.10. Collision with a fixed object

Let the final linear speed of the stick's CM be v, and let the final angular speed be ω (with counterclockwise positive). The momentum of the stick isn't conserved, because there is an external force from the fixed object. But energy is conserved because we are told that the collision is elastic. And angular momentum around the corner of the fixed object is conserved, because the external force at the corner produces no torque around that point.

Conservation of energy gives (using Eq. (7.8))

$$\frac{1}{2}mv_0^2 + 0 = \frac{1}{2}mv^2 + \frac{1}{2}\left(\frac{m\ell^2}{12}\right)\omega^2 \implies v_0^2 = v^2 + \frac{\ell^2\omega^2}{12}. \quad (8.39)$$

And conservation of angular momentum around the corner gives (using Eq. (8.5))

$$mv_0\frac{\ell}{2} + 0 = mv\frac{\ell}{2} + \frac{m\ell^2}{12}\omega \implies v_0 = v + \frac{\ell\omega}{6}. \quad (8.40)$$

We have two equations and two unknowns (v and ω). Plugging the v from Eq. (8.40) into Eq. (8.39) gives

$$v_0^2 = \left(v_0 - \frac{\ell\omega}{6}\right)^2 + \frac{\ell^2\omega^2}{12}$$

$$\implies 0 = -\frac{v_0\ell\omega}{3} + \frac{\ell^2\omega^2}{36} + \frac{\ell^2\omega^2}{12} \implies \omega = \frac{3v_0}{\ell}. \quad (8.41)$$

And then $v = v_0 - \ell\omega/6 = v_0/2$.

The final velocity of the bottom end of the stick (right after the collision) is the sum of the translational velocity of the CM plus the rotational velocity around the CM (associated with a circle of radius $\ell/2$). So

$$v_{\text{end}} = v + \omega\left(\frac{\ell}{2}\right) = \frac{v_0}{2} + \left(\frac{3v_0}{\ell}\right)\left(\frac{\ell}{2}\right) = 2v_0. \quad (8.42)$$

REMARKS: The final velocity of the top end of the stick is $v - \omega(\ell/2) = -v_0$, because the rotational motion is now *subtracted* from the translational motion. As a double check, the average of the final velocities of the two ends, namely $2v_0$ and $-v_0$, equals $v_0/2$ which is correctly the final velocity of the center of the stick. Note that since the final velocity of the top end is $-v_0$, we see that the velocity of the point of contact on the stick simply reverses sign during the collision. This is consistent with the relative-speed theorem mentioned in the remarks in the solution to Problem 8.8.

8.11. **Sticking to a stick**

(a) Energy isn't conserved, because the collision is inelastic. And linear momentum won't be relevant in finding the angular velocity ω (at least for our choice of origin). So we need only consider angular momentum. We'll apply conservation of L around a dot painted on the table at the location of the CM of the entire system, at the moment when the ball hits the stick. Since the ball and the stick have equal mass, this dot is $x/2$ from the center of the stick.

Relative to the dot on the table, the initial L is $mv_0(x/2)$, with counterclockwise taken to be positive. To find an expression for the final L, we need to calculate the moment of inertia of the entire system around the CM. Using the parallel-axis theorem for the stick, we have

$$I_{\text{stick}} + I_{\text{ball}} = \left[\frac{m\ell^2}{12} + m\left(\frac{x}{2}\right)^2\right] + m\left(\frac{x}{2}\right)^2 = \frac{m\ell^2}{12} + \frac{mx^2}{2}. \qquad (8.43)$$

Conservation of angular momentum around the dot therefore gives

$$mv_0\left(\frac{x}{2}\right) = 0 + \left(\frac{m\ell^2}{12} + \frac{mx^2}{2}\right)\omega \implies \omega = \frac{v_0 x}{\ell^2/6 + x^2}. \qquad (8.44)$$

The zero in this equation comes from the fact that the CM of the system moves directly away from our chosen origin, so the first term in Eq. (8.5) is zero. If we had chosen another point as the origin, such as the center of the stick, then we would have needed to use conservation of momentum to find the speed of the CM of the system, which is just $v_0/2$ since the total mass is $2m$. The solution in this case would involve $mvx_0/2$ being added to both sides of Eq. (8.44).

LIMITS: The above ω goes to zero for both $x = 0$ and $x = \infty$; these results make sense. (In the latter case, we would need to attach a massless extension to the stick.) By a continuity argument, it follows that ω must achieve a maximum at some intermediate value of x; see part (b).

(b) To maximize ω, we need to set the derivative of the expression in Eq. (8.44) equal to zero. Ignoring the denominator of the derivative, since we're setting the result equal to zero, we find

$$\left(x^2 + \frac{\ell^2}{6}\right)\cdot 1 - x \cdot 2x = 0 \implies x^2 = \frac{\ell^2}{6} \implies x = \frac{\ell}{\sqrt{6}} \approx (0.41)\ell. \qquad (8.45)$$

This result of $(0.41)\ell$ is about 80% of the $\ell/2$ distance from the center of the stick to the end.

(c) The velocity of the top end equals the CM velocity minus the backward rotational velocity. Conservation of linear momentum tells us that

$$mv_0 = (2m)v_{\text{CM}} \implies v_{\text{CM}} = v_0/2. \qquad (8.46)$$

8.5. PROBLEM SOLUTIONS

The top end of the stick is a distance $r = \ell/2 + x/2$ from the CM, so we want

$$0 = v_{CM} - r\omega$$
$$\implies 0 = \frac{v_0}{2} - \left(\frac{\ell}{2} + \frac{x}{2}\right)\omega$$
$$\implies \frac{v_0}{2} = \left(\frac{\ell}{2} + \frac{x}{2}\right)\left(\frac{v_0 x}{\ell^2/6 + x^2}\right)$$
$$\implies \ell^2/6 + x^2 = (\ell + x)x$$
$$\implies x = \ell/6. \tag{8.47}$$

REMARK: This result implies that the collision is located a distance $\ell/2 + x = 2\ell/3$ from the top end. This location is the same as the "center of percussion" relevant to a quick strike (see Problem 8.19). The reason these two results are the same is that the stick doesn't care that the ball sticks to it. The stick simply feels some forces (positive or negative) from the ball, and the center-of-percussion result in Problem 8.19 doesn't depend on the sign of the force.

8.12. Collision with a tilted stick

(a) There are no external forces, so p and L are conserved. E might or might not be; we'll find out eventually. Conservation of p quickly gives the speed of the dumbbell's CM as $v/2$, although we won't need this to find ω.

The CM is halfway between the masses, which means $\ell/2$ from each. The moment of inertia of the resulting dumbbell is therefore $I = 2 \cdot m(\ell/2)^2 = m\ell^2/2$. So conservation of L around the point on the table that coincides with the CM at the moment of the collision gives (using the fact that the moving mass has an "impact parameter" of $\ell/(2\sqrt{2})$)

$$mv\left(\frac{\ell}{2\sqrt{2}}\right) = 0 + \left(\frac{m\ell^2}{2}\right)\omega \implies \omega = \frac{v}{\sqrt{2}\,\ell}, \tag{8.48}$$

where the zero here comes from the fact that the CM is moving directly away from the origin, so the first term in Eq. (8.5) is zero.

(b) The energy of the dumbbell is partly translational and partly rotational. So the energy lost to heat (if any) is

$$E_i - E_f = \frac{1}{2}mv^2 - \left(\frac{1}{2}(2m)v_{CM}^2 + \frac{1}{2}I\omega^2\right)$$
$$= \frac{1}{2}mv^2 - \left(\frac{1}{2}(2m)\left(\frac{v}{2}\right)^2 + \frac{1}{2}\left(\frac{m\ell^2}{2}\right)\left(\frac{v}{\sqrt{2}\,\ell}\right)^2\right)$$
$$= mv^2\left(\frac{1}{2} - \frac{1}{4} - \frac{1}{8}\right) = \frac{1}{8}mv^2. \tag{8.49}$$

This is nonzero, so some energy is indeed lost to heat.

(c) The velocity of the lower-right mass is the sum of the CM velocity plus the rotational velocity relative to the CM. The former is $v/2$ to the right. The latter is $\omega r = (v/\sqrt{2}\,\ell)(\ell/2) = v/(2\sqrt{2})$ down and to the left at a 45° angle (since the rotation is clockwise). The sum of these two vectors is shown in Fig. 8.32. The desired velocity is therefore $(v_x, v_y) = (v/4, -v/4)$. Equivalently, it points down to the right at 45°, with magnitude $v/(2\sqrt{2})$.

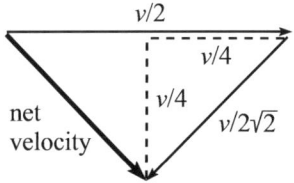

Figure 8.32

REMARK: The velocity we just found points along the stick, that is, directly away from the upper-left mass. There is a good reason for this. If we break up the initial velocity of the upper-left mass into components parallel and perpendicular to the stick, we can consider this mass to equivalently be two separate masses, each with mass m, with one moving perpendicular to the stick with speed $v/\sqrt{2}$, and the other moving along the direction of the stick with speed $v/\sqrt{2}$.

The former doesn't (initially) make the lower-right mass move (you can convince yourself of this by applying conservation of L around the upper-left end of the stick). But the latter gives the lower-right mass an impulse along the direction of the stick. This is why the velocity of the lower-right mass initially points along the stick.

This reasoning also tells us why the energy loss is $mv^2/8$. If we consider the same two hypothetical masses, the former leads to no energy loss, but the latter does. The stick is irrelevant as far as the energy loss from this latter hypothetical mass goes; this mass effectively just hits the lower-right mass and sticks to it. You can quickly use conservation of momentum to show that if a moving mass collides and sticks to an identical stationary one, then half of the initial energy is lost to heat. So the energy loss is $(1/2)(m(v/\sqrt{2})^2/2) = mv^2/8$.

8.13. Right-angled collision

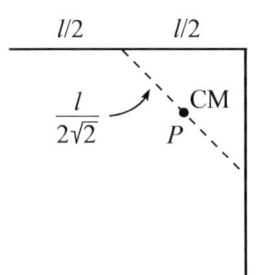

Figure 8.33

The CM of each stick is located a distance $\ell/(2\sqrt{2})$ from the CM of the entire system, as shown in Fig. 8.33. The parallel-axis theorem therefore gives the moment of inertia of the resulting rigid object, relative to the total CM, as

$$I_{\text{CM}} = 2\left(\frac{m\ell^2}{12} + m\left(\frac{\ell}{2\sqrt{2}}\right)^2\right) = \frac{5m\ell^2}{12}. \tag{8.50}$$

Conservation of angular momentum, around a dot on the table at the location of the total CM when the collision occurs (let's call this point P), gives

$$mv\left(\frac{\ell}{4}\right) = 0 + \left(\frac{5m\ell^2}{12}\right)\omega \implies \omega = \frac{3v}{5\ell}, \tag{8.51}$$

where we have used the fact that the "impact parameter" of the left stick relative to point P is $\ell/4$. The zero in Eq. (8.51) comes from the fact that the CM of the system moves directly away from point P, so the first term in Eq. (8.5) is zero.

To find the energy lost to heat, we first need to find the speed of the CM of the system. Conservation of linear momentum gives this speed as $mv = (2m)v_{\text{CM}} \implies v_{\text{CM}} = v/2$. The initial and final kinetic energies are then $K_i = mv^2/2$ and (using Eq. (7.8))

$$\begin{aligned}K_f &= \frac{1}{2}(2m)v_{\text{CM}}^2 + \frac{1}{2}I_{\text{CM}}\omega^2 \\ &= \frac{1}{2}(2m)\left(\frac{v}{2}\right)^2 + \frac{1}{2}\left(\frac{5m\ell^2}{12}\right)\left(\frac{3v}{5\ell}\right)^2 \\ &= \frac{13mv^2}{40}.\end{aligned} \tag{8.52}$$

The loss in energy is therefore $K_i - K_f = (1/2 - 13/40)mv^2 = 7mv^2/40$.

8.14. Stick on a table

Energy is conserved during the process, because we are told that all interactions (balls with stick, and stick with table) are elastic. Additionally, the angular momentum of the system relative to the corner of the table is conserved, because the external force from the corner provides no torque around it. (During the collision, the table applies a force to the stick only at the corner.)

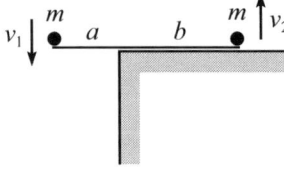

Figure 8.34

Let v_1 and v_2 be the velocities of the two balls right after the collision, with the positive directions chosen as shown in Fig. 8.34. If v_1 comes out to be negative, that means the left ball moves upward right after the collision. With counterclockwise taken to be positive, conservation of angular momentum relative to the corner of the table gives

$$mv_0 a = mv_1 a + mv_2 b \implies v_1 = v_0 - (b/a)v_2. \tag{8.53}$$

And conservation of energy gives

$$\frac{1}{2}mv_0^2 = \frac{1}{2}mv_1^2 + \frac{1}{2}mv_2^2 \implies v_0^2 = v_1^2 + v_2^2. \tag{8.54}$$

8.5. PROBLEM SOLUTIONS

Plugging the v_1 from Eq. (8.53) into Eq. (8.54) gives

$$v_0^2 = (v_0 - (b/a)v_2)^2 + v_2^2 \implies 0 = -2v_0v_2(b/a) + v_2^2(b^2/a^2 + 1)$$
$$\implies v_2 = v_0 \frac{2ab}{a^2 + b^2}. \qquad (8.55)$$

Either of Eq. (8.53) or Eq. (8.54) then gives

$$v_1 = v_0 \frac{a^2 - b^2}{a^2 + b^2}. \qquad (8.56)$$

LIMITS: If $b \to 0$ (more precisely, if $b \ll a$) then $v_1 \approx v_0$ and $v_2 \approx 0$. So the left ball keeps heading downward at speed v_0, and the right ball doesn't move. This makes sense intuitively.

If $a \to 0$ (more precisely, if $a \ll b$) then $v_1 \approx -v_0$ and $v_2 \approx 0$. So the left ball bounces *upward* (due to the minus sign in v_1) with the same speed v_0 that it originally had, and the right ball doesn't move. This also makes sense intuitively; the left ball is essentially just bouncing off the table.

If $a = b$ then $v_1 = 0$ and $v_2 = v_0$. So the left ball ends up at rest (instantaneously), and the right ball moves upward with speed v_0. It is believable, although not obvious, that $a = b$ leads to these velocities. However, by a continuity argument, it *is* obvious that there must be *some* relation between a and b that leads to the final speed of the left ball being zero, because v_1 is positive (downward) if $b \ll a$, and it is negative (upward) if $a \ll b$, so it must be zero for some intermediate value.

By conservation of energy, v_1 can't be any larger than v_0, of course. So the $a = b$ case yields the maximum possible speed of the right mass. (You can also deduce this from Eq. (8.55).) If the masses of the balls aren't equal, then the maximum possible speed of the right ball can quickly be found by using conservation of energy and letting the final speed of the left ball be zero; the right ball's final kinetic energy equals the left ball's initial kinetic energy. This method of finding the maximum speed of the right ball is *much* cleaner than finding the right ball's final speed and then taking the derivative to maximize it.

REMARKS: Note that v_0, v_1, and v_2 are in the ratio of $a^2 + b^2$ to $a^2 - b^2$ to $2ab$. These expressions are a standard way of generating Pythagorean triples. And the three velocities do indeed form a Pythagorean triple, in view of Eq. (8.54).

As an exercise, you can show that if you want both balls to bounce upward with the same speed, then a must be equal to $(\sqrt{2} - 1)b \approx (0.41)b$. Equivalently, $1/\sqrt{2} \approx 71\%$ of the stick must be lying on the table. Again, by a continuity argument, it follows that equal upward speeds must be obtained for some value of a between $a = b$ (where the left mass doesn't bounce up at all) and $a = 0$ (where the right mass doesn't bounce up at all).

8.15. Bouncing dumbbell

FIRST SOLUTION: We'll apply conservation of angular momentum around the contact point on the floor. Angular momentum is indeed conserved around this point, because the normal force from the floor has zero lever arm. For convenience, let $r \equiv \ell/2$. Also, we might as well work with a general moment of inertia $I = \beta mr^2$ around the center of the dumbbell. In the case at hand, $I = 2(m/2)r^2$, so $\beta = 1$. (A uniform stick would have $I = m(2r)^2/12 \implies \beta = 1/3$.) The post-bounce picture is shown in Fig. 8.35.

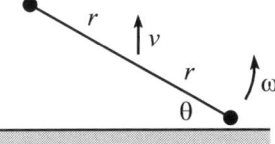

Figure 8.35

With counterclockwise L taken to be positive, and with the positive directions of the final v and ω as indicated in the figure, conservation of L around the contact point gives (using Eq. (8.5))

$$L_i = L_f \implies mv_0 r \cos\theta + 0 = -mvr\cos\theta + (\beta mr^2)\omega$$
$$\implies (v_0 + v)\cos\theta = \beta r\omega. \qquad (8.57)$$

Since the collision is elastic, we can also apply conservation of energy, which gives (using Eq. (7.8))

$$E_i = E_f \implies \frac{1}{2}mv_0^2 + 0 = \frac{1}{2}mv^2 + \frac{1}{2}(\beta mr^2)\omega^2$$
$$\implies v_0^2 - v^2 = \beta r^2\omega^2. \qquad (8.58)$$

The previous two equations involve two unknowns, v and ω. The easiest way to solve them is to divide Eq. (8.58) by Eq. (8.57). This gives $(v_0 - v)/\cos\theta = r\omega$. Equating this value of $r\omega$ with the one from Eq. (8.57) gives

$$\frac{v_0 - v}{\cos\theta} = \frac{(v_0 + v)\cos\theta}{\beta}$$

$$\implies v_0 \left(\frac{1}{\cos\theta} - \frac{\cos\theta}{\beta}\right) = v\left(\frac{1}{\cos\theta} + \frac{\cos\theta}{\beta}\right)$$

$$\implies v = v_0 \left(\frac{\beta - \cos^2\theta}{\beta + \cos^2\theta}\right)$$

$$\implies v = v_0 \left(\frac{1 - \cos^2\theta}{1 + \cos^2\theta}\right), \tag{8.59}$$

where we have used the fact that $\beta = 1$ for our dumbbell.

SECOND SOLUTION: Instead of applying conservation of L around the contact point, we can use the $\Delta L = l\Delta p$ relation from Eq. (8.10), with the dumbbell's CM taken to be the origin. (We're using l for the lever arm here, because we've already used r for half the length of the dumbbell.) Eq. (8.10) is applicable to a quick strike during which the lever arm is essentially constant, as it is here. In our setup, the lever arm (relative to the CM) of the normal force is $l = r\cos\theta$. The initial and final angular momenta relative to the CM are $L_i = 0$ and $L_f = I\omega$, so we have $\Delta L = I\omega$. Also, $\Delta p = mv - (-mv_0) = m(v + v_0)$. Therefore,

$$\Delta L = l\Delta p \implies (\beta m r^2)\omega = (r\cos\theta) \cdot m(v_0 + v)$$

$$\implies \beta r\omega = (v_0 + v)\cos\theta, \tag{8.60}$$

in agreement with the conservation-of-L statement in Eq. (8.57). This can be combined with the conservation-of-E statement in Eq. (8.58), as above.

LIMITS: If $\theta = 90°$ (the dumbbell is vertical), then Eq. (8.59) gives $v = v_0$. This makes sense; the dumbbell simply bounces back up with the same speed it initially had (and there is no rotation); the dumbbell may as well be a point mass in this case. If $\theta \to 0$ (the dumbbell is horizontal),[2] then Eq. (8.59) gives $v = 0$. What happens in this case is that the bouncing end heads upward with speed v_0, while the free end continues (at least briefly) to move downward with speed v_0. The CM moves with the average of the two velocities, which is zero.

In the $\theta \to 0$ case, the reason why the two resulting speeds are both v_0 is the following. When the dumbbell is horizontal (during the collision), the two masses behave like free particles. The stick effectively isn't there, for the following reason. Consider one of the masses. The stick can't apply a transverse force on this mass, because if it did, then by Newton's third law there would be a nonzero torque on the massless stick relative to the other mass, resulting in infinite angular acceleration. Therefore, since the stick effectively isn't there (during the collision), the right mass bounces up with speed v_0, and the left mass continues downward with speed v_0. See Problem 8.18(a) for another way to show that the stick effectively doesn't exist.

You should check that the expression for v in the third line of Eq. (8.59) behaves correctly in the $\beta \to 0$ and $\beta \to \infty$ limits. (Remember that we defined positive v to be upward.) The $\beta \to \infty$ case would require attaching extensions to the stick (with the extensions somehow being able to pass through the floor), and then attaching the masses to the ends of these extensions.

8.16. **Colliding with a wall**

FIRST SOLUTION: We've already done all of the work needed for this problem in Problem 8.15, because we solved that problem for a general moment of inertia, $I = \beta m r^2$ (where $r = \ell/2$). A uniform stick has $I = m(2r)^2/12$, so $\beta = 1/3$. The third line of

---

[2] Imagine that the right mass of the dumbbell hits near the left edge of a table, so that we don't have to worry about the left mass immediately running into the table.

8.5. PROBLEM SOLUTIONS

Eq. (8.59) then tells us that the final speed v of the CM is zero when $\cos\theta = 1/\sqrt{3}$, which means that $\theta \approx 54.7°$.

SECOND SOLUTION: If you want to solve the problem from scratch by invoking the fact that the final speed of the CM is zero, then conservation of L around the contact point gives (using Eq. (8.5) with counterclockwise L taken to be positive)

$$L_i = L_f \implies mv_0 \frac{\ell}{2}\cos\theta + 0 = 0 + \frac{m\ell^2}{12}\omega \implies \omega = \frac{6v_0\cos\theta}{\ell}. \tag{8.61}$$

There is only translational energy before the collision, and only angular energy after, so conservation of energy gives (using the value of ω we just found)

$$\frac{1}{2}mv_0^2 = \frac{1}{2}I\omega^2 \implies \frac{1}{2}mv_0^2 = \frac{1}{2}\left(\frac{m\ell^2}{12}\right)\left(\frac{6v_0\cos\theta}{\ell}\right)^2$$

$$\implies 1 = \frac{36}{12}\cos^2\theta \implies \cos\theta = \frac{1}{\sqrt{3}}. \tag{8.62}$$

THIRD SOLUTION: Instead of applying conservation of L around the contact point, we can use the $\Delta L = l\Delta p$ relation from Eq. (8.10), with the stick's CM taken to be the origin. We'll take the positive linear and angular directions to be leftward and counterclockwise (so that the force and torque provided by the wall are both positive). If the CM ends up at rest, then using $l = (\ell/2)\cos\theta$, the $\Delta L = l\,\Delta p$ relation gives

$$L_f - L_i = l(p_f - p_i) \implies I(\omega - 0) = \left(\frac{\ell}{2}\cos\theta\right)\cdot m(0 - (-v_0))$$

$$\implies \frac{m\ell^2}{12}\omega = \frac{\ell}{2}\cos\theta\cdot mv_0 \implies \omega = \frac{6v_0\cos\theta}{\ell}, \tag{8.63}$$

in agreement with the conservation-of-L statement in Eq. (8.61). This can be combined with the conservation-of-E statement in Eq. (8.62), as in the second solution.

REMARKS: In the post-collision motion of the stick, after it rotates through an angle of 2θ (with the CM at rest), the bottom end of the stick will collide with the wall. The resulting motion will then be a simple translation to the left with the original speed v_0, with no rotation (essentially the opposite of the initial motion); see Fig. 8.36. You can demonstrate this by performing the calculation for the second collision with the wall, similar to the one above for the first collision. Or you can simply note that if you consider the instant when the stick is parallel to the wall, and then look at the time-reversed motion, the stick will end up reversing the motion we analyzed above, and therefore head off to the left with speed v_0 and no rotation (the time reversal of the given initial motion).

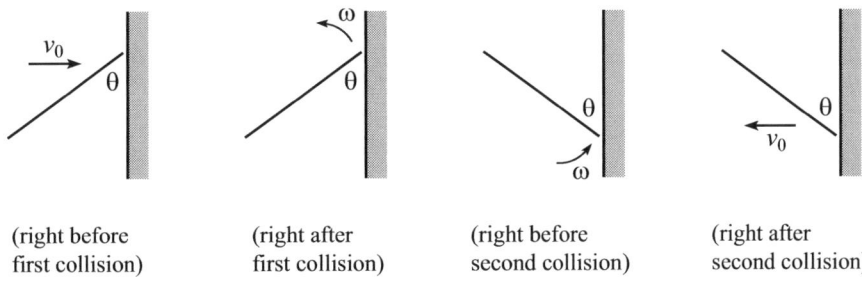

(right before first collision) (right after first collision) (right before second collision) (right after second collision)

Figure 8.36

A continuity argument makes it clear why there must exist an angle θ for which the final speed of the CM is zero. If $\theta = 90°$, then the CM (and the whole stick) simply bounces backwards. And if

$\theta \to 0$, then it is fairly intuitive that the CM will keep moving forward (until the other end of the stick hits the wall a split second later). So for some value of θ in between, the CM must end up at rest. Note that if the CM is at rest a moment after the first collision, then it must remain at rest until the second collision, because the net linear momentum of the stick (which equals mv_{CM}) can't change from zero to nonzero without the presence of an external force.

8.17. **Stick hitting a pole**

(a) We'll use conservation of angular momentum relative to the pole (the external force from the pole produces zero torque around the pole), and also conservation of energy (the collision is elastic). Conservation of energy gives

$$\frac{1}{2}mv_0^2 + 0 = 0 + \frac{1}{2}\left(\frac{1}{12}m\ell^2\right)\omega^2 \implies \omega = \sqrt{12}\,\frac{v_0}{\ell}, \tag{8.64}$$

where the first zero comes from the fact that the stick is initially not rotating, and the second zero comes from the fact that the final CM speed is zero.

Conservation of angular momentum relative to the pole (with counterclockwise taken to be positive) gives

$$mv_0 x + 0 = 0 + \left(\frac{1}{12}m\ell^2\right)\omega \implies x = \frac{\ell^2 \omega}{12 v_0}, \tag{8.65}$$

where the zeros arise for the same reasons as above. Substituting the ω from Eq. (8.64) into the x in Eq. (8.65) gives

$$x = \frac{\ell^2(\sqrt{12}\,v_0/\ell)}{12 v_0} = \frac{\ell}{\sqrt{12}} \approx (0.29)\ell. \tag{8.66}$$

This is larger than $\ell/4$, so the collision point is closer to the end of the stick than to the center.

REMARK: It is easy to see by continuity why there must exist a value of x that makes the final CM speed zero. If $x = 0$, then the stick bounces backward. But if $x = \ell/2$ (which corresponds to the stick's end), then intuitively we expect that the stick will keep moving forward (while rotating). So for some value of x in between, the CM must be at rest.

(b) FIRST SOLUTION: When the stick makes a half rotation and collides with the pole again, it will simply head off to the left with speed v_0, with no rotation. That is, the final state is exactly the same as the initial state. This obviously satisfies conservation of E and L with the initial conditions, so it must be what happens. (The other solution to the quadratic conservation equations is the rotational motion that we found in part (a).)

SECOND SOLUTION: The second collision is exactly the mirror image of the collision that would arise if we ran time backward through the first collision. And the latter scenario simply leads to the stick heading rightward with speed v_0 and no rotation (the time reversal of the initial state).

THIRD SOLUTION: If you actually want to apply conservation of E and L to the second collision, then with $\omega_0 = \sqrt{12}\,v_0/\ell$ being the ω we found in part (a), the conservation equations are

$$E: \quad \frac{1}{2}I\omega_0^2 = \frac{1}{2}mv_f^2 + \frac{1}{2}I\omega_f^2,$$
$$L: \quad I\omega_0 = mv_f x + I\omega_f, \tag{8.67}$$

where $x = \ell/\sqrt{12}$. There are various ways to solve for the unknowns ω_f and v_f here. One way is to simply solve for v_f in the second equation and plug the result into the

8.5. PROBLEM SOLUTIONS

first. This yields $\omega_f = \omega_0$ or $\omega_f = 0$, as you can show. The former of these must of course be a solution, because it corresponds to the initial conditions. The latter is the nontrivial solution we are concerned with. With $\omega_f = 0$, the E equation then gives $v_f = \omega_0 \sqrt{I/m} = (\sqrt{12}\, v_0/\ell)(\ell/\sqrt{12}) = v_0$.

8.18. Striking objects

(a) FIRST SOLUTION: Let the masses each be m, and let the stick have length $2r$. Then the moment of inertia of the dumbbell around its center is $2 \cdot mr^2$. So for a quick strike on one of the masses, Eq. (8.10) gives (with our origin chosen to be the center of the stick)

$$\Delta L = r\Delta p \implies (2mr^2)\omega = r \cdot (2m)v_{\rm CM} \implies \omega = \frac{v_{\rm CM}}{r}. \tag{8.68}$$

The motion of the end that isn't struck is the forward CM motion minus the backward rotational motion. So the speed of that end is

$$v_{\rm CM} - \omega r = v_{\rm CM} - \frac{v_{\rm CM}}{r}r = 0, \tag{8.69}$$

as desired.

SECOND SOLUTION: We can apply conservation of angular momentum around the fixed point P in space that coincides (initially) with the mass that you strike. Since the initial angular momentum is zero, and since you apply no torque around P, the final angular momentum is also zero. This implies that the other mass can't move (right after the strike), because if it did, it would contribute a nonzero angular momentum around P, and there is nothing else that can cancel this L; the stick is massless, and the mass you strike moves directly away from P (right after the strike).

The above reasoning applies only to the transverse (perpendicular to the stick) motion of the other mass. So to be complete, we should technically also say that there can be no motion in the longitudinal direction (along the stick), because the impulse you apply is perpendicular to the stick, so the longitudinal momentum must be zero.

Another (very similar) way to show that the other mass has no transverse motion (right after the strike) is the following. As mentioned near the end of the solution to Problem 8.15, the stick can't apply a transverse force on the other mass, because if it did, then by Newton's third law the mass would apply a force on the stick. There would then be a finite torque on the massless stick relative to the mass you strike, resulting in infinite angular acceleration.

(b) Let the total mass of the stick be m, and again let the length be $2r$. Then the moment of inertia around the center is $m(2r)^2/12 = mr^2/3$. For a quick strike at one of the ends, Eq. (8.10) gives (with our origin again chosen to be the center of the stick)

$$\Delta L = r\Delta p \implies \left(\frac{mr^2}{3}\right)\omega = r \cdot mv_{\rm CM} \implies \omega = \frac{3v_{\rm CM}}{r} \equiv \frac{3v_0}{r}. \tag{8.70}$$

The motion of the end that isn't struck is the forward CM motion minus the backward rotational motion. So the velocity of that end is

$$v_{\rm CM} - \omega r = v_0 - \frac{3v_0}{r}r = -2v_0. \tag{8.71}$$

This is negative, which means that this end moves backwards.

For the end that you hit, the motion is obtained by *adding* the forward rotational motion to the CM motion. So the velocity of this end is $v_{\rm CM} + \omega r = v_0 + (3v_0/r)r = 4v_0$. The average of the velocities of the ends (which are $-2v_0$ and $4v_0$) is correctly equal to the velocity v_0 of the center of the stick.

REMARKS: As an exercise, you can also derive the above $\omega = 3v_0/r$ result by applying conservation of angular momentum around the fixed point in space that coincides (initially) with the end that you strike. The initial (and hence final) L is zero. There are two canceling contributions to the final L, one from the translational motion, and the other from the rotational motion.

If you flick the end of a pencil with your finger, the pencil as a whole quickly moves forward in the direction of the strike. This happens fast enough to make it difficult to determine if the non-struck end actually moves backward, as specified in Eq. (8.71). But you can unequivocally settle this issue by placing an object on the back side of the non-struck end, very close to it. The object will move if and only if this end initially moves backward and hits the object.

8.19. Center of percussion

First note that a continuity argument quickly shows that such a strike point must exist. If you strike the middle of the stick, then the entire stick will of course simply move forward. But if you strike the far end of the stick, then from Problem 8.18(b) we know that the near end will move backward. Therefore, somewhere in between there must exist a strike point for which the near end doesn't move.

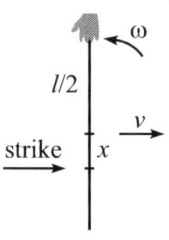

Figure 8.37

We'll use the $\Delta L = x\Delta p$ result from Eq. (8.10), with the CM of the stick chosen as the origin. So ΔL is measured with respect to the CM, and x is the distance from the strike to the CM; see Fig. 8.37. Both L and p start at zero, so $\Delta L = x\Delta p$ tells us that the resulting values of L and p are related by

$$L = xp \implies \left(\frac{1}{12}m\ell^2\right)\omega = x(mv) \implies \omega = \frac{12xv}{\ell^2}. \tag{8.72}$$

Due to the resulting rotation with angular speed ω, the end that you are holding will initially move backward with speed $(\ell/2)\omega$ with respect to the CM. We want this motion to be canceled by the forward speed v of the CM. The end that you are holding will then be (initially) at rest. Using the ω from above, this condition yields

$$\left(\frac{\ell}{2}\right)\omega = v \implies \frac{\ell}{2}\left(\frac{12xv}{\ell^2}\right) = v \implies x = \frac{\ell}{6}. \tag{8.73}$$

This is the distance from the center, so the strike should occur $\ell/2 + \ell/6 = 2\ell/3$ from your hand.

REMARK: As an exercise, you can also solve this problem by applying $\Delta L = r\Delta p$ with your hand chosen as the origin. The solution is fairly quick, due to the fact that (at least initially) the end that you are holding is assumed to be at rest, which means that the stick can be considered to be (instantaneously) rotating around your hand.

As another exercise, you can consider the more general case of an object with mass m and arbitrary moment of inertia I_{CM} around the CM. If you are holding the object at a distance d from the CM, you can show that the center of percussion is located a distance $x = I_{CM}/md$ from the CM, on the opposite side. In the above case of a uniform stick held at an end, this correctly gives $x = (m\ell^2/12)/(m \cdot \ell/2) = \ell/6$.

The center of percussion corresponds roughly to the "sweet spot" on a baseball bat when hitting a ball. However, there are a number of real-life complications at play, so the correspondence generally isn't exact.

8.20. Dumbbell and pole

First note that there does indeed exist a value of x that leads to no rotational motion of the dumbbell, due to the following continuity argument. If x is equal to r, then the pole is right next to the top mass. So this mass simply bounces off the pole and heads rightward with speed v_0. And since the bottom mass doesn't move right away, this means that the dumbbell rotates clockwise. On the other hand, if x equals zero (so the pole is right next to the center of the stick), then the dumbbell rotates counterclockwise, because it does so

8.5. PROBLEM SOLUTIONS

right after the swift strike to the top mass, and the collision with the pole doesn't change the angular momentum around the CM; so ω stays the same. Therefore, by continuity, there must exist a value of x between 0 and r for which the stick doesn't rotate after the collision with the pole. Now let's be quantitative.

The dumbbell's energy is conserved during the collision with the pole, because we are told that the collision is elastic. Also, the dumbbell's angular momentum around the pole is conserved, because the force provided by the pole provides no torque around it. But the dumbbell's linear momentum is *not* conserved, because the pole applies a force.

The initial energy is $E_i = mv_0^2/2$, because only one mass is moving. The final energy comes only from the CM motion of the dumbbell, because it isn't rotating. So $E_f = (2m)v_{CM}^2/2$. Conservation of E therefore gives

$$\frac{1}{2}mv_0^2 = \frac{1}{2}(2m)v_{CM}^2 \implies v_{CM} = \frac{v_0}{\sqrt{2}}. \tag{8.74}$$

We'll now apply conservation of angular momentum relative to the pole. The initial L comes from the moving mass, which is a distance $r - x$ from the pole. So $L_i = mv_0(r - x)$, in a counterclockwise sense. The final L comes only from the CM motion of the dumbbell, because there is no rotation. The CM is a distance x from the pole, so $L_f = (2m)v_{CM}x$, and this also has a counterclockwise sense. Conservation of L therefore gives

$$mv_0(r - x) = (2m)v_{CM}x. \tag{8.75}$$

Using $v_{CM} = v_0/\sqrt{2}$ from above, this becomes

$$v_0(r - x) = \sqrt{2}v_0 x \implies x = \frac{r}{\sqrt{2} + 1} = r(\sqrt{2} - 1) \approx (0.41)r. \tag{8.76}$$

Since half the length of the stick was defined to be r, we see that the pole is slightly closer to the dumbbell's center than to its end.

REMARKS: Another way of obtaining the result in Eq. (8.75) is to use Eq. (8.10). The pole applies a quick strike to the dumbbell, so Eq. (8.10) is applicable during this collision. Let's pick our origin to be a dot on the table that coincides with the dumbbell's CM. Then with clockwise L and rightward v taken to be positive, $\Delta L = x\Delta p$ gives

$$L_f - L_i = x(p_f - p_i) \implies 0 - (-mv_0 r) = x(2mv_{CM} - (-mv_0)), \tag{8.77}$$

where the zero here comes from the fact that the CM of the (nonrotating) dumbbell is moving directly away from our chosen origin. This equation is equivalent to Eq. (8.75).

8.21. **One full revolution**

Let the final linear and angular speeds be v and ω. Since the lever arm x is constant during the quick strike, we can use the $\Delta L = x\Delta p$ relation from Eq. (8.10), with the sticks's CM taken to be the origin. And since the stick starts at rest, the change in angular momentum ΔL is just $I\omega$, and the change in linear momentum Δp is just mv. $\Delta L = x\Delta p$ therefore gives

$$\frac{m\ell^2}{12}\omega = x(mv) \implies \frac{\omega}{v} = \frac{12x}{\ell^2} \tag{8.78}$$

Another expression for ω/v is obtained from the given information that the stick makes one complete revolution by the time the center reaches the dot. This implies that if t is the time to reach the dot, then $\omega t = 2\pi$. And we also know that $vt = d$ from the linear motion. Dividing these equations yields $\omega/v = 2\pi/d$. Equating our two expressions for ω/v gives

$$\frac{12x}{\ell^2} = \frac{2\pi}{d} \implies x = \frac{\pi\ell^2}{6d}. \tag{8.79}$$

Since x can be no larger than $\ell/2$, the minimum value of d is given by

$$\frac{\pi \ell^2}{6d} \le \frac{\ell}{2} \implies \frac{\pi}{3}\ell \le d. \tag{8.80}$$

So the minimum d is about $(1.05)\ell$, and in this case you should strike the stick at an end. Smaller values of d would require attaching a massless extension to the stick, which would allow x to be larger than $\ell/2$. A very long massless extension would allow you to make the stick essentially spin in place.

LIMITS: : If $d \to \infty$, then the expression for x in Eq. (8.79) correctly goes to zero. You want the stick to have a very small ω (which means that you must give it a very small angular impulse), because the time to reach the dot will be large.

8.22. **Sliding to rolling**

The key point in this problem is that the friction force F_f from the ground does *two* things. It decreases the ball's v via $F = ma$, and it also increases the ball's ω via $\tau = I\alpha$. The changes Δv and $\Delta \omega$ are therefore related. To see exactly how they are related, note that the lever arm (relative to the center of the ball) of the friction force is always the same (namely, the radius R of the ball). We can therefore use the $\Delta L = R\Delta p$ result from Eq. (8.10) to say that $I\Delta\omega = Rm\Delta v$. This equation relates the magnitudes of Δv and $\Delta\omega$. But since we know that v is decreasing and ω is increasing (given the sign conventions shown in Fig. 8.38), the actual relation between the signed quantities takes the form of (using $I = 2mR^2/5$ for a solid sphere)

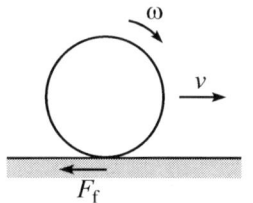

Figure 8.38

$$I\Delta\omega = -Rm\Delta v \implies \left(\frac{2}{5}mR^2\right)\Delta\omega = -Rm\Delta v$$

$$\implies \Delta\omega = -\frac{5}{2R}\Delta v. \tag{8.81}$$

When the ball eventually rolls without slipping, the final v_f and ω_f values are related by the non-slipping condition,

$$v_f = R\omega_f \implies \omega_f = \frac{v_f}{R}. \tag{8.82}$$

What happens physically is the following. At the start, the initial linear and angular velocities are $v_i = v_0$ and $\omega_i = 0$. So at the start, v_i is too large compared with ω_i, as far as the non-slipping $v = R\omega$ condition goes, so the ball is slipping. As time goes on, v decreases and ω increases (with the changes related by Eq. (8.81)) until finally the $v = R\omega$ condition is satisfied. At that time, the friction force suddenly drops to zero, and the ball then rolls indefinitely with the constant values of v_f and ω_f.

Let's be quantitative about the preceding paragraph. The changes in the linear and angular velocities are given by

$$\Delta v = v_f - v_i = v_f - v_0$$
$$\Delta\omega = \omega_f - \omega_i = \omega_f - 0 = \frac{v_f}{R}. \tag{8.83}$$

Plugging these expressions into Eq. (8.81) gives

$$\frac{v_f}{R} = -\frac{5}{2R}(v_f - v_0) \implies \frac{7v_f}{2} = \frac{5v_0}{2} \implies v_f = \frac{5v_0}{7}. \tag{8.84}$$

REMARKS: In deriving this result, note that nowhere did we say anything about the exact nature of the friction force. The force could be large (in which case the slipping ends quickly) or small (in which case the slipping lasts for a long time), or it could even vary with position. But in any case the

8.5. PROBLEM SOLUTIONS

final velocity will still be $5v_0/7$. This is true because the relation between $\Delta\omega$ and Δv in Eq. (8.81) follows from Eq. (8.10), which makes no mention of the exact nature of the force.

If you let the moment of inertia of the ball take the general form of $I = \beta m R^2$ (so $\beta = 2/5$ in the case of a solid ball), then as an exercise you can show that $v_f = v_0/(1 + \beta)$. This correctly gives $v_f = 5v_0/7$ when $\beta = 2/5$. And it also correctly gives $v_f = v_0$ when $\beta = 0$ (when all the mass is at the center of the ball); the angular acceleration is infinite for a split second, and the ball immediately rolls without slipping.

An another exercise, you can calculate the final speed in the case where the ball is initially only rotating with angular speed ω_0, and not translating. (Answer: $v_f = 2R\omega_0/7$, with the 2/7 more generally being $\beta/(1 + \beta)$.)

In our original setup, if you calculate the total final kinetic energy (translational plus rotational), you will find that it is less than the initial kinetic energy (all translational). This is expected, because some energy is lost to heat as the ball slips with respect to the ground; friction does work. However, when calculating this work as a force times a distance, the relevant distance is *not* the distance the ball moves. Rather, it is the distance moved by the contact point on the ball *relative* to the ground. The rotation of the ball makes this distance be smaller than the horizontal distance traveled by the ball. You can work everything out quantitatively, but it gets a bit messy.

8.23. Tipping a block

(a) We can use the parallel-axis theorem to calculate the moment of inertia of the block around the pivot:

$$I_{\text{block}} = I_{\text{CM}} + md^2 = \frac{m\ell^2}{6} + m\left(\frac{\ell}{\sqrt{2}}\right)^2 = \frac{2m\ell^2}{3}. \tag{8.85}$$

The moment of inertia of the ball (when it is at the corner) around the pivot is $m(\sqrt{2}\ell)^2 = 2m\ell^2$. The total moment of inertia of the system around the pivot is therefore $I = 2m\ell^2/3 + 2m\ell^2 = 8m\ell^2/3$.

During the collision, the angular momentum around the pivot is conserved, because the external force at the pivot produces no torque around it. Since the initial angular momentum is $mv\ell$ (the ball has an impact parameter of ℓ), conservation of L gives

$$mv\ell = \left(\frac{8m\ell^2}{3}\right)\omega \implies \omega = \frac{3v}{8\ell}. \tag{8.86}$$

(b) We'll now use conservation of energy to determine whether the CM of the system rises up to where it is directly over the pivot. This will give the cutoff value of v. Any value of v larger than this will result in the block falling over to the right.

Since the block and the ball have the same mass, the CM of the system is located halfway between the ball and the center of the block. So it is located $(3/4)(\sqrt{2}\ell)$ from the pivot. The CM therefore needs to rise from its starting height of $3\ell/4$ to its height of $(3/4)(\sqrt{2}\ell)$ when it is directly over the pivot. The initial kinetic energy of the system (right after the collision) is $I\omega^2/2$. The block will tip over if this kinetic energy is greater than or equal to the necessary increase in potential energy, that is, if

$$\frac{1}{2}I\omega^2 \geq (2m)g\,\Delta h$$

$$\implies \frac{1}{2}\left(\frac{8m\ell^2}{3}\right)\left(\frac{3v}{8\ell}\right)^2 \geq (2m)g\left(\frac{3}{4}\sqrt{2}\ell - \frac{3}{4}\ell\right)$$

$$\implies \frac{3}{16}mv^2 \geq \frac{3(\sqrt{2}-1)}{2}mg\ell$$

$$\implies v \geq \sqrt{8(\sqrt{2}-1)g\ell}, \tag{8.87}$$

which is about $(1.8)\sqrt{g\ell}$. For comparison, this is slightly larger than the $\sqrt{2g\ell} \approx (1.4)\sqrt{g\ell}$ speed attained by an object dropped a height ℓ from rest.

Note that because energy is lost in the (completely) inelastic collision, the relevant initial kinetic energy in the above conservation-of-E statement for the rising up of the block is the $3mv^2/16$ energy right *after* the collision, not the $mv^2/2$ energy right before it.

Chapter 9

Statics

9.1 Introduction

Zero force and torque

In this chapter we will deal with objects that are at rest, or *static*. If an object is at rest, then it has zero linear acceleration and zero angular acceleration. So the net force and net torque must both be zero:

$$\sum \mathbf{F} = 0 \quad \text{and} \quad \sum \tau = 0. \tag{9.1}$$

The converse isn't true, of course. That is, if the net force and torque are both zero (which means that the object is in *equilibrium*, by definition), then the object need not be static. It can be moving with constant linear velocity and/or constant angular velocity. But we will deal only with static objects here. So this entire chapter can be summarized by saying that if an object is static, then the net force and net torque must both be zero. However, a few things should be noted:

Planar setups

We will invariably deal with planar situations, as opposed to general 3-D ones. Therefore, since both the position $\mathbf{r}$ and the force $\mathbf{F}$ will lie in the plane of the page, the torque $\tau = \mathbf{r} \times \mathbf{F}$ will always point perpendicular to the page. So if the plane of the page is the x-y plane, then we are interested only in τ_z; the full vector nature of τ isn't important. Similarly, we are interested only in the force components F_x and F_y lying in the plane of the page. So there are only three equations, instead of the six in theory, that we need to write down when applying the equations in Eq. (9.1) to a given object (or more generally a given subsystem).

Choosing subsystems

Statics problems often involve making choices, and these choices often make solving problems a little less straightforward than the simplicity of Eq. (9.1) might suggest. If a system consists of various objects, then you need to decide which subsystem (a single object, or a combination of objects) you should apply Eq. (9.1) to. You will often need to apply it to a few different subsystems to obtain a sufficient amount of information to solve for whatever you're trying to solve for (often the various forces). Remember that Eq. (9.1) involves only the *external* forces (and torques) acting on a subsystem. So if a particular force appears as an internal force in a subsystem you've chosen, then there's no way you're going to be able to solve for it by using that subsystem. You will need to pick a different subsystem for which the desired force appears as an *external* force.

Many different sets of subsystems will get the job done by giving you enough information to solve for your unknowns; there isn't a single "right" way to pick the subsystems. However, some sets might make the calculations easier than others, although at the start it usually isn't obvious which ones these are. When solving a problem, it's best just to keep an eye on what you're trying to solve for, and to try to pick subsystems accordingly. If you get stuck, you can always just apply Eq. (9.1) to every possible subsystem. This will inevitably lead to the correct answer, even if there might have been a quicker way to go about things. If you write down every possible force and torque equation, there will be a good deal of redundancy in the equations, assuming that there is more than one object in the system.

Choosing an origin

In addition to choosing subsystems, another choice that you will need to make for each subsystem is the choice of origin around which the torque is calculated. You can pick different origins for different subsystems. It is often a good idea to choose your origin as the point at which the most forces act, because these forces will then provide zero torque around that origin (because the lever arm is zero), so there will be fewer unknowns in the $\tau_z = 0$ equation.

You might be worried that even if you have demanded that the torque is zero around one particular choice of origin, it might not be zero around another choice. The following fact dispels this worry:

- Given an object for which $\sum \mathbf{F} = 0$, then if $\sum \tau = 0$ around one choice of origin, then $\sum \tau = 0$ around any other choice of origin.

This means that you are free to pick the most convenient point as your origin. This fact is proved in Problem 9.1, although it is intuitively clear in the case of static objects: If an object is static, which implies that it isn't angularly accelerating around a given origin, then it isn't angularly accelerating around any other choice of origin either. Said in another way, a static object is static in a given frame no matter how you look at it.

9.2 Multiple-choice questions

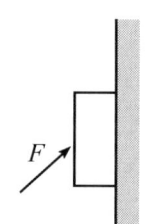

Figure 9.1

9.1. You hold a book at rest against a vertical wall by applying a force upward at an angle, as shown in Fig. 9.1. The static friction force from the wall on the book

(a) points upward

(b) points downward

(c) is zero

(d) The direction cannot be determined from the given information.

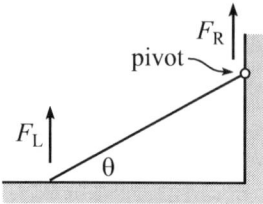

Figure 9.2

9.2. One end of a stick with mass m and length ℓ is pivoted on a wall, and the other end rests on a *frictionless* floor, as shown in Fig. 9.2. Let F_L and F_R be the vertical forces on the left and right ends of the stick, respectively. Then

(a) $F_L > F_R$

(b) $F_L < F_R$

(c) $F_L = F_R = mg/2$

(d) $F_L = F_R = mg$

(e) $F_L = F_R = mg \cos\theta$

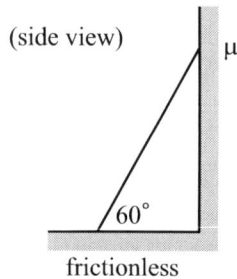

Figure 9.3

9.3. A ladder leans against a wall at a 60° angle, as shown in Fig. 9.3. The floor is frictionless, but there is friction with the wall. Assume that the coefficient of friction is large (say, $\mu = 10$). Is it possible for this setup to be static?

Yes   No

9.3. PROBLEMS

9.4. A uniform stick with mass m rests on two supports at its ends, as shown in Fig. 9.4. The forces exerted by the supports on the stick are equal because (circle all that apply)

(a) the total upward force from the supports must be mg

(b) the torques around the center of the stick must cancel

(c) the setup has left-right symmetry

Figure 9.4

9.5. Two uniform sticks with masses M and m are connected to each other and to a wall by pivots, as shown in Fig. 9.5. The left ends of both sticks are higher than their common right end. The vertical component of the force from the wall (acting via the pivot) on the left end of the bottom stick

(a) points upward

(b) points downward

(c) is zero

(d) More information about how m and M are related is required.

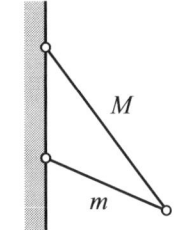

Figure 9.5

9.3 Problems

The first problem is a foundational problem.

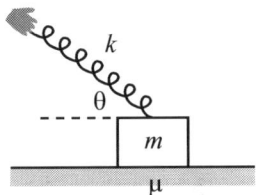

Figure 9.6

9.1. Any choice of origin

Prove the following statement: Given an object for which $\sum \mathbf{F} = 0$, then if $\sum \tau = 0$ around one choice of origin, then $\sum \tau = 0$ around any other choice of origin.

9.2. Not moving a block

A block with mass m rests on a horizontal table. The coefficient of static friction between the block and the table is μ. You push down on the block with a spring (with spring constant k) that is inclined at an angle θ, as shown in Fig. 9.6. What is the maximum distance you can compress the spring without having the block move?

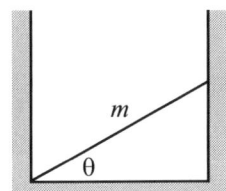

Figure 9.7

9.3. Stick in a well

A uniform stick with mass m is placed in a frictionless well. The angle the stick makes with the horizontal is θ, as shown in Fig. 9.7. What are the forces the well exerts on the stick at its two ends?

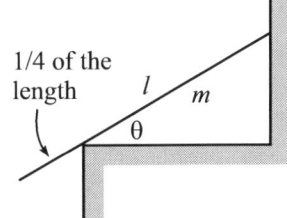

Figure 9.8

9.4. Stick on a corner

A stick with mass m and length ℓ leans against a frictionless wall, with a quarter of its length hanging over a corner, as shown in Fig. 9.8. It makes an angle θ with the horizontal. Assuming that there is sufficient friction at the corner to keep the stick at rest, what is the total force that the corner exerts on the stick?

9.5. Two sticks and a wall

Two sticks are connected with pivots to each other and to a wall, as shown in Fig. 9.9. The top stick is horizontal and has length ℓ, and the angle between the sticks is θ. Both sticks have the same linear mass density λ (kg/m). Find the force (give the horizontal and vertical components) that the lower sticks applies to the upper one.

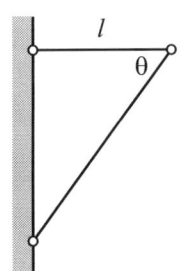

Figure 9.9

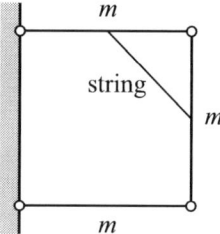

Figure 9.10

9.6. Supporting a square

Three sticks with mass m form three sides of a square, as shown in Fig. 9.10. They are connected with pivots to each other and to a wall. The midpoints of the top and right sticks are connected by a massless string. What is the tension in the string?

9.7. Corner on a plane 1

Two sticks, each with mass m and length ℓ, are rigidly connected to form a right angle. The bottom end of the object is pivoted on a plane inclined at a $45°$ angle, and the top end rests on the plane, as shown in Fig. 9.11. Assume that the plane is frictionless. What is the force on the object at its top end? At the pivot?

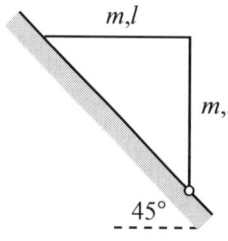
Figure 9.11

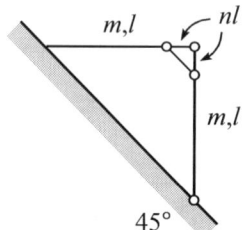
Figure 9.12

9.8. Corner on a plane 2

In the previous problem, the force from the plane on the left (top) stick has a rightward component. Since the setup is static, the right stick must therefore apply an equal and opposite leftward force on the left stick at the corner. By Newton's third law, the left stick must therefore apply a rightward force on the right stick at the corner. However, this then seems to imply that there is a nonzero clockwise torque on the right stick around the pivot at its bottom end. But the torque must be zero since the setup is static. What's going on here? How can a nonzero force acting on the right stick at the corner produce zero torque around the pivot?

Hint: Zoom in on the rigid corner, and imagine that it can be modeled as a pivot plus a tiny massless stick inclined at $45°$, as shown in Fig. 9.12. Let this tiny stick be connected (with pivots) to each of the original sticks at a distance $n\ell$ from the corner, where n is a small numerical factor. Find the force exerted by the tiny stick on each of the original sticks, and also the force that the pivot at the corner applies to each stick.

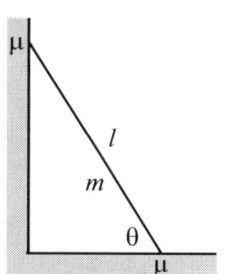

Figure 9.13

9.9. Leaning ladder

A ladder leans against a vertical wall, making an angle θ with the horizontal, as shown in Fig. 9.13. The coefficient of static friction between the ladder and *both* the floor and the wall is μ. What is the minimum value of θ (in terms of μ) for which it is possible for the ladder not to fall? *Hint*: You can assume that in the cutoff case where the ladder is about to fall, the friction forces take on their maximum allowed values (which is what you would intuitively expect), assuming that $\mu \leq 1$.

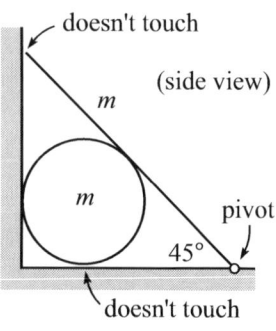

Figure 9.14

9.10. Stick on a cylinder

In Fig. 9.14 a stick with mass m rests on a cylinder also with mass m. The stick is inclined at a $45°$ angle. The stick's bottom end is connected to the ground by a pivot, and the top end is infinitesimally close to the wall, but *doesn't touch it*. Similarly, the cylinder is infinitesimally close to the ground, but *doesn't touch it*. What is the minimum coefficient of friction between the stick and the cylinder that allows the system to be static? (Assume that there is sufficient friction between the cylinder and the wall.)

9.11. Stick on a disk

A stick with mass m and length $2R$ is pivoted at one end on a vertical wall. It is held horizontal, and a disk with mass m and radius R is placed beneath it, in contact with both it and the wall, as shown in Fig. 9.15. The coefficient of friction between the disk and the wall is μ_w, and the coefficient of friction between the disk and the stick is μ_s. If the objects are released, what are the minimum values of μ_w and μ_s for which the system doesn't fall?

9.12. Cylinder on a stick

A massless stick with length $\sqrt{2}\ell$ is situated at a 45° angle in a box whose base has length ℓ, as shown in Fig. 9.16. A cylinder with mass m rests on the stick, and its radius is chosen so that the top of the cylinder is at the same height as the top of the stick. There is no friction anywhere in the setup. Find the force (give the horizontal and vertical components) that the box exerts on the stick at its lower end.

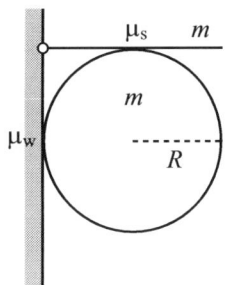

Figure 9.15

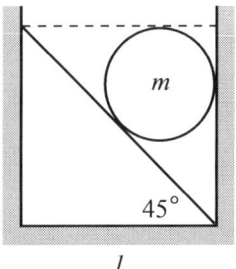

Figure 9.16

9.4 Multiple-choice answers

9.1. $\boxed{d}$ If the vertical component of your force is smaller than mg, then the friction force must point upward, to make the net vertical force be zero. On the other hand, if the vertical component of your force is larger than mg, then the friction force must point downward.

REMARK: If you push on the book too gently, then it will fall, of course, because the maximum upward static friction force (which is μN) won't be large enough, when combined with your upward F_y force, to balance the downward mg force. As an exercise, you can calculate the minimum F in terms of the angle θ that your force makes with the horizontal (along with m, g, and μ).

If you push on the book too hard, then it *might* rise up, if the maximum downward static friction force isn't large enough, when combined with the downward mg force, to balance your upward F_y force. However, if θ is small enough (more precisely, if $\tan\theta \leq \mu$), then the book won't rise up, no matter how hard you push. You will see how this scenario arises if you calculate the maximum F that allows the book to remain rest. Physically, what's going on is that the harder you push, the larger the normal force is, so the larger the maximum friction force is. If θ is small enough, then this effect wins out over the fact that you are also increasing the vertical component F_y of your force.

9.2. $\boxed{c}$ The vertical forces must add up to mg. And they must be equal so that there is no torque around the CM. (They are the only forces providing torques, because there is no horizontal force from the wall, because there is no horizontal force from the frictionless floor.) So both of the vertical forces must equal $mg/2$.

9.3. $\boxed{\text{No}}$ Since the floor is frictionless, it can't provide a horizontal force. The (horizontal) normal force from the wall must therefore be zero if the system is static. But $F_f \leq \mu N$ then says that the friction force from the wall must also be zero. So the wall can't apply a vertical force either. The normal force from the floor (which must be mg) is therefore the only force from the floor and wall. This force produces a nonzero torque around the center of the ladder, which means the system can't be static. Note that the specific angle of 60° is irrelevant.

REMARK: If you release the ladder from rest, what happens is that it rotates clockwise, with the bottom end sliding leftward and the top end sliding downward. There *is* a nonzero normal force from the wall in this nonstatic scenario, and this normal force is what causes the CM to accelerate leftward. Due to this nonzero normal force, there is now also a nonzero kinetic friction force from the wall.

9.4. $\boxed{b,c}$ Choice (a) is a true statement, but it deals only with the sum of the two forces, and not with their individual values.

Choice (b) is a valid reason, although technically it should be supplemented with the standard fact that the CM of a uniform stick is located at the center, which means that the gravitational force produces no torque around the center. Note that this torque argument tells us only that the forces are equal, and not what their common value is. We would need to use the force statement in choice (a) if we wanted to say that both forces are equal to $mg/2$.

Choice (c) is a valid reason because if you claimed that, say, the right force is larger than the left force, then imagine flipping the setup over (reversing right and left). You would then logically have to claim two different things: You would have to claim that the *left* force is now larger (because that's the same support that you previously said produced the larger force). But you would also have to claim that the *right* force is larger, because we have exactly the same setup as before (due to the symmetry), so you must make the same statement that you originally made, namely, that the right force is larger. You would therefore necessarily have to contradict yourself. The only way to escape this contradiction is to say (correctly) that the two forces are equal.

9.5. $\boxed{\text{d}}$ More information is required, because in the $M \gg m$ limit the wall pushes down on the left end of the bottom stick, whereas in the $m \gg M$ limit the wall pushes up on the left end of the bottom stick. We can justify these claims in the following way.

In general, the simplest way to determine which direction a force points is to imagine removing the force and then thinking about which way the system will move. In the present setup, let's replace the pivot at the left end of the bottom stick with a bead that is free to slide up and down a frictionless pole. So the left end is still constrained in the x direction, but now not the y direction. In the $M \gg m$ limit, it is intuitively clear that the top stick will simply swing down, in which case the left end of the bottom stick must rise up. Returning to our original scenario with the pivot, we see that the pivot must therefore provide a downward force on the left end of the bottom stick to keep it motionless. Conversely, in the $m \gg M$ limit, it is intuitively clear that left end of the bottom stick will swing down. In the original scenario with the pivot, the pivot must therefore provide an upward force on the left end of the bottom stick to keep it motionless.

9.5 Problem solutions

9.1. **Any choice of origin**

Even though we will generally deal only with planar setups, where the full vector nature of τ isn't important, it is easiest to prove the given statement by working in full generality with vectors.

Assume that $\sum \tau = 0$ around a given point, which we will call the origin. If a force $\mathbf{F}_i$ is applied at position $\mathbf{r}_i$, then the torque is $\tau_i = \mathbf{r}_i \times \mathbf{F}_i$. So if there are N forces, the $\sum \tau = 0$ assumption can be written as

$$0 = \sum \tau_{\text{origin}} = (\mathbf{r}_1 \times \mathbf{F}_1) + (\mathbf{r}_2 \times \mathbf{F}_2) + \cdots + (\mathbf{r}_N \times \mathbf{F}_N). \tag{9.2}$$

Now consider the torque around a different point P located at position $\mathbf{r}_P$ (relative to the above origin). Relative to P, each force $\mathbf{F}_i$ is applied at position $\mathbf{r}_i - \mathbf{r}_P$; see Fig. 9.17. So the total torque relative to P equals

$$\sum \tau_P = (\mathbf{r}_1 - \mathbf{r}_P) \times \mathbf{F}_1 + (\mathbf{r}_2 - \mathbf{r}_P) \times \mathbf{F}_2 + \cdots + (\mathbf{r}_N - \mathbf{r}_P) \times \mathbf{F}_N$$
$$= (\mathbf{r}_1 \times \mathbf{F}_1) + (\mathbf{r}_2 \times \mathbf{F}_2) + \cdots + (\mathbf{r}_N \times \mathbf{F}_N)$$
$$\quad - \mathbf{r}_P \times (\mathbf{F}_1 + \mathbf{F}_2 + \cdots + \mathbf{F}_N)$$
$$= \sum \tau_{\text{origin}} - \mathbf{r}_P \times \left(\sum \mathbf{F}\right)$$
$$= 0 - 0, \tag{9.3}$$

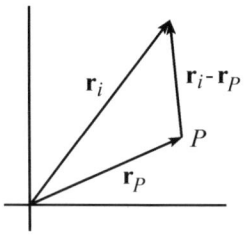

Figure 9.17

because $\sum \tau_{\text{origin}}$ and $\sum \mathbf{F}$ are both zero, by assumption. More generally, we just showed that if $\sum \mathbf{F} = 0$ then $\sum \tau_P = \sum \tau_{\text{origin}}$, even if the common value isn't zero.

9.5. PROBLEM SOLUTIONS

9.2. Not moving a block

The free-body diagram is shown in Fig. 9.18. In the cutoff case where the block barely doesn't move, the friction force takes on its maximum possible value, which is $F_f = \mu N$. Setting the net force equal to zero in the cutoff case gives

$$\sum F_x = 0 \implies kx\cos\theta - \mu N = 0,$$
$$\sum F_y = 0 \implies N - kx\sin\theta - mg = 0. \quad (9.4)$$

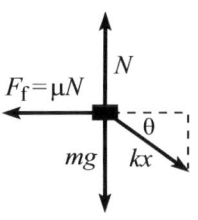

Figure 9.18

Solving for N in the second equation and substituting the result into the first equation gives the maximal compression as

$$kx\cos\theta - \mu(kx\sin\theta + mg) = 0 \implies x = \frac{\mu mg}{k(\cos\theta - \mu\sin\theta)}. \quad (9.5)$$

REMARK: If $\cos\theta - \mu\sin\theta = 0$, or equivalently if $\tan\theta = 1/\mu$, then our result for x is infinite. This means that you can compress the spring an infinite amount (assuming it's a long spring), and the block won't move. (Likewise if $\tan\theta > 1/\mu$.) In other words, you can apply an arbitrarily large force. The reason for this is that no matter how large the spring force is, the upper bound on the friction force (which is $\mu N = \mu(kx\sin\theta + mg) > \mu kx\sin\theta$) is always larger than the horizontal component of the spring force (which is $kx\cos\theta$). But if $\tan\theta < 1/\mu$, then the expression in Eq. (9.5) gives the maximum compression for which the block doesn't move.

LIMITS: If $\theta = 0$, then Eq. (9.5) gives $x = \mu mg/k$, which implies $kx = \mu mg$. This makes sense; the rightward spring force equals the maximum possible leftward friction force. If $\theta = 90°$, then we are automatically in the $\tan\theta > 1/\mu$ regime, so the block will stay at rest for any value of x. This makes sense, because you aren't pushing sideways at all.

9.3. Stick in a well

The various forces are shown in Fig. 9.19. The three non-gravitational forces are all normal forces because the well is frictionless. We can quickly say that the vertical force B at the left end of the stick is

$$B = mg, \quad (9.6)$$

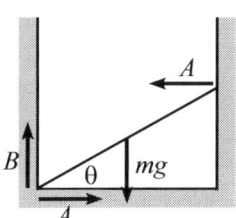

Figure 9.19

because the net vertical force on the stick must be zero. And the two A's are indeed equal as shown, because the net horizontal force must be zero. A can be found by (among other ways) balancing the torques around the left end. If the stick has length ℓ, then the lever arm of the mg force is $(\ell/2)\cos\theta$, and the lever arm of the upper A force is $\ell\sin\theta$, so we have

$$mg\left(\frac{\ell}{2}\cos\theta\right) = A(\ell\sin\theta) \implies A = \frac{mg}{2\tan\theta}. \quad (9.7)$$

REMARK: The total force on the left end of the stick has magnitude $\sqrt{A^2 + B^2}$ and makes an angle of $\tan\alpha = B/A = 2\tan\theta$ with the horizontal. This value of α implies that the total force does *not* point along the stick; the slope of the force is twice the $\tan\theta$ slope of the stick. The line of the force therefore passes through the point P that lies above the center of the stick, with twice the height; see Fig. 9.20. In retrospect, this must be the case, because the lines of both the stick's weight and the force on the right end pass through P, which means that they provide no torque around P. The line of the total force on the left end must therefore also pass through P so that it provides no torque around P, because the setup is static.

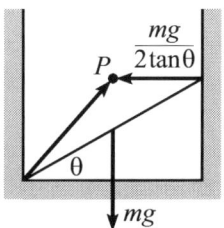

Figure 9.20

We have basically just proved the following general result: If a static object has three forces acting on it, then these three forces must be concurrent (that is, they must intersect at a common point). This is true because otherwise one force would cause a torque around the intersection of the lines of the other two forces. (Technically, the three forces could also be parallel. But then they may be considered to be concurrent at the "point" at infinity.)

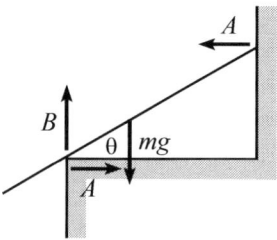

Figure 9.21

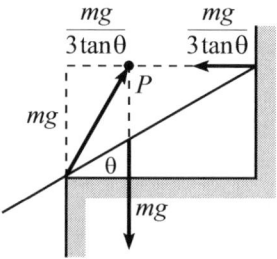

Figure 9.22

9.4. Stick on a corner

The various forces are shown in Fig. 9.21. As in Problem 9.3, we can quickly say that the vertical force B at the corner is

$$B = mg, \tag{9.8}$$

because the net vertical force on the stick must be zero. And the two A's are indeed equal as shown, because the net horizontal force must be zero. A can be found by (among other ways) balancing the torques around the corner. The lever arm of the mg force is $(\ell/4)\cos\theta$, and the lever arm of the upper A force is $(3\ell/4)\sin\theta$, so we have

$$mg\left(\frac{\ell}{4}\cos\theta\right) = A\left(\frac{3\ell}{4}\sin\theta\right) \implies A = \frac{mg}{3\tan\theta}. \tag{9.9}$$

As you can see, this problem is very similar to Problem 9.3.

REMARK: The total force from the corner, which has components A and B, is shown in Fig. 9.22. Note that the $3\tan\theta$ slope of this force is three times the $\tan\theta$ slope of the stick. The line of the force therefore passes through the point P shown (defined by the intersection of the lines of the other two forces). This has to be the case, due to the "concurrent forces" result we proved in the remark in the solution to Problem 9.3.

9.5. Two sticks and a wall

The free-body diagrams for the top stick and bottom stick, respectively, are shown in Fig. 9.23. In the second diagram, we have used the fact that the length of the bottom stick is $\ell/\cos\theta$.

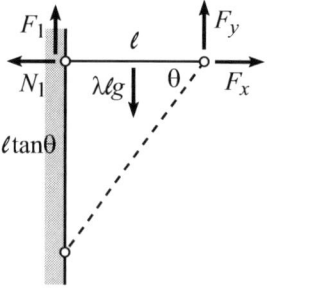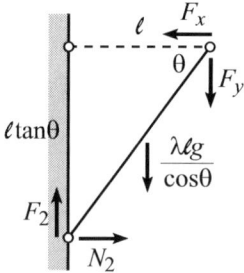

Figure 9.23

Note that we can't assume that the force between the sticks points along either stick (because why along one stick and not the other?). So we have labeled the magnitudes of the two components as F_x and F_y in the first diagram. By Newton's third law, the corresponding components point in the opposite directions in the second diagram. There are various ways to find F_x and F_y. Here is one:

- Balancing torques on the upper stick around the left end gives

$$\lambda\ell g\frac{\ell}{2} = F_y\ell \implies F_y = \frac{\lambda\ell g}{2}. \tag{9.10}$$

- Balancing horizontal forces on the upper stick gives $N_1 = F_x$. So the goal is to find N_1.

- Let's balance torques on the entire system around the bottom end of the bottom stick. F_x and F_y are internal forces, so they don't come into play. The only forces that provide nonzero torques around the bottom end of the bottom stick are N_1 and the

9.5. PROBLEM SOLUTIONS

weights of the two sticks. The mass of the bottom stick is $\lambda(\ell/\cos\theta)$, so balancing the counterclockwise and clockwise torques on the entire system gives

$$N_1(\ell\tan\theta) = \lambda\ell g\frac{\ell}{2} + \frac{\lambda\ell g}{\cos\theta}\frac{\ell}{2}$$
$$\implies N_1 = \frac{\lambda\ell g}{2}\left(1 + \frac{1}{\cos\theta}\right)\frac{1}{\tan\theta}$$
$$\implies F_x = N_1 = \frac{\lambda\ell g}{2}\left(\frac{1}{\tan\theta} + \frac{1}{\sin\theta}\right). \tag{9.11}$$

Having found F_y above, we alternatively could have found F_x by balancing torques on the bottom stick around the bottom end. The resulting equation would be basically the same as the first line of Eq. (9.11), as you can check.

LIMITS: If $\theta \to 0$ then $F_x \to \infty$, which makes intuitive sense. If $\theta \to \pi/2$ then $F_x \to \lambda\ell g/2$, which isn't so obvious (there are competing effects: the bottom stick is very long and massive, but it is barely tilted). This value of F_x equals the F_y we found in Eq. (9.10) (which is the F_y for any value of θ). So in the $\theta \to \pi/2$ limit, the total force that the bottom stick exerts on the top stick points up and to the right at a 45° angle, which is rather interesting.

9.6. Supporting a square

Consider the torque on the entire system, relative to the upper pivot on the wall. The relevant external forces are shown in Fig. 9.24. (The tension in the string is an internal force, so it doesn't matter here. And the various other forces from the pivots on the wall provide no torque around the upper pivot.) Let the length of each stick be ℓ (it will cancel out). Then balancing the torques on the entire system around the upper pivot on the wall gives

$$2 \cdot mg(\ell/2) + mg\ell = N\ell \implies N = 2mg. \tag{9.12}$$

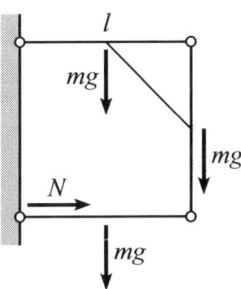

Figure 9.24

The sum of the horizontal forces on the bottom stick must be zero, so the vertical stick must exert a leftward force of $2mg$ on the bottom stick, to cancel the rightward N from the wall. By Newton's third law, the bottom stick therefore exerts a rightward force of $2mg$ on the vertical stick. Now look at torques on the vertical stick, relative to the pivot at its top end. The relevant forces are shown in Fig. 9.25, where T is the tension in the string. Balancing the torques gives

$$(T\cos 45°)(\ell/2) = (2mg)\ell \implies T = 4\sqrt{2}\,mg. \tag{9.13}$$

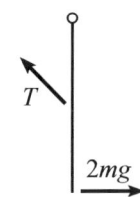

Figure 9.25

REMARK: If more generally the string is attached to the sticks at points that are a distance $n\ell$ from the upper right corner of the square ($n = 1/2$ in the above setup), then it quickly follows that the tension in the string equals $T = 2\sqrt{2}\,mg/n$. So T is inversely proportional to n. If the string is located very close to the corner, then the tension must be very large.

9.7. Corner on a plane 1

The various forces on the object are shown in Fig. 9.26. Since the plane is frictionless, we know that the force at the upper end is just a normal force perpendicular to the plane. However, we don't yet know the direction of the force at the pivot, so we have labeled the components as F_2 and F_3.

Let's look at torques on the entire object around the pivot. The only nonzero torques come from F_1 and the gravitational force mg on the horizontal stick. The lever arm for F_1 is the $\sqrt{2}\ell$ distance along the plane, and the lever arm for the mg force is a horizontal $\ell/2$. So balancing the two torques gives

$$F_1 \cdot \sqrt{2}\ell = mg \cdot \frac{\ell}{2} \implies F_1 = \frac{mg}{2\sqrt{2}}. \tag{9.14}$$

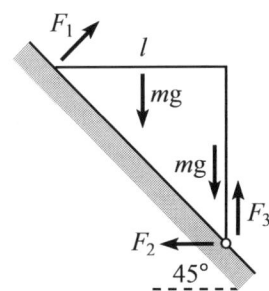

Figure 9.26

If we now look at the horizontal forces on the entire object, we see that the x component of F_1, which is $(mg/2\sqrt{2})\sin 45°$, must balance F_2. Hence

$$F_2 = \frac{mg}{4}. \qquad (9.15)$$

Finally, if we consider the vertical forces on the entire object, we see that the y component of F_1, which is $(mg/2\sqrt{2})\cos 45°$, plus F_3 must balance the total weight $2mg$. Hence

$$\frac{mg}{4} + F_3 = 2mg \implies F_3 = \frac{7mg}{4}. \qquad (9.16)$$

The force from the pivot therefore points upward and leftward with a slope of $F_3/F_2 = 7$.

REMARK: We can also work with a general angle θ for the plane's inclination, instead of 45°. If we let the horizontal stick have a given length ℓ (which means that the vertical stick has length $\ell\tan\theta$), and if both masses are still m, then you can show, among other things, that F_2 equals $(mg/2)\cos\theta\sin\theta$. This achieves a maximum when $\theta = 45°$. Intuitively, F_2 should go to zero for both $\theta \to 0$ and $\theta \to 90°$. It should therefore achieve a maximum at some intermediate angle, which happens to be 45°.

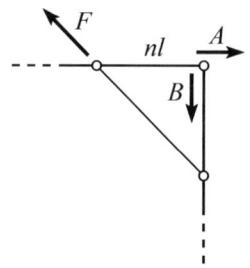

Figure 9.27

9.8. **Corner on a plane 2**

Fig. 9.27 shows a closeup view of the corner. Let F be the force that the tiny massless stick applies to the top stick, with the positive direction taken to be as shown. This force must point along the direction of the tiny massless stick; see Problem 7.7. We will find that F is positive, which means that the tiny stick is under compression. In other words, if it weren't there, the 90° angle at the corner would decrease if you held the sticks and then let them go; the left end of the top stick would slide down the plane, not up. You can check that this motion causes the height of the CM of the entire system to decrease. So the motion is consistent with conservation of energy.

Let A and B be the components of the force that the bottom stick applies to the top stick at the pivot, with the positive directions taken to be as shown in Fig. 9.27. We will find below that both A and B are positive, so these components do indeed point in the directions shown. (Technically we'll find that B is positive only if $n < 1/3$. But we're assuming n is small.) The forces on the right stick due to the tiny stick and the pivot are equal and opposite to the above F, A, and B forces.

Let's look at the torque on the top stick around the pivot at the corner. The total torque comes from three forces: (1) the vertical component of the force from the plane at the left end (which from Problem 9.7 equals $mg/4$) acting at a lever arm of ℓ, (2) the weight mg of the top stick acting at a lever arm of $\ell/2$, and (3) the vertical component $F/\sqrt{2}$ of the tiny stick's force, acting at a lever arm of $n\ell$. The second of these torques is counterclockwise, and the other two are clockwise, so setting the total torque equal to zero gives

$$\frac{mg}{4}\cdot\ell - mg\cdot\frac{\ell}{2} + \frac{F}{\sqrt{2}}\cdot n\ell = 0 \implies F = \frac{mg}{2\sqrt{2}\,n}. \qquad (9.17)$$

If n is very small (which it is, since our model represents the forces acting in the interior of the sticks at the corner), then F is very large. This is a general result for rigid objects that contain thin parts. In order to remain rigid, there is an inevitable need for internal torques, to keep the different parts of the system from rotating with respect to each other. And since the lever arms associated with the internal torques are very small if the object is thin, the internal forces must be very large.

Now let's find the A and B components of the force from the pivot on the top stick. Looking at the horizontal forces on the top stick tells us that the leftward $F/\sqrt{2}$ component of the force from the tiny stick (which from Eq. (9.17) equals $mg/4n$) must balance the rightward sum of A and the horizontal component of the force from the plane at the left end (which from Problem 9.7 equals $mg/4$). So we have

$$\frac{mg}{4n} = A + \frac{mg}{4} \implies A = \frac{mg}{4n} - \frac{mg}{4}. \qquad (9.18)$$

Looking at the vertical forces on the top stick tells us that the upward $F/\sqrt{2}$ component of the force from the tiny stick (which equals $mg/4n$) plus the vertical component of the force from the plane at the left end (which equals $mg/4$) must balance the downward sum of B and the weight mg. So we have

$$\frac{mg}{4n} + \frac{mg}{4} = B + mg \implies B = \frac{mg}{4n} - \frac{3mg}{4}. \qquad (9.19)$$

We've found all of the desired forces, so let's now check that the torque on the right stick, relative to the pivot at its bottom end, equals zero. Fig. 9.28 shows the relevant forces. The torque comes only from A (which acts at a lever arm of ℓ) and the horizontal component of F, which is $mg/4n$ (which acts at a lever arm of $\ell - n\ell$). (The force from the bottom pivot, the gravitational force, and the upward B force don't produce any torque around the bottom pivot.) The total clockwise torque is therefore

$$\frac{F}{\sqrt{2}}(\ell - n\ell) - A\ell = \frac{mg}{4n}(\ell - n\ell) - \left(\frac{mg}{4n} - \frac{mg}{4}\right)\ell = 0, \qquad (9.20)$$

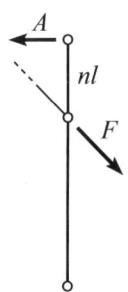

Figure 9.28

as desired. Using the results from Problem 9.7 for the F_2 and F_3 forces at the pivot, you can also quickly verify that the net vertical and horizontal forces on the right stick are zero.

REMARK: The point of all this is that since a rigid structure necessarily has a nonzero thickness, the internal forces don't all act at a single point. This makes it possible for a nonzero net force to produce a zero net torque. This is accomplished by having two (or more) large forces acting at slightly different lever arms. The ratio of the lever arms is the inverse of the ratio of the forces; hence the zero torque. *Multiplicatively*, the forces are nearly equal (as are the lever arms); the ratio is very close to 1. But *additively*, the forces differ by the given nonzero net force.

9.9. **Leaning ladder**

Let N be the normal force from the floor. Then the friction force from the floor is μN, because we are assuming that it takes on the maximum possible value in the cutoff case. Balancing horizontal forces on the ladder then quickly gives the normal force from the wall as μN. The friction force from the wall is therefore $\mu(\mu N)$, because again we are assuming that it takes on the maximum possible value. All of the forces are shown in Fig. 9.29.

Consider torques around the center of the ladder. The lever arms of the four force components at the two ends are all either $(\ell/2)\cos\theta$ or $(\ell/2)\sin\theta$. Three of the torques are clockwise, and one is counterclockwise. Demanding that the total torque be zero, and canceling the common factor of $\ell/2$, gives the cutoff condition on θ as

$$\mu^2 N \cos\theta + \mu N \sin\theta - N\cos\theta + \mu N \sin\theta = 0 \implies 2\mu\sin\theta = \cos\theta(1-\mu^2)$$

$$\implies \tan\theta = \frac{1-\mu^2}{2\mu}. \qquad (9.21)$$

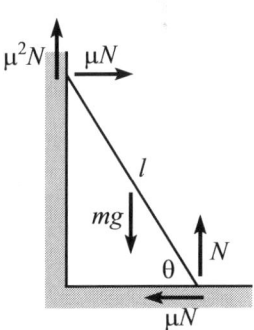

Figure 9.29

This is the desired minimum value of θ. Note that we never needed to use the fact that there is zero net force in the vertical direction. This condition just serves to tell us what N is:

$$mg = N + \mu^2 N \implies N = \frac{mg}{1+\mu^2}. \qquad (9.22)$$

However, if we had balanced torques around, say, the top or bottom end of the ladder, then we would have needed to use this equation, as you can check.

LIMITS: If $\mu \to 0$ then Eq. (9.21) tells us that $\theta \to 90°$, which makes sense; the ladder needs to be nearly vertical. And if $\mu \to 1$ then $\theta \to 0$; all of the force components at the ends of the ladder are equal to $mg/2$ in this case.

REMARK: Let's show that in the cutoff case where the ladder is about to fall, the friction forces do indeed take on their maximum allowed values (assuming $\mu \leq 1$), as we claimed in the statement of the problem. If we *don't* assume that they take on their maximum possible values, then we can label the above μN friction force from the floor more generally as F_1 (which is the same as the normal force from the wall), and we can label the $\mu^2 N$ friction force from the wall as F_2. The torque equation in Eq. (9.21) then takes the form of

$$F_2 \cos\theta + F_1 \sin\theta - N\cos\theta + F_1 \sin\theta = 0. \tag{9.23}$$

Since both F_1 and F_2 appear with positive coefficients, the left-hand side will be increased if we replace them with their upper limits. (Or it will remain the same, if F_1 and F_2 are already at their maximum values.) Therefore,

$$\mu^2 N \cos\theta + \mu N \sin\theta - N\cos\theta + \mu N \sin\theta \geq 0. \tag{9.24}$$

This inequality carries through in the steps in Eq. (9.21), so the condition on θ becomes $\tan\theta \geq (1-\mu^2)/2\mu$. In other words, the ladder cannot stay up if $\tan\theta < (1-\mu^2)/2\mu$. But we showed in Eq. (9.21) that it *can* stay up if $\tan\theta = (1-\mu^2)/2\mu$. And this scenario entails both friction forces taking on their maximum allowed values.

Note that if $\mu \geq 1$, the $\tan\theta \geq (1-\mu^2)/2\mu$ condition tells us that the ladder will stay up for any angle θ. If $\mu > 1$, the friction forces will *not* take on their maximum allowed values, because if they did, then Eq. (9.21) would imply that θ would have to be negative.

9.10. Stick on a cylinder

The free-body diagrams for the stick and the cylinder are shown in Fig. 9.30. All of the forces are drawn, but the following solution won't require finding f_g, n_g, or n_w. We will find F and N (and also f_w) and then invoke the $F \leq \mu N$ condition. The torque and force equations we will use are:

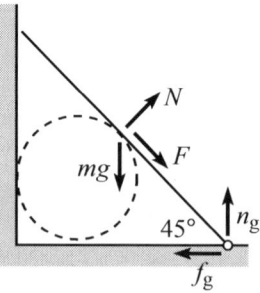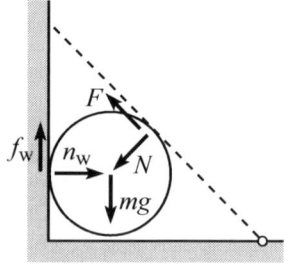

Figure 9.30

- If the radius of the cylinder is R, then balancing the torques on the cylinder around the center gives

$$f_w R = FR \implies f_w = F. \tag{9.25}$$

- If the stick has length ℓ, then balancing the torques on the stick around the pivot gives

$$N \cdot \frac{\ell}{2} = mg \cdot \frac{\ell}{2} \cos 45° \implies N = \frac{mg}{\sqrt{2}}. \tag{9.26}$$

- Balancing the vertical forces on the cylinder gives (using the above results for f_w and N):

$$f_w + F \sin 45° = mg + N \cos 45° \implies F(1 + 1/\sqrt{2}) = mg(1 + 1/2)$$

$$\implies F = \frac{3mg}{2 + \sqrt{2}}. \tag{9.27}$$

9.5. PROBLEM SOLUTIONS

Having found F and N, we can now invoke the $F \leq \mu N$ condition. This gives

$$\frac{3mg}{2+\sqrt{2}} \leq \mu \frac{mg}{\sqrt{2}} \implies \mu \geq \frac{3}{\sqrt{2}+1} = 3(\sqrt{2}-1) \approx 1.24. \tag{9.28}$$

REMARK: As an exercise, you can show that the minimum coefficient of friction between the cylinder and the wall is $3(2\sqrt{2}-1)/7 \approx 0.78$. (This requires finding n_w.) So we can get by with a smaller coefficient of friction with the wall than with the stick.

9.11. Stick on a disk

The free-body diagrams are shown in Fig. 9.31. Let's first determine how F_1 is related to N_1 and then invoke the $F_1 \leq \mu_w N_1$ condition. Balancing torques on the disk around the center gives $F_1 R = F_2 R \implies F_1 = F_2$. Balancing horizontal forces on the disk gives $N_1 = F_2$. Combining these relations gives $N_1 = F_1$. So the $F_1 \leq \mu_w N_1$ condition becomes

$$F_1 \leq \mu_w F_1 \implies \mu_w \geq 1. \tag{9.29}$$

Note that this result has nothing to do with the exact nature of the stick. It could have a different length (as long as it is at least the radius of the disk), be nonuniform, etc.

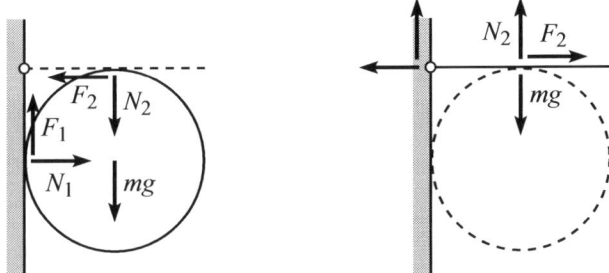

Figure 9.31

Now let's determine how F_2 is related to N_2 and then invoke the $F_2 \leq \mu_s N_2$ condition. Balancing torques on the stick around the pivot gives $N_2 = mg$. Balancing vertical forces on the disk gives $F_1 = N_2 + mg \implies F_1 = 2mg$. (Basically, F_1 is the vertical force supporting the whole system. There is no vertical force at the pivot even though we drew one in the figure to be general, because otherwise there would be a nonzero torque on the stick around its center.) But $F_1 = F_2$ from above, so we have $F_2 = 2mg$. The $F_2 \leq \mu_s N_2$ condition therefore becomes

$$2mg \leq \mu_s(mg) \implies \mu_s \geq 2. \tag{9.30}$$

We see that we need a larger coefficient of friction with the stick than with the wall. The entire set of forces in this problem is $N_1 = F_1 = F_2 = 2mg$ and $N_2 = mg$. But the actual values weren't necessary for the $\mu_w \geq 1$ result.

9.12. Cylinder on a stick

The free-body diagrams for the stick and the cylinder are shown in Fig. 9.32.

There are many ways to go about obtaining F_x and F_y. Here is one:

- Balancing vertical forces on the cylinder-plus-stick system gives $F_y = mg$.
- Balancing vertical forces on the cylinder gives

$$B \cos 45° = mg \implies B = \sqrt{2} mg. \tag{9.31}$$

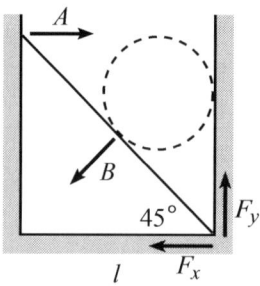
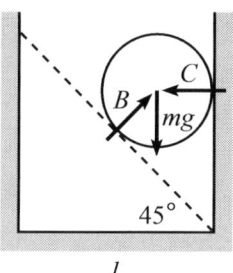

Figure 9.32

- Balancing torques on the massless stick around its bottom end gives

$$B \cdot \frac{\ell}{\sqrt{2}} = A\ell \implies A = \frac{B}{\sqrt{2}} = mg. \tag{9.32}$$

- Balancing horizontal forces on the stick gives

$$A = B\sin 45° + F_x \implies mg = \sqrt{2}mg \cdot \frac{1}{\sqrt{2}} + F_x \implies F_x = 0. \tag{9.33}$$

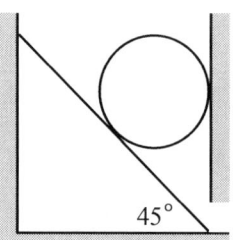

Figure 9.33

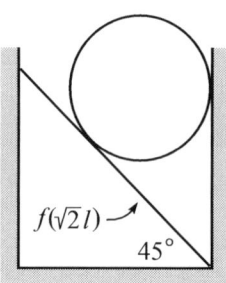

Figure 9.34

The force from the box on the stick at its lower end is therefore purely vertical, with magnitude mg. This means that if the box had a hole in its side, as shown in Fig. 9.33, the system would still be in equilibrium. (But it turns out it wouldn't be stable; that is, a small displacement would turn into a larger one.)

If you also want to find the force C from the right wall, then balancing horizontal forces on the cylinder gives

$$C = B\sin 45° = \sqrt{2}mg \cdot \frac{1}{\sqrt{2}} = mg. \tag{9.34}$$

REMARK: Let's change the size of the cylinder so that its contact point with the stick is a fraction f of the way from the bottom of the stick to the top, as shown in Fig. 9.34. (So $f = 1/2$ in the original problem.) As an exercise, you can show that the horizontal component of the force that the box exerts on the stick at its lower end is $F_x = mg(2f - 1)$. This correctly equals zero when $f = 1/2$.

If $f > 1/2$ (as in Fig. 9.34) then F_x is positive, which means that the box pushes to the left on the stick (we defined positive F_x to point leftward in Fig. 9.32). Equivalently, the stick wants to slide to the right through the wall of the box. But it can't, so the setup is static.

If $f < 1/2$ (so we have a small cylinder down closer to the bottom of the box) then F_x is negative, that is, the force on the stick is directed to the right. But the force can't be directed to the right, because we are assuming that there is no friction anywhere in the setup. The stick wants to slide to the left, away from the right wall, and since there is nothing to keep it from doing so, it slides to the left and the cylinder falls. The setup can't be static.

Chapter 10

Oscillations

10.1 Introduction

Hooke's law

Consider a force that depends on position according to

$$F(x) = -kx. \tag{10.1}$$

A force of this form (proportional to $-x$) is said to obey *Hooke's law*. The force is negative if x is positive, and positive if x is negative. So it is a restoring force; it is always directed back toward the equilibrium point (the origin). Since $F = -dU/dx$, the associated Hooke's-law potential energy is

$$U(x) = \frac{1}{2}kx^2. \tag{10.2}$$

Plots of $F(x)$ and $U(x)$ are shown in Fig. 10.1. Hooke's-law forces are extremely important because they are ubiquitous in nature, due to the fact that near an equilibrium point (which is where systems generally hang out), any potential-energy function looks essentially like a parabola; see Problem 10.1. So we can always approximate $U(x)$ as $kx^2/2$ for some value of k (although this approximation will, of course, break down for sufficiently large x). We'll often use a spring (see page 70) as an example of a Hooke's-law force, but there are countless other examples – pendulums, objects floating in water, electrical circuits, etc.

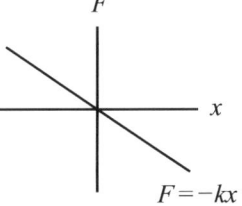
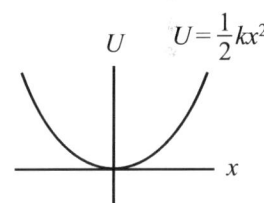

Figure 10.1

Simple harmonic motion

If we have a Hooke's-law force, $F = -kx$, then Newton's second law becomes

$$F = ma \implies -kx = ma \implies m\ddot{x} = -kx. \tag{10.3}$$

What is the solution, $x(t)$, to this equation? There are many ways to solve it, but the easiest way is to just note that we want to find a function whose second derivative is proportional to the negative of itself. And we know that sines and cosines have this property. So let's try a solution of the form,

$$x(t) = A\cos(\omega t + \phi). \tag{10.4}$$

Plugging this into $m\ddot{x} = -kx$ gives

$$m(-\omega^2 A\cos(\omega t + \phi)) = -kA\cos(\omega t + \phi) \implies \omega = \sqrt{\frac{k}{m}}, \tag{10.5}$$

where we have canceled the common factor of $A\cos(\omega t + \phi)$. We see that the expression for $x(t)$ in Eq. (10.4) is a solution to $F = ma$, provided that $\omega = \sqrt{k/m}$. A and ϕ can take on arbitrary values, and the solution is still valid. The sinusoidal motion in Eq. (10.4) is called *simple harmonic motion*. The quantity ω is a very important one:

265

- ω is the *angular frequency* of the oscillatory motion. The argument $\omega t + \phi$ of the cosine in Eq. (10.4) is an angle measured in radians (not degrees), and ω is the rate at which this angle increases. (So you could also call ω the "angular speed" or "angular velocity." But "angular frequency" is more common.)

You can quickly verify that $x(t + 2\pi/\omega) = x(t)$, which means that the motion repeats itself after every time interval of

$$T = \frac{2\pi}{\omega} = 2\pi \sqrt{\frac{m}{k}}. \tag{10.6}$$

The time T is the *period* of the oscillation. The *frequency* of the oscillation, in cycles per second (that is, in "hertz") is

$$\nu = \frac{1}{T} = \frac{\omega}{2\pi} = \frac{1}{2\pi} \sqrt{\frac{k}{m}}. \tag{10.7}$$

We see that ω is larger than ν by a factor of 2π. This is due to the fact that there are 2π radians in each cycle. *Note*: The term "frequency," instead of the full "angular frequency," is often used when referring to ω. So the word "frequency" is somewhat ambiguous in practice. But as long as you write down the symbol ω or ν, it will be clear what you're referring to.

Initial conditions

Aside from the time t, there are three parameters in the expression for $x(t)$ in Eq. (10.4), namely A, ω, and ϕ. The angular frequency ω is determined by k and m via Eq. (10.5), and these are in turn determined by the setup; someone has to give you a particular spring and a particular mass. In contrast, the values of A and ϕ are *not* determined by k and m; that is, they are not determined by the setup. The $x(t)$ in Eq. (10.4) is a solution to Eq. (10.3) for *any* arbitrary values of A and ϕ. If we want to determine what the actual values of these two parameters are, we must specify two *initial conditions,* most commonly the initial position x_0 and the initial velocity v_0 at time $t = 0$. Differentiating Eq. (10.4), we see that the velocity is given by $v(t) = dx/dt = -\omega A \sin(\omega t + \phi)$. Letting $t = 0$ in this expression for $v(t)$ and also in the expression for $x(t)$ in Eq. (10.4), we find that the initial conditions at $t = 0$, namely $x_0 = x(0)$ and $v_0 = v(0)$, can be written as

$$x_0 = A\cos\phi \qquad \text{and} \qquad v_0 = -\omega A \sin\phi. \tag{10.8}$$

These two equations can be solved for A and ϕ (see Problem 10.2). These two parameters are:

- A is the *amplitude* of the motion. It is the largest value that $x(t)$ achieves; $x(t)$ bounces back and forth between A and $-A$.

- ϕ is the *phase* of the motion. It dictates where in the cycle of motion the mass is at $t = 0$. A larger value of ϕ means that the mass is further along in the cycle at $t = 0$. The phase ϕ depends on the choice of origin for time. If we pick a different instant when we set our clock equal to zero, we obtain a different value of ϕ for the same motion.

Other forms of x(t)

Equation (10.4) is one way to write the solution to Eq. (10.3). But there are others, for example,

$$x(t) = B\sin(\omega t + \psi) \qquad \text{or} \qquad x(t) = C\cos\omega t + D\sin\omega t. \tag{10.9}$$

These satisfy Eq. (10.3), provided that we again have $\omega = \sqrt{k/m}$. If $x(t)$ is written in the $C\cos\omega t + D\sin\omega t$ form, it isn't obvious what the amplitude of the motion is. But you can show in Problem 10.3 that the amplitude is $\sqrt{C^2 + D^2}$.

Note that no matter which form of $x(t)$ we use, there are always two free parameters (A and ϕ; or B and ψ; or C and D). These two parameters are determined by the two initial conditions. The two parameters in one expression are related to the two in another. For example, using

10.1. INTRODUCTION

the trig sum formula for sine, and comparing the two expressions in Eq. (10.9), we see that $C = B\sin\psi$ and $D = B\cos\psi$.

The $C\cos\omega t + D\sin\omega t$ form of $x(t)$ is the most amenable to the application of initial conditions. The $x(0) = x_0$ condition quickly gives $C = x_0$. And using $v(t) = dx/dt = -\omega C\sin\omega t + \omega D\cos\omega t$, the $v(0) = v_0$ condition quickly gives $\omega D = v_0 \Longrightarrow D = v_0/\omega$. Depending on the setup and goal of a given problem, a particular form of $x(t)$ might make things easier than the other forms.

We know that sines and cosines are solutions to the $F = ma$ equation in Eq. (10.3). The fact that a sine plus a cosine (the second form given in Eq. (10.9)) is again a solution is due to the *linearity* of the $F = ma$ equation; x appears only to the first power (the number of derivatives doesn't matter). This linearity implies that any linear combination of two solutions is again a solution, as you can show in Problem 10.4.

Energy

Consider the standard system of a mass on a spring oscillating back and forth with amplitude A and angular frequency ω. The energy will slosh back and forth between being only potential (at the points of maximal stretch or compression, where the mass is instantaneously at rest) and being only kinetic (when the mass passes through the equilibrium point, where the spring is neither stretched nor compressed). Assuming that we have an ideal system with no damping forces, energy should be conserved. To determine what the constant value of the energy is, we can conveniently look at the $x = \pm A$ points, where the energy is all potential; the energy is therefore $kA^2/2$. And indeed, you can show in Problem 10.5 that the equation

$$\frac{kx^2}{2} + \frac{mv^2}{2} = \frac{kA^2}{2} \tag{10.10}$$

is satisfied for all values of t. That is, the sum of the potential and kinetic energies takes on the constant value of $kA^2/2$.

Generality

As mentioned above, Hooke's law is ubiquitous in nature because near an equilibrium point, any potential-energy function looks basically like a parabola. A spring is one example. Another common example is a pendulum. You can show in Problem 10.7 that the equation for the angle θ (measured relative to the vertical) of a point-mass pendulum with length ℓ is $\ddot\theta = -(g/\ell)\theta$, assuming that the amplitude of the oscillations is small. Having derived this equation, there is no need to waste any time solving it, because it takes exactly the same form as Eq. (10.3), if we rewrite that equation as $\ddot x = -(k/m)x$. The only difference is that k/m is replaced by g/ℓ (and x is replaced by θ). Therefore, since the $\ddot x = -(k/m)x$ equation has a solution of the form given in Eq. (10.4), with the condition that ω must be equal to $\sqrt{k/m}$, we conclude that the $\ddot\theta = -(g/\ell)\theta$ equation must also have a solution of the form in Eq. (10.4), but with ω now equal to $\sqrt{g/\ell}$.

In general, if you solve a problem that involves some variable z (by using $F = ma$, $\tau = I\alpha$, a conservation statement, or whatever), and if at the end of the day you arrive at an equation of the form,

$$\ddot z = -(\text{something})z, \tag{10.11}$$

then you can simply write down the answer with no further work: you know that $z(t)$ undergoes oscillatory motion given by Eq. (10.4), with an angular frequency ω equal to

$$\omega = \sqrt{\text{something}}. \tag{10.12}$$

10.2 Multiple-choice questions

10.1. The expression for the angle of an oscillating pendulum (with a small amplitude) is $\theta(t) = \theta_0 \cos(\omega t + \phi)$.

True or false: The angular velocity $\dot\theta \equiv d\theta/dt$ is equal to the angular velocity ω.

T   F

10.2. The position of a particle is given by $x(t) = x_0 \cos(\omega t + \pi/6)$, where $x_0 = 6\,\text{m}$ and $\omega = 2\,\text{s}^{-1}$. The maximum speed the particle achieves is

(a) 3 m/s   (b) 6 m/s   (c) 12 m/s   (d) 24 m/s   (e) 36 m/s

10.3. If you plot the functions $\cos \omega t$ and $\cos(\omega t + \pi/4)$, the $\cos(\omega t + \pi/4)$ curve will be shifted relative to the $\cos \omega t$ curve. Which way is it shifted?

(a) rightward   (b) leftward

10.4. Which of the pendulums shown below has the largest frequency of small oscillations? The objects all have the same mass, and the CM's are all the same distance from the support. The (massless) rods are glued to each object to form rigid systems. The result from Problem 10.8 will come in handy (but don't look at the remark in the solution).

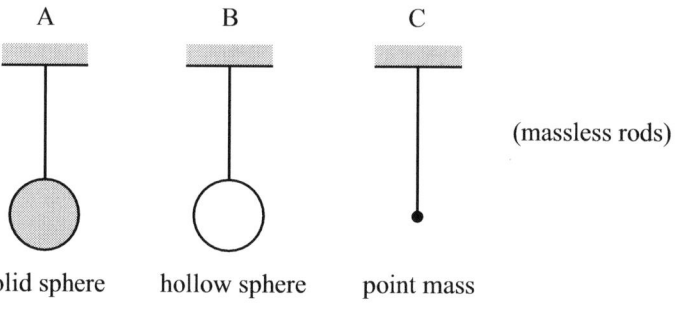

(a) A   (b) B   (c) C   (d) They all have the same frequency.

10.3 Problems

The first eight problems are foundational problems.

10.1. Quadratic potential near a minimum

Use a Taylor series to show that near a local minimum, any (well-behaved) function, in particular the potential energy $U(x)$, looks essentially like a quadratic function (a parabola).

10.2. Initial conditions

Solve the equations in Eq. (10.8) for A and ϕ.

10.3. Amplitude

Show that the amplitude of the motion given by $x(t) = C \cos \omega t + D \sin \omega t$ equals $\sqrt{C^2 + D^2}$. You can do this by taking the derivative of $x(t)$ to find the maximum.

10.4. Linearity

Homogeneous linear differential equations have the property that the sum, or any linear combination, of two solutions is again a solution. ("Homogeneous" means that there is a zero on one side of the equation.) Consider, for example, the second-order linear differential equation (although the property holds for any order),

$$A\ddot{x} + B\dot{x} + Cx = 0. \tag{10.13}$$

10.3. PROBLEMS

Show that if $x_1(t)$ and $x_2(t)$ are solutions, then the sum $x_1(t) + x_2(t)$ is also a solution. Show that this property does *not* hold for the *nonlinear* differential equation, $A\ddot{x} + B\dot{x}^2 + Cx = 0$.

10.5. Conservation of energy

Using the form of $x(t)$ given in Eq. (10.4) (along with the derivative, $v(t) = -\omega A \sin(\omega t + \phi)$), show that the total energy (kinetic plus potential) of simple harmonic motion takes on the constant value of $kA^2/2$.

10.6. Circular motion and simple harmonic motion

(a) In the x-y plane, a particle moves in a circular path with radius A, centered at the origin, as shown in Fig. 10.2. The angular velocity ω is constant, so the angle with respect to the x axis is given by ωt (or technically $\omega t + \phi$, but the phase won't be important here). Using what you know about circular motion, determine the projections onto the x axis of the position **r**, velocity **v**, and acceleration **a** vectors.

(b) Show that the three projections you found are equal, respectively, to the $x(t)$, $v(t)$, and $a(t)$ functions for a particle undergoing simple harmonic motion, with position given by $x(t) = A \cos \omega t$. In other words, show that the projection of uniform circular motion onto a diameter is simple harmonic motion. (Imagine shining a light downward and looking at the shadow of the particle on the x axis.)

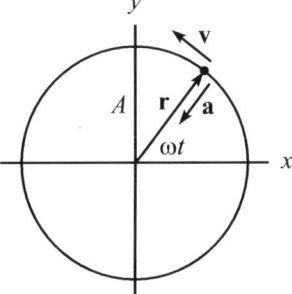

Figure 10.2

10.7. Simple pendulum

Consider the simple pendulum shown in Fig. 10.3. (The word "simple" refers to the fact that the mass is a point mass, as opposed to an extended mass in the "physical" pendulum in Problem 10.8.) The mass hangs on a massless string and swings back and forth in a vertical plane. Let ℓ be the length of the string, and let $\theta(t)$ be the angle the string makes with the vertical. Show that θ satisfies $\ddot{\theta} = -(g/\ell)\theta$, assuming that θ is small enough so that the $\sin \theta \approx \theta$ small-angle approximation is a good one.

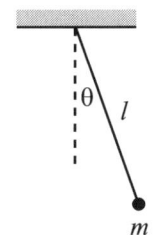

Figure 10.3

10.8. Physical pendulum

Consider an extended pendulum whose CM is a distance ℓ from the pivot, as shown in Fig. 10.4. The pendulum's mass is m, and its moment of inertia around the pivot is I_p. Find the frequency of small oscillations, under the same small-angle approximation as in Problem 10.7.

10.9. Adding on some mass

A mass m is attached to a spring with spring constant k and oscillates on a horizontal table with amplitude A. At an instant when the spring is stretched by $A/2$, a second mass m is dropped vertically onto the original mass and immediately sticks to it. What is the amplitude of the resulting motion?

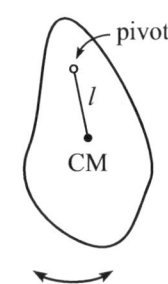

Figure 10.4

10.10. Oscillating in phase

The system in Fig. 10.5 lies on a frictionless horizontal table. Both masses are m, and the two spring constants are nk (where n is a numerical factor) and k. The springs have the same relaxed length. Assuming that it is possible to set up initial conditions so that the masses oscillate back and forth with the two springs always having equal lengths at any given instant, what is n?

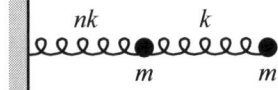

Figure 10.5

10.11. Two masses and a spring

In Fig. 10.6, masses M and m are connected by a spring with spring constant k and relaxed length ℓ. The system lies on a frictionless table. The masses are pulled apart so that the spring is stretched a distance d relative to equilibrium, and then the masses are simultaneously released from rest. Find the position of m as a function of time (let $x = 0$ be its equilibrium position). *Hint*: What is the CM of the system doing?

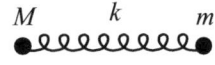

Figure 10.6

10.12. Hanging on springs

Two identical horizontal springs with spring constant k and relaxed length *zero* are connected to each other and to two walls a distance 2ℓ apart, as shown in Fig. 10.7. A mass m is then attached to the midpoint and is slowly lowered down to its equilibrium position, where it sits at rest. At this equilibrium position, the mass is a distance z_0 below the initial line of the springs, as shown (so z_0 is defined to be a positive number).

(a) What is z_0?

(b) If the mass is given a kick in the vertical direction, what is the frequency of the resulting oscillations?

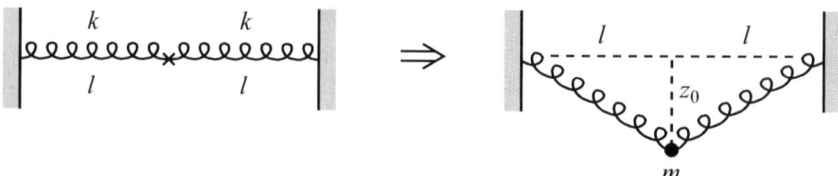

Figure 10.7

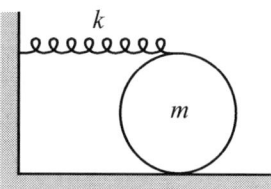

Figure 10.8

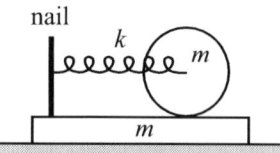

Figure 10.9

10.13. Cylinder and spring

A uniform solid cylinder with mass m and radius R is connected at its highest point to a spring (at its relaxed length) with spring constant k, as shown in Fig. 10.8. If the cylinder rolls without slipping on the ground, what is the frequency of small oscillations? Careful, the top of the cylinder moves more than the center!

10.14. Wheel on a board

A uniform wheel is free to roll without slipping on a board, which in turn is free to slide on a frictionless table. A spring with spring constant k connects the axle of the wheel to a nail stuck in the board, as shown in Fig. 10.9. Both the wheel and the board have mass m, and the nail is massless. The wheel is moved away from the equilibrium position, and then the system is released from rest. What is the frequency of the oscillatory motion?

10.15. Board on a cylinder

The axle of a uniform cylinder with mass m and radius R is connected to a spring with spring constant k, as shown in Fig. 10.10. A horizontal board with mass m rests on top of the cylinder, and the board also rests on top of a frictionless support near its left end.

(a) The system is displaced from equilibrium. If there is no slipping between the cylinder and the board, or between the cylinder and the ground, what is the frequency of the oscillatory motion?

(b) If the amplitude of the oscillation of the center of the cylinder is A, what is the maximum value of the friction force between the cylinder and the board?

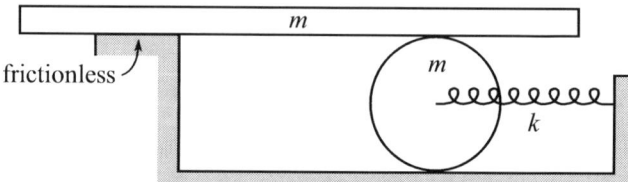

Figure 10.10

10.3. PROBLEMS

10.16. Oscillating disk

A uniform disk with mass m and radius R lies on a frictionless horizontal table and is free to rotate about a pivot at its center. A spring with spring constant k and relaxed length *zero* has one end attached to a point on the rim of the disk and the other end bolted down on the table at a distance R from the rim, as shown in Fig. 10.11. If the disk is initially in the equilibrium position shown and is then given a tiny angular kick, what is the frequency of small oscillations?

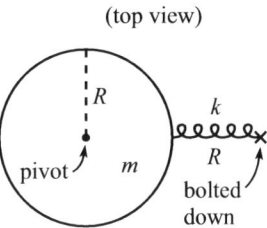

Figure 10.11

10.17. Maximum frequency 1

A coin with radius R is pivoted at a point that is a distance d from its center. The coin is free to swing back and forth in the vertical plane defined by the plane of the coin. What value of d yields the largest frequency of small oscillations?

10.18. Maximum frequency 2

A uniform stick with mass m and length ℓ lies on a frictionless horizontal surface (so you can ignore gravity in this problem). It is pivoted at a point a distance x from its center. A spring (at its relaxed length) with spring constant k is attached to the far end of the stick, perpendicular to the stick, as shown in Fig. 10.12. If the stick is given a tiny kick, what value of x yields the largest frequency of small oscillations?

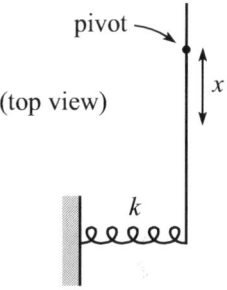

Figure 10.12

10.19. Oscillating board

A uniform board with mass M and length L is pivoted at its center, as shown in Fig. 10.13. Its right end is connected to a spring with spring constant k. The relaxed length is such that the system is in equilibrium when the board is horizontal. The board is given a tiny vertical kick at one of its ends. Throughout this problem, assume that all oscillations are small, and make suitable approximations ($\cos\theta \approx 1$, $\sin\theta \approx \theta$, where θ is the angle the board makes with the horizontal).

(a) What is the frequency of the resulting small oscillations?

(b) Assume that the (small) angular amplitude is θ_0. A small block with mass $m \ll M$ is placed on top of the board near the center, as shown in the right diagram in Fig. 10.13. Assume that the top of the board is frictionless, so that the block is free to slide back and forth. And assume that m is so small that the presence of the block doesn't affect the motion of the board. If initial conditions have been set up so that the block oscillates back and forth between the points with horizontal coordinates $x = \pm x_0$ (with the pivot at $x = 0$), what is x_0? *Hint*: What is the $F = ma$ equation for the block?

(c) Describe what the collective motion of the block and the board looks like by drawing four pictures: when the board is (1) tilted upward the most, then (2) horizontal, then (3) tilted downward the most, then (4) horizontal again.

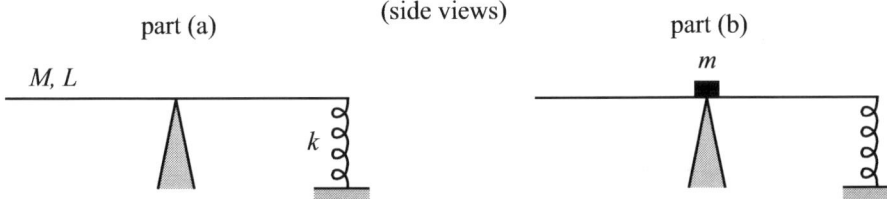

Figure 10.13

10.4 Multiple-choice answers

10.1. $\boxed{F}$ The derivative $d\theta/dt$ is equal to $-\omega\theta_0 \sin(\omega t + \phi)$, which is an oscillating function of time with a maximum value of $\omega\theta_0$ and a minimum value of $-\omega\theta_0$. $d\theta/dt$ is an "angular velocity" in the sense that it is the rate of change of the angle $\theta(t)$ that the pendulum makes with the vertical.

In contrast, the parameter ω is a fixed quantity. It is an "angular velocity" in the sense that it is the rate of change of the angle $\omega t + \phi$ (the argument of the cosine function). This angle increases without bound as time goes on (or at least keeps running from 0 to 2π, if you want to strip off irrelevant integral multiples of 2π), as opposed to just bouncing back and forth between the small angles θ_0 and $-\theta_0$, as the angle $\theta(t)$ does.

10.2. $\boxed{c}$ The velocity is $v(t) = -\omega x_0 \sin(\omega t + \pi/6)$. The maximum value of the sine term is 1 (the phase doesn't matter here), so the maximum magnitude of the velocity (that is, the maximum speed) is ωx_0, which equals $(2\,\text{s}^{-1})(6\,\text{m}) = 12\,\text{m/s}$

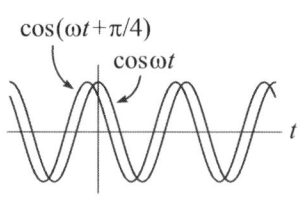

Figure 10.14

10.3. $\boxed{b}$ The two curves are shown in Fig. 10.14, and we see that the $\cos(\omega t + \pi/4)$ curve is shifted *leftward* relative to the $\cos \omega t$ curve. The reason for this is that since $\pi/4$ is a positive phase, $\cos(\omega t + \pi/4)$ reaches a given value *before* $\cos \omega t$ does. For example, $\cos \omega t$ is maximum when $\omega t = 0$ (that is, when $t = 0$), whereas $\cos(\omega t + \pi/4)$ is maximum when $\omega t = -\pi/4$ (that is, when t takes on the negative value of $t = -\pi/4\omega$).

REMARK: We say that the $\cos(\omega t + \pi/4)$ curve is *ahead* of the $\cos \omega t$ curve, because the former reaches a given value before the latter. If you look at Fig. 10.14, however, you might think that the $\cos(\omega t + \pi/4)$ curve is *behind* the $\cos \omega t$ curve, considering that it is shifted slightly leftward. But this spatial appearance isn't important. The leftward shift means that $\cos(\omega t + \pi/4)$ reaches a given value first, so it is ahead. If we had instead posed the question with the function $\cos(\omega t - \pi/4)$, then the negative phase would mean that this curve is shifted to the *right* and hence *behind* the $\cos \omega t$ curve.

10.4. $\boxed{c}$ For a given displacement angle, all three objects experience the same torque from gravity relative to the pivot on the ceiling, because gravity effectively acts at the CM. But Problem 10.8 tells us that the frequency of small oscillations equals $\omega = \sqrt{mg\ell/I}$, where ℓ is the common distance from the pivot to the CM, and I is the moment of inertia relative to the pivot. So the question reduces to which object has the smallest I. From the parallel-axis theorem, the I relative to the pivot is given by $I = m\ell^2 + I_{\text{CM}}$. Since the two spheres have a nonzero I_{CM}, the point mass has the smallest I and hence the largest ω. The hollow sphere has the largest I_{CM}, and hence the smallest ω, because its mass is farthest from the CM.

10.5 Problem solutions

10.1. **Quadratic potential near a minimum**

Let the local minimum occur at x_0, and expand $U(x)$ in a Taylor series around this point. (See Appendix B for a review of Taylor series.) With a prime denoting differentiation, we have

$$U(x) = U(x_0) + U'(x_0)(x - x_0) + \frac{U''(x_0)}{2!}(x - x_0)^2 + \frac{U'''(x_0)}{3!}(x - x_0)^3 + \cdots . \quad (10.14)$$

This is valid for any value of x_0. But in the case where x_0 corresponds to a minimum, we can use the fact that $U'(x_0) = 0$ to eliminate the second term. Additionally, the first term, $U(x_0)$ is a constant, so it is irrelevant. (Only differences in energy matter. Equivalently, the force is given by $F = -dU/dx$, and the derivative of a constant is zero.) Furthermore, the fourth term (along with all higher-order terms) is very small compared with the third

10.5. PROBLEM SOLUTIONS

term because it has an extra power of $(x - x_0)$, and we are assuming that this quantity is very small (because we are assuming that x is very close to x_0). We are therefore left with only the second term:[1]

$$U(x) \approx \frac{1}{2} U''(x_0)(x - x_0)^2. \quad (10.15)$$

This is the desired quadratic form of $U(x)$ near the local minimum. The $U''(x_0)$ term is a constant, so the x dependence comes solely from the $(x - x_0)^2$ term. In short, any arbitrary (well-behaved) function looks like a parabola in the vicinity of a minimum.

If we define y by $y \equiv x - x_0$ (in other words, if we shift our origin so that it is located at x_0), then the potential energy can be written as

$$U(y) = \frac{1}{2} U''(0) y^2 \implies U(y) \equiv \frac{1}{2} k y^2, \quad (10.16)$$

where $k \equiv U''(0)$. This $U(y)$ looks just like the $U(x)$ in Eq. (10.2), so we see that near a stable equilibrium point,[2] everything looks like a Hooke's-law spring. And the effective spring constant is simply the second derivative of the potential energy, evaluated at the equilibrium point. From Eq. (10.5) the frequency of small oscillations is therefore

$$\omega = \sqrt{\frac{U''(0)}{m}}. \quad (10.17)$$

10.2. Initial conditions

We can solve for ϕ by dividing the second of the given equations by the first. The result is

$$\tan \phi = -\frac{v_0}{\omega x_0}. \quad (10.18)$$

And we can solve for A by dividing the second of the given equations by ω, and then squaring both equations and adding them. The identity $\sin^2 \phi + \cos^2 \phi = 1$ allows us to eliminate ϕ, and the result for A is

$$A = \sqrt{x_0^2 + \frac{v_0^2}{\omega^2}}. \quad (10.19)$$

Technically there should be a "$\pm$" in front of the square root, but the usual convention is to take A as a positive quantity.

There is an ambiguity in the above result for ϕ, because there are two distinct angles (differing by π) that have their tangent equal to $-v_0/\omega x_0$. The correct angle can be determined by looking at the sign of either x_0 or v_0. If, for example, x_0 is positive, then the first of the equations in Eq. (10.8) tells us that $\cos \phi$ is positive, which means that ϕ must lie in the first or fourth quadrants. Only one of the two possible angles specified by Eq. (10.18) satisfies this condition.

10.3. Amplitude

FIRST SOLUTION: Setting the derivative of $C \cos \omega t + D \sin \omega t$ equal to zero gives

$$0 = \frac{dx}{dt} = -\omega C \sin \omega t + \omega D \cos \omega t \implies \tan \omega t_0 = \frac{D}{C}. \quad (10.20)$$

This tells us the time t_0 at which $x(t)$ is maximum (or minimum). The angle ωt_0 that satisfies $\tan \omega t_0 = D/C$ is shown in Fig. 10.15. We have drawn the triangle with C and

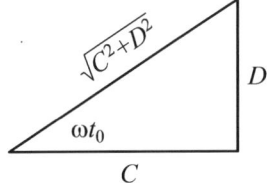

Figure 10.15

---

[1] Even if $U'''(x_0)$ is much larger than $U''(x_0)$, we can always make $(x - x_0)$ small enough so that the third-order term is negligible. The one exception is the case where $U''(x_0)$ is exactly equal to zero. But even in this case the $\omega = \sqrt{U''(0)/m}$ result given below in Eq. (10.17) is still valid.

[2] "Stable" means that $k \equiv U''(0)$ is positive, so that we have a rightside-up parabola, instead of an upside-down one. More physically, if an object is given a small displacement away from a stable equilibrium point, it always feels a force directed back toward that point. In contrast, although the force at an *unstable* equilibrium point is likewise zero (that's the definition of "equilibrium"), a small displacement yields a force directed *away* from that point.

274   CHAPTER 10. OSCILLATIONS

D positive, but they need not be. From this figure we can quickly read off the values of $\cos\omega t_0$ and $\sin\omega t_0$. Plugging these values into $x(t_0) = C\cos\omega t_0 + D\sin\omega t_0$ gives the maximum value of $x(t)$ as

$$x_{\max} = x(t_0) = C \cdot \frac{C}{\sqrt{C^2 + D^2}} + D \cdot \frac{D}{\sqrt{C^2 + D^2}} = \sqrt{C^2 + D^2}. \tag{10.21}$$

This $x_{\max}$ is the amplitude, because the motion bounces back and forth between $x_{\max}$ and $-x_{\max}$. (The minimum of $x(t)$ is obtained from the triangle whose legs are $-C$ and $-D$, instead of the C and D in Fig. 10.15. This triangle also satisfies $\tan\omega t_0 = D/C$.)

SECOND SOLUTION: If $x(t)$ is written in the $A\cos(\omega t + \phi)$ form, we know that the amplitude is simply A. Using the trig sum formula for cosine, we can rewrite $A\cos(\omega t + \phi)$ as

$$x(t) = A\cos\omega t\cos\phi - A\sin\omega t\sin\phi. \tag{10.22}$$

Comparing this with the $x(t) = C\cos\omega t + D\sin\omega t$ expression, we see that the C and D parameters in one expression for $x(t)$ are related to the A and ϕ parameters in the other by

$$C = A\cos\phi \quad\text{and}\quad D = -A\sin\phi. \tag{10.23}$$

We then quickly have

$$C^2 + D^2 = A^2(\cos^2\phi + \sin^2\phi) = A^2. \tag{10.24}$$

So the amplitude, which we know is A, is also $\sqrt{C^2 + D^2}$.

10.4. **Linearity**

If $x_1(t)$ and $x_2(t)$ are solutions to the given linear equation, then

$$A\ddot{x}_1 + B\dot{x}_1 + Cx_1 = 0, \quad\text{and}$$
$$A\ddot{x}_2 + B\dot{x}_2 + Cx_2 = 0. \tag{10.25}$$

If we add these two equations, and switch from the dot notation to the d/dt notation, then we have (using the fact that the sum of the derivatives is the derivative of the sum)

$$A\frac{d^2(x_1 + x_2)}{dt^2} + B\frac{d(x_1 + x_2)}{dt} + C(x_1 + x_2) = 0. \tag{10.26}$$

But this is just the statement that $x_1 + x_2$ is a solution to our differential equation, as we wanted to show. This technique clearly works for any linear combination of x_1 and x_2, not just their sum. Note that the right-hand side of the equation needs to be zero, otherwise we wouldn't end up with the same term on the right-hand side when we add the equations. Hence the "homogeneous" qualifier in the statement of the problem.

Now consider the nonlinear equation,

$$A\ddot{x} + B\dot{x}^2 + Cx = 0. \tag{10.27}$$

If x_1 and x_2 are solutions to this equation, and if we add the differential equations applied to each of them, we obtain

$$A\frac{d^2(x_1 + x_2)}{dt^2} + B\left[\left(\frac{dx_1}{dt}\right)^2 + \left(\frac{dx_2}{dt}\right)^2\right] + C(x_1 + x_2) = 0. \tag{10.28}$$

This is *not* the statement that $x_1 + x_2$ is a solution, which is instead the (incorrect) statement that

$$A\frac{d^2(x_1 + x_2)}{dt^2} + B\left(\frac{d(x_1 + x_2)}{dt}\right)^2 + C(x_1 + x_2) = 0. \tag{10.29}$$

10.5. PROBLEM SOLUTIONS

The preceding two equations differ by the cross term in the middle term in Eq. (10.29), namely $2B(dx_1/dt)(dx_2/dt)$. In general this term is nonzero, so Eq. (10.29) isn't valid, and we conclude that $x_1 + x_2$ isn't a solution. No matter what the order of the differential equation is, we see that these cross terms will arise if and only if the equation isn't linear.

REMARK: This property of homogeneous linear differential equations – that the sum of two solutions is again a solution – is extremely useful. It means that we can build up solutions from other solutions. Systems that are governed by linear equations are *much* easier to deal with than systems that are governed by nonlinear equations. In the latter type of system, the various solutions aren't related in an obvious way. Each one sits in isolation, in a sense. General Relativity is an example of a theory that is governed by nonlinear equations, and solutions are indeed very hard to come by.

10.5. Conservation of energy

Plugging the $x(t)$ and $v(t)$ functions into the energy gives

$$\begin{aligned}
E &= \frac{1}{2}kx^2 + \frac{1}{2}mv^2 \\
&= \frac{1}{2}k(A\cos(\omega t + \phi))^2 + \frac{1}{2}m(-\omega A \sin(\omega t + \phi))^2 \\
&= \frac{1}{2}A^2\big(k\cos^2(\omega t + \phi) + m\omega^2 \sin^2(\omega t + \phi)\big) \\
&= \frac{1}{2}kA^2\big(\cos^2(\omega t + \phi) + \sin^2(\omega t + \phi)\big) \qquad \text{(using } \omega^2 \equiv k/m) \\
&= \frac{1}{2}kA^2,
\end{aligned} \qquad (10.30)$$

where we have used the identity $\sin^2\theta + \cos^2\theta = 1$.

10.6. Circular motion and simple harmonic motion

(a) The **r** vector has length A and makes an angle of ωt with respect to the x axis. So the projection of **r** onto the x axis (that is, the x coordinate of **r**) is

$$r_x(t) = A\cos\omega t. \qquad (10.31)$$

The length of the **v** vector is the speed of the particle, which equals $A\omega$ from the standard $v = r\omega$ result for circular motion. Since **v** is perpendicular to **r**, you can show that **v** makes an angle of $\pi/2 - \omega t$ with respect to the negative x axis at the position shown in Fig. 10.2.[3] So the projection of **v** onto the x axis is

$$v_x(t) = -(A\omega)\cos(\pi/2 - \omega t) = -A\omega \sin\omega t, \qquad (10.32)$$

where the minus sign comes from the fact that **v** points partially leftward at the given position.

From the standard $a = v^2/r$ result for circular motion, the length of the **a** vector is $v^2/A = (A\omega)^2/A = A\omega^2$. Since **a** points radially inward (opposite to **r**), its projection onto the x axis is obtained by multiplying by $-\cos\omega t$, which gives

$$a_x(t) = -A\omega^2 \cos\omega t. \qquad (10.33)$$

(b) If the position of a particle undergoing simple harmonic motion is given by $x(t) = A\cos\omega t$, then taking the derivative of $x(t)$ gives the velocity as $v(t) = -A\omega\sin\omega t$, and taking another derivative gives the acceleration as $a(t) = -A\omega^2\cos\omega t$. These results equal the projections we found in part (a), as desired. We therefore see that the physical interpretation of the angle ωt in the $x(t) = A\cos\omega t$ expression for simple

---
[3]You can work out the geometry, or in the spirit of Multiple-Choice Question 1.12, you know that the angle is either ωt or $\pi/2 - \omega t$, and you can determine the correct choice by taking a limit, for example, $\omega t \to 0$.

harmonic motion is the angle of a particle running around in a circle with the same angular frequency ω.

REMARKS: There was nothing special about the x axis, of course, so the projection onto any diameter yields simple harmonic motion. Depending on when we start our clock, there will in general be a phase in the angle, $\omega t + \phi$. But the ϕ just goes along for the ride in the above calculations and doesn't affect the equalities we found.

Once we demonstrated above that the projection of the circular-motion position **r** equals the simple-harmonic position $x(t) = A\omega \cos \omega t$, there was actually no need to demonstrate the same thing for the velocity and acceleration, because if the x coordinates of the circular and simple harmonic motions match up for all time, then all derivatives of these common x values (in particular, v and a) will necessarily also match up for all time. But it didn't hurt to check anyway.

It is intuitively believable that the projection of uniform circular motion yields simple harmonic motion, because in simple harmonic motion the particle spends more time in a given, say, centimeter near the extremes of the oscillation than it does in a given centimeter near the origin, because it is going slower near the extremes. This agrees with the projection of circular motion, because the particle is moving nearly vertically on the right and left sides of the circle, so it hangs out longer in a given horizontal span than it does when it is moving horizontally as it crosses the y axis (which corresponds to crossing the origin in simple harmonic motion).

10.7. Simple pendulum

The tangential component of the gravitational force on the mass is $-mg \sin\theta$, with the negative sign indicating that the force always points back to the equilibrium position at $\theta = 0$, where the string is vertical. Since the tangential acceleration can be written as $a = \ell\alpha \equiv \ell\ddot\theta$, the $F = ma$ equation in the tangential direction is

$$F = ma \implies -mg\sin\theta = m(\ell\ddot\theta). \tag{10.34}$$

The tension in the string combines with the radial component of gravity to produce the centripetal acceleration, but the radial $F = ma$ equation serves only to tell us the tension, which we won't need here.

Eq. (10.34) isn't solvable in closed form. But for small oscillations, we can use the $\sin\theta \approx \theta$ approximation to write

$$-mg\theta \approx m\ell\ddot\theta \implies \ddot\theta = -(g/\ell)\theta, \tag{10.35}$$

as desired. As stated on page 267, the pendulum therefore undergoes simple harmonic motion with a frequency of $\omega = \sqrt{g/\ell}$. The true motion is arbitrarily close to this, for sufficiently small amplitudes.

REMARKS: For general (not necessarily small) values of the amplitude, the actual frequency is *smaller* than the idealized small-amplitude result of $\omega = \sqrt{g/\ell}$. Equivalently, the period $T = 2\pi/\omega$ is *larger* than $2\pi\sqrt{\ell/g}$. However, the difference is generally negligible. For example, even if the amplitude is 20°, it can be shown numerically that the period is larger than $2\pi\sqrt{\ell/g}$ by less than 1%. If the amplitude is 45°, the discrepancy is still only about 4%. But if the amplitude is 90°, the discrepancy grows to 18%.

The negative sign in Eq. (10.34) is critical. Without it, the force would be a repulsive force instead of a restoring one. And the solution to Eq. (10.35) would be a growing or decaying exponential (you can quickly verify that $Ae^{\pm\alpha t}$ is a solution, where $\alpha = \sqrt{g/\ell}$) instead of an oscillatory trig function.

10.8. Physical pendulum

We were able to solve the simple pendulum in the preceding problem by using only the tangential $F = ma$ equation. In the present case of an extended pendulum, we'll need to use $\tau = I\alpha$ relative to the pivot. By using the torque relative to the pivot, we will be able to ignore the force acting at the pivot, which generally points in a messy direction (not along the line from the pivot to the CM). In the case of the simple pendulum, the

10.5. PROBLEM SOLUTIONS

force pointed nicely along the string, which made the solution easy, in that the tangential $F = ma$ equation sufficed.

Let θ be defined as shown in Fig. 10.16. Relative to the pivot, the lever arm of the gravitational force (which effectively acts at the CM) is $\ell \sin\theta$. So the gravitational torque is $-mg(\ell \sin\theta)$, where the minus sign indicates that the torque always acts to bring the pendulum back to vertical. The $\tau = I\alpha$ equation, relative to the pivot, is then (using $\sin\theta \approx \theta$)

$$\tau = I\alpha \implies -mg\ell \sin\theta = I_p\ddot\theta \implies \ddot\theta = -\left(\frac{mg\ell}{I_p}\right)\theta. \tag{10.36}$$

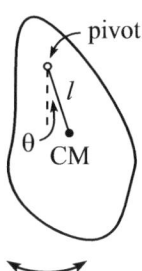

Figure 10.16

From Eqs. (10.11) and (10.12), the frequency of small oscillations is therefore

$$\omega = \sqrt{\frac{mg\ell}{I_p}}. \tag{10.37}$$

REMARK: In the special case where the extended pendulum is actually a point mass, I_p is simply $m\ell^2$, so Eq. (10.37) gives $\omega = \sqrt{g/\ell}$, as it should. For a given length ℓ from the pivot to the CM, a point mass has the largest frequency of any possible shape, because the parallel-axis theorem tells us that $I_p = I_{CM} + m\ell^2$, which means that I_p is always greater than or equal to $m\ell^2$ (with equality holding in the case of a point mass for which $I_{CM} = 0$). So the $\sqrt{mg\ell/I_p}$ result in Eq. (10.37) is always less than or equal to $\sqrt{g/\ell}$.

10.9. Adding on some mass

FIRST SOLUTION: The total energy of the system is $kA^2/2$, because that is the potential energy when the spring is stretched maximally. Right before the second mass is added on, conservation of energy for the initial oscillatory motion gives the speed of the oscillating mass as

$$\frac{1}{2}kA^2 = \frac{1}{2}k\left(\frac{A}{2}\right)^2 + \frac{1}{2}mv^2 \implies \frac{3}{8}kA^2 = \frac{1}{2}mv^2 \implies v = \sqrt{\frac{3k}{4m}}A. \tag{10.38}$$

When the second mass is added on, conservation of momentum gives the speed of the resulting mass $2m$ as $v/2$. (We can't use conservation of mechanical energy during the inelastic collision, because some of the energy is converted into heat.)

Having found the speed of the mass $2m$ right after the collision, we can use conservation of energy for the resulting oscillatory motion to determine the new amplitude B:

$$\frac{1}{2}kB^2 = \frac{1}{2}k\left(\frac{A}{2}\right)^2 + \frac{1}{2}(2m)\left(\frac{1}{2}\sqrt{\frac{3k}{4m}}A\right)^2$$

$$\implies \frac{1}{2}kB^2 = \frac{1}{8}kA^2 + \frac{3}{16}kA^2 \implies B = \sqrt{\frac{5}{8}}A. \tag{10.39}$$

SECOND SOLUTION: We can also solve this problem by treating it as an "initial conditions" problem. The new oscillatory motion starts with position $x_0 = A/2$ and speed $v_0 = (A/2)\sqrt{3k/4m}$ (from the first solution above). If we write the position in the form of $x(t) = C\cos\omega t + D\sin\omega t$, then the discussion on page 267 tells us that $C = x_0$ and $D = v_0/\omega$. But the ω of the new motion is $\omega = \sqrt{k/2m}$ because the mass is now $2m$. So

$$D = \frac{v_0}{\omega} = \frac{(A/2)\sqrt{3k/4m}}{\sqrt{k/2m}} = \sqrt{\frac{3}{8}}A. \tag{10.40}$$

With $C = x_0 = A/2$ and $D = \sqrt{3/8}\,A$, Problem 10.3 gives the amplitude B of the new motion as

$$B = \sqrt{C^2 + D^2} = A\sqrt{\frac{1}{4} + \frac{3}{8}} = \sqrt{\frac{5}{8}}A, \tag{10.41}$$

in agreement with the first solution.

10.10. Oscillating in phase

At a given instant, let the springs each be stretched by x (or compressed, if x is negative). Then relative to the equilibrium positions, the positions of the left and right masses are x and $2x$, respectively. The forces from the springs have magnitudes $(nk)x$ and kx, so the two $F = ma$ equations are (paying attention to the signs)

$$\text{left mass}: \quad -nkx + kx = m\ddot{x} \implies \ddot{x} = -\frac{(n-1)k}{m}x,$$

$$\text{right mass}: \quad -kx = m(2\ddot{x}) \implies \ddot{x} = -\frac{k}{2m}x. \tag{10.42}$$

(Note that whereas the force on the left mass is due to both springs, the force on the right mass is due only to the right spring, because the right mass touches only the right spring.) Since the springs always have equal lengths, the masses must have the same frequency of oscillation. So the coefficients of the x's in the above two equations must be equal. Hence

$$\frac{(n-1)k}{m} = \frac{k}{2m} \implies n - 1 = \frac{1}{2} \implies n = \frac{3}{2}. \tag{10.43}$$

REMARK: This factor of 3/2 makes sense, because it makes the net force on the left mass be $-(3/2)kx + kx = -(1/2)kx$, which is half as large as the $-kx$ force on the right mass. This factor of 1/2 is correct, because the left mass moves half as far as the right mass, which means that its acceleration is half as large (given that it has the same frequency).

10.11. Two masses and a spring

The CM is initially at rest, so it is always at rest because there are no external forces. We'll present three solutions.

FIRST SOLUTION: In this solution we'll use $F = ma$. To do this, we'll need to calculate how the stretching distance of the spring is related to the displacements of the masses. The distances that the masses move are inversely proportional to their masses, because otherwise the CM would move. So if the spring stretches a distance z (or compresses, if z is negative), then mass m moves a distance $Mz/(M + m)$ to the right, and mass M moves a distance $mz/(M + m)$ to the left. These two distances correctly add up to z and are in the correct ratio.

Let's work in terms of the displacement x of mass m. If m moves by $x = Mz/(M + m)$, then the spring stretches by $z = (M + m)x/M$. This means that the spring force acting on m is $-kz = -k(M + m)x/M$. The $F = ma$ equation for m is therefore

$$-\frac{k(M+m)}{M}x = m\ddot{x} \implies \ddot{x} = -\frac{k(M+m)}{Mm}x. \tag{10.44}$$

The frequency of the oscillatory motion is then

$$\omega = \sqrt{\frac{k(M+m)}{Mm}}. \tag{10.45}$$

This ω is symmetric in m and M, so we would have obtained the same result (as we know we must) if we had instead looked at the mass M; you can quickly verify this.

The spring is initially stretched by d, so the initial position of m (relative to its equilibrium position) is $Md/(M + m)$. This is the amplitude of m's motion. And m is initially at rest, so if we write the position with a cosine then we don't need a phase. The desired position of m as a function of time is therefore

$$x(t) = \frac{Md}{M+m}\cos\omega t, \tag{10.46}$$

where ω is given above. The position of M is similar, except with an overall minus sign and an m instead of an M in the numerator of the amplitude.

10.5. PROBLEM SOLUTIONS

LIMITS: If $M \gg m$ then Eq. (10.46) gives $x(t) \approx d\cos\omega t$, with $\omega \approx \sqrt{k/m}$ from Eq. (10.45). This makes sense, because M is effectively a brick wall. If $m \gg M$ then we have $x(t) \approx 0$, with $\omega \approx \sqrt{k/M}$. This makes sense because m is now effectively a brick wall and hence barely moves, but whatever tiny motion it undergoes has frequency $\sqrt{k/M}$, because that is the frequency of M's motion on the other end of the spring.

SECOND SOLUTION: The idea of this solution is to note that since the CM is at rest, we can imagine bolting down the spring at the location of the CM. This won't affect the system, because this point on the spring doesn't move anyway. We then effectively have two masses each connected by shortened springs to a wall located at the CM.

Let the original spring have a relaxed length of ℓ. Then the effective shortened spring that is connected to m has a relaxed length of $M\ell/(M+m)$, because the distances to the CM are inversely proportional to the masses (as we noted in the first solution). Since this is a factor $M/(M+m)$ shorter than the original spring, Problem 4.2 tells us that the spring constant of this shortened spring is larger by a factor $(M+m)/M$, that is, $k_{\text{eff}} = (M+m)k/M$. The frequency of m's motion is therefore

$$\omega = \sqrt{\frac{k_{\text{eff}}}{m}} = \sqrt{\frac{k(M+m)}{Mm}}, \qquad (10.47)$$

in agreement with Eq. (10.45). The solution proceeds as above.

THIRD SOLUTION: In this solution we'll use conservation of energy. The initial potential energy of the spring, $kd^2/2$, shows up at a later time as potential energy $kz^2/2$ (where z is the stretching distance) plus kinetic energy. From the first solution, we know how the positions, and hence velocities, of the two masses relate to $\dot{z}$. So conservation of energy gives

$$\frac{1}{2}kd^2 = \frac{1}{2}kz^2 + \frac{1}{2}m\left(\frac{M\dot{z}}{M+m}\right)^2 + \frac{1}{2}M\left(\frac{m\dot{z}}{M+m}\right)^2$$

$$= \frac{1}{2}kz^2 + \frac{1}{2}\left(\frac{mM^2 + Mm^2}{(M+m)^2}\right)\dot{z}^2$$

$$= \frac{1}{2}kz^2 + \frac{1}{2}\left(\frac{mM}{M+m}\right)\dot{z}^2. \qquad (10.48)$$

But this is simply the conservation-of-energy equation associated with *one* particle with mass $Mm/(M+m)$ on the end of a spring with spring constant k. Therefore,

$$\omega = \sqrt{\frac{k}{\frac{Mm}{M+m}}} = \sqrt{\frac{k(M+m)}{Mm}}. \qquad (10.49)$$

This is the frequency of the oscillation of the variable z (the stretching distance), so it is also the frequency of the oscillation of each mass. The solution proceeds as above.

If you want to be a little more formal with Eq. (10.48), you can take the time derivative of that equation to obtain

$$0 = \frac{1}{2}k \cdot 2z\dot{z} + \frac{1}{2}\frac{Mm}{M+m} \cdot 2\dot{z}\ddot{z} \implies \ddot{z} = -\frac{k(M+m)}{Mm}z, \qquad (10.50)$$

which tells us that the frequency is given by Eq. (10.49).

REMARK: The quantity $m_{\text{eff}} \equiv Mm/(M+m)$ is known as the *reduced mass* of the system. As we saw above, in many respects we can pretend that our system consists of a single mass m_{eff} attached to the given spring (with the other end held fixed). This reduced mass comes up often in systems involving two masses. For example, the earth-sun system (a rotational system involving gravity) behaves in many respects like a mass m_{eff} orbiting around a fixed point. If one mass is

10.12. Hanging on springs

(a) If θ is the angle that the springs make with the horizontal, then each spring has length $\ell/\cos\theta$. So each spring exerts a force of $k\ell/\cos\theta$ (remember that the relaxed length is zero). The upward component of this force is $(k\ell/\cos\theta)\sin\theta = k\ell\tan\theta$. But $\ell\tan\theta$ equals z_0. So the total upward force from the two springs is $2kz_0$. This simple $2kz_0$ form tells us that we effectively have two zero-length springs hanging vertically.[4] In equilibrium, the upward force from the springs must balance the weight of the mass, so $mg = 2kz_0 \implies z_0 = mg/2k$. Note that this doesn't depend on ℓ.

(b) If the mass is given a vertical kick, let y be the displacement from equilibrium, with upward taken to be positive. The upward force from the spring is then $2k(z_0 - y)$. The $2kz_0$ part of this will always cancel the weight mg. So the net upward force as a function of y is $F = -2ky$. The $F = ma$ equation that governs the motion is therefore $-2ky = m\ddot{y}$, so the frequency of the oscillations is $\omega = \sqrt{2k/m}$.

REMARK: The reason for this simple $\omega = \sqrt{2k/m}$ result is that, as we noted above, our tilted springs behave effectively like a single vertical spring with spring constant $2k$; and then from Problem 5.6 we know that this hanging spring behaves like a spring with relaxed length z_0 (which doesn't affect the frequency) in a world with no gravity.

10.13. Cylinder and spring

From Problem 7.8(b), we know that if the center of the cylinder moves a distance x, then the point at the top moves a distance $2x$. So the spring force is $F_s = -k(2x)$. If positive directions are defined as shown in Fig. 10.17, then the $F = ma$ equation for the cylinder is

$$F_f + (-2kx) = ma, \tag{10.51}$$

where $a \equiv \ddot{x}$. With clockwise torque taken to be positive, the $\tau = I\alpha$ equation around the center of the cylinder is (using the non-slipping condition $\alpha = a/R$)

$$-F_f R + (-2kx)R = \left(\frac{mR^2}{2}\right)\left(\frac{a}{R}\right) \implies -F_f - 2kx = \frac{1}{2}ma. \tag{10.52}$$

Adding this to the above $F = ma$ equation eliminates the friction force F_f, and we obtain

$$-4kx = \frac{3}{2}m\ddot{x} \implies \ddot{x} = -\frac{8k}{3m}x. \tag{10.53}$$

The frequency of small oscillations is therefore $\omega = \sqrt{8k/3m}$.

REMARK: This frequency of $\omega = \sqrt{8k/3m}$ is *larger* than the frequency of $\sqrt{k/m}$ for a mass on the end of a spring. There are two reasons for this. First, for a given displacement x of the center of the cylinder, the spring stretches by $2x$. So the spring force is larger than it is for a mass on a spring stretched by x; we effectively have a stronger spring. Second, the friction force F_f always points in the *same* direction as the spring force; as an exercise, you can show that F_f always equals $F_s/3$. So the net force is increased compared with the case where the ground is frictionless.

As an exercise, you can show that if the spring were attached to the *center* of the cylinder, the frequency would be $\sqrt{2k/3m}$, which is *smaller* than the frequency of $\sqrt{k/m}$ for a mass on the end of a spring. This is due to the fact that the friction force always points *opposite* to the spring force, as you can verify.

---

[4]This is a general property of a tilted spring with relaxed length zero. The spring is equivalent to two zero-length sub-springs, with the same spring constant, lying along the horizontal and vertical projections of the original spring. (The above reasoning works the same way for the horizontal component.) In the present case, the horizontal sub-springs cancel, and we are left with two vertical sub-springs.

10.5. PROBLEM SOLUTIONS

In view of the previous two paragraphs, a continuity argument dictates that there must exist an attachment point for which the friction force is always zero (which is the cutoff case between always pointing in the same direction as the spring force and always pointing opposite to it). You can show that this point is located at a height of $3R/2$ above the ground, or equivalently halfway between the center and the top.[5] Also by continuity, there must be an attachment point for which the frequency equals $\sqrt{k/m}$ (because attaching at the top yields a higher frequency, and attaching at the center yields a lower frequency). You can show that this point is located at a height of $\sqrt{3/2}\,R$ above the ground, or equivalently about $(0.22)R$ above the center.

10.14. Wheel on a board

Let x be the amount the spring is stretched (or compressed, if x is negative). Then the free-body diagrams for the horizontal forces (spring and friction) on the board and the wheel are shown in Fig. 10.18. We have chosen positive F_f on each object to point in the directions indicated, although F_f will be negative half the time.

Figure 10.18

Note that the net forces on the board and the wheel are always equal and opposite. Therefore, since the masses of the objects are the same, their accelerations must always be equal and opposite. And since both objects start at rest, this implies that the velocities are always equal and opposite. And likewise for the positions relative to equilibrium. So if the spring stretches by x, the wheel must move $x/2$ to the right and the board must move $x/2$ to the left.

The $F = ma$ equation for the wheel (with rightward taken to be positive) is

$$F = ma \implies F_f - kx = m\left(\frac{\ddot{x}}{2}\right). \tag{10.54}$$

The $F = ma$ equation for the board is simply the negative of this equation. With clockwise taken to be positive, the $\tau = I\alpha$ equation for the wheel, around its center, is

$$\tau = I\alpha \implies -F_f R = \left(\frac{mR^2}{2}\right)\alpha. \tag{10.55}$$

The final equation we need is the non-slipping condition. This is $\ddot{x} = R\alpha$, which is most easily seen by working in the reference frame of the board. If the cylinder moves rightward by a distance x with respect to the board (which is how we defined x), it does so by rolling through an angle θ, where $x = R\theta$. Taking two derivatives of this relation gives $\ddot{x} = R\alpha$. We now have three equations and three unknowns: F_f, $\ddot{x}$, and α.

Using $\ddot{x} = R\alpha$ in the $\tau = I\alpha$ equation in Eq. (10.55), we obtain

$$-F_f R = \left(\frac{mR^2}{2}\right)\frac{\ddot{x}}{R} \implies F_f = -\frac{m\ddot{x}}{2}. \tag{10.56}$$

Plugging this into the $F = ma$ equation in Eq. (10.54) gives

$$-\frac{m\ddot{x}}{2} - kx = \frac{m\ddot{x}}{2} \implies \ddot{x} = -\frac{k}{m}x \implies \omega = \sqrt{\frac{k}{m}}. \tag{10.57}$$

---

[5]This is also the center of percussion for the cylinder, relative to the bottom point (see Problem 8.19 for the definition of the center of percussion). You should think about why this is the case.

REMARK: If we instead had two point masses m on the ends of the spring (or equivalently, if the wheel were replaced by a mass sliding without friction on the board), the $F = ma$ equation for each mass would be $-kx = m(\ddot{x}/2) \implies \ddot{x} = -(2k/m)x$. So the frequency would be $\omega = \sqrt{2k/m}$, which is larger than the above result of $\omega = \sqrt{k/m}$ for the original setup. (Since the midpoint of the spring doesn't move in our new setup, each mass is effectively on the end of a half-length spring with twice the spring constant; see Problem 4.2. Hence the $2k$ term in the frequency.) The frequency in the original setup with the rolling wheel is smaller because the friction always acts in the direction opposite to the spring force (because the spring force is $-kx$, while the friction force is $F_f = kx/2$ by substituting the $\ddot{x}$ from Eq. (10.57) into Eq. (10.56)). So the net force on each object is reduced.

10.15. Board on a cylinder

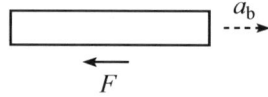

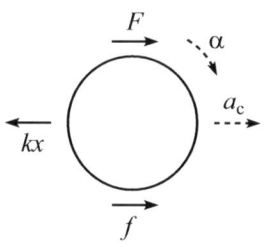

Figure 10.19

(a) FIRST SOLUTION: Let x be the position of the center of the cylinder relative to equilibrium, with rightward taken to be positive. The free-body diagrams (for just the horizontal forces) for the board and the cylinder are shown in Fig. 10.19. We've arbitrarily chosen the positive directions for the friction forces F and f as shown. If they come out to be negative (and half the time they are in fact negative), then they simply point in the other direction. The directions of F in the two diagrams must be opposite, though, by Newton's third law. We'll take positive a_b and a_c to be rightward, and positive α to be clockwise, as shown.

From Problem 7.8(a), the non-slipping condition between the cylinder and the ground is $a_c = R\alpha$. And from Problem 7.8(b), the non-slipping condition between the cylinder and the board is $a_b = 2a_c$. So the various force and torque equations are:

- $F = ma$ for the cylinder:
$$F + f - kx = ma_c. \tag{10.58}$$

- $\tau = I\alpha$ for the cylinder (using $R\alpha = a_c$):
$$FR - fR = \left(\frac{mR^2}{2}\right)\alpha \implies F - f = \frac{ma_c}{2}. \tag{10.59}$$

- $F = ma$ for the board (using $a_b = 2a_c$):
$$-F = ma_b \implies -F = 2ma_c. \tag{10.60}$$

We have three equations in three unknowns (F, f, and a_c, where $a_c \equiv \ddot{x}$). Adding the first two equations to eliminate f, and then plugging $F = -2ma_c$ into the result gives

$$2(-2ma_c) - kx = \frac{3ma_c}{2} \implies -kx = \frac{11}{2}ma_c$$

$$\implies \ddot{x} = -\left(\frac{2k}{11m}\right)x. \tag{10.61}$$

The frequency of the oscillatory motion is therefore $\omega = \sqrt{2k/11m}$.

REMARK: This frequency is smaller than $\sqrt{k/m}$. This makes sense, because $\sqrt{k/m}$ would be the frequency if the board didn't exist and if the cylinder were instead a block sliding on the ground. In our actual setup, both of the friction forces F and f point opposite to the spring force at all times (as you can verify), and this effectively reduces the spring constant, making the frequency be smaller than it would be if we just had a sliding block.

SECOND SOLUTION: We can also find ω by using conservation of energy. Again let x be the position of the center of the cylinder relative to equilibrium. Then from the non-slipping condition between the cylinder and the board (which says that the board's acceleration is always twice the cylinder's), the position of the board is $2x$. Using Eq. (7.8) for the energy of the cylinder, the total energy of the system is

$$E = \left[\frac{1}{2}m\dot{x}^2 + \frac{1}{2}\left(\frac{1}{2}mR^2\right)\dot{\theta}^2\right] + \frac{1}{2}m(2\dot{x})^2 + \frac{1}{2}kx^2. \tag{10.62}$$

The non-slipping condition between the cylinder and the ground, $R\alpha = a_c$, can alternatively be written in terms of velocities as $R\dot\theta = \dot x$. Using this to eliminate $\dot\theta$ in favor of $\dot x$, the energy can be rewritten as

$$E = \frac{1}{2}\left(\frac{11m}{2}\right)\dot x^2 + \frac{1}{2}kx^2. \tag{10.63}$$

But this is simply the energy of a mass $11m/2$ on a spring with spring constant k. And we know that the frequency of such a system is $\omega = \sqrt{k/(11m/2)} = \sqrt{2k/11m}$, in agreement with the first solution.

If you want to be a little more systematic, you can take the derivative of Eq. (10.63) to obtain (using the fact that E is constant)

$$0 = \left(\frac{11m}{2}\right)\dot x \ddot x + kx\dot x \implies \ddot x = -\left(\frac{2k}{11m}\right)x, \tag{10.64}$$

which reproduces Eq. (10.61).

(b) If the amplitude is A, then $x(t)$ takes the form of $x = A\cos\omega t$ (or technically $A\cos(\omega t + \phi)$, but the phase won't matter here). The acceleration of the cylinder is then $a_c = \ddot x = -\omega^2 A\cos\omega t$. From Eq. (10.60), the friction force F between the cylinder and the board is

$$F = -2ma_c = 2m\omega^2 A\cos\omega t$$
$$= 2m\left(\frac{2k}{11m}\right)A\cos\omega t = \frac{4kA}{11}\cos\omega t. \tag{10.65}$$

The maximum value of F is therefore $4kA/11$. This grows with k and A, which makes sense. And the units of kA are correctly the units of force, because the spring force takes the form of $-kx$ (k times a distance).

10.16. **Oscillating disk**

Let θ be the angular displacement of the disk. For small θ, the length of the spring is still essentially equal to the radius R. There are two consequences of this. First, the squat triangle in Fig. 10.20 with base $2R$ is essentially isosceles. This implies that the angle on the right is equal to the angle on the left (which we defined to be θ), as shown. Second, the spring force has magnitude essentially equal to $F = kR$ (remember that the relaxed length is zero).

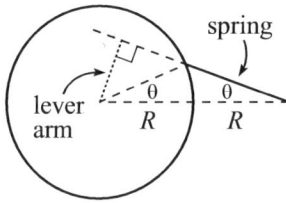

Figure 10.20

By looking at the thin right triangle with hypotenuse $2R$ in Fig. 10.20, the lever arm of the spring force, relative to the center of the disk, is $\ell = (2R)\sin\theta \approx 2R\theta$, where we have used the $\sin\theta \approx \theta$ small-angle approximation. The effect of the torque from the spring is to decrease θ if θ is positive, so the torque is $\tau = -F\ell = -(kR)(2R\theta)$. The $\tau = I\alpha$ equation around the center of the disk is therefore

$$-(kR)(2R\theta) = \left(\frac{mR^2}{2}\right)\ddot\theta \implies \ddot\theta = -\frac{4k}{m}\theta \implies \omega = 2\sqrt{\frac{k}{m}}. \tag{10.66}$$

Another way to determine the torque is to note that the spring makes an angle of 2θ with respect to the radial, because the spring and the radial each make angles of θ with respect to the horizontal, but in opposite directions. So the tangential spring force is $F_\theta = -(kR)\sin 2\theta \approx -(kR)(2\theta)$, which gives a torque of $\tau = F_\theta R = -(kR \cdot 2\theta)R$, as above.

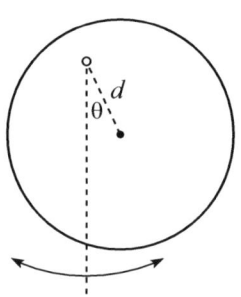

Figure 10.21

10.17. Maximum frequency 1

The parallel-axis theorem gives the moment of inertia of the coin around the pivot in Fig. 10.21 as $I = I_{CM} + md^2 = mR^2/2 + md^2$. The gravitational force mg has a lever arm of $d\sin\theta$ relative to the pivot, so the $\tau = I\alpha$ equation is

$$-mgd\sin\theta = (mR^2/2 + md^2)\ddot\theta \implies \ddot\theta = -\left(\frac{gd}{R^2/2 + d^2}\right)\theta, \qquad (10.67)$$

where we have used the small-angle approximation, $\sin\theta \approx \theta$. The frequency of small oscillations is therefore

$$\omega = \sqrt{\frac{gd}{R^2/2 + d^2}}. \qquad (10.68)$$

(We've just rederived Eq. (10.37) here.) We want to maximize ω as a function of d. Maximizing ω is equivalent to maximizing ω^2, so setting the derivative of $d/(R^2/2 + d^2)$ equal to zero gives (ignoring the denominator of the derivative)

$$(R^2/2 + d^2)(1) - d(2d) = 0 \implies R^2/2 = d^2 \implies d = R/\sqrt{2}. \qquad (10.69)$$

We therefore want the pivot to be at a radius of about $(0.71)R$.

REMARKS: With $d = R/\sqrt{2}$, the frequency ω in Eq. (10.68) is $\omega = (1/2^{1/4})\sqrt{g/R} \approx (0.84)\sqrt{g/R}$. This is correctly larger than the frequency at the two extreme values of d, namely $\omega = 0$ at $d = 0$, and $\omega = \sqrt{2g/3R} \approx (0.82)\sqrt{g/R}$ at $d = R$. If we furthermore want to allow a massless extension to be attached to the coin, then we can take the $d \to \infty$ limit, in which case $\omega \to 0$.

If you want to work with a general moment of inertia I_{CM}, you can show that the maximum frequency is obtained when the distance d from the pivot to the CM is $d = \sqrt{I_{CM}/m}$. This correctly reduces to $d = R/\sqrt{2}$ in the above case of the coin with $I_{CM} = mR^2/2$. Note that any pivot point on a circle (in the plane of the page) of a given radius d around the CM yields the same frequency, even if the object isn't a nice symmetric coin. The direction from the CM to the pivot doesn't matter, because it doesn't matter in the parallel-axis theorem; only the distance d is important.

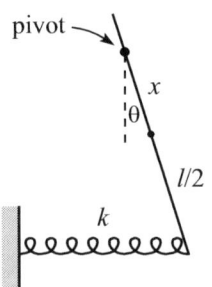

Figure 10.22

10.18. Maximum frequency 2

If θ is small, then we can use the $\sin\theta \approx \theta$ approximation in Fig. 10.22 to say that the spring is stretched (or compressed, if θ is negative) by essentially $(\ell/2 + x)\theta$. So the spring provides a force of $F = -k(\ell/2 + x)\theta$. This force produces a torque with respect to the pivot. For small θ, the lever arm of the force is essentially equal to $d = \ell/2 + x$, so the torque (with counterclockwise taken to be positive, since that's how we've defined θ) is

$$\tau = Fd = \left(-k\left(\frac{\ell}{2} + x\right)\theta\right)\left(\frac{\ell}{2} + x\right) = -k\left(\frac{\ell}{2} + x\right)^2\theta. \qquad (10.70)$$

Using the parallel-axis theorem, the moment of inertia of the stick relative to the pivot is $I = I_{CM} + mx^2 = m\ell^2/12 + mx^2$. So $\tau = I\alpha$ gives

$$-k\left(\frac{\ell}{2} + x\right)^2\theta = \left(\frac{m\ell^2}{12} + mx^2\right)\ddot\theta. \qquad (10.71)$$

Solving for $\ddot\theta$, we can read off the oscillation frequency as the square root of the (negative of the) coefficient of the θ term. So we have

$$\omega = \sqrt{\frac{k}{m}}\sqrt{\frac{(\ell/2 + x)^2}{\ell^2/12 + x^2}}. \qquad (10.72)$$

10.5. PROBLEM SOLUTIONS

Maximizing this (or equivalently maximizing ω^2) by setting the derivative equal to zero gives

$$0 = \left(\frac{\ell^2}{12} + x^2\right) \cdot 2\left(\frac{\ell}{2} + x\right) - \left(\frac{\ell}{2} + x\right)^2 \cdot 2x$$

$$\implies \left(\frac{\ell}{2} + x\right)x = \frac{\ell^2}{12} + x^2 \implies x = \frac{\ell}{6}. \tag{10.73}$$

This value of x means that the pivot is a third of the way along the stick. With $x = \ell/6$, the ω in Eq. (10.72) turns out to be $\sqrt{4k/m}$, compared with $\sqrt{3k/m}$ when $x = 0$ and again $\sqrt{3k/m}$ when $x = \ell/2$.

LIMITS: In the $x \to -\ell/2$ limit (the above reasoning is still valid if x is negative), we have $\omega \to 0$. This makes sense because the pivot is right where the spring is attached, so the spring force has no lever arm, which means the torque is zero. In the $x \to \infty$ limit (which involves a long massless extension attached to the stick), we have $\omega \to \sqrt{k/m}$. This makes sense because the stick simply oscillates back and forth without rotating, so it's effectively just a point mass m.

10.19. **Oscillating board**

(a) Let y be the height of the top end of the spring, relative to equilibrium. Then $y = (L/2)\sin\theta \approx (L/2)\theta$, where θ is the angle of the board. The spring force is therefore $-ky = -k(L/2)\theta$. The lever arm of this force relative to the pivot is essentially $L/2$, so the $\tau = I\alpha$ equation for the board, with counterclockwise taken to be positive, is

$$-\left(k \cdot \frac{L}{2}\theta\right)\frac{L}{2} = \left(\frac{1}{12}ML^2\right)\ddot{\theta} \implies \ddot{\theta} = -\frac{3k}{M}\theta \implies \omega = \sqrt{\frac{3k}{M}}. \tag{10.74}$$

We see that ω grows with k and decreases with M. This behavior makes sense.

(b) If the angular amplitude is θ_0, then $\theta(t)$ takes the form,

$$\theta(t) = \theta_0 \cos\omega t, \tag{10.75}$$

where $\omega = \sqrt{3k/M}$. (A possible additional phase ϕ of the cosine is irrelevant for the present purposes.) The forces on the block are the downward gravitational force and the slightly tilted normal force N from the board. N is essentially equal to mg because the block is only negligibly accelerating, since we are assuming that θ_0 is very small. The horizontal component of N is therefore $N_x = -N\sin\theta \approx -(mg)\sin\theta \approx -mg\theta$. The horizontal $F = ma$ equation for the block is then

$$-mg\theta = m\ddot{x} \implies \ddot{x} \approx -g\theta. \tag{10.76}$$

Using the $\theta(t)$ from Eq. (10.75), we can write this as

$$\ddot{x} = -g\theta_0 \cos\omega t. \tag{10.77}$$

To solve this equation for $x(t)$, we can just integrate twice. This gives

$$x(t) = \frac{g\theta_0}{\omega^2}\cos\omega t = \frac{Mg\theta_0}{3k}\cos\omega t, \tag{10.78}$$

where we have used the ω from Eq. (10.74). The amplitude is therefore

$$x_0 = \frac{Mg\theta_0}{3k}. \tag{10.79}$$

REMARK: Technically, the most general solution to Eq. (10.77) has an $At + B$ added on to the solution we found in Eq. (10.78). This doesn't mess up the equality in Eq. (10.77), because the second derivative of $At + B$ is zero. But since we are told that initial conditions have been set up so that the block oscillates back and forth symmetrically around the pivot, both A and B must be zero. For general initial conditions, A isn't zero if the initial velocity (that is, the velocity when the board is tilted maximally) isn't zero, so the block heads off to infinity.

LIMITS: The amplitude x_0 grows with M (which is the mass of the board; the mass m of the block doesn't appear) and decreases with k; these behaviors make sense, because M and k influence the frequency of the oscillation. For example, if the frequency is large (which arises if M is small and/or k is large), then the block doesn't have much time to move, so the amplitude must be small. The amplitude also grows with g and θ_0, which makes sense, because these influence the force at a given point in the oscillation. If the force is large, then the block can be instantaneously at rest at a large position x_0 and still be dragged back to the origin in time (that is, after a quarter of the given period).

(c) Comparing the $x(t)$ in Eq. (10.78) with the $\theta(t)$ in Eq. (10.75), we see that they are both proportional to (positive) $\cos \omega t$. So they are in phase. This means that when θ is maximum (when the board is tilted upward the most), x is also maximum (the block is farthest to the right). Likewise, x is minimum when θ is minimum. And x is zero when θ is zero; that is, the block is right over the pivot whenever the board is horizontal. The pictures are shown in Fig. 10.23.

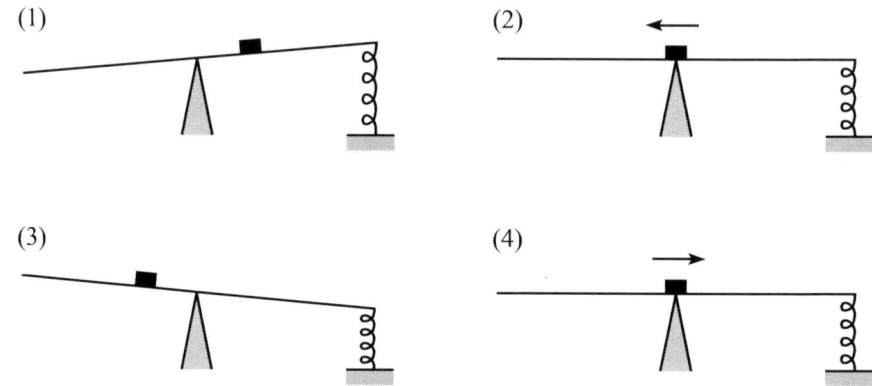

Figure 10.23

Chapter 11

Gravity

11.1 Introduction

Universal law of gravitation

Newton's *universal law of gravitation*, which is the basis for everything we will do in this chapter, gives the magnitude of the (attractive) gravitational force between two point masses m_1 and m_2, separated by a distance r, as

$$F = \frac{Gm_1m_2}{r^2}. \tag{11.1}$$

Newton first wrote down this law in 1687. The important ingredients in the law are that the force is proportional to the product of the masses and inversely proportional to the square of the separation. The gravitational constant G depends on the system of units used. In our standard mks system, the value is

$$G = 6.674 \cdot 10^{-11} \frac{\text{m}^3}{\text{kg s}^2}. \tag{11.2}$$

In 1798, Henry Cavendish measured G to the impressive accuracy of one percent. His very delicate experiment involved determining the force between two everyday-sized objects with known masses. (The earth couldn't be used as one of the masses because its mass wasn't known until Cavendish did his experiment.) The smallness of G means that the force between two everyday-sized masses is barely noticeable, so Cavendish needed to take great pains to isolate the tiny force he was trying to measure. See Section 5.4.2 in Morin (2008) for a discussion of the experiment.

Relation between *G* and *g*

It turns out that Eq. (11.1) holds for spheres as well as point masses. That is, from the outside, a sphere can be treated like a point mass at the center, as far as the gravitational force is concerned (more on this below). So for a point mass m located near the surface of the earth, Eq. (11.1) gives the force between the mass and the earth as $F = GmM_\text{E}/R_\text{E}^2$. However, we already know what this force is; it is the standard $F = mg$ gravitational force, where $g = 9.8\,\text{m/s}^2$. Equating these two expressions for the force tells us that

$$g = \frac{GM_\text{E}}{R_\text{E}^2}. \tag{11.3}$$

And indeed, if we plug in the values of the various constants, we obtain

$$g = \frac{GM_\text{E}}{R_\text{E}^2} = \frac{(6.67 \cdot 10^{-11}\,\text{m}^3/\text{kg s}^2)(5.97 \cdot 10^{24}\,\text{kg})}{(6.37 \cdot 10^6\,\text{m})^2} = 9.81\,\text{m/s}^2, \tag{11.4}$$

as expected.

Equation (11.3) isn't exact, in that the actual value of g varies over the surface of the earth, from about 9.78 at the equator to about 9.83 at the poles. It depends in a fairly complicated way on the nonspherical shape of the earth, which bulges at the equator due to the rotation of the earth (which itself causes a slight centrifugal-force correction to the effective value of g for a dropped ball; see Multiple-Choice Question 12.4). Additionally, there are local variations due to variations in the mass density of the earth's crust. Measuring these small variations with a *gravimeter* (which has a remarkable sensitivity) can indicate what substances (oil, minerals, etc.) might be lurking deep in the ground.

Note that the relation in Eq. (11.3) by itself isn't enough to determine G. That is, Cavendish could not have determined the value of G by simply measuring the acceleration of a dropped ball. Equation (11.3) determines only the product GM_E. (More precisely, it determines GM_E/R_E^2, but the value of R_E was known reasonably accurately in Cavendish's time.) And for all we know, the value of G might be 10 times larger than what we thought, with M_E being 10 times smaller. Maybe the interior of the earth is made of styrofoam (not likely!). But once Cavendish determined G via an experiment with two *known* masses, he could then determine the value of M_E. This is why his experiment was known as "weighing the earth" (or rather, "massing the earth"). It is fascinating how the behavior of a few lead balls in a tabletop experiment can tell us what the mass of the earth is. The average density of the earth turns out to be about 5.5 times the density of water. So we conclude that the earth is certainly *not* made of styrofoam. Iron is a much better bet. Cavendish's tabletop experiment therefore tells us something about the earth's core, which we have no hope of investigating by direct access.

Potential energy

We know from Eqs. (5.4) and (5.7) that the potential energy associated with a force is $U(r) = -\int F\, dr$. (The radial direction is all that matters here.) Consider the potential energy U of a mass m located a distance r from a mass M. We can arbitrarily choose the reference point where we define U to be zero, but it is customary to take this reference point to be at infinity. So with $U = 0$ at $r = \infty$, the value of U at any other distance r from the given mass M is

$$U(r) = -\int_\infty^r F\, dr' = -\int_\infty^r \left(-\frac{GmM}{r'^2}\right) dr' = -\frac{GmM}{r'}\bigg|_\infty^r = -\frac{GmM}{r}. \quad (11.5)$$

We have used the fact that the gravitational force is attractive, which means that it points in the direction of decreasing r; hence the minus sign in F. We have put primes on the integration variable so that it isn't confused with the specific value of r that is the limit of integration. We see that the potential energy is negative. However, this negative nature isn't important; only differences in potential energy matter. Someone could add, say, 13 joules to the energy everywhere, and it would describe the same system. All that matters is that the potential energy decreases as we move in toward the mass M.

As mentioned in Chapter 5, a potential energy can be defined only if a force is conservative. The $F = -GmM/r^2$ force is indeed conservative, because when calculating the work done between two given points, only the starting and ending values of r matter; the path taken is irrelevant. (You can consider any path to consist of radial and tangential displacements, and no work is done along the tangential ones.) The particular $1/r^2$ nature of the (radial) force isn't important here. Any other function of r would still yield a conservative force.

Force from a sphere

The nice thing about the inverse-square law in Eq. (11.1) is that it leads to the following two facts:

- Outside a uniform hollow spherical shell with total mass m, the shell can be treated like a point mass m at the center, as far as the gravitational force is concerned.

- Inside a uniform hollow spherical shell with total mass m, the shell produces zero gravitational force; the shell effectively doesn't exist.

The first of these facts requires a bit of a calculation to demonstrate; see Problem 11.1. The second fact can be proved in a much simpler way; see Problem 11.2. Note that the shell needs to be spherical; the above facts don't apply to a cubical shell, for example.

If we have a solid sphere instead of a hollow shell, we can still make use of the above facts, because we can treat the solid sphere as the superposition of many thin hollow shells. We quickly conclude that if we are located at radius r inside a solid sphere of radius R, then the ball of mass inside radius r looks like a point mass at the center (with the same mass as the ball of radius r), while the mass between r and R effectively doesn't exist. This result holds even if the mass density isn't uniform throughout the sphere, provided that it depends only on radius (so that each thin hollow shell is uniform).

Kepler's laws

The motion of the planets around the sun is completely governed by the universal law of gravitation, Eq. (11.1) (assuming we neglect effects of general relativity). This law can be used to derive three other laws, knows as Kepler's laws:

1. The planets move in elliptical orbits, with the sun at one focus.

2. A planet sweeps out equal areas in equal times; see Fig. 11.1.

3. The square of the period of revolution, T, is proportional to the cube of the semimajor axis, a, of the elliptical orbit. More precisely:

$$T^2 = \frac{4\pi^2 a^3}{GM_{sun}}.\tag{11.6}$$

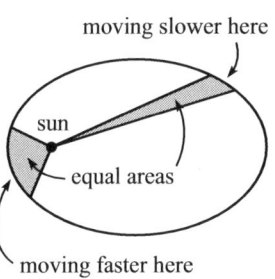

Figure 11.1

If we look at the historical order in which things developed, Kepler wrote down these laws in the early 1600's, about 75 years before Newton wrote down Eq. (11.1). Kepler arrived at his laws empirically (that is, by looking at observational data), which is quite a remarkable feat. But Newton showed that they are all consequences of a single universal law.

Kepler's second law can be derived quickly by showing that it is equivalent to the statement of conservation of angular momentum; see Problem 11.5. In contrast, the first and third laws require a lengthy calculation to derive (see Section 7.4 in Morin (2008)), although the third can be easily demonstrated in the special case of a circular orbit; see Problem 11.6.

Technically, Kepler's three laws hold only in the approximation where the sun is much more massive than the planets. Since this is a good approximation, we won't worry about the minor corrections to the laws.

11.2 Multiple-choice questions

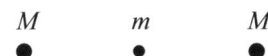

Figure 11.2

11.1. In the setup shown in Fig. 11.2, the two outside masses are glued in place. The center mass is initially located at the midway point. If it is displaced slightly to the right, the resulting net force on it is

   (a) rightward   (b) leftward   (c) zero

11.2. Two planets have the same mass density, but one has twice the radius of the other. What is the ratio of the acceleration due to gravity on the larger planet to that on the smaller planet?

   (a) 1/4   (b) 1/2   (c) 1   (d) 2   (e) 4

11.3. Two planets have the same mass density, but one has twice the radius of the other. What is the ratio of the potential energy (relative to infinity) of a mass m on the surface of the larger planet to that on the surface of the smaller planet?

   (a) 1/4   (b) 1/2   (c) 1   (d) 2   (e) 4

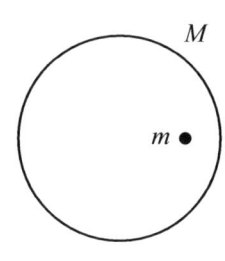

Figure 11.3

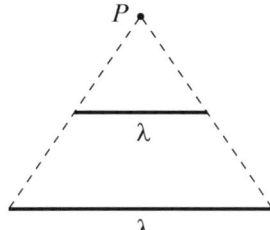

Figure 11.4

11.4. Two identical planets have radius R and uniform mass density ρ. Their centers are a distance d apart. If all distances (both R and d) are scaled up by a factor of 2, while ρ is held constant, what is the ratio of the force between the planets in the new setup to the force in the original setup?

(a) 1/4   (b) 1   (c) 4   (d) 16   (e) 64

11.5. A mass m is located off-center in the interior of a uniform ring with mass M, as shown in Fig. 11.3. What is the direction of the gravitational force on m due to the ring? (You may want to look at the reasoning in Problem 11.2 first.)

(a) leftward

(b) rightward

(c) upward

(d) downward

(e) The force is zero inside the ring.

11.6. The two sticks shown in Fig. 11.4 have the same linear mass density λ, and they subtend the same angle from a given point P, as indicated. Their distances from P are ℓ and 2ℓ. Which stick creates a larger force on a mass at P? (This question is in the same spirit as the preceding one.)

(a) the top stick

(b) the bottom stick

(c) They produce equal forces at P.

(d) The relative size of the forces depends on ℓ.

11.7. A mass m is located very close to a very large (essentially infinite) flat sheet with surface mass density σ. Another mass m is located very close to a hollow spherical shell with radius R and the same surface mass density σ. What is the ratio of the gravitational force on the first m to the gravitational force on the second m? You can use the result from Problem 11.8 that the force from an infinite sheet is $F = 2\pi G\sigma m$.

(a) 1/4   (b) 1/2   (c) 1   (d) 2   (e) 4

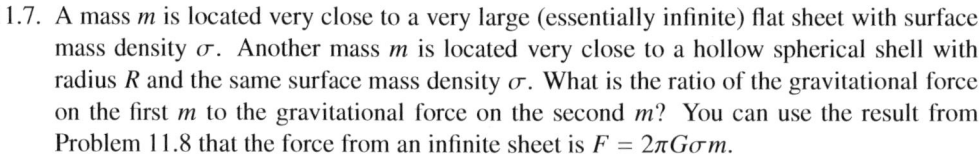

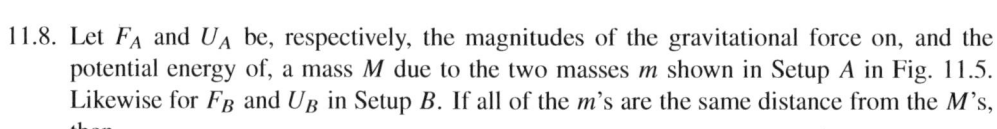

Figure 11.5

11.8. Let F_A and U_A be, respectively, the magnitudes of the gravitational force on, and the potential energy of, a mass M due to the two masses m shown in Setup A in Fig. 11.5. Likewise for F_B and U_B in Setup B. If all of the m's are the same distance from the M's, then

(a) $F_A = F_B$ and $U_A = U_B$

(b) $F_A = F_B$ and $U_A < U_B$

(c) $F_A < F_B$ and $U_A = U_B$

(d) $F_A < F_B$ and $U_A < U_B$

(e) $F_A > F_B$ and $U_A < U_B$

11.9. Relative to infinity, the gravitational potential energy of an object at the center of a hollow spherical shell is

(a) zero

(b) less than zero, but larger than the potential energy at the surface of the shell

(c) equal to the potential energy at the surface

(d) less than the potential energy at the surface, but greater than negative infinity

(e) negative infinity

11.3 Problems

The first nine problems are foundational problems.

11.1. Force from a spherical shell

A uniform hollow spherical shell has mass M and radius R. Find the force on a mass m at radius r due to the shell, both inside and outside. Do this by calculating the potential energy due to the shell, and then taking the derivative to obtain the force. *Hint*: Slice the shell into the rings shown in Fig. 11.6, and then integrate over the rings. You will need to use the law of cosines.

Note: The reason for finding the force via the potential energy is that the potential energy is a scalar quantity, whereas the force is a vector. If we tried to directly calculate the force, we would have to worry about forces pointing in all sorts of different directions. With the potential energy, we just need to add up some numbers.

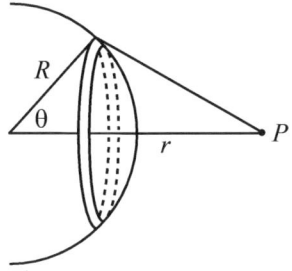

Figure 11.6

11.2. Zero force inside a spherical shell

Show that the gravitational force inside a uniform hollow spherical shell is zero by showing that the pieces of mass at the ends of the thin cones in Fig. 11.7 give canceling forces on a given mass at point P.

11.3. Force inside a solid sphere

What is the gravitational force on a mass m located at radius r inside a planet with radius R, mass M, and uniform mass density?

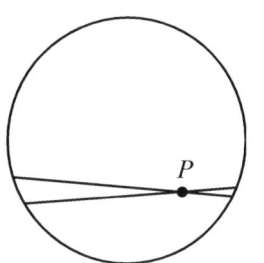

Figure 11.7

11.4. Low-orbit and escape velocities

(a) What is the speed of a low-orbit satellite circling around a planet with mass M and radius R, just above the surface?

(b) What is the escape velocity (or rather, escape speed) from the surface of a planet with mass M and radius R? That is, what minimum initial speed is required for a mass to end up infinitely far away from the planet (and thus refute the "What goes up must come down" claim)?

11.5. Kepler's 2nd law

Show that Kepler's second law (which says that a planet sweeps out equal areas in equal times) is equivalent to the statement of conservation of angular momentum.

11.6. Kepler's 3rd law for circles

For a circular orbit, use $F = ma$ to derive Kepler's third law, Eq. (11.6), from scratch.

11.7. Force from a line

A mass m is placed a distance ℓ away from an infinite straight line with mass density λ (kg/m). Show that the force on the mass is $F = 2G\lambda m/\ell$. *Hint*: If you use an angle as your variable, then you'll have to do some geometry, but the resulting integral should be easy. If you use the distance along the line as your variable, then a $\tan\theta$ trig substitution should simplify things.

11.8. Force from a plane

A mass m is placed a distance ℓ away from an infinite plane with mass density σ (kg/m²). Show that the force on the mass is $F = 2\pi G\sigma m$. *Hint*: Imagine building up the plane from an infinite number of adjacent rods, and then integrate over these rods, using the result from Problem 11.7.

11.9. Two expressions for potential energy

From Eq. (11.5) we know that the gravitational potential energy is given by $U(r) = -GmM/r$. However, we also know that near the surface of the earth the potential energy is given by $U(h) = mgh$. Verify that these two expressions are consistent, assuming that h is much smaller than the radius R of the earth. You will need to make a Taylor-series approximation.

11.10. Attracting particles

Two particles with masses m and M are initially at rest, a very large (essentially infinite) distance apart. They are attracted to each other due to gravity. What is their relative speed when they are a distance r apart?

11.11. Sphere and stick

A uniform sphere with mass m and radius R is placed a distance $2R$ from a uniform stick with mass m and length $2R$, as shown in Fig. 11.8. The objects are attracted to each other due to gravity. If they are released from rest, what are their speeds right before they collide?

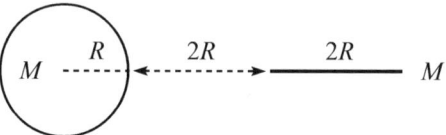

Figure 11.8

11.12. Dismantling a planet

Consider a spherical planet with mass M and radius R, with uniform mass density. Assume that the planet is made up of little grains of sand. How much energy is required to take all of the grains and move them out to infinity, all very far away from each other? *Hint*: How much energy is required to take the grains of sand in a very thin outer shell of the sphere and move them out to infinity? You can then repeat this process by removing a thin outer shell from successively smaller spheres, until you are left with nothing.

11.13. Planet with variable density

(a) A spherical planet has mass M and radius R. If its density depends on the radial position according to $\rho(r) = b/r$, what is b in terms of M and R?

(b) What is the difference between the potential energy of a mass m located on the surface of the planet, and the potential energy of a mass m located at the center?

11.14. Descending in a mine shaft

(a) Assuming that the density of the earth is constant (call it ρ), find an expression for the gravitational force on a mass m, as a function of the radial position r inside the earth. You should find that the (magnitude of the) gravitational force decreases as you descend in a mine shaft.

(b) However, the density of the earth is *not* constant, and in fact the gravitational force *increases* as you descend in a mine shaft. It turns out that the general condition under which this is true for a planet is $\rho_c < A\rho_{avg}$, for a certain value of A. Here ρ_{avg} is the average density of the planet, and ρ_c is the density of the crust at the surface. Find A. (Assume that the density is spherically symmetric, that is, independent of r.)

11.3. PROBLEMS

11.15. Orbiting objects

(a) A planet with mass M is orbited by a satellite with negligible mass. The radius of the orbit is R. What is the angular frequency of the motion?

(b) Two planets, both with mass $M/2$, orbit around their CM. They both travel in a circle (the same circle), and they are always a distance R apart. What is the angular frequency of the motion?

(c) Your answers to the above two questions should be the same. This might make you think that if two planets orbit around their CM, with each one moving in a circle (two different circles in general now), and with the separation always being R, then the angular frequency of the motion depends on the masses only through their sum. Show that this is indeed the case.

11.16. Grabbing a mass

Two equal masses m are a distance ℓ apart and interact via gravity. They are given the proper tangential speed v_0 so that they both travel in a circle of radius $\ell/2$ around their CM.

(a) What is v_0?

(b) If one of the masses is grabbed and held at rest, what is the closest distance the other mass comes to it? *Hints:* What is the direction of the velocity at closest approach? What quantities are conserved?

11.17. Orbiting satellite

(a) A satellite with mass m is in a circular orbit with radius R around a planet with mass M, which is assumed to be fixed in place. What is the total energy of the satellite (in terms of G, m, M, and R)? As usual, take the gravitational potential energy to be zero at infinity.

(b) Assume now that the satellite has been given a kick so that its orbit looks like the one shown in Fig. 11.9, where the closest and farthest approaches to M are R and nR, where n is a given numerical factor. What is the total energy of the satellite (in terms of n, G, m, M, and R)? Check the $n \to 1$, $n \to \infty$, and $n \to 0$ limits.

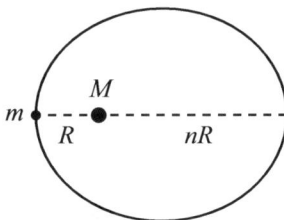

Figure 11.9

11.18. Projectile in a planet

(a) Consider a planet with radius R and uniform mass density ρ. In terms of G, R, and ρ, what is the speed of an object in a circular orbit just above the surface of the planet?

(b) Consider the gravitational potential energy of a mass m at radius r *inside* the planet. Calculate the potential energy for $r < R$, relative to the center of the planet (not relative to infinity).

(c) An object is projected tangentially at the surface of the planet with a speed equal to half of the speed you found in part (a). Under normal circumstances, the object will simply hit the ground. But assume that someone has calculated the unhindered trajectory of the object and has (hypothetically!) dug a tunnel into the planet along the predicted path, so that the object travels through this tunnel instead of hitting the ground. What is the smallest value of the radius r that the object achieves?

11.19. Oscillating stick in a planet 1

Consider the following highly unrealistic setup. A rigid stick with mass m and length ℓ is located in the interior of a planet with mass M and radius R. The center of the stick is located at the midpoint of the chord on which it lies (but the stick need not be as long as the entire chord). The center of the stick is a distance h from the center of the planet, as shown in Fig. 11.10.

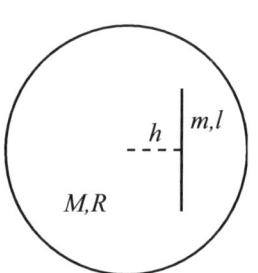

Figure 11.10

(a) What is the gravitational force from the planet on a small piece of the stick with mass dm located at radius r? Assume that the planet has uniform density.

(b) Assume that somehow the stick is able to move freely within the planet (you can pretend that a thin "sheet" of the planet has been hollowed out). The stick is released from rest. Show that it undergoes simple harmonic motion, and find the frequency.

11.20. Oscillating stick in a planet 2

A narrow tube is drilled from one point to a diametrically opposite point through a planet with mass M and radius R. A thin stick with length $2R$ and linear mass density λ (kg/m) is placed in the tube, with one end a distance x above the surface of the planet, as shown in Fig. 11.11. Assume that there is no friction between the stick and the tube, and assume that the planet has uniform density. The stick is released.

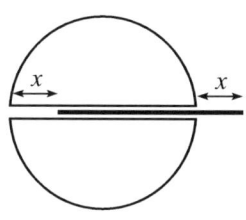

Figure 11.11

(a) As a function of x, find an *exact* expression for the net gravitational force on the stick.

(b) For small x (more precisely, for $x \ll R$), find an *approximate* expression for the force on the stick, to first order in x. (If you got stuck in part (a), it's possible to solve this part on its own.)

(c) Find the frequency of small oscillations of the stick.

11.21. Asteroid slab

(a) Consider a spherical planet with radius R and uniform volume mass density ρ. A thin tube is drilled along a diameter, as shown in Fig. 11.12(a). If an object is dropped from rest into the tube, what is the frequency of its oscillatory motion?

(b) Consider now an asteroid in the shape of a slab with thickness D, with the same uniform volume mass density ρ as the sphere in part (a). Assume that the two dimensions perpendicular to the thickness D are essentially infinite (much larger than D). A thin tube is drilled along the "D" direction, as shown in Fig. 11.12(b). If an object is dropped from rest into the tube, what is the frequency of its oscillatory motion? *Hint*: You'll want to use the result from Problem 11.8 that the force on a mass m due to a thin *sheet* with infinite extent and uniform *surface* mass density σ (kg/m^2) equals $2\pi\sigma Gm$, which is independent of the distance from the sheet.

(c) You will find that the frequencies in parts (a) and (b) aren't equal. Give a qualitative physical argument why the larger frequency is in fact larger.

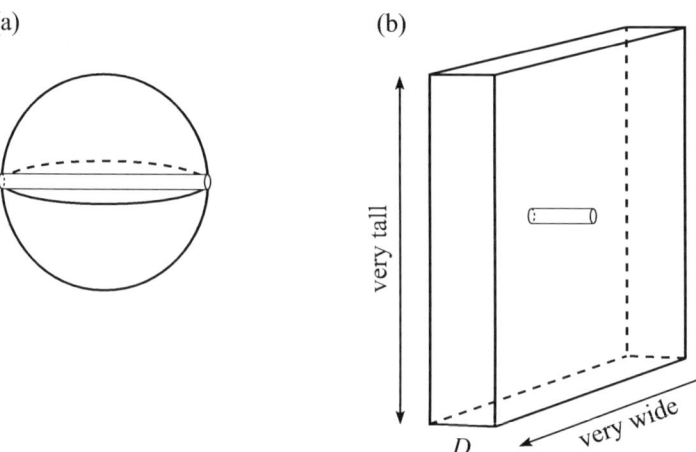

Figure 11.12

11.4 Multiple-choice answers

11.1. [a] Since the gravitational force falls off with r, and since m is now closer to the right M, the attractive rightward force from the right M is larger than the attractive leftward force from the left M. The net force is therefore rightward.

11.2. [d] The gravitational force has magnitude $F = GMm/R^2$. (Equivalently, g is given by GM/R^2.) And a planet's mass M is proportional to its volume (for a given density), which is proportional to R^3. So $F \propto R^3/R^2 = R$. Doubling the radius of a planet therefore means doubling the force on a given object with mass m at the surface. So the acceleration is doubled.

REMARK: A fact that you may be familiar with is that the acceleration due to gravity on the moon is about 1/6 that on the earth. If the densities were the same, then from the above reasoning, this would imply that the radius of the moon is 1/6 the radius of the earth. But in fact the moon's average density is only about 3/5 the earth's (apparently green cheese is less dense than iron!). This means (as you can show) that the ratio of moon to earth radii is a factor of 5/3 larger than the equal-density result of 1/6. So the moon's radius is $5/18 \approx 0.28$ of the earth's radius. The actual radii of about 1740 km and 6370 km yield a ratio of 0.27, which agrees with 0.28, at least to the accuracy of the numbers we used here.

11.3. [e] The gravitational potential energy is $U = -GMm/R$. And a planet's mass M is proportional to its volume, which is proportional to R^3. So $U \propto R^3/R = R^2$. Doubling the radius of a planet therefore means quadrupling the potential energy of a given object with mass m on the surface.

REMARK: Using the numbers given in the solution to the preceding question, you can show that the ratio of the potential energies of a mass on the surface of the earth vs. on the moon is roughly $(5/3)/(0.27)^2 \approx 23$.

11.4. [d] If the radius of a planet is doubled, then the volume $4\pi R^3/3$ increases by a factor of $2^3 = 8$. And since the density ρ is held constant, the mass also increases by a factor of 8. The gravitational force is Gm_1m_2/r^2, which equals Gm^2/d^2 here. Since m increases by a factor of 8, and d increases by a factor of 2, the force therefore increases by a factor of $8^2/2^2 = 16$.

11.5. [b] This is a lower-dimensional analog of Problem 11.2. (And Multiple-Choice Question 11.1 is an even lower dimensional analog!) In the spirit of Problem 11.2, draw two lines through the mass m (with a very small angle between them), and look at the forces from the two short arcs they define on the ring. Let the distances a and b be as shown in Fig. 11.13.

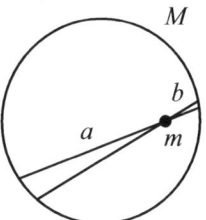

Figure 11.13

The arc on the left is longer by a factor a/b, so there is a/b more mass there (instead of a^2/b^2 as in Problem 11.2, because we're dealing with lengths here instead of areas). But the force from a small bit of mass in the left arc is smaller by a factor $1/(a/b)^2$ than the force from an equal bit in the right arc, due to the inverse-square nature of the gravitational force. The force from the whole left arc is therefore smaller by a factor $(a/b)/(a/b)^2 = b/a$ than the force from the whole right arc. So the net force has a rightward component. This reasoning holds for any such corresponding arcs, with one arc to the right of P and the other to the left. The net force is therefore directed to the right. (The vertical components of the forces cancel by symmetry.)

11.6. [a] Consider two corresponding pieces of the sticks subtending the same infinitesimal angle. These pieces are essentially point masses. There is twice as much mass in the larger piece (in the bottom stick), but it is twice as far from P. The dm/r^2 factor in the $Gm(dm)/r^2$ force law is therefore $2/2^2 = 1/2$ times as large for the bottom stick as for the top stick. This holds for all corresponding pieces, so the top stick creates twice as large a force as the bottom stick.

REMARK: What if we had 2-D objects like disks instead of 1-D sticks? (Imagine that the two horizontal lines in Fig. 11.4 represent the cross sections of two disks.) In this case we want to consider corresponding *patches* of the disks that are defined by the same infinitesimally thin *cone*. There is now *four* times as much mass in the larger patch, because areas are proportional to lengths squared. This factor of 4 cancels the factor of 4 from the r^2 in the denominator of the force law, so the forces at P from the two patches are equal. (This is the same reasoning as in Problem 11.2.) The answer in the 2-D case is therefore (c). Note that the 2-D objects don't have to be disks; they can be any similar shapes defined by the same cone-like surface emanating from P, and the forces will still be equal. And the ratio of the distances need not be 2 to 1. And the objects need not be horizontal, as long as they are parallel.

11.7. $\boxed{\text{b}}$ The mass of the spherical shell is $(4\pi R^2)\sigma$, so the force from it is

$$F_{\text{shell}} = \frac{G(4\pi R^2 \sigma)m}{R^2} = 4\pi G\sigma m. \tag{11.7}$$

And since we know that the force from the sheet is $F_{\text{sheet}} = 2\pi G\sigma m$, the desired ratio of F_{sheet} to F_{shell} is $1/2$.

REMARK: Another way of deriving this factor of 1/2 is the following. (This method also gives a slick way of deriving the $F_{\text{sheet}} = 2\pi G\sigma m$ result in Problem 11.8.) Consider a mass m that is very close to a hollow spherical shell. For example, pretend that the earth is a hollow shell, and place a mass m a meter above the surface. Consider the disk on the earth's surface consisting of all points within, say, a kilometer of the mass m. Any distance that is very large compared with a meter (so that the disk looks effectively like an infinite sheet, from m's point of view) but very small compared with the radius of the earth (so that the disk is essentially flat) will do.

The earth's (hypothetical) hollow shell may be divided into the disk plus the remaining part, which is a shell with a hole in it. The force from the entire shell can therefore be written as

$$\mathbf{F}_{\text{shell}} = \mathbf{F}_{\text{disk}} + \mathbf{F}_{\text{rem}}, \tag{11.8}$$

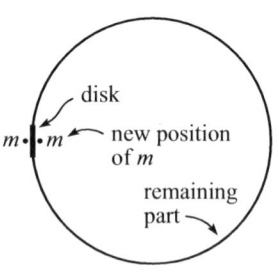

Figure 11.14

where "rem" stands for "remaining." What are the relative sizes of $\mathbf{F}_{\text{disk}}$ and $\mathbf{F}_{\text{rem}}$? We claim that they are equal. This can be seen in the following way. Take the mass m and move it to the other side of the disk, as shown in Fig. 11.14. The mass m is now *inside* a complete spherical shell, and we know that the force inside a spherical shell is zero. So $\mathbf{F}'_{\text{disk}}$ (the new force from the disk) and $\mathbf{F}_{\text{rem}}$ must be equal and opposite:

$$\mathbf{F}'_{\text{disk}} = -\mathbf{F}_{\text{rem}}. \tag{11.9}$$

Now, the only difference between this new setup and the original one where the mass was outside the shell is that the force from the (essentially flat) disk has changed sign, that is,

$$\mathbf{F}'_{\text{disk}} = -\mathbf{F}_{\text{disk}}. \tag{11.10}$$

(The mass m is still in essentially the same location relative to the remaining part, so $\mathbf{F}_{\text{rem}}$ hasn't changed.) The preceding two equations imply that in the original setup where the mass is outside the shell, the $\mathbf{F}_{\text{disk}}$ and $\mathbf{F}_{\text{rem}}$ forces must be equal:

$$\mathbf{F}_{\text{disk}} = \mathbf{F}_{\text{rem}}. \tag{11.11}$$

Eq. (11.8) then tells us that $\mathbf{F}_{\text{shell}} = 2\mathbf{F}_{\text{disk}}$. But $\mathbf{F}_{\text{disk}}$ is essentially the same as the force from an infinite sheet, so we have $\mathbf{F}_{\text{shell}} = 2\mathbf{F}_{\text{sheet}} \implies \mathbf{F}_{\text{sheet}} = (1/2)\mathbf{F}_{\text{shell}}$, as desired. Additionally, since we showed above that $F_{\text{shell}} = 4\pi G\sigma m$, we see that we have reproduced the $F_{\text{sheet}} = 2\pi G\sigma m$ result in Problem 11.8.

11.8. $\boxed{\text{c}}$ $F_A = 0$ because the forces from the two m's cancel. And $F_B > 0$ because the forces from both m's point rightward. So the answer must be either (c) or (d). The potential energy is the same in both setups, because potential energies are scalars (and not vectors like forces) which simply add, without any worry about directions of components. So it doesn't matter where the m's are in relation to M, as long as they are the same distance from M. The answer is therefore (c).

11.9. [c] The force inside a spherical shell is zero, so all points inside and on the surface have the same potential energy, because $U = -\int F\,dr$ (no work is done when moving an object around inside the shell), or equivalently because $F = -dU/dr$ (the derivative of a constant is zero).

REMARK: If we replace the hollow spherical shell with a solid sphere, then there is now a nonzero force inside (directed radially inward), which means that the potential energy decreases as we march in from the surface to the center. The total decrease is finite (you just need to integrate the force in Problem 11.3), so the answer for a solid sphere is (d). The potential energy at the center turns out to be 3/2 times the (negative) value at the surface, as you can show.

11.5 Problem solutions

11.1. Force from a spherical shell

As suggested in the hint, we'll slice the shell into rings as indicated in Fig. 11.6. The distance ℓ from an arbitrary point P to the ring is a function of R, r, and θ. It can be found as follows. In Fig. 11.15, segment AB has length $R\sin\theta$, and segment BP has length $r - R\cos\theta$. So the length ℓ in right triangle ABP is

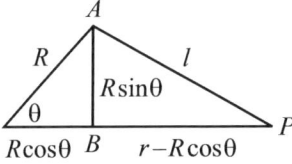

Figure 11.15

$$\ell = \sqrt{(R\sin\theta)^2 + (r - R\cos\theta)^2} = \sqrt{R^2 + r^2 - 2rR\cos\theta}, \qquad (11.12)$$

where we have used $\sin^2\theta + \cos^2\theta = 1$. What we've done here is just prove the law of cosines.

The area of a ring lying between θ and $\theta + d\theta$ on the shell is the width (which is $R\,d\theta$) times the circumference (which is $2\pi R\sin\theta$). In terms of the mass density of the shell, $\sigma = M/(4\pi R^2)$, the potential energy of a mass m at point P due to a thin ring is therefore $-Gm(\sigma(R\,d\theta)(2\pi R\sin\theta))/\ell$. This is true because the gravitational potential energy,

$$U(\ell) = -\frac{Gm_1 m_2}{\ell}, \qquad (11.13)$$

is a scalar quantity, so the contributions from the little mass pieces simply add. Every piece of the ring is the same distance from P, and this distance is all that matters; the direction from P is irrelevant (unlike it would be with the force). Using the ℓ from Eq. (11.12), and integrating over the entire shell, the total potential energy of a mass m at P is

$$\begin{aligned} U(r) &= -\int_0^\pi \frac{2\pi\sigma GR^2 m\sin\theta\,d\theta}{\sqrt{R^2 + r^2 - 2rR\cos\theta}} \\ &= -\frac{2\pi\sigma GRm}{r}\sqrt{R^2 + r^2 - 2rR\cos\theta}\,\Big|_0^\pi. \end{aligned} \qquad (11.14)$$

The $\sin\theta$ in the numerator is what makes this integral nice and doable. The $\theta = 0$ lower limit of the integral involves the quantity $\sqrt{(R-r)^2}$. A square root is defined to be a positive quantity, so this equals either $R - r$ or $r - R$, depending on which radius is larger. So we must consider two cases:

- If $r > R$ (that is, if m is outside the shell), then we have

$$U(r) = -\frac{2\pi\sigma GRm}{r}\Big((r+R) - (r-R)\Big) = -\frac{G(4\pi R^2\sigma)m}{r} = -\frac{GMm}{r}, \qquad (11.15)$$

which is simply the potential energy due to a point mass M located at the center of the shell. In other words, a uniform spherical shell with mass M can be treated like point mass M at the center, as far as the potential energy of an external mass is concerned.

- If $r < R$ (that is, if m is inside the shell), then we have

$$U(r) = -\frac{2\pi\sigma GRm}{r}\Big((r+R) - (R-r)\Big) = -\frac{G(4\pi R^2 \sigma)m}{R} = -\frac{GMm}{R}, \quad (11.16)$$

which is independent of r. So the potential energy inside the shell is constant.

Having found $U(r)$, we can obtain $F(r)$ by taking the negative of the derivative of $U(r)$. This gives

$$\begin{aligned} F(r) &= -\frac{GMm}{r^2} \quad &&\text{if } r > R, \\ F(r) &= 0 \quad &&\text{if } r < R. \end{aligned} \quad (11.17)$$

The minus sign in the first of these forces indicates radially inward.

REMARK: A *solid* sphere may be considered to be built up from many concentric spherical shells. This implies that if m is outside a given solid sphere, then the force on m is $-GMm/r^2$, where M is the total mass of the sphere. So from the outside, a sphere looks like a point mass. This result holds even if the shells have different mass densities (but each shell must have uniform density, that is, the density can be a function of only r). Note that the gravitational force between two spheres is the same as if they were both replaced by point masses. This follows from two applications of our "point mass" result.

Inside a *hollow* spherical shell, the force is zero, so it's just like the shell doesn't exist. This implies that at a point at radius r inside a *solid* sphere with radius R, only the mass that lies inside radius r is relevant (and this mass looks like a point mass, from the previous paragraph). This is true because we can build up the solid sphere from many thin shells, and all of the shells between r and R effectively don't exist. See Problem 11.3.

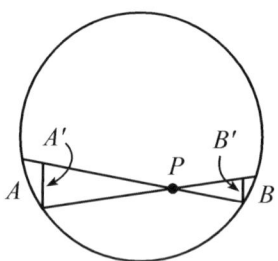

Figure 11.16

11.2. Zero force inside a spherical shell

In Fig. 11.16 let a be the distance from P to piece A, and let b be the distance from P to piece B. Draw the "perpendicular" bases of the cones, and call them A' and B'. The ratio of the areas of A' and B' is a^2/b^2, because areas are proportional to lengths squared. The key point to now note is that the angle between the planes of A and A' is the same as the angle between the planes of B and B'. This is true because the chord between A and B meets the circle at equal angles at its ends. This equality of the angles implies that the ratio of the areas (and hence masses) of A and B is also a^2/b^2.

However, the gravitational force decreases like $1/r^2$, and this effect exactly cancels the a^2/b^2 ratio of the masses; A is larger, but it is farther away. Therefore, the forces on a given mass m at P due to A and B (which can be treated like point masses, because the cones are assumed to be thin) are equal in magnitude; and opposite in direction, of course. If you want to write things out explicitly, the force due to A is

$$\frac{GmM_A}{a^2} = \frac{Gm(M_B \cdot a^2/b^2)}{a^2} = \frac{GmM_B}{b^2}, \quad (11.18)$$

which is the force due to B. If we draw enough cones to cover the whole shell, then the contributions to the force from little pieces over the whole shell will cancel in pairs, so we are left with zero force at P. This holds for any point P inside the shell.

11.3. Force inside a solid sphere

The force at radius r inside a planet is effectively due only to the mass that lies inside radius r (call it M_r). This is true because Problem 11.2 tells us that the gravitational force inside a uniform hollow spherical shell is zero, and because the mass outside radius r may be considered to be built up from many concentric spherical shells. We can find the mass M_r by noting that since mass is proportional to volume (because the mass density is uniform here), and since volume is proportional to radius cubed, we have $M_r/M =$

11.5. PROBLEM SOLUTIONS

$r^3/R^3 \implies M_r = (r^3/R^3)M$. The desired force is therefore (including the minus sign to indicate that the force is attractive)

$$F(r) = -\frac{GM_r m}{r^2} = -\frac{G(r^3 M/R^3)m}{r^2} = -\frac{GMmr}{R^3}. \tag{11.19}$$

This equals zero at the center of the planet, as it must, by symmetry. If you want to write $F(r)$ in terms of the mass density ρ, then substituting $4\pi R^3 \rho/3$ for M gives $F(r) = -4\pi G\rho mr/3$.

REMARK: $F(r)$ is a Hooke's-law force, because it is proportional to $-r$. The motion of an object hypothetically moving radially in the interior of a planet (inside a narrow tube) is therefore simple harmonic motion. (This holds even for motion along a non-radial chord.) Problems 11.19–11.21 are related to this fact.

11.4. Low-orbit and escape velocities

(a) Let the mass of the satellite be m. Since the radius of the satellite's circular orbit is essentially equal to the planet's radius R, the gravitational force on the satellite is GMm/R^2. This force must account for the centripetal acceleration v^2/R, so we have

$$F = ma \implies \frac{GMm}{R^2} = \frac{mv^2}{R} \implies v = \sqrt{\frac{GM}{R}}. \tag{11.20}$$

Using $g = GM/R^2$ from Eq. (11.3), this can also be written as $v = \sqrt{gR}$.

REMARK: The v in Eq. (11.20) implies that the angular frequency of the orbital motion is $\omega = v/R = \sqrt{GM/R^3}$. If we assume that the density ρ is uniform, then M is proportional to the volume, which is proportional to R^3. Hence ω is independent of R. That is, the orbit time around a small spherical asteroid is the same as the orbit time around a large planet (assuming the same density). The satellite's speed is much larger in the latter case; it is proportional to R.

(b) This is a task for conservation of energy (as opposed to $F = ma$ in part (a)). The object starts off with kinetic energy $mv^2/2$ and potential energy $-GMm/R$. And it ends up with essentially zero kinetic energy very far away (because the escape velocity corresponds to the case where the object barely makes it to infinity) and zero potential energy (because $-GMm/r = 0$ when $r = \infty$). Conservation of energy therefore gives

$$K_i + U_i = K_f + U_f$$

$$\implies \frac{mv^2}{2} + \left(-\frac{GMm}{R}\right) = 0 + 0$$

$$\implies v = \sqrt{\frac{2GM}{R}} = \sqrt{2gR}. \tag{11.21}$$

REMARK: The escape velocity in part (b) is $\sqrt{2}$ times the low-orbit speed in part (a). Equivalently, since $K = mv^2/2$, it takes twice as much energy to send an object to infinity as it does to send it into a low orbit. (However, this assumes unrealistically that the object is abruptly given a large initial speed and then allowed to move freely.) The rough numerical values of the two speeds for the case of the earth are

$$v_{\text{orbit}} = \sqrt{gR} \approx \sqrt{(10\,\text{m/s}^2)(6.4 \cdot 10^6\,\text{m})} = 8{,}000\,\text{m/s} = 8\,\text{km/s}, \tag{11.22}$$

$$v_{\text{escape}} = \sqrt{2gR} \approx \sqrt{2(10\,\text{m/s}^2)(6.4 \cdot 10^6\,\text{m})} = 11{,}300\,\text{m/s} = 11.3\,\text{km/s}.$$

For comparison, the speed of sound is about 340 m/s, and some high-powered rifles have a muzzle velocity of about 1000 m/s. The International Space Station orbits at about 250 miles above the surface of the earth (as did the Space Shuttle). Since this distance is small compared with the 4000-mile radius of the earth, the orbit radius r is only slightly larger than R, so the speed is only slightly smaller than the low-orbit speed of $\sqrt{GM/R} = 8\,\text{km/s}$. It is indeed slightly smaller, because all of the R's in Eq. (11.20) get replaced with r's. Note that we *cannot* simply replace R with r in the $\sqrt{gR}$ expression for the speed, because $g = GM/R^2$ has the earth's radius R built into it.

11.5. Kepler's 2nd law

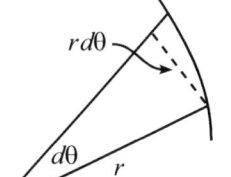

Figure 11.17

During a short period of time, the region swept out by the radius vector to the planet is essentially a very thin triangle, as shown in Fig. 11.17. (The slight curvature of the base is irrelevant if the triangle is very thin.) The small area dA of this triangle is $dA = r(r\,d\theta)/2$, because $r\,d\theta$ is the length of the base, and r is essentially equal to the altitude. Dividing both sides of this equation by dt, we have (using $L = mr v_{\tan} = mr(r\dot\theta) = mr^2\dot\theta$)

$$\frac{dA}{dt} = \frac{r^2\dot\theta}{2} = \frac{L}{2m}. \tag{11.23}$$

So a constant dA/dt (as claimed by Kepler's 2nd law) is equivalent to a constant L (that is, to conservation of angular momentum). And the angular momentum is indeed conserved, because the gravitational force always points toward the mass M, so it can never apply a torque around M.

11.6. Kepler's 3rd law for circles

If a mass m orbits in a circle of radius r around a fixed mass M, the radial $F = ma$ equation is (using $v = r\omega$)

$$F = ma \implies \frac{GmM}{r^2} = \frac{mv^2}{r} \implies \frac{GmM}{r^2} = mr\omega^2. \tag{11.24}$$

Since the angular frequency ω and the period T are related by $\omega = 2\pi/T$, we can rewrite Eq. (11.24) as

$$\frac{GM}{r^3} = \omega^2 \implies \frac{GM}{r^3} = \left(\frac{2\pi}{T}\right)^2 \implies T^2 = \frac{4\pi^2 r^3}{GM}. \tag{11.25}$$

This agrees with Eq. (11.6), because the semimajor axis a of a circle is simply the radius r.

11.7. Force from a line

In Fig. 11.18, the distance from the mass m to a little piece of the line is $r = \ell/\cos\theta$. In the magnified view, we therefore have the length $(\ell/\cos\theta)\,d\theta$ as shown, because that segment spans an angle $d\theta$ of a circle with radius $\ell/\cos\theta$ centered at the mass m.

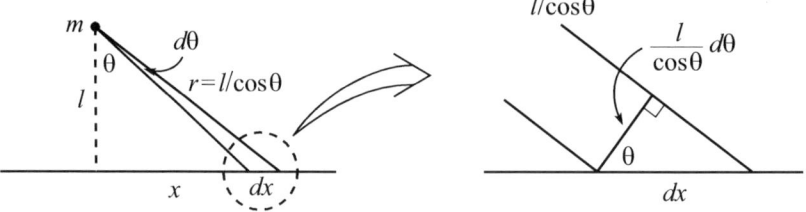

Figure 11.18

The length dx and mass dM of the little piece on the line are therefore

$$dx = \frac{1}{\cos\theta}\left(\frac{\ell\,d\theta}{\cos\theta}\right) \implies dM = \lambda\,dx = \frac{\lambda\ell\,d\theta}{\cos^2\theta}. \tag{11.26}$$

11.5. PROBLEM SOLUTIONS

The expression for dx could alternatively be derived in a much quicker manner by taking the differential of the relation $x = \ell \tan\theta$.

The horizontal components of the forces from all the little dM's in the line cancel in pairs, so we need only worry about the vertical components. This brings in a factor of $\cos\theta$. The magnitude of the (downward) vertical force due to a given dM is then

$$dF = \frac{Gm(dM)}{r^2}\cos\theta = \frac{Gm\left(\dfrac{\lambda\ell\,d\theta}{\cos^2\theta}\right)}{\left(\dfrac{\ell}{\cos\theta}\right)^2}\cos\theta = \frac{G\lambda m}{\ell}\cos\theta\,d\theta. \qquad (11.27)$$

The total downward force on m therefore has magnitude

$$F = \int_{-\pi/2}^{\pi/2} \frac{G\lambda m}{\ell}\cos\theta\,d\theta = \frac{G\lambda m}{\ell}\sin\theta\bigg|_{-\pi/2}^{\pi/2} = \frac{2G\lambda m}{\ell}. \qquad (11.28)$$

REMARKS: How much work does it take to drag the mass m to a position far away from the line? In terms of a general distance y from the line, the upward force that we must apply to m is $F(y) = 2G\lambda m/y$. Work is force times distance, so we need to calculate the integral $\int_\ell^\infty F(y)\,dy$. But $F(y) \propto 1/y$, so this integral diverges (like a log). The work required is therefore infinite. This should be contrasted with the finite amount of work, GmM/ℓ, in the case where the line is replaced by a point mass M.

If you want to solve this problem by working in terms of the position x along the line instead of the angle θ, then the distance from the mass m to a little piece of the line is $r = \sqrt{\ell^2 + x^2}$ (with $x = 0$ being at the foot of the perpendicular from m). And the little piece has mass $dM = \lambda\,dx$. The differential force in Eq. (11.27) is then

$$dF = \frac{Gm(dM)}{r^2}\cos\theta = \frac{Gm(\lambda\,dx)}{\ell^2 + x^2}\cdot\frac{\ell}{\sqrt{\ell^2 + x^2}}. \qquad (11.29)$$

To integrate this from $x = -\infty$ to $x = \infty$, you can either look up the integral, plug it into a computer, or use the suggested $x = \ell\tan\theta$ trig substitution.

11.8. Force from a plane

In Fig. 11.19, the horizontal line represents the plane, which extends into and out of the page (and also to the left and right). The short segment represents a rod extending into and out of the page, with small width dx. From Problem 11.7, the gravitational force from the rod on the mass m is $2G\lambda m/r$, where the effective linear mass density of the rod is $\lambda = \sigma\,dx$. This relation follows from the fact that the amount of mass in a length L of the rod (into and out of the page) can be written as both λL (by definition) and $\sigma L\,dx$, because $L\,dx$ is the area of the rod. Hence $\lambda = \sigma\,dx$.

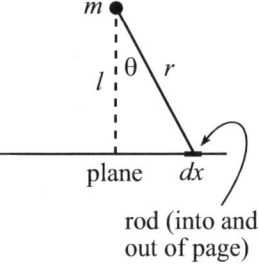

Figure 11.19

The horizontal component of the force from the rod on m cancels with the horizontal component of the force from the rod located symmetrically on the left side of m. So we care only about the vertical component (as expected). This brings in a factor of $\cos\theta$. And since $x = \ell\tan\theta$, we have $dx = \ell\,d\theta/\cos^2\theta$. (Or you can use the geometrical reasoning in the solution to Problem 11.7.) The total vertical force at m is therefore

$$\begin{aligned}F &= \int_{-\infty}^{\infty} \frac{2G(\sigma\,dx)m}{r}\cos\theta \\ &= \int_{-\pi/2}^{\pi/2} \frac{2G\sigma m(\ell\,d\theta/\cos^2\theta)}{\ell/\cos\theta}\cos\theta \\ &= 2G\sigma m\int_{-\pi/2}^{\pi/2} d\theta \\ &= 2\pi G\sigma m,\end{aligned} \qquad (11.30)$$

as desired. Note that this is independent of ℓ (more on this below).

REMARKS: As an exercise, you can also find the force from the plane by considering the plane to be built up from an infinite set of concentric rings, and integrating over the rings. You will first need to show that the force on a mass m located a distance ℓ from the center of a ring (along the axis of the ring) with linear mass density λ and radius r is $2\pi G m \lambda \ell r/(\ell^2 + r^2)^{3/2}$.

Note the following r dependence of the force due to objects with different dimensions. From Newton's universal law of gravitation, the force from a zero-dimensional point mass is proportional to $1/r^2$. From Problem 11.7, the force from an infinite one-dimensional line is proportional to $1/r$. And from the present problem, the force from an infinite two-dimensional plane is proportional to $1/r^0$ (that is, it is independent of r).

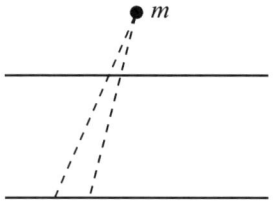

Figure 11.20

Is there a quick intuitive way of seeing why the force from a plane is independent of the distance from the plane? Indeed there is. Consider the two planes in Fig. 11.20, and imagine drawing a thin cone emanating from m. This cone defines a small patch on each plane. If a and b are the distances from m to the patches, then the areas of the patches are proportional to a^2 and b^2. These factors exactly cancel the effect of the r^2 in the denominator of the Gm_1m_2/r^2 force law (the patches are small enough to be treated like point masses). The two patches therefore produce equal forces on m. And since both planes can be covered by corresponding patches, the forces from the two planes are equal, as we wanted to show. (This reasoning is the same as the reasoning we gave in the remark in the solution to Multiple-Choice Question 11.6.) Physically, if you move farther away from a plane, the force on you from any given bit of mass in the plane is smaller. But there is now a larger area of the plane pulling you generally toward the plane. For example, in Fig. 11.21 the little piece of the plane shown applies a larger downward force on the top m than on the bottom m, because the angle to the bottom m is so shallow (and the distances aren't drastically different, at least for the little piece shown).

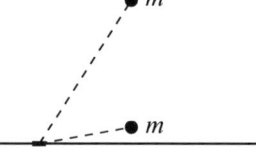

Figure 11.21

11.9. Two expressions for potential energy

From Eq. (11.5), the difference between the potential energy at height h above the ground (that is, at radius $R + h$) and the potential energy at the ground (that is, as radius R) is

$$U(R+h) - U(R) = -\frac{GMm}{R+h} - \left(-\frac{GMm}{R}\right), \qquad (11.31)$$

where M is the mass of the earth. This is the exact expression for the difference, but let's generate an approximate expression, valid in the $h \ll R$ limit. We have (using the $1/(1+\epsilon) \approx 1 - \epsilon$ approximation)

$$\begin{aligned} U(R+h) - U(R) &= GMm\left(\frac{1}{R} - \frac{1}{R+h}\right) = \frac{GMm}{R}\left(1 - \frac{1}{1+h/R}\right) \\ &\approx \frac{GMm}{R}\left(1 - \left(1 - \frac{h}{R}\right)\right) = m\left(\frac{GM}{R^2}\right)h = mgh, \end{aligned} \qquad (11.32)$$

where we have used the $g = GM/R^2$ relation from Eq. (11.3).

REMARK: We see that $U(h) = mgh$ is simply an approximation to the exact difference given in Eq. (11.31). Physically, the essence of the approximation is that if h is small compared with R, then the gravitational force is essentially constant over the interval between R and $R+h$. This then implies that the work done by the gravitational force over this interval can be obtained by simply multiplying the force mg by the distance h, as opposed to actually evaluating the integral $W = \int F\,dr$. Of course, we've basically gone in circles here. In Eq. (11.5) we integrated the gravitational force to obtain the potential energy, and then in Eq. (11.32) we effectively differentiated the potential energy (by taking the difference between two nearby values) to get back to the force (the GMm/R^2 part of the result).

11.10. Attracting particles

FIRST SOLUTION: Energy and momentum are both conserved during the motion. The initial values of E and p are both zero, so they are zero at all times. If v and V are the two

11.5. PROBLEM SOLUTIONS

speeds at a later time (with the velocities directed toward each other), then conservation of momentum tells us that the individual momenta must be equal and opposite at all times:

$$mv = MV. \tag{11.33}$$

And conservation of energy gives

$$\frac{1}{2}mv^2 + \frac{1}{2}MV^2 - \frac{GMm}{r} = 0. \tag{11.34}$$

That is, the loss in potential energy shows up as kinetic energy. Using Eq. (11.33) to write v in terms of V, we obtain

$$\frac{1}{2}m\left(\frac{MV}{m}\right)^2 + \frac{1}{2}MV^2 = \frac{GMm}{r}$$

$$\implies \left(\frac{M}{m} + 1\right)V^2 = \frac{2Gm}{r}$$

$$\implies V = m\sqrt{\frac{2G}{(M+m)r}}. \tag{11.35}$$

The speed of m is then $v = (M/m)V = M\sqrt{2G/(M+m)r}$. The relative speed of the two masses when they are a distance r apart is therefore

$$v + V = (M+m)\sqrt{\frac{2G}{(M+m)r}} = \sqrt{\frac{2G(M+m)}{r}}. \tag{11.36}$$

REMARK: Interestingly, this result depends only on the total mass $M + m$, and not on each mass separately. So if we have a given total mass, then no matter how we split it up into two masses, they will always have the same relative speed at a given separation. If one mass is much larger than the other, then it remains essentially at rest, so the relative speed comes solely from the motion of the smaller mass. If the masses are equal, then they each have half the speed of the smaller mass in the preceding case.

SECOND SOLUTION: Let's work in terms of the desired relative speed, call it u. Eq. (11.33) tells us that the speeds of the two masses are inversely proportional to the masses. (Equivalently, the given frame in which the masses are initially at rest is the CM frame, because the CM is initially, and hence always, at rest in this frame. And we know that in the CM frame the speeds of the masses are inversely proportional to the masses; this makes the total momentum be zero, as it must be in the CM frame.) You can then show that the individual speeds in the CM frame are

$$v_{\text{CM}} = \frac{Mu}{M+m} \quad \text{and} \quad V_{\text{CM}} = \frac{mu}{M+m}. \tag{11.37}$$

These speeds are inversely proportional to the masses and correctly add up to u. The conservation-of-energy statement in the CM frame (the lab frame) is therefore

$$0 = \frac{1}{2}m\left(\frac{Mu}{M+m}\right)^2 + \frac{1}{2}M\left(\frac{mu}{M+m}\right)^2 - \frac{GMm}{r}$$

$$\implies 0 = \frac{1}{2}\left(\frac{Mm}{M+m}\right)u^2 - \frac{GMm}{r}$$

$$\implies u = \sqrt{\frac{2G(M+m)}{r}}, \tag{11.38}$$

in agreement with the first solution. This solution was actually the same as the first one, with the only difference being that we chose to work from the start with the relative speed u.

REMARK: The equation in the second line above indicates that when computing the kinetic energy in the CM frame, we can pretend that we just have one particle with mass $Mm/(M+m)$ moving with the relative speed u of the actual two particles. This mass $Mm/(M+m)$ is called the *reduced mass* of the system. It comes up often in systems involving two masses. See, for example, Problem 10.11.

11.11. Sphere and stick

We will use conservation of energy. The sphere looks like a point mass from the outside, so we need to find the potential energy of a system consisting of a point mass m and a stick with mass m and length $2R$, initially a distance $3R$ away (with the far end a distance $5R$ away). The potential energy (due to the interaction with the point mass) of a little piece of the stick with mass $dm = \lambda\, dr$ (where the density λ equals $m/2R$) is $-Gm\,dm/r$, where r runs from $3R$ to $5R$. The total initial potential energy is therefore

$$U_i = -\int_{3R}^{5R} \frac{Gm(\lambda\, dr)}{r} = -Gm\left(\frac{m}{2R}\right)\ln r \Big|_{3R}^{5R}$$
$$= -\frac{Gm^2}{2R}\Big[\ln(5R) - \ln(3R)\Big] = -\frac{Gm^2}{2R}\ln\left(\frac{5}{3}\right). \tag{11.39}$$

At the end of the process, right before the stick and the sphere collide, the stick extends from $r = R$ to $r = 3R$. So to obtain the final potential energy U_f, the only modification we need to make in Eq. (11.39) is to replace $\ln(5/3)$ with $\ln(3/1)$. The change in potential energy is therefore

$$U_f - U_i = -\frac{Gm^2}{2R}\left[\ln\left(\frac{3}{1}\right) - \ln\left(\frac{5}{3}\right)\right] = -\frac{Gm^2}{2R}\ln\left(\frac{9}{5}\right). \tag{11.40}$$

This loss in potential energy shows up as kinetic energy. Since the masses are equal, conservation of momentum tells us that the final speeds are equal. So conservation of energy gives

$$2 \cdot \frac{mv^2}{2} = \frac{Gm^2}{2R}\ln\left(\frac{9}{5}\right) \implies v = \sqrt{\frac{Gm}{2R}\ln\left(\frac{9}{5}\right)}. \tag{11.41}$$

11.12. Dismantling a planet

Consider the planet at an intermediate stage of the dismantling, when it has radius r. What is the potential energy (due to the interaction with the rest of the planet) of a thin shell with thickness dr at the surface? The mass of the rest of the planet (which can be treated like a point mass) is $m_r = (4\pi r^3/3)\rho$,[1] and the mass of the thin shell is $dm = (4\pi r^2\, dr)\rho$, where $\rho = M/(4\pi R^3/3)$. The potential energy of the shell, due to the gravitational force from the rest of the planet, is therefore

$$dU = -\frac{Gm_r\, dm}{r} = -\frac{G(4\pi r^3 \rho/3)(4\pi r^2 \rho\, dr)}{r} = -\frac{16G\pi^2 \rho^2 r^4\, dr}{3}. \tag{11.42}$$

In order to move the shell out to infinity (with all of its bits of mass far away from each other), we must give it an energy of $|dU|$.[2] To successively remove shells of smaller and smaller radius r (from the initial radius R down to zero), we need to add up $|dU|$ for all of the thin shells that make up the whole sphere. That is, we need to integrate $|dU|$ from $r = 0$ to $r = R$. The total energy required (that is, the total work we must do) to dismantle the planet is therefore

$$E = \int |dU| = \frac{16G\pi^2 \rho^2}{3} \int_0^R r^4\, dr = \frac{16G\pi^2 \rho^2}{3} \frac{R^5}{5}. \tag{11.43}$$

---

[1] Technically the radius here should be $r - dr$, but this correction is negligible in the $dr \to 0$ limit.

[2] There is no need to worry about the internal potential energy of the thin shell (due to the gravitational interaction between different bits of the shell), because this energy is of order $(dm)^2$, which is a second-order small quantity and therefore negligible compared with the first-order small quantity $dU \propto dm$.

11.5. PROBLEM SOLUTIONS

We can eliminate ρ in favor of the total mass M by using $\rho = M/(4\pi R^3/3)$. This yields

$$E = \frac{16G\pi^2}{3}\left(\frac{3M}{4\pi R^3}\right)^2 \frac{R^5}{5} = \frac{3}{5}\frac{GM^2}{R}. \qquad (11.44)$$

From dimensional analysis, we know that the answer must be proportional to GM^2/R, because that is the only way to produce an energy from the given quantities M and R (along with G). But we need to do a calculation to show that the factor out front is 3/5.

REMARK: This is a picky point, but we'll address it anyway. You might say that since the removal of the shells starts at radius R and ends at radius zero, the limits of the integral in Eq. (11.43) should be reversed. But that can't be correct, because it would yield a negative result for E. The reason for this sign error is that we tacitly assumed that dr was a positive quantity in Eq. (11.42). If you want to treat dr as a negative quantity (which would automatically be the case if r runs from R down to zero), then the minus sign in Eq. (11.42) should be removed from dU (so that dU is still a negative quantity), in which case a minus sign should be added to the $|dU|$ in Eq. (11.43) (so that $|dU|$ is still a positive quantity). With the integration limits in Eq. (11.43) reversed, the final result for E will then correctly be positive. At every intermediate stage in a calculation, it's always good to do a quick check and make sure that at least the sign is correct.

11.13. Planet with variable density

(a) If we slice the planet into concentric shells with thickness dr, the mass of each shell is $dM = (4\pi r^2\, dr)\rho$. Since $\rho = b/r$, this equals $4\pi br\, dr$. The mass of the entire planet is obtained by integrating dM from $r = 0$ to $r = R$, so

$$M = \int dM = 4\pi b \int_0^R r\, dr = 2\pi b R^2 \implies b = \frac{M}{2\pi R^2}. \qquad (11.45)$$

The units of b are correct, because they make $\rho = b/r$ correctly have units of mass per length-cubed (that is, mass per volume).

(b) FIRST SOLUTION: We'll find the force on a mass m at radius r inside the planet, and then integrate this force to find the difference in potential energy. The gravitational force at radius r is determined by the mass m_r inside radius r, which is given by the same type of integral as in part (a):

$$m_r = 4\pi b \int_0^r r'\, dr' = 2\pi b r^2 = 2\pi\left(\frac{M}{2\pi R^2}\right)r^2 = M\frac{r^2}{R^2}. \qquad (11.46)$$

Note that if the density ρ were instead constant, then m_r would be proportional to volume, that is, to r^3. But the given density $\rho = b/r$ falls off with r, so m_r doesn't grow as quickly with r; the dependence in this case happens to be r^2.

Using the m_r from Eq. (11.46), the inward gravitational force on a mass m at radius r is

$$F(r) = \frac{Gm_r m}{r^2} = \frac{G(Mr^2/R^2)m}{r^2} = \frac{GMm}{R^2}. \qquad (11.47)$$

This is independent of r, so the magnitude of the radially inward force is the same everywhere inside the planet. $\rho(r) \propto 1/r$ is the special form of the density that creates this nice result. Since the force is independent of r, the potential energy difference ΔU, which in general is $\int F\, dr$, is simply the force times the distance here. Hence

$$\Delta U = FR = \left(\frac{GMm}{R^2}\right)R = \frac{GMm}{R}, \qquad (11.48)$$

with the energy at the surface being larger than the energy at the center. This result looks similar to the standard expression for the gravitational potential energy due to a point mass or a sphere. So we see that it takes GMm/R energy to bring the mass m

from the center to the surface, and then another GMm/R to bring it from the surface out to infinity.

SECOND SOLUTION: In this solution we'll find the potential energy of a mass m at the center of the planet, relative to infinity. And then we'll subtract off the potential energy at the surface, relative to infinity.

The potential energy U of a mass m at the center of the planet equals to the sum of the dU's at the center due to all the thin concentric spherical shells that make up the planet. Now, the dU at the center of a thin shell equals the dU at the surface, because there is no gravitational force inside the shell and hence no change in potential energy inside. And the dU at the surface of a shell of radius r is simply $-G(dM)m/r$, where dM is the mass of the shell. As in part (a), this dM equals $4\pi br\,dr$. So the dU at the center of a shell is $-G(4\pi br\,dr)m/r = -4\pi Gmb\,dr$. Note that this is independent of r. Integrating dU over all of the thin shells gives the total U at the center of the planet as (using the b from part (a))

$$U_{\text{center}} = -\int_0^R 4\pi Gmb\,dr = -4\pi GmbR$$
$$= -4\pi Gm\left(\frac{M}{2\pi R^2}\right)R = -\frac{2GMm}{R}. \qquad (11.49)$$

But the U at the surface of the planet is the standard $U_{\text{surface}} = -GMm/R$. So we see that U_{surface} is GMm/R larger than U_{center}, in agreement with the first solution.

11.14. Descending in a mine shaft

(a) (This part of the problem is just Problem 11.3 again.) We know that the gravitational force inside a hollow spherical shell is zero. So all of the mass of the earth outside radius r, which can be thought of as built up from shells, produces zero force on a mass m at radius r. Only the mass inside radius r matters, and it can be treated like a point mass at the center. This mass is $M_r = (4\pi r^3/3)\rho$, so the radially inward gravitational force on m has magnitude

$$F = \frac{GM_r m}{r^2} = \frac{G(4\pi r^3 \rho/3)m}{r^2} = \frac{4\pi G\rho mr}{3}. \qquad (11.50)$$

This decreases as r decreases, so the gravitational force decreases as you descend in a mine shaft in a planet with uniform mass density.

(b) We want to find the condition under which the force $F = GM_r m/r^2$ increases as r decreases, or equivalently, decreases as r increases. So we want to find the condition under which

$$\frac{d}{dr}\left(\frac{GM_r m}{r^2}\right) < 0 \implies \frac{d}{dr}\left(\frac{M_r}{r^2}\right) < 0$$
$$\implies r^2 \frac{dM_r}{dr} - M_r(2r) < 0, \qquad (11.51)$$

where we have ignored the denominator of the derivative quotient rule, because we're only comparing the result with zero. Since the mine shaft is located within the crust of the earth, we have $dM_r/dr = 4\pi R^2 \rho_c$, where R is the radius of the earth. This is true because if we travel outward a distance r, an additional shell with volume $(4\pi R^2)\,dr$ is now located inside our radial position. Technically, the R here should be replaced with our radial position r, but r is essentially equal to R since we're close to the surface of the earth. Any correction would involve a multiplicative factor that is essentially equal to 1.

The mass M_r is essentially the mass of the entire earth, so it equals $(4\pi R^3/3)\rho_{\text{avg}}$. Again, the issue of whether we use the entire radius of the earth R, or a slightly

11.5. PROBLEM SOLUTIONS

smaller r, is irrelevant. So Eq. (11.51) becomes (letting the other factors of r there also be equal to R)

$$R^2(4\pi R^2 \rho_c) - \left(\frac{4}{3}\pi R^3 \rho_{avg}\right)(2R) < 0 \implies \rho_c < \frac{2}{3}\rho_{avg}. \quad (11.52)$$

So the desired value of A is $2/3$.

REMARK: It makes sense that depending on the ratio of the two densities ρ_c and ρ_{avg}, it is possible for the force to either increase or decrease as you descend in a mine shaft. This can be seen by looking at two extreme cases. In the limit where the planet consists of a thin massive shell surrounding a styrofoam interior (so that $\rho_c \gg \rho_{avg}$), the gravitational force decreases to zero by the time you descend all the way to the inner surface of the shell. So the force definitely decreases as you descend. The decreasing mass M_r in $F = GM_r m/r^2$ is where all of the change in F comes from in this limit.

In the opposite limit where the planet consists of a styrofoam shell surrounding a massive interior (so that $\rho_c \ll \rho_{avg}$), the force definitely increases as you descend, because you are getting closer to the massive solid sphere (which behaves like a point mass at the center). The decreasing amount of styrofoam inside your radius r is inconsequential, so the decreasing r in $F = GM_r m/r^2$ is where all of the change in F comes from in this limit.

11.15. Orbiting objects

(a) Let the mass of the satellite be m. The gravitational force on the satellite is responsible for its centripetal acceleration, so the radial $F = ma$ equation is (writing the centripetal acceleration v^2/R as $R\omega^2$)

$$\frac{GMm}{R^2} = mR\omega^2 \implies \omega = \sqrt{\frac{GM}{R^3}}. \quad (11.53)$$

Since G has units of $m^3/(kg\,s^2)$, the units of ω are correctly s^{-1}.

(b) The gravitational force on each planet is $G(M/2)^2/R^2$, and both planets move in the circle of radius $R/2$ shown in Fig. 11.22. So the radial $F = ma$ equation for one of the masses is

$$\frac{G(M/2)^2}{R^2} = \left(\frac{M}{2}\right)\left(\frac{R}{2}\right)\omega^2 \implies \omega = \sqrt{\frac{GM}{R^3}}, \quad (11.54)$$

which is the same result as in part (a). Note that the full R (the distance between the planets) appears in the gravitational force, whereas $R/2$ (the radius of the circle) appears in the centripetal acceleration.

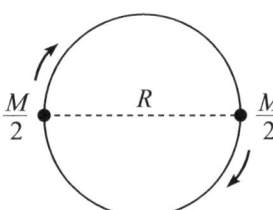

Figure 11.22

(c) Label the planets as A and B, and let them have masses a and b, where $a + b$ is constrained to be a given mass M. The distances from A and B to the CM are, respectively, $bR/(a + b)$ and $aR/(a + b)$. These two distances correctly add up to R and are inversely proportional to the masses. Planet A therefore moves in a circle of radius $bR/(a + b)$, as shown in Fig. 11.23. So the radial $F = ma$ equation for A is (the equation for B would give the same result)

$$\frac{Gab}{R^2} = a\left(\frac{bR}{a+b}\right)\omega^2 \implies \omega = \sqrt{\frac{G(a+b)}{R^3}} = \sqrt{\frac{GM}{R^3}}, \quad (11.55)$$

as desired. The masses a and b cancel, and only the sum $a + b = M$ remains.

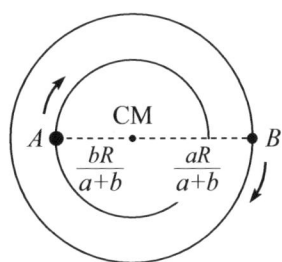

Figure 11.23

11.16. Grabbing a mass

(a) Each mass travels in a circle of radius $\ell/2$, so the centripetal acceleration is $v_0^2/(\ell/2)$. The gravitational force on each mass due to the other mass is Gm^2/ℓ^2. So the $F = ma$ equation for each mass is

$$\frac{Gm^2}{\ell^2} = \frac{mv_0^2}{\ell/2} \implies v_0 = \sqrt{\frac{Gm}{2\ell}}. \quad (11.56)$$

(b) After one mass is grabbed, the other mass will orbit around it in a noncircular orbit. During this motion, both the energy and the angular momentum of the orbiting mass are conserved. (The gravitational force from the other mass produces no torque around itself.) Also, the velocity is tangential at closest approach, because if it had a radial component, this would contradict the fact that the mass is at closest approach. (The mass would either be getting closer, or it would have been closer a moment earlier.)

Let the distance at closest approach be r, and let v be the speed there. Then conservation of angular momentum (around the grabbed mass) gives

$$\ell v_0 = rv \implies v = \frac{\ell}{r} v_0. \tag{11.57}$$

Conservation of energy then gives

$$\frac{1}{2} m v_0^2 - \frac{Gm^2}{\ell} = \frac{1}{2} m v^2 - \frac{Gm^2}{r}$$

$$\implies \frac{1}{2} v_0^2 - \frac{Gm}{\ell} = \frac{1}{2} \left(\frac{\ell}{r} v_0\right)^2 - \frac{Gm}{r}$$

$$\implies Gm \left(\frac{1}{r} - \frac{1}{\ell}\right) = \frac{1}{2} v_0^2 \left(\frac{\ell^2}{r^2} - 1\right)$$

$$\implies 2Gm \left(\frac{\ell - r}{\ell r}\right) = v_0^2 \left(\frac{\ell^2 - r^2}{r^2}\right)$$

$$\implies \frac{2Gmr}{\ell(\ell + r)} = v_0^2. \tag{11.58}$$

But we know from part (a) that $v_0 = \sqrt{Gm/2\ell}$. So we have

$$\frac{2Gmr}{\ell(\ell + r)} = \frac{Gm}{2\ell} \implies 4r = \ell + r \implies r = \frac{\ell}{3}. \tag{11.59}$$

REMARK: It makes sense that this distance of closest approach is smaller than ℓ, because you can quickly show that the orbiting mass would need a speed of $\sqrt{Gm/\ell}$ to orbit in a circle of radius ℓ around the grabbed mass. Since the initial speed of $v_0 = \sqrt{Gm/2\ell}$ is smaller than this, the orbiting mass will fall somewhat inward toward the grabbed mass and therefore achieve a separation smaller than ℓ.

11.17. Orbiting satellite

(a) The potential energy at radius R is

$$U = -\frac{GMm}{R}. \tag{11.60}$$

The radial $F = ma$ equation is $GMm/R^2 = mv^2/R$, which gives $v^2 = GM/R$. So the kinetic energy is

$$K = \frac{mv^2}{2} = \frac{GMm}{2R}. \tag{11.61}$$

The total energy is therefore

$$E = K + U = -\frac{GMm}{2R}. \tag{11.62}$$

This is negative, which means that work would have to be done on the object to move it far from the planet, out to infinity.

11.5. PROBLEM SOLUTIONS

(b) Let the speeds at the leftmost and rightmost points in the orbit in Fig. 11.9 be v_1 (the speed right after the kick) and v_n. The velocities are tangential at these points. Both energy and angular momentum are conserved throughout the motion. (The gravitational force from the planet produces no torque around itself.) Conservation of angular momentum gives

$$mv_1 R = mv_n(nR) \implies v_n = \frac{v_1}{n}. \tag{11.63}$$

Conservation of energy then gives

$$\frac{mv_1^2}{2} - \frac{GMm}{R} = \frac{mv_n^2}{2} - \frac{GMm}{nR}$$

$$\implies \frac{v_1^2}{2} - \frac{GM}{R} = \frac{1}{2}\left(\frac{v_1}{n}\right)^2 - \frac{GM}{nR}$$

$$\implies \frac{1}{2}v_1^2\left(1 - \frac{1}{n^2}\right) = \frac{GM}{R}\left(1 - \frac{1}{n}\right)$$

$$\implies \frac{1}{2}v_1^2\left(1 + \frac{1}{n}\right) = \frac{GM}{R}$$

$$\implies \frac{mv_1^2}{2} = \frac{n}{n+1}\frac{GMm}{R}. \tag{11.64}$$

The total energy is therefore

$$E = \frac{mv_1^2}{2} - \frac{GMm}{R} = \frac{n}{n+1}\frac{GMm}{R} - \frac{GMm}{R} = -\frac{GMm}{(n+1)R}. \tag{11.65}$$

LIMITS: If $n \to 1$ then $E \to -GMm/2R$, in agreement with the result in part (a). Note, interestingly, that setting $n = 1$ too soon in the conservation-of-energy statement would just give the trivial statement that $0 = 0$. We need to wait to take the $n \to 1$ limit if we want to obtain something useful.

If $n \to \infty$ then $E \to 0$. This makes sense because the satellite is moving very slowly when it is at the very large radius of nR (because $v_n = v_1/n$), so at this distant point we have $K + U = 0 + 0$.

If $n \to 0$ then $E \to -GMm/R$. This corresponds to the satellite being dropped essentially radially inward to the planet, starting from radius R. This is true because the kinetic energy K is essentially zero at the start, since the last line of Eq. (11.64) tells us that $v_1 \to 0$ when $n \to 0$. (So the required kick is an opposing one that essentially stops the mass, at least briefly.) The initial energy therefore comes only from the initial potential energy of $-GMm/R$. The satellite just "bounces" in and out in a very thin elliptical orbit.

11.18. **Projectile in a planet**

(a) The mass of the planet is $M = (4\pi R^3/3)\rho$, so the $F = ma$ equation for the circular motion is

$$\frac{G(4\pi R^3 \rho/3)m}{R^2} = \frac{mv_{\text{circ}}^2}{R} \implies v_{\text{circ}} = \sqrt{\frac{4\pi G\rho R^2}{3}}. \tag{11.66}$$

The units of G are $m^3/(kg\, s^2)$, and the units of ρ are kg/m^3. So v correctly has units of m/s. This problem is just Problem 11.4(a) in terms of ρ instead of M.

(b) Only the mass that is inside radius r (call it M_r) contributes to the force at r. So we have

$$F(r) = -\frac{GM_r m}{r^2} = -\frac{G(4\pi r^3 \rho/3)m}{r^2} = -\frac{4\pi G\rho m r}{3}, \tag{11.67}$$

where the minus sign indicates radially inward. The potential energy, relative to the center, is the negative of the integral of $F(r)$. So

$$U(r) = -\int_0^r \frac{-4\pi G\rho m r'}{3} dr' = \frac{2\pi G\rho m r^2}{3}. \tag{11.68}$$

This is correctly positive; the potential energy at radius r is larger than the potential energy at the center. You can check that the units of U are correctly kg m^2/s^2.

(c) We'll apply conservation of energy and angular momentum between the initial point on the surface and the point with the minimum value of r. Using the form of the potential energy in Eq. (11.68), conservation of energy gives

$$U_i + K_i = U_f + K_f$$
$$\implies \frac{2\pi G\rho m R^2}{3} + \frac{mv_i^2}{2} = \frac{2\pi G\rho m r^2}{3} + \frac{mv_f^2}{2}. \qquad (11.69)$$

Angular momentum around the center of the planet is conserved because the gravitational force provides no torque around the center. At the minimum value of r in the motion, the velocity is tangential because any radial motion would contradict the fact that r is the minimum radius. So conservation of angular momentum gives

$$mRv_i = mrv_f \implies v_f = \frac{R}{r}v_i. \qquad (11.70)$$

Plugging this value of v_f into Eq. (11.69) and rearranging yields

$$\frac{2\pi G\rho}{3}(R^2 - r^2) = \frac{v_i^2}{2}\left(\frac{R^2}{r^2} - 1\right)$$
$$\implies \frac{2\pi G\rho r^2}{3} = \frac{v_i^2}{2}. \qquad (11.71)$$

This relation gives the minimum radius r in terms of a general (tangential) initial speed v_i. We will now invoke the given information that $v_i = (1/2)v_{\text{circ}}$, which yields $v_i^2 = (1/4)v_{\text{circ}}^2$. Using the value of v_{circ} from part (a), Eq. (11.71) becomes

$$\frac{2\pi G\rho r^2}{3} = \frac{1}{2} \cdot \frac{1}{4} \frac{4\pi G\rho R^2}{3}$$
$$\implies r^2 = \frac{R^2}{4} \implies r = \frac{R}{2}. \qquad (11.72)$$

REMARK: If we more generally have $v_i = nv_{\text{circ}}$ instead of $v_i = (1/2)v_{\text{circ}}$, where n is a numerical factor less than 1, then the $1/4$ in the first line of Eq. (11.72) becomes n^2, which means that the minimum value of r is $r = nR$. If $n = 1$ then we just have the original circular orbit at the surface of the planet. And if $n \approx 0$ then the object starts at rest and falls in to the center of the planet (and then pops back out, in a very thin elliptical orbit).

11.19. **Oscillating stick in a planet 1**

(a) (This part of the problem is just Problem 11.3 again.) The force at radius r inside the planet is due to the mass inside radius r. Since mass is proportional to volume, this mass is $M_r = M(r^3/R^3)$. The radially inward force on a mass dm at radius r therefore has magnitude

$$F_r = \frac{GM_r\,dm}{r^2} = \frac{G(Mr^3/R^3)\,dm}{r^2} = \frac{GM\,dm\,r}{R^3}. \qquad (11.73)$$

(b) F_r points radially, so if we add up the forces on all the little dm's that make up the stick, the sum of the components parallel to the stick is zero (they cancel in pairs). We therefore need only look at the components perpendicular to the stick. If x is the displacement of the stick relative to the center of the planet (so the initial value of x is h), then the perpendicular component brings in a factor of x/r. The perpendicular force therefore has magnitude

$$\frac{GM\,dm\,r}{R^3} \cdot \frac{x}{r} = \frac{GM\,dm\,x}{R^3}. \qquad (11.74)$$

This is independent of the position along the stick, so when we integrate over the whole stick, the dm simply turns into m. The total force on the stick is therefore

$$F_x = -\frac{GMmx}{R^3}, \quad (11.75)$$

where the minus sign indicates that the force is pulling the stick toward the center of the planet. The $F_x = ma_x$ equation for the stick is then

$$-\frac{GMmx}{R^3} = m\ddot{x} \implies \ddot{x} = -\left(\frac{GM}{R^3}\right)x. \quad (11.76)$$

This is a simple-harmonic-oscillator equation, and the frequency is $\omega = \sqrt{GM/R^3}$. Since G has units of $m^3/(kg\,s^2)$, the units of ω are correctly s^{-1}.

REMARKS: Since $g = GM/R^2$ from Eq. (11.3), this frequency can alternatively be written as $\omega = \sqrt{g/R}$, which is as nice a result as we could have hoped for. In terms of the (uniform) density ρ of the planet, the mass is $M = (4\pi R^3/3)\rho$, so $\omega = \sqrt{GM/R^3}$ can also be written as $\omega = \sqrt{4\pi G\rho/3}$, which is independent of R. This makes sense; the force isn't affected if we increase the size of the planet by adding on shells, because a shell produces zero force in its interior.

The frequency of $\omega = \sqrt{GM/R^3}$ shows up in many different setups involving planets. It is the frequency of a low-orbit satellite; see Problem 11.4(a). And related to this, it is also the maximum frequency of a planet's rotation before it flies apart. It is also the frequency of the oscillations of a mass dropped through a diametrical tube in a planet; see Problem 11.21(a). This must be the case, because the result of this problem is independent of the length of the stick. So we could make the stick be very short, in which case we would essentially have a point mass falling through a tube.

The preceding result for a mass dropped through a tube actually holds more generally. Even if the tube lies along an arbitrary chord instead of a diameter, the frequency is still $\omega = \sqrt{GM/R^3} = \sqrt{g/R}$. You can demonstrate this as an exercise; the calculation is very similar to the one in this problem. The period of the oscillation is $2\pi/\omega = 2\pi\sqrt{R/g}$, which you can show equals about 5000 seconds in the case of the earth, or about 84 minutes (assuming, incorrectly, that the earth has uniform density). In other words, if you drop something through a straight frictionless tube from one point on the earth to *any* other point, the time to reach that point will always be 42 minutes, whether it is on the other side of the earth, or only a mile away, or even just at the other end of a (perfectly frictionless!) table, provided that the perpendicular bisector of the table points exactly toward the center of the earth.

11.20. Oscillating stick in a planet 2

(a) The part of the stick marked as A (the part within radius $R - x$) in Fig. 11.24 experiences zero net force, by symmetry. The unbalanced force comes from parts B and C. Let's find these forces.

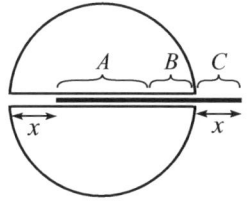

Figure 11.24

Force on C: The gravitational force on a mass m that lies outside the planet is simply $-GMm/r^2$, where the minus sign indicates radially inward. So the force on a little piece of the stick with mass $dm = \lambda\,dr$ is $-GM(\lambda\,dr)/r^2$. Integrating this from R to $R + x$ gives the force on C as

$$F_C = -\int_R^{R+x} \frac{GM\lambda\,dr}{r^2} = \left.\frac{GM\lambda}{r}\right|_R^{R+x}$$
$$= GM\lambda\left(\frac{1}{R+x} - \frac{1}{R}\right) = -\frac{GM\lambda x}{R(R+x)}. \quad (11.77)$$

Force on B: From Problem 11.3, the force on a mass m at radius r inside the planet is $-GMmr/R^3$. Replacing m with $dm = \lambda\,dr$ and integrating from $R - x$ to x gives the force on B as

$$F_B = -\int_{R-x}^{R} \frac{GM\lambda r\, dr}{R^3} = -\frac{GM\lambda}{R^3}\frac{r^2}{2}\bigg|_{R-x}^{R}$$
$$= -\frac{GM\lambda}{2R^3}(R^2 - (R-x)^2) = -\frac{GM\lambda}{2R^3}(2Rx - x^2). \qquad (11.78)$$

The total force on the stick is

$$F_B + F_C = -\frac{GM\lambda}{2R^3}(2Rx - x^2) - \frac{GM\lambda x}{R(R+x)}. \qquad (11.79)$$

(b) If $x \ll R$, then in F_B in Eq. (11.78) we can ignore the x^2 compared with the $2Rx$. And in F_C in Eq. (11.77) we can ignore the x compared with the R in the denominator. So we have

$$F_B + F_C \approx -\frac{GM\lambda x}{R^2} - \frac{GM\lambda x}{R^2} = -\frac{GM}{R^2}(\lambda \cdot 2x). \qquad (11.80)$$

This makes sense, because the gravitational force on a mass m near the surface of the planet is GMm/R^2. (If the mass is slightly inside or outside the planet, the force is slightly different from this, but the correction is negligible if x is small.) And the mass of the B+C part of the stick is $m = \lambda(2x)$, which can be treated like a point mass if x is small.

(c) The radial $F = ma$ equation for the stick is

$$F = m_{\text{stick}}a \implies -\frac{GM(\lambda \cdot 2x)}{R^2} = (\lambda \cdot 2R)\ddot{x} \implies \ddot{x} = -\frac{GM}{R^3}x, \qquad (11.81)$$

where we have used the fact that the *total* mass of the stick is $m_{\text{stick}} = \lambda(2R)$. Eq. (11.81) is a simple-harmonic-oscillator equation, and the frequency of small oscillations is $\omega = \sqrt{GM/R^3}$. This is the same as the frequency in Problem 11.19.

REMARKS: As in Problem 11.19, the frequency can be written as as $\omega = \sqrt{4\pi G\rho/3}$, which is independent of R. The reason for this independence in the present setup is that the $F = ma$ equation for the stick is $m_{B+C}\, g = m_{\text{stick}}a$, and both g (as you can check) and m_{stick} grow linearly with R, so these effects cancel.

As an exercise, you can show more generally that $\sqrt{GM/R^3}$ is the frequency of small oscillations in the analogous setup where the tube lies along an arbitrary chord through the earth, not necessarily a diameter.

11.21. **Asteroid slab**

(a) From Problem 11.3, the force on a mass m at radius r inside the planet, when written in terms of the density ρ, is $-4\pi G\rho m r/3$. This force is directed radially inward, so the radial $F = ma$ equation yields simple harmonic motion:

$$-\frac{4\pi G\rho m}{3}r = m\ddot{r} \implies \ddot{r} = -\frac{4\pi G\rho}{3}r \implies \omega = \sqrt{\frac{4\pi G\rho}{3}}. \qquad (11.82)$$

This frequency is the same as the frequency we found in the preceding two problems.

REMARK: The frequency is independent of the radius of the sphere. Equivalently, it is independent of the radius r_0 where the object is released, because the mass outside r_0 is effectively not present (since a hollow shell produces zero force inside), so we effectively just have a smaller sphere with radius r_0.

(b) Consider a thin sheet (that is, a thin sub-slab of the given slab) with area A and tiny thickness ϵ. The volume of this sheet is $A\epsilon$, so the mass is $\rho(A\epsilon)$, by the definition of ρ. But the mass is also $A\sigma$, by the definition of σ. Equating these two expressions for the mass gives the mass per unit *area* of the thin sheet as $\sigma = \rho\epsilon$.

Consider the force on a mass m located *outside* (or on the surface of) a wide slab with thickness ℓ. We can think of this slab as built up from many thin sheets. And since Problem 11.8 tells us that the force from a sheet is independent of the distance from it, we can imagine squashing the slab down to a thin sheet. The slab has thickness ℓ, so the reasoning in the previous paragraph tells us that the mass per unit area is $\rho\ell$, which then implies that the surface mass density σ of the squashed sheet is $\rho\ell$. From Problem 11.8, the force due to this sheet is therefore $2\pi(\rho\ell)Gm$. And this is also the force due to the slab with thickness ℓ.

When the object is located a distance x from the center plane of the slab, the forces from the two sub-slabs on either side of it (with thicknesses $D/2 + x$ and $D/2 - x$) point in opposite directions. So the net force on the object is

$$-2\pi\rho\left(\frac{D}{2} + x\right)Gm + 2\pi\rho\left(\frac{D}{2} - x\right)Gm = -(4\pi\rho Gm)x. \quad (11.83)$$

The $F = ma$ equation in the x direction then yields simple harmonic motion:

$$-(4\pi Gm\rho)x = m\ddot{x} \implies \ddot{x} = -(4\pi G\rho)x \implies \omega = \sqrt{4\pi G\rho}. \quad (11.84)$$

REMARK: This frequency is independent of the thickness of the slab. Equivalently, it is independent of the x_0 where the object is released, because the mass outside the initial $|x_0|$ value is effectively not present (since the two outer slabs in Fig. 11.25 produce equal and opposite forces on the mass m), so we effectively just have a thinner slab with thickness $2x_0$. This logic actually provides another way of obtaining the force in Eq. (11.83). Since we effectively have only the middle slab with thickness $2x$ in Fig. 11.25 (which is equivalent to a sheet with $\sigma = \rho \cdot 2x$), Problem 11.8 gives the force as $-2\pi(\rho \cdot 2x)Gm = -4\pi\rho Gmx$.

(c) The frequency in the slab is larger than the frequency in the sphere, by a factor of $\sqrt{3}$. The physical reason for this is that for a given displacement, the restoring force is larger in the case of the slab, because more matter is pulling the object back toward equilibrium. In more detail: we know that in the case of a sphere, only the mass inside radius r matters. Similarly, in the case of a slab, only the mass inside the range from $-x$ to x matters. Therefore, since there is more mass in a slab with thickness $2x$ than in a sphere with diameter $2x$, the restoring force from the slab is larger.

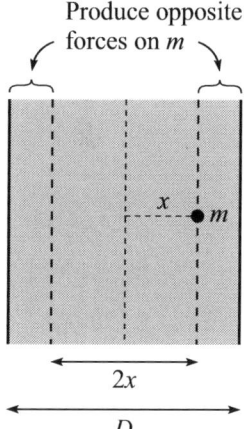

Figure 11.25

Chapter 12

Fictitious forces

12.1 Introduction

The concept of fictitious forces

Newton's second law, $\mathbf{F} = m\mathbf{a}$, holds only in inertial frames. It does *not* hold in accelerating frames. Is there a way to modify $\mathbf{F} = m\mathbf{a}$ so that it holds in an accelerating frame, for example, a merry-go-round or an accelerating train? (Here $\mathbf{a}$ is the acceleration with respect to the accelerating frame.) The answer is "Yes," provided that we introduce some new *fictitious forces* in the accelerating frame.

It turns out that there are four different types of fictitious forces. They are called the *translational, centrifugal, Coriolis*, and *azimuthal* forces. When these are added on to whatever real forces exist (gravity, friction, normal, etc., which are the same in any (nonrelativistic) frame, inertial or not), then the $\mathbf{F} = m\mathbf{a}$ equation becomes valid in the accelerating frame. We won't go through the derivation of the fictitious forces, but see Section 10.1 in Morin (2008) if you are interested; the derivation is a bit involved. We'll simply state the result here.

To specify an accelerating frame, we must state two things: (1) the position of the origin, $\mathbf{R}(t)$, with respect to a given inertial frame, and (2) the angular velocity $\omega(t)$ with respect to the given inertial frame.[1] It can then be shown that the acceleration $\mathbf{a}$ in the accelerating frame is given by

$$m\mathbf{a} = \mathbf{F}_{\text{real}} - m\frac{d^2\mathbf{R}}{dt^2} - m\omega \times (\omega \times \mathbf{r}) - 2m\omega \times \mathbf{v} - m\frac{d\omega}{dt} \times \mathbf{r}$$
$$\equiv \mathbf{F}_{\text{real}} + \mathbf{F}_{\text{translational}} + \mathbf{F}_{\text{centrifugal}} + \mathbf{F}_{\text{Coriolis}} + \mathbf{F}_{\text{azimuthal}}. \qquad (12.1)$$

Note that the minus signs are included in the definitions of the fictitious forces. There are two types of quantities in Eq. (12.1). The quantities $\mathbf{r}$, $\mathbf{v} \equiv d\mathbf{r}/dt$, and $\mathbf{a} \equiv d^2\mathbf{r}/dt^2$ are all measured internally with respect to the accelerating frame. In contrast, the quantities $\mathbf{R}$ and ω are properties of the accelerating frame as a whole. Someone living in the frame would need to be told what $\mathbf{R}$ and ω are.

The whole point of all this is that if you are in an accelerating frame, and if you add up all the terms on the right-hand side of Eq. (12.1) to obtain the total force (real plus fictitious), then you can divide by m to obtain the acceleration $\mathbf{a}$, just as you would do with $\mathbf{F} = m\mathbf{a}$ in an inertial frame.

---

[1] The angular velocity *vector* ω is defined to be the vector whose magnitude equals the angular speed ω and whose direction is along the axis of rotation, with the orientation determined by the right-hand rule: if you curl your fingers in the direction of the rotation, your thumb will point in the direction of ω.

12.1. INTRODUCTION

The various fictitious forces

Let's look at each of the four fictitious forces in turn.

- $\mathbf{F}_{\text{translational}} = -m d^2\mathbf{R}/dt^2$

  This fictitious force is intuitive. If you are standing on a train that is accelerating rightward with acceleration $d\mathbf{R}^2/dt^2$ (see Fig. 12.1), then you feel like you are getting flung leftward (consistent with the minus sign in the definition of the translational force). If you don't brace yourself, you will indeed move leftward with respect to the train. Of course, from the perspective of someone on the ground, if the floor of the train is frictionless then you will simply remain at rest while the train accelerates rightward. The magnitude of the translational force can be written more concisely as

  $$F_{\text{translational}} = m a_{\text{frame}}, \qquad (12.2)$$

  where $\mathbf{a}_{\text{frame}} \equiv d^2\mathbf{R}/dt^2$ is the acceleration of the origin of the accelerating frame (which is the acceleration of all points in the frame in the case of a non-rotating object like a train).

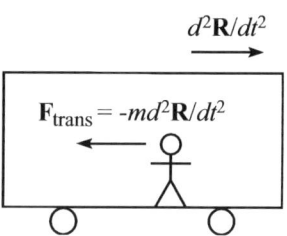

Figure 12.1

- $\mathbf{F}_{\text{centrifugal}} = -m\boldsymbol{\omega} \times (\boldsymbol{\omega} \times \mathbf{r})$

  This fictitious force is also intuitive. Consider the case of the rotating carousel in Fig. 12.2, where the counterclockwise nature of ω implies that the ω vector points out of the page. At the location marked by the dot, $\boldsymbol{\omega} \times \mathbf{r}$ points upward in the plane of the page, with magnitude ωr. Hence $\boldsymbol{\omega} \times (\boldsymbol{\omega} \times \mathbf{r})$ points leftward with magnitude $\omega^2 r$, which means that $\mathbf{F}_{\text{centrifugal}}$ points rightward, due to the minus sign in the definition. In other words, the centrifugal force points radially outward with magnitude

  $$F_{\text{centrifugal}} = m\omega^2 r. \qquad (12.3)$$

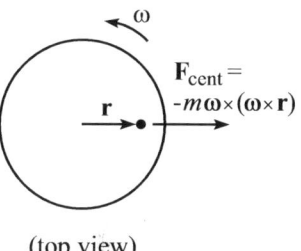

(top view)

Figure 12.2

  The outward direction makes sense; you feel like you are getting flung outward. And the magnitude also makes sense; see Problem 12.2.

  There is rarely any need to use the full definition of $\mathbf{F}_{\text{centrifugal}}$ with the two cross products. The $m\omega^2 r$ form is usually quite sufficient. And indeed, it is always sufficient, even in 3-D cases (such as the rotating earth), if r is taken to be the distance from the axis of rotation.[2] The centrifugal force always points away from the axis.

- $\mathbf{F}_{\text{Coriolis}} = -2m\boldsymbol{\omega} \times \mathbf{v}$

  This fictitious force is less intuitive, because it involves the velocity $\mathbf{v}$ in the accelerating frame. Rounding a corner in a car (in which the centrifugal force can be quite appreciable) is a common experience, whereas moving appreciably with respect to the car (so that $\mathbf{v}$ and the Coriolis force are appreciable) isn't so common. However, if you happen to have access to a carousel and spend some time walking around on it at a decent clip, the Coriolis force will become more intuitive.

  In the case where $\mathbf{v}$ is perpendicular to ω (as it is when walking on a carousel), the magnitude of the Coriolis force is

  $$F_{\text{Coriolis}} = 2m\omega v. \qquad (12.4)$$

  The direction is determined by the right-hand rule in the cross product $\boldsymbol{\omega} \times \mathbf{v}$; the result will always be perpendicular to $\mathbf{v}$. Depending on the direction of rotation of the carousel (that is, whether ω points vertically upward or downward), the force always points to your right, no matter which way you walk; or it always points to your left, no matter which way you walk. See Fig. 12.3 for the case where ω points out of the page and $\mathbf{F}_{\text{cor}}$ always points to your right.

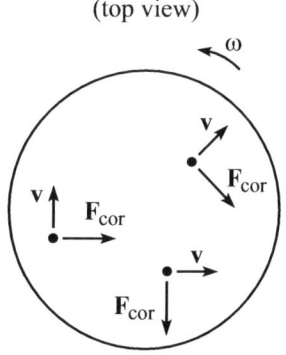

(top view)

$\mathbf{F}_{\text{cor}} = -2m\boldsymbol{\omega} \times \mathbf{v}$

Figure 12.3

---

[2]In the 3-D case, the $\mathbf{r}$ vector from a given origin to a particular point may have a component along the axis of rotation. This component is irrelevant in the centrifugal force; it is correctly eliminated when taking the cross product $\boldsymbol{\omega} \times \mathbf{r}$.

- $\mathbf{F}_{\text{azimuthal}} = -m(d\omega/dt) \times \mathbf{r}$

Although this force looks a little complicated since it involves $d\omega/dt$, it is actually fairly intuitive, at least in a common special case. The vector ω can change in two basic ways. It can change because its length changes (that is, the speed of rotation changes), or because its direction changes (that is, the axis of rotation changes), or both. The latter of these cases gets a bit complicated, so let's deal only with the former (which is more common anyway), where the rotation simply either speeds up or slows down. In this case, this azimuthal force is quite intuitive. It's really no different from the translational force. If you are standing on a carousel and the rate of rotation increases, then you feel like you are getting flung backwards tangentially. If you're enclosed in a box and can't see outside, then (ignoring the centrifugal force) it's just like you're on a train that accelerates in the tangential direction; you get flung backwards in the accelerating frame. The magnitude of the azimuthal force is

$$F_{\text{tangential}} = m\dot\omega r = m\ddot\theta r = ma_{\text{tangential}}, \tag{12.5}$$

in agreement with the form of the translational force.

A frequent error involving fictitious forces is to mix up which frame you're solving a problem in, and to solve it partially in one frame (say, a rotating frame where a centrifugal force exists) and partially in another (say, an inertial frame where a centrifugal force doesn't exist). So be careful to pick one frame and stay there. If you want to solve the problem again by picking another frame and staying in that one, then by all means do so. But keep the calculations separate.

Remember that fictitious forces have nothing whatsoever to do with inertial frames. Equivalently, they are all zero in any inertial frame. So if you ever find yourself using the word "centrifugal" when solving a problem in a nonrotating frame, you know that something is amiss.

12.2 Multiple-choice questions

12.1. It is possible to experience "zero gravity" on board an airplane if the plane travels up and down with the appropriate $v(t)$ along a path that looks roughly like the one shown below. Which region of the path causes the passengers to experience a feeling of "weightlessness"?

(a) A: where the path is approximately a straight line with negative slope

(b) B: the bottom part of the dips

(c) C: where the path is approximately a straight line with positive slope

(d) D: the top part of the bumps

(e) E: the entire path

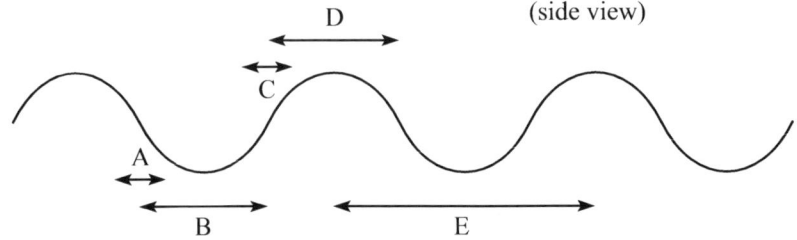

12.2. MULTIPLE-CHOICE QUESTIONS

12.2. A pendulum hangs from the ceiling of an elevator. Which of the following scenarios yields the largest frequency of oscillations?

(a) The elevator accelerates upward at 5 m/s^2.

(b) The elevator accelerates downward at 5 m/s^2.

(c) The elevator moves upward at constant speed 5 m/s.

(d) The elevator moves downward at constant speed 5 m/s.

12.3. You are riding on a subway train and having trouble standing because the train is bouncing around. If x is the position of the train, then a nonzero value of which of the following quantities is the most relevant to your difficulty in standing? (We'll deal with just one dimension here, although it's often the train's sideways motion that makes it difficult to stand.)

(a) x (b) $\dot{x}$ (c) $\ddot{x}$ (d) $\dddot{x}$

12.4. You stand on a scale at the equator and record the reading. If the earth then hypothetically stopped spinning but (doubly hypothetically) kept its same shape, the reading on the scale would

(a) increase   (b) decrease   (c) remain the same

12.5. You are holding a two-minute hourglass that happens to have one minute of sand in each half. You would like to move all of the sand from one half to the other in less than one minute. Which fictitious force can you most easily use to your advantage?

(a) translational

(b) centrifugal

(c) Coriolis

(d) azimuthal

(e) There is no possible way to move all of the sand from one half to the other in less than one minute.

12.6. A race car is driving around a circular track. Consider the following statement: "$F = ma$ tells us that the centrifugal force of $m\omega^2 r$ causes the centripetal acceleration of $\omega^2 r$." This statement is true in

(a) the ground frame

(b) the car frame

(c) both frames

(d) neither frame

12.7. A race car is driving around a circular track. Fill in the blanks: In the ground frame the _____ force(s) cause(s) an acceleration equal to _____ , whereas in the car frame the _____ force(s) cause(s) an acceleration equal to _____ .

(a) friction and centrifugal, $\omega^2 r$, friction and centrifugal, 0

(b) friction and centrifugal, $\omega^2 r$, centrifugal, $\omega^2 r$

(c) friction, $\omega^2 r$, friction and centrifugal, 0

(d) friction, 0, friction and centrifugal, $\omega^2 r$

(e) friction, $\omega^2 r$, centrifugal, 0

12.8. A mass on the end of a string is given the proper initial velocity so that it swings around in a horizontal circle. Which of the following diagrams correctly indicates the forces acting on the mass in the reference frame rotating along with the pendulum? (The forces are *not* drawn to scale.)

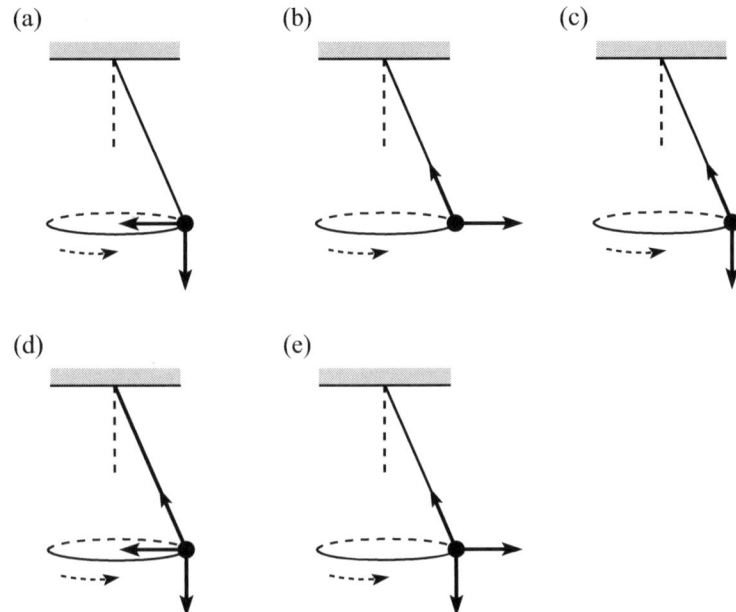

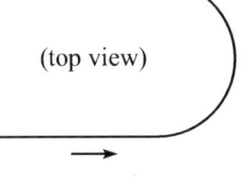

Figure 12.4

12.9. You drive with constant speed along a road that begins straight and then becomes part of a circle, as shown in Fig. 12.4. When you enter the circular part,

(a) you don't feel anything different as time goes on

(b) you initially don't feel anything different, but then you gradually feel like you are getting pulled more and more toward the right side of the car

(c) you abruptly feel like you are getting pulled, by a constant amount, toward the right side of the car

(d) you abruptly feel like you are getting pulled, by an amount that increases with time, toward the right side of the car

12.10. A carousel sits on the bed of a truck. The truck accelerates to the right with acceleration a, and the carousel rotates with constant angular speed ω, as shown below. An object is at rest with respect to the carousel at one of the points shown. Assuming that a and ω are related in a well-chosen way, at which of the five points can the sum of the fictitious forces (in the frame of the carousel) on the object be zero?

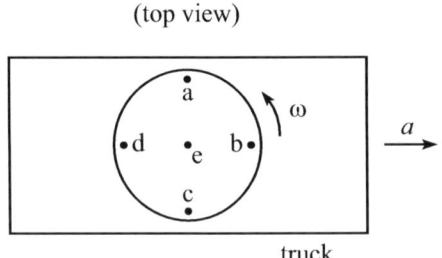

12.11. A futuristic space station might take the form of a large cylinder that rotates around its axis. In the rotating frame, the centrifugal force at locations on the inner surface of the cylinder will feel like an artificial gravity force. If the cylinder has a radius of 50 m, what should the period of revolution be if the goal is to mimic earth's gravity, $g \approx 10\,\text{m/s}^2$?

(a) 0.07 s   (b) 2.2 s   (c) 5 s   (d) 14 s   (e) 31 s

12.3. PROBLEMS

12.12. A person walks inward along a radial line painted on a rotating carousel. In the rotating frame, the Coriolis force points in the

(a) inward radial direction

(b) outward radial direction

(c) forward tangential direction

(d) backward tangential direction

(e) There is no Coriolis force.

12.13. A skater spins around with her arms outstretched. She then draws her arms in, which causes her angular speed to increase. At a given instant, her rotating frame coincides with a frame S rotating with constant speed. In frame S, which fictitious force produces the torque that causes her to angularly accelerate out of S?

(a) translational   (b) centrifugal   (c) Coriolis   (d) azimuthal

12.14. You are walking around in an arbitrary manner on a large rotating platform. At a given instant, under which of the following conditions is the Coriolis force on you much smaller than the centrifugal force (in magnitude)?

(a) The angular velocity of the platform is small.

(b) You are at a small radius on the platform.

(c) The radial component of your velocity (with respect to the platform) is small.

(d) The tangential component of your velocity (with respect to the platform) is small.

(e) Your speed with respect to the platform is small compared with the speed of the platform (at your location) with respect to the inertial frame of the ground.

12.3 Problems

12.1. Only one force

For each of the four fictitious forces (translational, centrifugal, Coriolis, azimuthal) give a scenario in which at a given instant, only that one fictitious force is nonzero.

12.2. Standing on a carousel

A person stands at rest with respect to a carousel, a distance r from the center. The carousel rotates with angular speed ω. Write down the horizontal $F = ma$ equations in both the ground frame and the carousel frame, and compare the two equations.

12.3. Circular pendulum

Consider a "circular pendulum" with mass m and length ℓ, as shown in Fig. 12.5. The mass swings around in a horizontal circle with the (massless) string always making an angle θ with the vertical. Find mass's angular speed ω by drawing the free-body diagram for the mass and writing down the $F = ma$ equations in the vertical and horizontal directions. Do this in (a) the lab frame, and (b) the rotating frame of the pendulum.

12.4. Circling stick

A uniform stick with mass m and length ℓ is attached by a pivot to a ceiling. Initial conditions have been set up so that the stick rotates around the vertical axis, making a constant angle θ with respect to the vertical, as shown in Fig. 12.6. Find the angular speed ω by considering the setup in the reference frame that rotates along with the stick (at frequency ω) around the vertical axis.

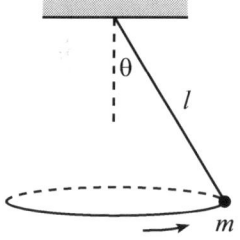

Figure 12.5

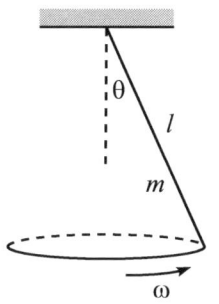

Figure 12.6

12.5. Swirling down a drain

Is the swirling of water as it goes down a drain caused by the Coriolis force arising from the earth's rotation? Said in another way, does the water swirl in different directions in the northern and southern hemispheres? A rough order-of-magnitude calculation should give you the answer.

12.6. Walking radially on a carousel

(a) A radial line is painted on a carousel that rotates with constant angular frequency ω. If you walk inward along this radial line with constant speed v, a tangential friction force must act on you at your feet (in addition to a radial friction force). By considering which fictitious force this tangential friction force opposes, find the value of the friction force.

(b) Calculate the rate of change of your angular momentum, as measured with respect to the ground frame. Show that this dL/dt is exactly accounted for by the torque due to the friction force you found in part (a).

12.7. Walking tangentially on a carousel

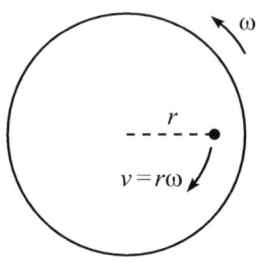

Figure 12.7

A carousel rotates counterclockwise with constant angular frequency ω with respect to the ground. A person runs clockwise in a circle of radius r (centered at the center of the carousel) with speed ωr with respect to *the ground*; see Fig. 12.7. (So the person's speed with respect to the carousel is $2\omega r$.) Find the required friction force by drawing the free-body diagram (for the horizontal forces) for the person and writing down the radial $F = ma$ statement. Do this in (a) the ground frame, and (b) the carousel frame.

12.8. Bead on a rod

Here is a problem that is much easier to solve in a rotating frame than in the lab frame: A frictionless rod is pivoted at one end and rotates around in a horizontal plane with a constant angular frequency ω. A bead on the rod starts out at rest with respect to the rod, at a distance r_0 from the pivot.

(a) At a general later time, draw the free-body diagram on the bead, in the rotating frame of the rod.

(b) What is $r(t)$? *Hint*: Write down the radial $F = ma$ equation, and then guess an exponential solution of the form $r(t) = Ae^{\alpha t}$. You can solve for α and then apply the initial conditions.

(c) What is the horizontal normal force from the rod on the bead, as a function of time?

12.4 Multiple-choice answers

12.1. $\boxed{d}$ By "weightless" we mean that the total force (both real and fictitious) on you in the accelerating reference frame of the plane is zero, just as it would be if you were floating in deep space. The gravitational force always points downward, so if the total force on you is to be zero in the frame of the plane, then we need a fictitious force pointing *upward*. This force is the translational force. It points upward with magnitude mg if the plane accelerates *downward* with acceleration g, due to the minus sign in $\mathbf{F}_{\text{trans}} = -md^2\mathbf{R}/dt^2$. Choice D is the only part of the path for which it is possible for the acceleration to point exactly downward. (In A and C, any nonzero acceleration will point along the diagonal straight line. And in B, the acceleration will necessarily have an upward v^2/R component at the bottom of the dip.)

REMARK: We can also answer this question by working in the inertial reference frame of the ground and not mentioning anything about fictitious forces. Imagine shooting a projectile through the air, and imagine enclosing the projectile in a box and moving the box along with the projectile, so that the projectile always stays in the middle of the box (and hence doesn't touch it). Due to air resistance, it will take some effort to move the box along the idealized parabolic trajectory with an acceleration of exactly g downward. But inside the box, the projectile won't feel any air resistance, so it will take the idealized parabolic trajectory. This is just the scenario we want: You are the projectile, and the plane is the box. You will float freely inside the plane, with no need for any normal force to keep you in the plane. This is exactly what it would feel like if you were floating in the middle of a (non-accelerating) box in deep space.

12.2. $\boxed{a}$ Since the elevator is accelerating upward in choice (a), there is a translational fictitious force directed downward. This gets added to the downward gravitational force mg, making the total downward force larger. The pendulum therefore effectively lives in a world where $g_{\text{eff}} = g + 5 \text{ m/s}^2 \approx 15 \text{ m/s}^2$. Since the pendulum's frequency is proportional to $\sqrt{g_{\text{eff}}/\ell}$ by dimensional analysis, and since g_{eff} is largest for choice (a), this choice has the largest frequency.

REMARKS: Choices (c) and (d) have the same frequency as in the case of a stationary elevator. The constant speed doesn't change the frequency; only the acceleration matters. Choice (b) has the smallest frequency among the given options. In the special case where the elevator accelerates downward at $g \approx 10 \text{ m/s}^2$, the frequency is zero. This makes sense because the elevator is in freefall, so the pendulum just floats at rest (if it was initially at rest) with respect to the elevator. If the elevator accelerates downward faster than g, then the pendulum will experience a net force *upward* in the accelerating frame, so the pendulum string will go limp as the mass "falls" upward and hits the ceiling.

12.3. $\boxed{d}$ The position x certainly doesn't matter; the train could be in New York or it could be in Boston, and the ride would feel the same (local idiosyncracies aside). Nor does the velocity $\dot{x}$ matter; if you are enclosed in a windowless train, then $\dot{x}$ can take on any constant value, and you will never be able to tell what that value is.

In contrast, the acceleration $\ddot{x}$ is noticeable. You will have to brace yourself, or perhaps lean, to avoid falling over. A constant $\ddot{x}$ means that due to the translational fictitious force, you effectively live in a world in which gravity points downward at an angle; we encountered this earlier in Multiple-Choice Question 4.6. (This is all consistent with Einstein's "principle of equivalence" in General Relativity, which states that locally you can't tell the difference between acceleration and gravity.) However, if $\ddot{x}$ is constant then this effective gravitational force is also constant. So as long as you initially brace yourself properly, you can continue doing the same thing as time goes on, and you will have a peaceful ride. In short, the acceleration $\ddot{x}$ is noticeable, but not annoying (assuming that it isn't too large).

The third derivative $\dddot{x}$ is what gets you. If $\dddot{x}$ is nonzero, then the effective gravitational force in your reference frame is changing, so you have to keep changing how you brace yourself. And if this change is unpredictable (which means that some higher derivatives are nonzero, so technically those derivatives are the ones that get you), then you will have trouble standing; imagine suddenly slamming on the brakes. The third derivative $\dddot{x}$ is called the "jerk," an appropriate name considering that the train is getting jerked around.

12.4. $\boxed{a}$ In the case of the spinning earth, let's look at things in the rotating frame of the earth. The forces on you are the inward gravitational, outward centrifugal, and outward normal from the scale; this normal force is what the scale reads. You aren't accelerating in the rotating frame of the earth, because you are standing at rest on the scale. So the total force on you must be zero. In terms of the magnitudes of the three forces, we therefore have

$$-F_{\text{grav}} + F_{\text{cent}} + F_N = 0 \implies F_N = F_{\text{grav}} - F_{\text{cent}}. \quad (12.6)$$

We see that the normal force (that is, the reading on the scale) is smaller than the gravitational force. But the gravitational force is what the normal force from the scale equals in the case where the earth isn't spinning, because then the F_{cent} term doesn't exist in Eq. (12.6). Putting it all together, we have $F_N^{\text{no spin}} = F_{\text{grav}} > F_N^{\text{spin}}$. So the answer is (a).

Note that our assumption that the earth keeps the same shape (which it actually wouldn't do) allows us to say that F_{grav} is the same in both scenarios.

Alternatively, we can answer this question by working in a background inertial frame, without making any reference to fictitious forces. In the case of the spinning earth, you *are* accelerating, because you are traveling around in a circle as the earth rotates with respect to the given inertial frame. The $F = ma$ equation for your circular motion is

$$F_{\text{grav}} - F_N = \frac{mv^2}{R}. \tag{12.7}$$

This tells us that F_{grav} is larger than F_N (which is the reading on the scale). But F_{grav} is what the scale would read if the earth stopped spinning. So the answer is (a).

REMARK: The difference in the readings of the scales in the two scenarios is $F_{\text{cent}} = mR\omega^2$ (or mv^2/R, which is the same thing). The ratio of this difference to mg is $R\omega^2/g$, which you can show is about 0.3%. This is a small but measurable correction. For more discussion of this, see page 463 in Morin (2008).

12.5. $\boxed{\text{b}}$ If you swing your arm around in a circular windmill motion, holding the hourglass such that it points along the direction of your arm, then in the rotating frame of your arm, the centrifugal force will push the sand from the inside half to the outside half. (For fast motion, the centrifugal force dominates the gravitational force.) Try it. It's easy to cut the time in half. Just be sure to hold the hourglass tightly, lest there be an additional and less desirable effect of the centrifugal force!

REMARKS: Let's make a rough estimate of the size of the effect. If you make one revolution of your arm per second, then $\omega = 2\pi \, \text{s}^{-1}$. If your arm is a meter long, then $\omega^2 r = (2\pi)^2 \, \text{m/s}^2 \approx 40 \, \text{m/s}^2 \approx 4g$. This means that at the bottom of the circular path, the hourglass feels a centrifugal force of $m\omega^2 r = 4mg$ downward. When this is added to the downward gravitational force, the hourglass effectively lives in a world where $g_{\text{eff}} = 5g$. By similar reasoning, $g_{\text{eff}} = 3g$ (upward) at the top of the circular path. These numbers imply that even though we can't assume that the sand's flow rate should be strictly proportional to g_{eff}, it is certainly reasonable to expect that the rate should be significantly larger than in the normal "g" case where the hourglass just sits on a table.

You can also make the sand flow faster by using the translational force. If you accelerate the hourglass upward, then the downward translational force makes g_{eff} be larger than g. So the sand will flow faster. However, since you're holding the hourglass, you can obviously accelerate it upward for only so long. The overall effect will therefore be small. The important advantage of the centrifugal force is that because the motion is circular instead of linear, you can continue it indefinitely.

12.6. $\boxed{\text{d}}$ Since the word "centrifugal" is involved, we can't be talking about the ground frame, because there are no fictitious forces in an inertial frame. So the answer must be (b) or (d). And similarly, since the acceleration $\omega^2 r$ is involved, we can't be talking about the car frame, because there is no acceleration in the car frame (the car is at rest in its frame, by definition). So the answer must be (a) or (d). The correct answer is therefore (d).

REMARK: In short, the given statement mixes apples and oranges. In the ground frame there is no centrifugal force, but there is a nonzero acceleration; whereas in the car frame there is a centrifugal force, but there is zero acceleration.

12.7. $\boxed{\text{c}}$ This question is similar to the preceding one. In the ground frame the only (horizontal) force is the friction between the tires and the ground, and the acceleration is $v^2/r = \omega^2 r$. In the accelerating frame of the car there is also the centrifugal force (in addition to the friction force; a real force is present in every frame), and the acceleration is zero because the car is at rest in the car frame.

12.4. MULTIPLE-CHOICE ANSWERS

12.8. $\boxed{e}$ The forces are tension, gravity, and centrifugal, with the centrifugal force pointing outward.

REMARK: The net force (both horizontal and vertical) must be zero in the rotating frame, because the mass is at rest in that frame. Choice (e) is the only one for which the forces have any possibility of adding up to zero. If we drew the forces to scale, then assuming that the tension force stays as it is, the gravitational force would have to be a hair shorter, so that it balances the vertical component of the tension. And the centrifugal force would have to be cut roughly in half, so that it balances the horizontal component of the tension.

12.9. $\boxed{c}$ There is no acceleration on the straight part of the road. Your acceleration on the circular part takes on the constant value of v^2/r. So assuming that you remain at rest with respect to the car, you feel a horizontal force (perhaps friction from your seat) that starts at zero and then abruptly jumps to the constant value of mv^2/r. Equivalently, in the reference frame of the car, you suddenly feel a centrifugal force with the constant value of $m\omega^2 r = mv^2/r$ pointing outward (rightward). This is balanced by the friction force, assuming that you remain at rest in the car.

REMARK: Note that even though the circular arc joins smoothly (with no kink; the slope is continuous) with the straight part of the road, the acceleration does indeed have a discontinuity. This is due to the fact that the acceleration is related to the curvature of the road, which involves the *second* derivative of the path, as opposed to the first derivative. And the second derivative in our setup has a discontinuity. (For another example, imagine joining the negative y axis with the right half of the parabola $y = ax^2$. The derivative jumps abruptly from zero to $2a$.) In actual roads that you drive on, a sudden jerking motion is avoided by having the curvature change gradually. The desired shape happens to be a so-called "Cornu spiral." Entrance and exit ramps on highways take this general shape.

12.10. $\boxed{b}$ The two fictitious forces are the translational and centrifugal. The translational force points to the left, because the truck is accelerating to the right. So we need the centrifugal force to point to the right, if the two forces are to cancel. The centrifugal force always points radially outward, so we need radially outward to be the same as rightward. This is the case at point (b).

REMARK: What this means physically is that if you are holding onto the carousel as you rotate around, and if you release your grip when you are at point (b), then you won't accelerate away from this point (at least briefly). To see how this works when viewed in the reference frame of the ground, consider for simplicity (although this assumption isn't necessary) the case where the truck starts accelerating from rest (but is accelerating). Then after you release your grip, you travel in a straight line, upward in the given figure. If the truck weren't accelerating, a dot painted on the carousel at point (b) would curve to the left as it rotates around in a circle. But since the truck *is* accelerating, the whole carousel, including the dot, accelerates to the right. This keeps the dot from curving left, and it stays right below you (if a is chosen to be $\omega^2 r$), at least briefly. In other words, you don't accelerate away from the dot.

12.11. $\boxed{d}$ The centrifugal force is $\omega^2 r$, so we want

$$\omega^2 r = g \implies \omega^2 = \frac{g}{r} = \frac{10 \text{ m/s}^2}{50 \text{ m}} \implies \omega = 0.45 \text{ s}^{-1}. \quad (12.8)$$

The period is then $T = 2\pi/\omega = 14$ s.

REMARK: This result of 14 s is longer (that is, the cylinder is spinning slower) than most people might guess. (However, the speed of a point on the cylinder is $\omega r \approx 22$ m/s ≈ 50 mph, which is reasonably brisk). Even if the radius were only 10 m, the period would be about 6 s, which is still fairly large. And a radius of 1 m (which is far too small for a space station, of course) would require a period of 2 s. Given that your arm is about a meter long, this last case implies (as you can verify) that if you want to spin a bucket of water in a vertical circle at roughly constant speed (with your rotating arm being the radius), and if you want the water to stay inside the bucket even at the top of the circle where the bucket is upside down, then the period of the motion should be no more than 2 s, which again is probably a longer period than you'd think. See Problem 4.19.

12.12. $\boxed{\text{c}}$ If the carousel lies in the plane of the page and rotates counterclockwise, then ω points out of the page, by the right-hand rule. Since **v** points radially inward, the Coriolis force $\mathbf{F}_{\text{cor}} = -2m\boldsymbol{\omega} \times \mathbf{v}$ points in the forward tangential direction, as you can check with the right-hand rule (don't forget the minus sign). If the carousel instead rotates clockwise, then the Coriolis force points in the opposite direction in space, compared with what we found above (since ω now points into the page). But since the rotation is now opposite too, the force still points in the *forward* tangential direction. So the "forward tangential" result is independent of the direction of rotation.

REMARKS: In the carousel frame there is also a radially outward centrifugal force, of course. So the *total* fictitious force (Coriolis plus centrifugal) points in a diagonal direction between tangentially forward and radially outward. The exact angle depends on the ratio of $2m\omega v$ to $m\omega^2 r$.

The above "forward tangential" result implies that if you stand on a rotating carousel and aim a ball at the center and throw it, it will drift away from the target in the forward tangential direction. This outcome is quite obvious when observed in the inertial reference frame of the ground: When you throw the ball, it has a nonzero tangential velocity (the same tangential velocity that you have), in addition to the radial velocity you give it. So it certainly won't end up at the center. It takes the same path taken by a ball thrown diagonally (radially inward and tangentially forward) by a stationary observer on the ground.

12.13. $\boxed{\text{c}}$ For the radially inward motion of the arms, you can quickly check that the Coriolis force, $-2m\boldsymbol{\omega} \times \mathbf{v}$, points in the forward tangential direction (as in the preceding question). This force produces the desired torque.

REMARK: The correct answer (c) can also be obtained by eliminating the other choices. The translational force is zero because the frame S is only rotating, the azimuthal force is zero because S has constant ω, and the centrifugal force points radially so it can't provide a torque.

If you want to work instead with the angularly accelerating frame S' of the skater, then the net torque must be zero (because the skater remains in S', by definition). This zero net torque comes about because the Coriolis torque is exactly canceled by the azimuthal torque. As an exercise, you can verify that this works out quantitatively for a point mass by setting the derivative of the (constant) angular momentum $mr^2\omega$ (as measured in the lab frame) equal to zero.

12.14. $\boxed{\text{e}}$ The Coriolis force has magnitude $2m\omega v$, and the centrifugal force has magnitude $m\omega^2 r$. The former is much smaller than the latter if

$$2m\omega v \ll m\omega^2 r \implies v \ll \frac{\omega r}{2}. \tag{12.9}$$

But ωr is simply the (tangential) speed of the platform at your location, due to the platform's rotation. This factor of 2 isn't critical since we're just making a rough statement. So the answer is (e).

REMARK: Note that none of the other answers can possibly be correct, because they all involve saying that a *dimensionful* quantity is large or small. Such a statement is meaningless. Is a meter large or small? It is large on an atomic scale, but small on a planetary scale. It makes sense only to say either (1) that a *dimensionless* number is large or small, where it is understood that we mean in comparison with the number 1, or (2) that one dimensionful quantity is large or small compared with another quantity with the *same* dimensions. This is the type of statement in choice (e). These two types of statements are actually equivalent, because in (2) we're really just saying that the ratio of the two quantities is large or small compared with 1.

12.5 Problem solutions

12.1. Only one force

TRANSLATIONAL: A linearly accelerating frame has no ω or $d\omega/dt$, so only the translational force is nonzero. For the remaining three cases, we'll assume that the frame has no overall linear acceleration, so that the translational force is always zero.

CENTRIFUGAL: If you stand at rest at a nonzero **r** on a carousel rotating with constant ω, then your velocity **v** is zero (which makes the Coriolis force zero), and $d\omega/dt$ is zero (which makes the azimuthal force zero).

CORIOLIS: If you walk with nonzero velocity **v** through the origin of a rotating carousel, then since **r** = 0, both the centrifugal and azimuthal forces are zero.

AZIMUTHAL: If you stand at rest at a nonzero **r** on a carousel that has nonzero angular acceleration (so $d\omega/dt \neq 0$) but instantaneously has zero angular velocity, then your velocity **v** is zero (which makes the Coriolis force zero), and ω is zero (which makes the centrifugal force zero).

REMARK: Which of the above scenarios can go on for an extended period of time? The translational and centrifugal scenarios certainly can, because you're simply standing at rest in a frame that has constant $d^2\mathbf{R}/dt^2$ or ω. But the Coriolis scenario works only at one instant (when you pass through the origin where **r** = 0), because $\mathbf{v} \equiv d\mathbf{r}/dt$ must be nonzero, so a split second later you will have a nonzero **r**, which makes the centrifugal force nonzero. Similarly, the azimuthal scenario works only at one instant (when $\omega = 0$), because $d\omega/dt$ must be nonzero, so a split second later ω will be nonzero, which makes the centrifugal force nonzero.

12.2. Standing on a carousel

In the ground frame, the only horizontal force on the person is the friction force at her feet. The person is traveling in a circle, so her acceleration is the standard v^2/r inward centripetal acceleration. The $F = ma$ equation in the ground frame is therefore

$$F_{\text{friction}} = \frac{mv^2}{r}. \tag{12.10}$$

The friction force points radially inward.

In the rotating carousel frame, the centrifugal force has magnitude $mr\omega^2$ from Eq. (12.3), and it points radially outward. The person is at rest, so there is zero acceleration. The friction force is the same as it is in the ground frame, so the $F = ma$ equation in the carousel frame is (with radially inward taken to be positive)

$$F_{\text{friction}} - F_{\text{cent}} = 0 \implies F_{\text{friction}} - mr\omega^2 = 0. \tag{12.11}$$

This is the same equation as in Eq. (12.10), because the mv^2/r term there can be written as $mr\omega^2$, since $v = \omega r$. The only difference in the two $F = ma$ equations is that one term has been moved from the right-hand side to the left-hand side.

REMARK: Although the two $F = ma$ equations differ trivially in a mathematical sense, they differ significantly in a physical sense: In the ground frame, the friction force produces the centripetal acceleration. In the rotating frame, the friction force balances the centrifugal force, in order to produce zero acceleration. To repeat: in the ground frame there is no fictitious force but a nonzero acceleration, whereas in the carousel frame there is a fictitious force but no acceleration.

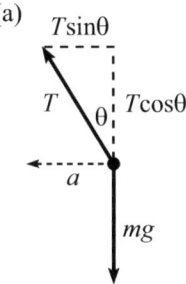

12.3. Circular pendulum

(a) The free-body diagram in the lab frame is shown in Fig. 12.8(a). The mass moves in a circle in this frame, and there are two forces acting on it: tension and gravity. The two $F = ma$ equations are (taking leftward to be positive for the x equation)

$$F_x = ma_x \implies T \sin\theta = \frac{mv^2}{r} = mr\omega^2 = m(\ell \sin\theta)\omega^2,$$
$$F_y = ma_y \implies T \cos\theta - mg = 0 \implies T = \frac{mg}{\cos\theta}, \qquad (12.12)$$

where we have used the fact that the radius of the circular motion is $\ell \sin\theta$. Plugging the T from the F_y equation into the F_x equation gives

$$\left(\frac{mg}{\cos\theta}\right)\sin\theta = m(\ell \sin\theta)\omega^2 \implies \omega = \sqrt{\frac{g}{\ell \cos\theta}}. \qquad (12.13)$$

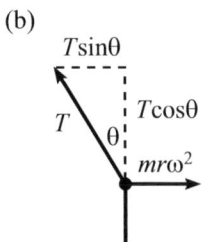

(b) The free-body diagram in the rotating frame is shown in Fig. 12.8(b). The mass is at rest in this frame, so there is no acceleration. In addition to the tension and gravity forces, we now have the centrifugal force. The sum of the forces must be zero, so the two $F = ma$ equations are (again taking leftward to be positive for the x equation, although this doesn't matter, since we're setting the force equal to zero)

$$F_x = ma_x \implies T \sin\theta - mr\omega^2 = 0 \implies T \sin\theta = m(\ell \sin\theta)\omega^2,$$
$$F_y = ma_y \implies T \cos\theta - mg = 0 \implies T = \frac{mg}{\cos\theta}. \qquad (12.14)$$

Figure 12.8

These are exactly the same equations as in Eq. (12.12), so we obtain the same result for ω. The only difference in the initial statements of the $F_x = ma_x$ equations in the two frames is that the acceleration term in the lab frame turns into a fictitious force term in the accelerating frame and appears on the other side of the equation.

12.4. Circling stick

In the rotating frame of the stick, the stick is at rest, so this problem reduces to a statics problem. We could try to solve it by using $\sum \mathbf{F} = 0$, but things would get complicated because we would have to figure out the nontrivial force applied by the pivot. (A common error is to assume that this force points along the stick; it doesn't.) But if we use $\sum \tau = 0$, with the pivot chosen as the origin, then we don't need to know the force from the pivot.

In order for the net torque around the pivot to vanish, the torque from gravity must be canceled by the torque from the centrifugal force. Gravity effectively acts at the CM, which means that it has a lever arm of $(\ell/2) \sin\theta$; see Fig. 12.9. So the clockwise gravitational torque is

$$\tau_{\text{grav}} = mg\frac{\ell}{2}\sin\theta. \qquad (12.15)$$

Figure 12.9

To find the torque from the centrifugal force, consider a small piece of the stick with mass dm, a distance z from the pivot, as shown. This piece is a distance $r = z \sin\theta$ from the axis of rotation. The mass dm can be written as $\lambda\, dz$, where $\lambda = m/\ell$ is the linear mass density. So the centrifugal force on the piece is $(dm)\omega^2 r = (\lambda\, dz)\omega^2(z \sin\theta)$. The associated lever arm is $z \cos\theta$, so the counterclockwise torque due to the centrifugal force on the piece is

$$d\tau_{\text{cent}} = (\lambda\, dz)\omega^2(z \sin\theta) \cdot (z \cos\theta) = (\lambda\omega^2 \sin\theta \cos\theta)z^2\, dz. \qquad (12.16)$$

Integrating this over the whole stick gives the total counterclockwise centrifugal torque as

$$\tau_{\text{cent}} = (\lambda\omega^2 \sin\theta \cos\theta)\int_0^\ell z^2\, dz = \lambda\omega^2 \sin\theta \cos\theta \frac{\ell^3}{3} = m\omega^2 \sin\theta \cos\theta \frac{\ell^2}{3}, \qquad (12.17)$$

where we have used $\lambda \ell = m$. Equating τ_{cent} with τ_{grav} gives

$$m\omega^2 \sin\theta \cos\theta \frac{\ell^2}{3} = mg\frac{\ell}{2}\sin\theta \implies \omega = \sqrt{\frac{3g}{2\ell \cos\theta}}. \tag{12.18}$$

LIMITS: If $\theta \to 90°$, then $\omega \to \infty$, which makes sense. If $\theta \approx 0$, then $\omega \approx \sqrt{3g/2\ell}$. This isn't obvious, although we do know from considerations of units that the answer must be proportional to $\sqrt{g/\ell}$. But it takes a calculation to show that the constant of proportionality is $\sqrt{3/2}$.

REMARKS: The frequency in Eq. (12.18) is larger by a factor of $\sqrt{3/2} \approx 1.22$ than the frequency we obtained in Eq. (12.13) in Problem 12.3 for the case where the stick is replaced by a massless string and a point mass. It makes sense that the present frequency is larger, because we can consider the stick to be built up from many point masses. If all of these point masses were connected to the pivot by massless strings, then from Problem 12.3 we know that the masses closer to the pivot would have larger ω's, because $\omega \propto 1/\sqrt{\ell}$. However, the actual stick is a rigid object, so all of the masses are constrained to lie on a line and rotate with the same ω. The end effect is that the higher masses drag the lower ones along a little faster than they would have otherwise moved, and the frequency ends up being $\sqrt{3/2}$ larger than the frequency with which a point mass at the end of the stick would naturally move.

If you want to solve this rotating-stick problem by working in the lab frame, you will need to use $\tau = d\mathbf{L}/dt$. And furthermore you will need to take into account the full vector nature of the angular momentum. This makes things more involved than the planar problems we dealt with in Chapter 8. For a discussion of this more general class of angular momentum problems, see Chapter 9 in Morin (2008), and Section 9.4.2 in particular.

12.5. Swirling down a drain

No, the swirling isn't caused by the Coriolis force. The Coriolis effect is tiny and is washed out by the motion arising from the inevitable initial speeds of different parts of the water. To see this, let's figure out the size of the Coriolis effect. We're just trying to get a rough idea, so we'll drop all factors of order 1.

Ignoring a trig factor of order 1 that depends on the location on the earth, the Coriolis force is $2m\omega v$. So the acceleration is $a = 2\omega v$. The Coriolis force is perpendicular to the velocity, so the force will cause a sideways deflection. A typical sideways deflection distance is

$$d \approx (1/2)at^2 \approx \omega v t^2 = \omega(vt)t \approx \omega(\ell)t, \tag{12.19}$$

where ℓ is the distance traveled. You can show that the earth's angular speed ω is about $7.3 \cdot 10^{-5}\,\text{s}^{-1}$. We'll take $\ell \approx 0.1$ m and $t \approx 10$ s, which seem like reasonable values; a water molecule might take 10 seconds to approach the drain in a sink that is 20 cm wide. Eq. (12.19) then gives a deflection distance on the order of 10^{-4} m = 0.1 mm, which is far smaller than the size of the drain. Even the tiniest initial velocities (caused by convection currents, etc.) will wash out this effect. A minuscule sideways speed of 10^{-5} m/s will give the same deflection of 10^{-4} m over the time of 10 s.

The swirling that you see must therefore be caused by the initial velocities, combined with conservation of angular momentum. If you manage to get the initial velocities *very* close to zero, then you might see a tiny Coriolis vortex. But this is definitely not responsible for the swirling you see day to day.

REMARKS: The Coriolis effect does cause *hurricanes* to swirl because (1) the speeds involved are larger, so the Coriolis acceleration is larger, and (2) the time scale is longer (on the order of a day), so there is more time for the force to act. In the northern hemisphere, the Coriolis force causes an initial rightward deflection of a stream of air when viewed from above (as you can check with $-2m\omega \times \mathbf{v}$), which causes a low pressure system on the left side of the stream, which the stream then ends up circling around. The direction of the hurricane's spinning is therefore opposite to what you might naively expect from considering the orientation of the Coriolis deflection. That is, hurricanes rotate counterclockwise (when viewed from above) in the northern hemisphere, whereas the Coriolis force causes objects to rotate in clockwise arcs of circles. A hurricane *as a whole* undergoes the standard rightward deflection in the northern hemisphere. So it generally bends through a west-then-north-then-east path.

Another example of an observable effect of the Coriolis force at the surface of the earth is Foucault's pendulum: the plane of a swinging pendulum will gradually precess as time goes on. The exact rate of precession depends on the latitude position on the earth, but at the poles the plane of the pendulum makes one complete revolution per day. This is clear if you view things from an outside inertial frame looking down on the earth; if a swinging pendulum is suspended over the north pole, then the earth rotates beneath the pendulum while the plane of the pendulum remains fixed. An earth observer therefore sees the plane of the pendulum rotate in the opposite direction.

If you want to explain this phenomenon by working in the earth's rotating frame, the relevant force is the Coriolis force. As mentioned in the paragraph following Eq. (12.4), the Coriolis force always points in the *same* direction (which is rightward at the north pole) relative to the velocity. During each half oscillation from apex to apex, the pendulum gets deflected slightly to the right. The overall effect is shown (in exaggerated form) in Fig. 12.10.[3] If the pendulum starts at A, it will eventually end up at B, and so on. (It does *not* simply run back and forth along a single slightly bent curve; that would involve getting deflected to the *left*, relative to the velocity, half the time.) The effect is too small to observe during each half oscillation, but it adds up over the course of a day. We see that there is a gradual clockwise (when viewed from above) precession of the pendulum's plane, when viewed in the earth's frame. This is consistent with the fact that in the external inertial frame, the earth rotates counterclockwise beneath the pendulum. Foucault's pendulum provides a low-tech way of demonstrating that the earth does indeed rotate.

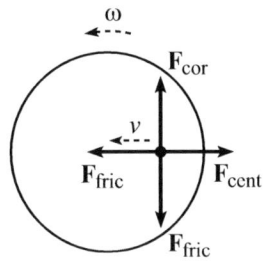

Figure 12.10

12.6. Walking radially on a carousel

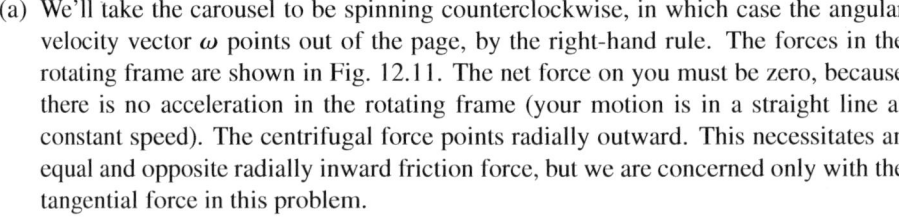

Figure 12.11

(a) We'll take the carousel to be spinning counterclockwise, in which case the angular velocity vector ω points out of the page, by the right-hand rule. The forces in the rotating frame are shown in Fig. 12.11. The net force on you must be zero, because there is no acceleration in the rotating frame (your motion is in a straight line at constant speed). The centrifugal force points radially outward. This necessitates an equal and opposite radially inward friction force, but we are concerned only with the tangential force in this problem.

The Coriolis force, $-2m\omega \times \mathbf{v}$, points in the positive tangential direction, as you can check with the right-hand rule. This necessitates an equal and opposite friction force pointing in the negative tangential direction. So the tangential friction force has magnitude $F_{\rm f} = |-2m\omega \times \mathbf{v}| = 2m\omega v$.

(b) Your angular momentum in the ground frame is $L = mvr = m(\omega r)r = mr^2\omega$. Its rate of change is therefore (using $\dot r = -v$ since your motion is radially inward, and using the fact that ω is assumed to be constant)

$$\frac{dL}{dt} = \frac{d(mr^2\omega)}{dt} = 2mr\dot r\omega = 2mr(-v)\omega = (-2m\omega v)r = (-F_{\rm f})r. \qquad (12.20)$$

And $-F_{\rm f}r$ is the torque due to the tangential friction force we found in part (a). The minus sign is present because the friction force points in the backward tangential direction, so its effect is to decrease your angular momentum. Intuitively, your radius is decreasing while ω is held constant, so L must decrease. Formally, $\boldsymbol{\tau} = \mathbf{r} \times \mathbf{F}_{\rm f}$ points into the page, while positive L was defined to point out of the page, in the direction of ω. So L is decreasing.

---

[3] This figure shows the path of a pendulum released from rest relative to the *earth*. If the pendulum is instead released from rest relative to the *inertial frame* (so that the motion is exactly planar in the inertial frame, as we discussed above), then the pendulum will have a tiny initial tangential velocity in the earth's frame, because the earth is rotating beneath the pendulum in the inertial frame. This means that the path in Fig. 12.10 will now have very thin loops at the apexes of the motion instead of the cusps shown. But the overall precession will be the same.

12.5. PROBLEM SOLUTIONS

12.7. Walking tangentially on a carousel

(a) The free-body diagram in the lab frame is shown in Fig. 12.12(a). The only horizontal force is the friction force, so the radial $F = ma$ equation for the circular motion with speed ωr is

$$F = ma \implies F_f = \frac{mv^2}{r} \implies F_f = mr\omega^2. \quad (12.21)$$

The rotation of the carousel is irrelevant in this calculation.

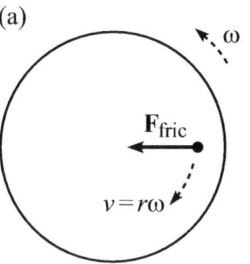

(b) The free-body diagram in the carousel frame is shown in Fig. 12.12(b). We now have the centrifugal and Coriolis fictitious forces, in addition to the friction force. $\mathbf{F}_{cent}$ points radially outward with the usual magnitude of $mr\omega^2$. $\mathbf{F}_{cor}$ points radially inward, as you can check with the right-hand rule, because ω points out of the page and $\mathbf{v}$ points downward at the point shown. The person's speed in the rotating frame is $\omega r + \omega r = 2\omega r$, so $\mathbf{F}_{cor}$ has magnitude $F_{cor} = 2m\omega v = 2m\omega(2\omega r) = 4mr\omega^2$. The radial $F = ma$ equation for the circular motion with speed $2\omega r$ is therefore

$$F_{cor} + F_f - F_{cent} = \frac{mv^2}{r} \implies 4mr\omega^2 + F_f - mr\omega^2 = \frac{m(2\omega r)^2}{r}$$

$$\implies F_f = mr\omega^2. \quad (12.22)$$

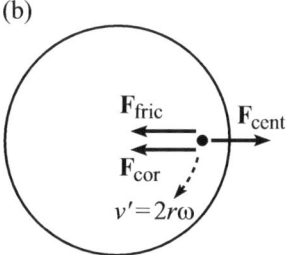

Figure 12.12

This agrees with the result in part (a), as it must. The friction force has the same value in any frame.

12.8. Bead on a rod

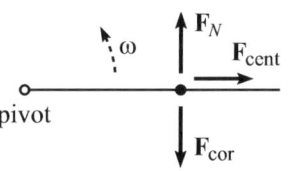

Figure 12.13

(a) Fig. 12.13 shows the forces acting on the bead in the rotating frame. The centrifugal force, $mr\omega^2$, is always present. The Coriolis force, $2m\omega v$, will be nonzero when $v \neq 0$ (in other words, at all times after the initial instant). The bead will be moving radially outward, so you can check with the right-hand rule that the Coriolis force points in the backward tangential direction. There must also be a normal force to balance the Coriolis force, because there is no acceleration in the tangential direction in the frame of the rod, since the bead is constrained to be on the rod. There are also gravitational and normal forces perpendicular to the plane of the page, but those aren't important in this problem.

(b) The radial $F = ma$ equation is

$$F_{cent} = ma \implies mr\omega^2 = m\ddot{r}. \quad (12.23)$$

If we guess a solution of the form $r(t) = Ae^{\alpha t}$, we obtain

$$m(Ae^{\alpha t})\omega^2 = m(\alpha^2 Ae^{\alpha t}) \implies \omega^2 = \alpha^2 \implies \alpha = \pm\omega. \quad (12.24)$$

We have therefore found two solutions. The most general solution is an arbitrary linear combination of these:

$$r(t) = Ae^{\omega t} + Be^{-\omega t}, \quad (12.25)$$

where A and B are determined by the initial conditions, which are $r(0) = r_0$ and $\dot{r}(0) = 0$. The $r(0) = r_0$ condition yields

$$Ae^0 + Be^0 = r_0 \implies A + B = r_0. \quad (12.26)$$

Taking the derivative of r gives $\dot{r} = \omega Ae^{\omega t} - \omega Be^{-\omega t}$, so the $\dot{r}(0) = 0$ condition yields

$$\omega Ae^0 - \omega Be^0 = 0 \implies A = B. \quad (12.27)$$

Eq. (12.26) then gives $A = B = r_0/2$. So our solution for $r(t)$ is

$$r(t) = \frac{r_0}{2}\left(e^{\omega t} + e^{-\omega t}\right). \tag{12.28}$$

REMARK: As a double check, this $r(t)$ does indeed satisfy $r(0) = r_0$ and $\dot{r}(0) = 0$. For large t, the $e^{-\omega t}$ term is negligible, so we have $r(t) \approx (r_0/2)e^{\omega t}$. That is, $r(t)$ grows exponentially. For small (but nonzero) t, we can use the Taylor series $e^x \approx 1 + x + x^2/2$ to write

$$r(t) \approx \frac{r_0}{2}\left(\left(1 + \omega t + \frac{\omega^2 t^2}{2}\right) + \left(1 - \omega t + \frac{\omega^2 t^2}{2}\right)\right) = r_0 + \frac{1}{2}r_0\omega^2 t^2. \tag{12.29}$$

This makes sense because right at the start, the centrifugal force points outward with magnitude $mr_0\omega^2$. The radial acceleration (which is essentially constant over a small time interval t) is therefore $a = r_0\omega^2$. So the distance traveled in time t is $at^2/2 = (r_0\omega^2)t^2/2$, in agreement with Eq. (12.29).

(c) The normal force is equal and opposite to the Coriolis force, which has magnitude $2m\omega v$. Since $v \equiv \dot{r} = (r_0/2)\omega(e^{\omega t} - e^{-\omega t})$, we have

$$F_{\text{normal}} = 2m\omega v = 2m\omega \cdot \frac{r_0\omega}{2}(e^{\omega t} - e^{-\omega t}) = mr_0\omega^2(e^{\omega t} - e^{-\omega t}). \tag{12.30}$$

For large t, the $e^{-\omega t}$ term is negligible, so the normal force (and the Coriolis force) is essentially equal to $mr_0\omega^2 e^{\omega t}$. This is twice as large as the centrifugal force, $mr\omega^2$, because $r(t) \approx (r_0/2)e^{\omega t}$ for large t.

Chapter 13

Appendices

13.1 Appendix A: Vectors

13.1.1 Basics

Although it is possible to define a *vector* in a more precise way, the definition that will suffice for our purposes is that a vector is a mathematical object that has both a *magnitude* (that is, a length) and a *direction*. For example, Fig. 13.1 shows a vector that points upward and rightward with a length of about 3 cm. In general, a vector can point in an arbitrary direction in 3-D space, but we'll deal mainly with 2-D vectors in the plane of the page. Vectors also exist in 1-D, but in contrast with 2-D and 3-D where a vector can point in an infinite number of directions, a vector in 1-D can point in only two directions – to the right or to the left, if we're dealing with a horizontal line. So the direction of a 1-D vector can be specified by simply writing a positive or negative sign.

Figure 13.1

A vector in any dimension is completely determined by its magnitude and direction. This means that all of the vectors in Fig. 13.2 are actually the same vector. It doesn't matter where the tail and the tip of a given vector are, as long as the segment joining them has a given length and points in a given direction.

We'll denote vectors with boldface letters, such as **r** or **v**, etc. And we'll denote the magnitude (length) with an italic latter, such as r or v, etc. The length is also sometimes denoted with absolute value bars around the vector, so $|\mathbf{r}|$ means the same thing as r. The length is a *scalar*, that is, just a number (with units).

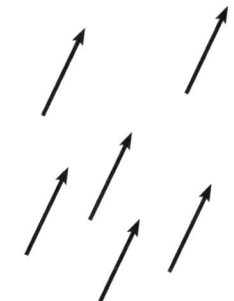

Figure 13.2

It is important to distinguish between a vector and its magnitude. The magnitude is only one piece of the complete information contained in the entire vector (with the other piece being the direction). The classic example of this distinction is the *velocity* **v** of an object versus the *speed* v. By "velocity" we mean the complete vector, and by "speed" we mean just the magnitude. A statement such as "My velocity is 10 m/s" uses incorrect terminology, whereas statements such as "My speed is 10 m/s" and "My velocity is 10 m/s in the northwest direction" use correct terminology.

Examples of vectors that come up in this book are: position **r**, velocity **v**, acceleration **a**, momentum **p**, force **F**, angular momentum **L**, and angular velocity $\boldsymbol{\omega}$. Examples of scalars are: length ℓ, speed v, energy E, mass m, and angular speed ω.

13.1.2 Cartesian coordinates

If you want to do anything quantitative with vectors, it is usually necessary to use a *coordinate system* to describe them. There are many different kinds of coordinate systems you can choose from, the most common being Cartesian, polar, cylindrical, and spherical. Additionally, you have the freedom to choose the orientation of your coordinate axes.

The simplest and most common type of coordinate system is a Cartesian one, which in 3-D space involves the three mutually perpendicular x, y, and z axes. The Cartesian *components* of

331

a vector **a** are written as (a_x, a_y, a_z). To understand what this notation means, we can write **a** in terms of the *basis* vectors $\hat{\mathbf{x}}$, $\hat{\mathbf{y}}$, $\hat{\mathbf{z}}$. These basis vectors are unit vectors (that is, they have length 1) pointing along the three coordinate axes. In terms of the basis vectors, we have

$$\mathbf{a} = (a_x, a_y, a_z) = a_x \hat{\mathbf{x}} + a_y \hat{\mathbf{y}} + a_z \hat{\mathbf{z}}. \tag{13.1}$$

So (a_x, a_y, a_z) is just shorthand for $a_x \hat{\mathbf{x}} + a_y \hat{\mathbf{y}} + a_z \hat{\mathbf{z}}$. We'll discuss the addition of vectors in Section 13.1.4 below, but in short, the $\mathbf{a} = a_x \hat{\mathbf{x}} + a_y \hat{\mathbf{y}} + a_z \hat{\mathbf{z}}$ relation says that if you march a distance a_x in the x direction, then a_y in the y direction, and then a_z in the z direction, you will end up at the tip of the **a** vector.

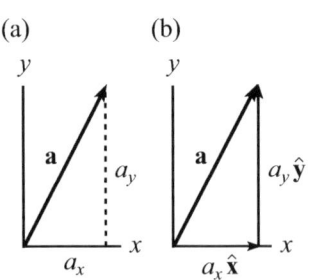

Figure 13.3

The word "component" can be used in two slightly different ways. If you ask someone for the x component of a vector **a**, the answer might be a_x (which is a number; see Fig. 13.3(a)), or the answer might be $a_x \hat{\mathbf{x}}$ (which is a vector; see Fig. 13.3(b)). The difference is just semantics. Either answer is acceptable, provided that it is used in the appropriate way, depending on whether it is a number or a vector. For example, if a person says, "A vector is the sum of its components," then she is clearly thinking of the components as vectors and making the correct statement that $\mathbf{a} = a_x \hat{\mathbf{x}} + a_y \hat{\mathbf{y}} + a_z \hat{\mathbf{z}}$, as opposed to the incorrect statement that $\mathbf{a} = a_x + a_y + a_z$. This latter equation certainly can't be correct, because it equates a vector with a nonvector. Additionally, the magnitude isn't even correct.

The length of a vector **a** whose Cartesian coordinates are (a_x, a_y, a_z) is

$$a \equiv |\mathbf{a}| = \sqrt{a_x^2 + a_y^2 + a_z^2}. \tag{13.2}$$

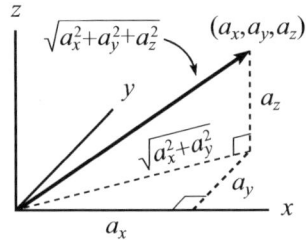

Figure 13.4

This follows from two applications of the Pythagorean theorem in Fig. 13.4, first with the right triangle in the x-y plane, and then with the vertical right triangle extending upward from the x-y plane to the point (a_x, a_y, a_z).

It should be emphasized that a vector has a meaning independent of any coordinate system (type, choice of origin, orientation of axes). Components are used to describe a vector, but a vector is what it is, independent of how we choose to describe it. The vector **a** we drew in Fig. 13.1 has length 3 cm and points up to the right. But we don't even need to give these bits of information to describe it. All we need to do is draw it on the page, as we did in Fig. 13.1. That arrow is the vector **a**, period.

However, the fact of the matter is that using components makes it much easier to deal with vectors. It would be a pain to have to keep drawing vectors whenever you needed to do something with them, such as adding them (see below). And if your drawing skills are shaky, then you might not trust the answer you obtain, anyway. Components provide a reliable and rigorous way of describing vectors. If the task of a given problem is to find a certain vector, then in practice the most common ways of stating the answer are to give the Cartesian coordinates or to give the magnitude and direction.

REMARK: Here is a somewhat subtle point, concerning the distinction between *points* and *vectors*. If someone gives you the triplet of numbers (a_x, a_y, a_z), is this a point in space, or is it a vector? Well, it could be either, so you would need to be told which it is. The main difference between a point (a_x, a_y, a_z) and a vector (a_x, a_y, a_z) is that whereas a vector doesn't depend on the choice of origin, a point very much does. The coordinates of the point (a_x, a_y, a_z) are measured with respect to a given origin, so if you don't know where that origin is, then you don't know where the point (a_x, a_y, a_z) is. In contrast, the components of the vector (a_x, a_y, a_z) give the position of the head with respect to the tail, so it doesn't matter where the origin is. In Fig. 13.4 we actually used (a_x, a_y, a_z) to mean both a point and a vector.

Given two points in space, the position of one with respect to the other is a vector, because the relative position doesn't depend on the choice of origin. If you are given the coordinates of a particular house, then you also need to be given the origin that these coordinates are measured with respect to. But if you walk from one house to another (in which case your displacement will be described by a vector), then you're going to take the same walk independent of the location of the point that someone arbitrarily decides to call the origin.

13.1.3 Polar coordinates

Consider a *point* in 2-D space (and yes, we're talking about a point here, not a vector). We can use Cartesian coordinates to describe the location of this point relative to a given origin, as we did above. But we can also use polar coordinates. The polar coordinates of a point are the distance r from the origin and the angle θ relative to the horizontal axis, as shown in Fig. 13.5. By looking at the right triangle in the figure, the Cartesian coordinates (x, y) of the point can be obtained from the polar coordinates (r, θ) by

$$x = r\cos\theta \quad \text{and} \quad y = r\sin\theta. \tag{13.3}$$

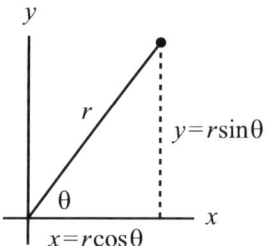

Figure 13.5

Equivalently, the polar coordinates (r, θ) can be obtained from the Cartesian coordinates (x, y) by

$$r = \sqrt{x^2 + y^2} \quad \text{and} \quad \theta = \arctan(y/x). \tag{13.4}$$

The second expression here doesn't quite determine θ, due to the $\tan(\theta + \pi) = \tan\theta$ ambiguity. But the correct angle can be determined by the sign of x (or y); this narrows down the quadrant the angle is in. In 3-D, the generalization of polar coordinates is *spherical coordinates*, which involve an additional angle associated with circling around a sphere (the longitudinal angle on the earth).

Let's now look at what form a *vector* (as opposed to a point) takes when written in polar coordinates.[1] Things aren't as simple as they are in Cartesian coordinates, because while the Cartesian basis vectors $\hat{\mathbf{x}}, \hat{\mathbf{y}}, \hat{\mathbf{z}}$ don't depend on position, the polar basis vectors $\hat{\mathbf{r}}$ and $\hat{\boldsymbol{\theta}}$ do. At a given point, the unit vector $\hat{\mathbf{r}}$ is defined to point in the radial direction, and the unit vector $\hat{\boldsymbol{\theta}}$ is defined to point in the counterclockwise tangential direction. These directions depend on where the point is located in the plane, as shown at points P_1 and P_2 in Fig. 13.6. The polar basis vectors $\hat{\mathbf{r}}$ and $\hat{\boldsymbol{\theta}}$ are undefined at the origin, because the radial and tangential directions aren't uniquely defined there.

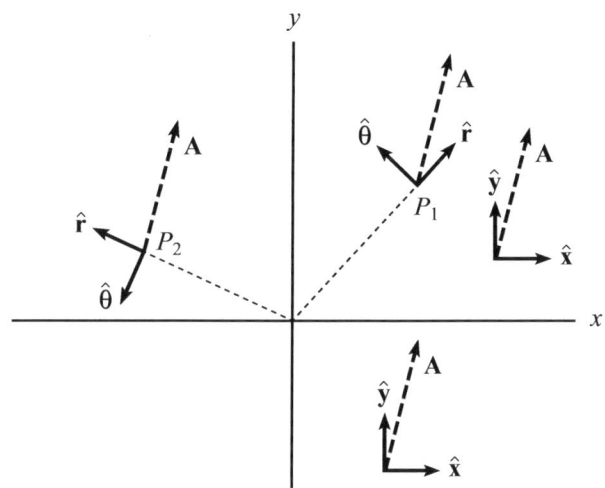

Figure 13.6

To be clear, the $\hat{\mathbf{r}}$ vector pointing up and to the right at point P_1 in Fig. 13.6 is the same as a similar vector pointing up and to the right, no matter where we place it in the plane (as in Fig. 13.2). However, this $\hat{\mathbf{r}}$ vector has nothing to do with what we define to be the $\hat{\mathbf{r}}$ vector at, say, point P_2, where $\hat{\mathbf{r}}$ points up and to the left. Although we use the same *symbols* $\hat{\mathbf{r}}$ and $\hat{\boldsymbol{\theta}}$ for the

---

[1] The rest of this subsection on polar coordinates isn't important for this book; the $\hat{\mathbf{r}}$ and $\hat{\boldsymbol{\theta}}$ basis vectors don't appear anywhere except in this appendix. But it is natural to include the following discussion, given that we've introduced polar coordinates. In more advanced topics in physics, this discussion is highly relevant.

basis vectors at any point, this doesn't mean that the basis vectors at different points are actually the same *vectors*.

So what does a vector look like in when written in polar coordinates? Consider the vector **A** shown above in Fig. 13.6. This vector is the same vector, no matter where we place it in the plane (as in Fig. 13.2). In Cartesian coordinates, let's say **A** is given by $(0.7)\hat{\mathbf{x}} + (2.3)\hat{\mathbf{y}}$ (just a rough estimate, by comparing **A** with the unit basis vectors). This expression describes **A** in Cartesian coordinates, no matter where we draw **A**. But in polar coordinates, the description depends on what basis vectors we use. That is, it depends on where we put **A** (more precisely, where we put the tail of **A**). At point P_1, **A** happens to be $(2.1)\hat{\mathbf{r}} + (1.2)\hat{\boldsymbol{\theta}}$, whereas at point P_2, **A** happens to be $(0.4)\hat{\mathbf{r}} - (2.4)\hat{\boldsymbol{\theta}}$. These are two different expressions for the same vector. Note that all of the above expressions for **A** have the same length (at least up to rounding errors), as they must. The length happens to be about 2.4.

Polar (or spherical) coordinates are very useful when describing vectors in setups that possess circular (or spherical) symmetry. For example, the full vector form of the gravitational force on a mass m due to a mass M located at the origin is

$$\mathbf{F} = -\frac{GMm}{r^2}\hat{\mathbf{r}}, \tag{13.5}$$

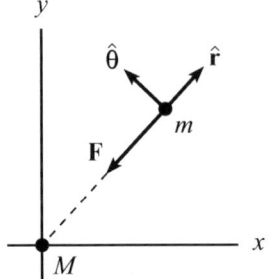

Figure 13.7

as shown in Fig. 13.7. Here we are using the $\hat{\mathbf{r}}$ basis vector at the most reasonable point to pick, which is the location of the mass m. (The only other special point in the plane is the origin, where M is. But as mentioned above, the polar basis vectors are undefined there.) The above polar expression for **F** is much cleaner than the Cartesian expression,

$$\mathbf{F} = -\frac{GMm}{x^2+y^2}\left(\frac{x}{\sqrt{x^2+y^2}}\hat{\mathbf{x}} + \frac{y}{\sqrt{x^2+y^2}}\hat{\mathbf{y}}\right). \tag{13.6}$$

This is indeed the same force as in Eq. (13.5), because when $\hat{\mathbf{r}}$ is written in terms of Cartesian coordinates, it equals the vector in parentheses in Eq. (13.6). This is true because this vector has length 1, and it is proportional to the vector (x, y) which points in the $\hat{\mathbf{r}}$ direction by definition.

The gravitational force **F** is an example of a *vector field*. In a vector field, every point in space has a vector associated with it. Another example of a vector field is wind velocity; every point is space has associated with it the local velocity of the wind. (In contrast, in a *scalar field*, every point has a number (with units) associated it. An example is the temperature.) When describing a vector field, it is understood that when writing a vector at a given point in terms of basis vectors, you are using the basis vectors at that point, although this specification isn't necessary in Cartesian coordinates because the basis vectors are the same everywhere.

REMARK: If someone working with polar coordinates gives you the pair of numbers (a_r, a_θ), what does it represent? As with Cartesian coordinates, the expression (a_r, a_θ) is ambiguous. It could mean a point in the plane, as in Fig. 13.5, with $a_r = r$ and $a_\theta = \theta$ (in which case you would need to be told where the origin is). Or it could mean the components of the vector $a_r\hat{\mathbf{r}} + a_\theta\hat{\boldsymbol{\theta}}$ (in which case you would need to be told what basis vectors $\hat{\mathbf{r}}$ and $\hat{\boldsymbol{\theta}}$ to use). In practice, most people would assume that (a_r, a_θ) represents a point in the plane. Note that technically the ambiguity is removed if you pay attention to units/dimensions. In the case where (a_r, a_θ) represents a point in the plane, a_r has dimensions of length and a_θ is a dimensionless angle. But in the case where (a_r, a_θ) represents a vector, a_r and a_θ must have the same dimensions. This is true because the dimensions of $\hat{\mathbf{r}}$ and $\hat{\boldsymbol{\theta}}$ are the same; they are both dimensionless unit vectors, just as $\hat{\mathbf{x}}$ and $\hat{\mathbf{y}}$ are. In an expression such as $x\hat{\mathbf{x}}$, the units of meters are in the x, not the $\hat{\mathbf{x}}$.

13.1.4 Adding and subtracting vectors

What is the sum of two vectors, for example, the two vectors **a** and **b** shown in Fig. 13.8(a)? To add two vectors, we put them "tail-to-head," as shown in Fig. 13.8(b). It doesn't matter which vector you start with; you can add them in either order and you will obtain the same result, as shown with the parallelogram in the figure. In other words, vector addition is *commutative*. In

13.1. APPENDIX A: VECTORS

the top triangle, the vector from the tail of **a** to the head of **b** is the desired sum; you can think of this as **a** + **b**. In the bottom triangle, the vector from the tail of **b** to the head of **a** is also the desired sum; you can think of this as **b** + **a**.

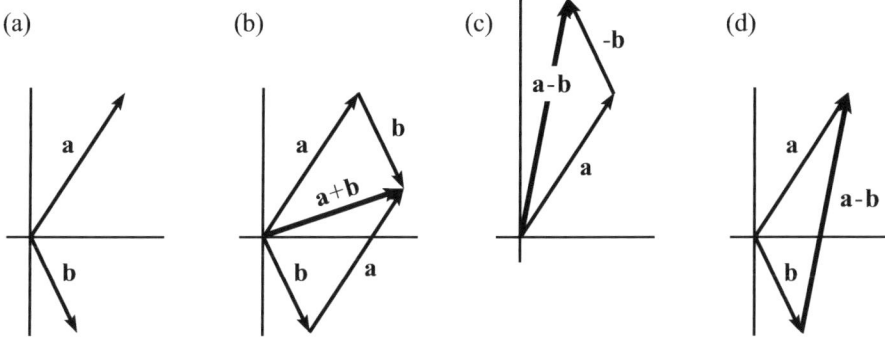

Figure 13.8

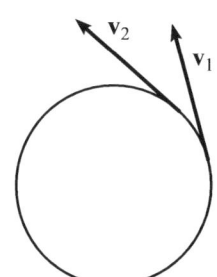

Figure 13.9

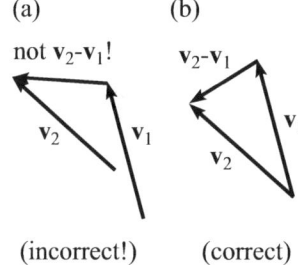

Figure 13.10

The tail-to-head procedure makes intuitive sense: If you go for a walk and end up with a displacement of **a**, and from there you go for another walk and end up with a displacement of **b** (relative to where you began your second walk), then your total displacement is given by the **a** + **b** vector in Fig. 13.8(b).

If you want to subtract two vectors, this operation is the same as adding on the negative of one of the vectors. That is, **a** − **b** = **a** + (−**b**). (The negative of a vector is the vector that has the same magnitude but points in the opposite direction.) If we consider the same vectors **a** and **b** as in Fig. 13.8(a), the difference **a** − **b** is shown in Fig. 13.8(c); we have added −**b** to **a**. Equivalently, **a** − **b** is the vector from the tip of **b** to the tip of **a** (with the tails of the vectors coinciding), because **a** − **b** is the vector that you need to add to **b** to obtain **a**. This equality **b** + (**a** − **b**) = **a** is shown in Fig. 13.8(d).

If you want to use the method in Fig. 13.8(d) to calculate the difference between two vectors, you must remember to put the tails of the two vectors at the same point. A common error is to draw the (incorrect) difference vector as the vector that connects the tips of two given vectors whose tails don't coincide. Consider, for example, a particle traveling around a circle at constant speed. The velocity vectors at two different times are shown in Fig. 13.9; they have the same length but different directions. The difference $\mathbf{v}_2 - \mathbf{v}_1$ between these vectors is *not* the vector shown in Fig. 13.10(a), obtained by simply connecting the tips of the vectors as they appear on the page in Fig. 13.9. The correct difference, obtained by having the tails of $\mathbf{v}_1$ and $\mathbf{v}_2$ coincide, is shown in Fig. 13.10(b). Of course, you can also obtain this difference by adding $-\mathbf{v}_1$ to $\mathbf{v}_2$, as you can check.

If you want to multiply a vector by a scalar n (a number), you just need to multiply the length by n and keep the direction the same. In the special case where n is an integer, this is the same as lining up n copies of the vector. For example, since $3\mathbf{b} = \mathbf{b} + \mathbf{b} + \mathbf{b}$, the vector $3\mathbf{b}$ is shown in Fig. 13.11. If you want to multiply a vector by another vector, there are two possible ways to do this, involving the *dot product* and the *cross product*. These are discussed below in Sections 13.1.6 and 13.1.7.

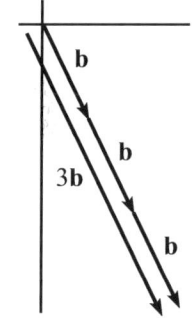

Figure 13.11

As mentioned earlier, although it is possible in theory to manipulate vectors by doing nothing other than drawing them (which has been our strategy thus far in this subsection), it is usually necessary to make use of coordinates if we want to be quantitative. If we are adding two vectors, we simply need to add the corresponding Cartesian components. For simplicity let's work in 2-D, where a vector is described by two Cartesian components (a_x, a_y). In the addition example in Fig. 13.8(b), let's say that the two vectors are $\mathbf{a} = (2,3)$ and $\mathbf{b} = (1,-2)$. Then the sum is

$$\mathbf{a} + \mathbf{b} = (2,3) + (1,-2) = (2+1, 3-2) = (3,1), \tag{13.7}$$

as shown in Fig. 13.12 (we've scaled up the size of the figure, relative to Fig. 13.8). It is clear

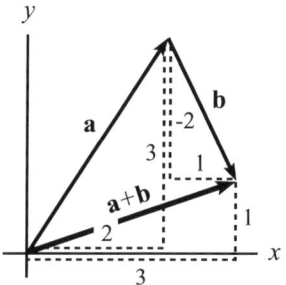

Figure 13.12

from this figure that the x component of the sum equals the sum of the x components of the two individual vectors. Likewise for the y components. A common mistake when adding vectors is to forget that they are vectors and simply add the magnitudes. Remember that vectors aren't just numbers, and that you always need to first break vectors into their Cartesian components, and then you can add the corresponding components. To subtract two vectors, you simply need to subtract the components.

13.1.5 Using components

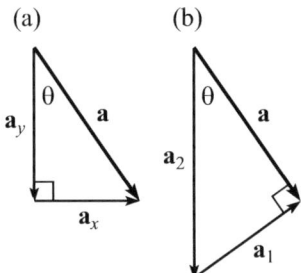

Figure 13.13

What is the vertical component of the vector **a** shown in Fig. 13.13? Is it the vertical vector $\mathbf{a}_y$ in Fig. 13.13(a) with length $a \cos\theta$, or is it the vertical vector $\mathbf{a}_2$ in Fig. 13.13(b) with length $a/\cos\theta$? The correct choice is the first one, $\mathbf{a}_y$. When breaking a given vector into components, the vector is always the hypotenuse of the triangle representing the components. The axes of our coordinate systems are always orthogonal, so the components must always be orthogonal, which isn't the case in Fig. 13.13(b). A vector is the sum of its components, and while it is certainly true that $\mathbf{a} = \mathbf{a}_2 + \mathbf{a}_1$ in Fig. 13.13(b), just as it is true that $\mathbf{a} = \mathbf{a}_y + \mathbf{a}_x$ in Fig. 13.13(a), the former of these decompositions isn't helpful in general (because $\mathbf{a}_1$ has a vertical component, whereas $\mathbf{a}_x$ does not).

When solving a problem, choosing your coordinate system wisely can make the problem much easier to solve. Cartesian and polar coordinates are the most common options. But additionally, if you are using Cartesian coordinates, you might find that one orientation of your axes makes things easier than another. For example, in setups involving objects on inclined planes, you can pick your axes to be horizontal and vertical, or you can pick them to be parallel and perpendicular to the plane. It often isn't obvious which is the better set of axes, so you might need to try both. When using the latter set, you will need to break the gravitational acceleration **g** (or equivalently, the gravitational force $m\mathbf{g}$) into components parallel and perpendicular to the plane. These components have lengths $g \sin\theta$ and $g \cos\theta$, as shown in Fig. 13.14. (You can do the geometry to determine which angles in the figure are equal to θ, or you can just use the limiting-case reasoning in Multiple-Choice Question 1.12.) The rectangle formed by the components is shown in Fig. 13.15. It's personal preference which of the two right triangles you draw to find the components.

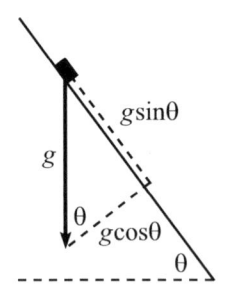

Figure 13.14

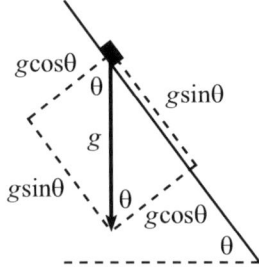

Figure 13.15

The downward $m\mathbf{g}$ force is a vector, and a vector is the sum of its components. So the sum of the $mg \sin\theta$ force along the plane and the $mg \cos\theta$ force perpendicular to the plane equals the original $m\mathbf{g}$ force. (There is also a normal force from the plane acting on the object, but that won't concern us here.) In other words, the object on the plane behaves in exactly the same way whether it is acted on by the single gravitational force or by two people pulling along the plane and perpendicular to the plane with forces of $mg \sin\theta$ and $mg \cos\theta$. The sum of the *magnitudes* of these two forces doesn't equal the magnitude of the original $m\mathbf{g}$ force (the sum is larger), but that is irrelevant. All that matters is that the sum of the two component *vectors* equals the original $m\mathbf{g}$ vector. When solving problems, it is often helpful to forget all about the original force vector and instead consider it to be two separate forces generated by two people pulling/pushing along the directions of the components.

Many equations in physics are vector equations. An example is $\mathbf{F} = m\mathbf{a}$. Another example is a conservation-of-momentum equation of the form $\mathbf{p}_{\text{initial}} = \mathbf{p}_{\text{final}}$. Since these are vector equations, they don't say simply that the magnitudes are the same on both sides; they say that the complete vectors (including the directions) are the same. Equivalently, they say that the corresponding components on each side are equal. In other words, a vector equation is really three equations, one for each component. So $\mathbf{F} = m\mathbf{a}$ says the same thing as the combination of the three $F_x = ma_x$, $F_y = ma_y$, and $F_z = ma_z$ equations. (If you're dealing with a 2-D system in the x-y plane, then you really have only two equations, because $F_z = ma_z$ is the trivial statement that $0 = 0$.) Therefore, when dealing with vectors, you will invariably need to break them up into components and deal with these components separately.

13.1. APPENDIX A: VECTORS

13.1.6 Dot product

As mentioned in Section 13.1.4, there are two possible ways to multiply two vectors together: the *dot product* and the *cross product*. These products don't come up too often in this book, but since they do appear, we'll provide a brief review of them. The dot product shows up in the definition of work, and the cross product shows up in the definitions of torque, angular momentum, and fictitious forces.

Consider two vectors whose Cartesian components are given by

$$\mathbf{a} = (a_x, a_y, a_z) \quad \text{and} \quad \mathbf{b} = (b_x, b_y, b_z). \tag{13.8}$$

The *dot product*, or *scalar product*, between these vectors is defined as

$$\mathbf{a} \cdot \mathbf{b} \equiv a_x b_x + a_y b_y + a_z b_z. \tag{13.9}$$

The dot product takes two vectors and produces a scalar, which is just a number (with units). You can quickly use Eq. (13.9) to show that the dot product is commutative (that is, $\mathbf{a} \cdot \mathbf{b} = \mathbf{b} \cdot \mathbf{a}$) and distributive (that is, $(\mathbf{a} + \mathbf{b}) \cdot \mathbf{c} = \mathbf{a} \cdot \mathbf{c} + \mathbf{b} \cdot \mathbf{c}$). Note that the dot product of a vector with itself is $\mathbf{a} \cdot \mathbf{a} = a_x^2 + a_y^2 + a_z^2$, which is just its length squared. So $\mathbf{a} \cdot \mathbf{a} = |\mathbf{a}|^2 \equiv a^2$.

Taking the sum of the products of the corresponding components of two vectors, as we did in Eq. (13.9), might seem like a silly and arbitrary thing to do. Why don't we instead look at, say, the sum of the cubes of the products of the corresponding components? The reason is that the dot product as we've defined it has many nice properties, the most useful of which is that it can be written as

$$\mathbf{a} \cdot \mathbf{b} = |\mathbf{a}||\mathbf{b}| \cos\theta \equiv ab\cos\theta, \tag{13.10}$$

where θ is the angle between the two vectors. We can demonstrate this as follows. Consider the dot product of the vector $\mathbf{c} \equiv \mathbf{a} + \mathbf{b}$ with itself, which is simply the square of the length of $\mathbf{c}$. Using the distributive and commutative properties, we have

$$\begin{aligned} c^2 = (\mathbf{a} + \mathbf{b}) \cdot (\mathbf{a} + \mathbf{b}) &= \mathbf{a} \cdot \mathbf{a} + 2\mathbf{a} \cdot \mathbf{b} + \mathbf{b} \cdot \mathbf{b} \\ &= a^2 + 2\mathbf{a} \cdot \mathbf{b} + b^2. \end{aligned} \tag{13.11}$$

But from the law of cosines applied to the triangle in Fig. 13.16, we have

$$c^2 = a^2 + b^2 - 2ab\cos\gamma = a^2 + b^2 + 2ab\cos\theta, \tag{13.12}$$

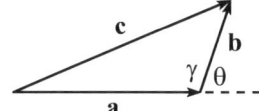

Figure 13.16

because $\gamma = \pi - \theta$. Equating the two results for c^2 in Eqs. (13.11) and (13.12) yields $\mathbf{a} \cdot \mathbf{b} = ab\cos\theta$, as desired. The angle between two vectors is therefore given by

$$\cos\theta = \frac{\mathbf{a} \cdot \mathbf{b}}{|\mathbf{a}||\mathbf{b}|}. \tag{13.13}$$

A nice corollary of this result is that if the dot product of two vectors is zero, then $\cos\theta = 0$, which means that the vectors are perpendicular. If someone gives you the vectors $(1, -2, 3)$ and $(4, 5, 2)$, it is by no means obvious visually that they are perpendicular. But we know from Eq. (13.13) that they indeed are.

The dot product $\mathbf{a} \cdot \mathbf{b}$ can be written as either $a(b\cos\theta)$ or $b(a\cos\theta)$. So $\mathbf{a} \cdot \mathbf{b}$ equals the length of $\mathbf{a}$ times the component of $\mathbf{b}$ along $\mathbf{a}$. Or vice versa, depending on which length you want to group with the $\cos\theta$ factor. If we rotate our coordinate system, the dot product of two vectors remains the same, because from Eq. (13.10) it depends only on their lengths and the angle between them; and these are unaffected by the rotation. In other words, the dot product is a scalar. (The technical definition of a scalar is something that is invariant under a rotation of the coordinate system.) This certainly isn't obvious from looking at the original definition in Eq. (13.9), because the coordinates get all messed up during the rotation.

13.1.7 Cross product

The *cross product*, or *vector product*, between two vectors is defined via a determinant as

$$\mathbf{a} \times \mathbf{b} \equiv \begin{vmatrix} \hat{\mathbf{x}} & \hat{\mathbf{y}} & \hat{\mathbf{z}} \\ a_x & a_y & a_z \\ b_x & b_y & b_z \end{vmatrix}$$
$$= \hat{\mathbf{x}}(a_y b_z - a_z b_y) + \hat{\mathbf{y}}(a_z b_x - a_x b_z) + \hat{\mathbf{z}}(a_x b_y - a_y b_x). \tag{13.14}$$

The cross product takes two vectors and produces another vector. As with the dot product, you can show that the cross product is distributive. However, it is *anti*-commutative (that is, $\mathbf{a} \times \mathbf{b} = -\mathbf{b} \times \mathbf{a}$), which is evident from Eq. (13.14); interchanging two rows of a determinant negates its value. This implies that the cross product of any vector with itself is zero.

As with the dot product, the reason why we study this particular combination of components is that it has many nice properties, the most useful of which are that the direction of $\mathbf{a} \times \mathbf{b}$ is perpendicular to both $\mathbf{a}$ and $\mathbf{b}$ (in the orientation determined by the right-hand rule; see below) and that the magnitude is

$$|\mathbf{a} \times \mathbf{b}| = |\mathbf{a}||\mathbf{b}|\sin\theta \equiv ab\sin\theta. \tag{13.15}$$

We'll derive this relation below, but let's first show that $\mathbf{a} \times \mathbf{b}$ is indeed perpendicular to both $\mathbf{a}$ and $\mathbf{b}$. We'll do this by making use of the fact that if the *dot* product of two vectors is zero, then the vectors are perpendicular. We have

$$\mathbf{a} \cdot (\mathbf{a} \times \mathbf{b}) = a_x(a_y b_z - a_z b_y) + a_y(a_z b_x - a_x b_z) + a_z(a_x b_y - a_y b_x) = 0, \tag{13.16}$$

as desired; you can check that all of the terms cancel in pairs. And likewise for $\mathbf{b} \cdot (\mathbf{a} \times \mathbf{b})$. So $\mathbf{a} \times \mathbf{b}$ is perpendicular to both $\mathbf{a}$ and $\mathbf{b}$.

However, there is still an ambiguity in the direction of $\mathbf{a} \times \mathbf{b}$, because although we know that $\mathbf{a} \times \mathbf{b}$ points along the direction perpendicular to the plane spanned by $\mathbf{a}$ and $\mathbf{b}$, there are two possible directions along this line. Assuming that our coordinate system has been chosen to be "right-handed" (which means that if you point the fingers of your right hand in the direction of $\hat{\mathbf{x}}$ and then swing them to $\hat{\mathbf{y}}$, your thumb points along $\hat{\mathbf{z}}$), then the direction of $\mathbf{a} \times \mathbf{b}$ is determined by the right-hand rule. That is, if you point the fingers of your right hand in the direction of $\mathbf{a}$ and then swing them to $\mathbf{b}$ (through the angle that is less than 180°), then your thumb points along $\mathbf{a} \times \mathbf{b}$. This is consistent with the fact that Eq. (13.14) gives $(1,0,0) \times (0,1,0) = (0,0,1)$, which is the statement that $\hat{\mathbf{x}} \times \hat{\mathbf{y}} = \hat{\mathbf{z}}$.

Let's now demonstrate the result in Eq. (13.15), which is equivalent to $|\mathbf{a} \times \mathbf{b}|^2 = a^2 b^2 (1 - \cos^2\theta)$, which in turn is equivalent to $|\mathbf{a} \times \mathbf{b}|^2 = a^2 b^2 - (\mathbf{a} \cdot \mathbf{b})^2$. Written in terms of the components, this last equation is

$$(a_y b_z - a_z b_y)^2 + (a_z b_x - a_x b_z)^2 + (a_x b_y - a_y b_x)^2 = (a_x^2 + a_y^2 + a_z^2)(b_x^2 + b_y^2 + b_z^2)$$
$$- (a_x b_x + a_y b_y + a_z b_z)^2. \tag{13.17}$$

If you stare at this long enough, you'll see that it's true. The three different types of terms agree on both sides. For example, both sides have an $a_y^2 b_z^2$ term, a $-2a_y b_y a_z b_z$ term, and no $a_x^2 b_x^2$ term.

13.2 Appendix B: Taylor series

13.2.1 Basics

Taylor series are extremely useful for checking limiting cases, in particular in situations where a given parameter is small. A Taylor series expresses a given function of x as a series expansion in powers of x. The general form of a Taylor series is (the primes here denote differentiation)

$$f(x_0 + x) = f(x_0) + f'(x_0)x + \frac{f''(x_0)}{2!} x^2 + \frac{f'''(x_0)}{3!} x^3 + \cdots . \quad (13.18)$$

This equality can be verified by taking successive derivatives of both sides of the equation and then setting $x = 0$. For example, taking the first derivative and then setting $x = 0$ gives $f'(x_0)$ on the left. And this operation also gives $f'(x_0)$ on the right, because the first term is a constant and gives zero when differentiated, the second term gives $f'(x_0)$, and all of the rest of the terms give zero once we set $x = 0$ because they all contain at least one power of x. Likewise, if we take the second derivative of each side and then set $x = 0$, we obtain $f''(x_0)$ on both sides. And so on for all derivatives. Therefore, since the two functions on each side of Eq. (13.18) are equal at $x = 0$ and also have their nth derivatives equal at $x = 0$ for all n, they must in fact be the same function (assuming that they're nicely behaved functions, as we generally assume in physics).

Some specific Taylor series that often come up are listed below. They are all expanded around $x = 0$; that is, $x_0 = 0$ in Eq. (13.18). We use these series countless times throughout this book when checking how expressions behave in the limit of some small quantity. The series are all derivable via Eq. (13.18), but sometimes there are quicker ways of obtaining them. For example, Eq. (13.20) is most easily obtained by taking the derivative of Eq. (13.19), which itself is simply the sum of a geometric series.

$$\frac{1}{1+x} = 1 - x + x^2 - x^3 + \cdots \quad (13.19)$$

$$\frac{1}{(1+x)^2} = 1 - 2x + 3x^2 - 4x^3 + \cdots \quad (13.20)$$

$$\ln(1+x) = x - \frac{x^2}{2} + \frac{x^3}{3} - \cdots \quad (13.21)$$

$$e^x = 1 + x + \frac{x^2}{2!} + \frac{x^3}{3!} + \cdots \quad (13.22)$$

$$\cos x = 1 - \frac{x^2}{2!} + \frac{x^4}{4!} - \cdots \quad (13.23)$$

$$\sin x = x - \frac{x^3}{3!} + \frac{x^5}{5!} - \cdots \quad (13.24)$$

$$\sqrt{1+x} = 1 + \frac{x}{2} - \frac{x^2}{8} + \cdots \quad (13.25)$$

$$\frac{1}{\sqrt{1+x}} = 1 - \frac{x}{2} + \frac{3x^2}{8} + \cdots \quad (13.26)$$

$$(1+x)^n = 1 + nx + \binom{n}{2}x^2 + \binom{n}{3}x^3 + \cdots \quad (13.27)$$

These series might look a little scary, but in most situations there is no need to include terms beyond the first-order term in x. For example, $\sqrt{1+x} \approx 1 + x/2$ is usually a good enough approximation. The smaller x is, the better the approximation is, because any term in the expansion is smaller than the preceding term by a factor of order x. Note that you can quickly verify that the $\sqrt{1+x} \approx 1 + x/2$ expression is valid to first order in x, by squaring both sides to obtain $1 + x \approx 1 + x + x^2/4$. Similar reasoning at second order shows that $-x^2/8$ is correctly the next term in the expansion.

As mentioned in Footnote 4 in Chapter 1, we won't worry about taking derivatives to rigorously derive all of the above Taylor series. We'll just take them as given, which means that if you haven't studied calculus yet, that's no excuse for not using Taylor series! Instead of deriving them, let's just check that they're believable. This can easily be done with a calculator. For example, consider what e^x looks like if x is a very small number, say, $x = 0.0001$. Your calculator (or a computer, if your want more digits) will tell you that

$$e^{0.0001} = 1.0001000050001666\ldots \tag{13.28}$$

This can be written more informatively as

$$\begin{aligned} e^{0.0001} =\ & 1.0 \\ & + 0.0001 \\ & + 0.000000005 \\ & + 0.0000000000001666\ldots \\ =\ & 1 + (0.0001) + \frac{(0.0001)^2}{2!} + \frac{(0.0001)^3}{3!} + \cdots. \end{aligned} \tag{13.29}$$

This last line agrees with the form of the Taylor series for e^x in Eq. (13.22). If you made x smaller (say, 0.000001), then the same pattern would form, but just with more zeros between the numbers than in Eq. (13.28). If you kept more digits in Eq. (13.28), you could verify the $x^4/4!$ and $x^5/5!$, etc., terms in the e^x Taylor series. But things aren't quite as obvious for these terms, because we don't have all the nice zeros as we do in the first 12 digits of Eq. (13.28).

Note that the left-hand sides of all of the Taylor series listed above involve only 1's and x's. So how do we make an approximation to an expression of the form, say, $\sqrt{N+x}$? We could of course use the general Taylor-series expression in Eq. (13.18) and generate the series from scratch by taking derivatives. But we can save ourselves some time by making use of the similar-looking series in Eq. (13.25). We can turn the N into a 1 by factoring out an N from the square root, which gives $\sqrt{N}\sqrt{1+x/N}$. Having generated a 1, we can now apply Eq. (13.25), with the only modification being that the small quantity x that appears in that equation is replaced by the small quantity x/N. This gives (to first order in x)

$$\sqrt{N+x} = \sqrt{N}\sqrt{1+\frac{x}{N}} \approx \sqrt{N}\left(1+\frac{1}{2}\frac{x}{N}\right) = \sqrt{N} + \frac{x}{2\sqrt{N}}. \tag{13.30}$$

Again, you can quickly verify that this expression is valid to first order in x by squaring both sides. If $N = 100$ and $x = 1$, then this approximation gives $\sqrt{101} \approx 10 + 1/20 = 10.05$, which is very close to the actual value of $\sqrt{101} = 10.0499\ldots$.

13.2.2 How many terms to keep?

When making a Taylor-series approximation, how do you know how many terms in the series to keep? For example, if the exact answer to a given problem takes the form of $e^x - 1$, then the Taylor series $e^x \approx 1 + x$ tells us that our answer is approximately equal to x. You can check this by picking a small value for x (say, 0.01) and plugging it in your calculator. This approximate form makes the dependence on x (for small x) much more transparent than the original expression $e^x - 1$ does.

But what if our exact answer had instead been $e^x - 1 - x$? The Taylor series $e^x \approx 1 + x$ would then yield an approximate answer of zero. And indeed, the answer *is* approximately zero. However, when making approximations, it is generally understood that we are looking for the *leading-order* term in the answer (that is, the smallest power of x with a nonzero coefficient). If our approximate answer comes out to be zero, then that means we need to go (at least) one term further in the Taylor series, which means $e^x \approx 1 + x + x^2/2$ in the present case. Our approximate answer is then $x^2/2$. (You should check this by letting $x = 0.01$.) Similarly, if the exact answer had instead been $e^x - 1 - x - x^2/2$, then we would need to go out to the term of order x^3 in the Taylor series for e^x.

Be sure to be consistent in the powers of x that you deal with. If the exact answer is, say, $e^x - 1 - x - x^2/3$, and if you use the Taylor series $e^x \approx 1 + x$, then you will obtain an approximate answer of $-x^2/3$. This is incorrect, because it is inconsistent to pay attention to the $-x^2/3$ term in the exact answer while ignoring the corresponding $x^2/2$ term in the Taylor series for e^x. Including both terms gives the correct approximate answer as $x^2/6$.

So what is the answer to the above question: How do you know how many terms in the series to keep? Well, the answer is that before you do the calculation, there's really no way of knowing how many terms to keep. The optimal strategy is probably to just hope for the best and start by keeping only the term of order x. This will often be sufficient. But if you end up with a result of zero, then you can go to order x^2, and so on. Of course, you could play it safe and always keep terms up to, say, fourth order. But that is invariably a poor strategy, because you will probably never need to go out that far in a series.

13.2.3 Dimensionless quantities

Note that whenever you use a Taylor series from the above list to make an approximation in a physics problem, the parameter x must be *dimensionless*. If it weren't dimensionless, then the terms with the various powers of x in the series would all have different units, and it makes no sense to add terms with different units.

As an example of an expansion involving a properly dimensionless quantity, consider the approximation made in going from Eq. (1.3) to Eq. (1.4) in the beach-ball example in Chapter 1. In this setup, the small dimensionless quantity x is the bt/m term that appears in the exponent in Eq. (1.3). This quantity is indeed dimensionless, because from the original expression for the drag force, $F_d = -bv$, we see that b has units of N/(m/s), or equivalently kg/s. Hence bt/m is dimensionless.

We can restate the above dimensionless requirement in a more physical way. Consider the question, "What is the velocity $v(t)$ in Eq. (1.3), in the limit of small t?" This question is meaningless, because t has dimensions. Is a year a large or small time? How about a hundredth of a second? There is no way to answer this without knowing what situation we're dealing with. A year is short on the time scale of galactic evolution, but a hundredth of a second is long on the time scale of a nuclear process. It makes sense only to look at the limit of a large or small *dimensionless* quantity. And by "large or small," we mean compared with the number 1.

Equivalently, in the beach-ball example the quantity m/b has dimensions of time, so the value of m/b is a time that is inherent to the system. It therefore *does* make sense to look at the limit where $t \ll m/b$ (that is, $bt/m \ll 1$), because we are comparing two things, namely t and m/b, that have the *same* dimensions. We will sometimes be sloppy and say things like, "In the limit of small t." But you know that we really mean, "In the limit of a small dimensionless quantity that has a t in the numerator," or, "In the limit where t is much smaller than a certain quantity that has dimensions of time."

After you make an approximation, how do you know if it is a "good" one? Well, just as it makes no sense to ask if a dimensionful quantity is large or small without comparing it to another quantity with the same dimensions, it makes no sense to ask if an approximation is "good" or "bad" without stating what accuracy you want. In the beach-ball example, let's say that we're looking at a value of t for which $bt/m = 1/100$. In Eq. (1.4) we kept the bt/m term in the Taylor series for $e^{-bt/m}$, and this directly led to our answer of $-gt$. We ignored the $(bt/m)^2/2$ term in the Taylor series. This is smaller than the bt/m term that we kept, by a factor of $(bt/m)/2 = 1/200$. So the error is roughly half a percent. (The corrections from the higher-order terms will be even smaller.) If this is enough accuracy for whatever purpose you have in mind, then the approximation is a good one. If not, then it's a bad one, and you need to add more terms in the series until you get your desired accuracy.

13.3 Appendix C: Limiting cases, scientific method

In Section 1.1.3 we discussed the strategy of checking limiting (or special) cases after solving a problem. It turns out that checking limiting cases is directly analogous to the scientific method (the procedure of testing hypotheses against experiments, and then modifying the hypotheses if needed). To see how this analogy comes about, recall that the point of checking limiting cases is that although it is often difficult to determine how a system behaves in general, your intuition usually gives you a very good idea of how a system behaves in certain limiting cases. Of course, it is quite possible that your intuition can lead you astray. But for the purposes of the present discussion, we'll assume that your intuition is always correct.

Once you've solved a problem and obtained an answer, checking a limiting case leads to two possible results:

1. Your answer (which you are hoping is correct) doesn't agree with what your intuition says (which we are assuming is correct). In this case you conclude that your answer must be *incorrect*.

2. Your answer *does* agree with what your intuition says. In this case all you can conclude is that your answer *might* be correct. Note well that checking a limiting case can never tell you that your answer is *definitely* correct.

The above two cases are exactly analogous to what happens in real life with the scientific method. In the real world, everything comes down to experiment. If you have a theory or hypothesis that you think is correct, then you need to check that its predictions are consistent with experiments. The specific experiments you do are the analog of the special cases you check after solving a problem. That is, when checking special cases, you check your answer to a problem (what you hope is true) against your intuition about a special case (what you know is true). And when using the scientific method, you check your theory (what you hope is true) against the result of an experiment (what you know is true). This is summarized in the following table.

	Physics class	*Science*
What you *hope* is true	Answer to problem	Theory
What you *know* is true	Intuition about special cases	Result of experiment

In the real world, if your theory *isn't* consistent with the experiments, then you need to go back and fix it, just as you would need to go back and fix your answer to a physics problem if it weren't consistent with the special cases. Your theory might need minor tweaks, or it might be total garbage and need a complete overhaul. If, on the other hand, your theory *is* consistent with the experiments, then although this is nice, the only thing it actually tells you is that your theory *might* be correct. And considering the way things usually turn out, the odds are that it actually isn't correct, but rather the limiting case of a more correct theory.

For example, Newtonian physics (which is what this book is about) is consistent with any "everyday" type of experiment that you might do. But that doesn't mean that Newtonian physics is correct. In fact, Newtonian physics is certainly *not* correct. It's a perfectly good theory for everyday scenarios, but it breaks down when things get very small (when quantum mechanics takes over) or very fast (when relativity takes over). Newtonian physics is simply a limiting theory of these *more* correct theories. And we use the term "more correct" here, because these theories (quantum mechanics and relativity) aren't correct either. They're both just limiting theories of a more correct theory (quantum field theory), which in turn is a limiting theory of something else. And so on and so forth. Turtles, turtles, all the way down – or at least much farther than we can presently see.

Every time you do an experiment, you don't actually prove anything. Instead, what you do is narrow down the possibilities of what *might* be true, by ruling out incorrect theories. And that's *all* you can do. You can solve a physics problem for your class, of course, but no one has solved the problem of the universe yet. So the only thing we can do is perform experiments and narrow things down. Imagine a multiple-choice question that has a million possible answers to choose from. And imagine that there is no possibility of directly deducing which answer is correct. Then all we can do is start chipping away and eliminating answers by looking at special cases. It's a slow and daunting process, but it's the only process we have.

The collective endeavor of science is therefore to squeeze down the set of possible theories until there's only one left. We have a long way to go, of course! At any stage in the process, the theories that are still on the table are the ones we haven't been able to disprove. So that's how science works. You can't actually *prove* anything, so you learn to settle for the things you can't *disprove*.

Consider, when seeking gestalts,
The theories that science exalts.
It's not that they're known
To be written in stone.
It's just that we can't say they're false.

13.4 Appendix D: Problems requiring calculus

The problems listed below require calculus. "1 M" means "Chapter 1 Multiple Choice," and "1 P" means "Chapter 1 Problems," etc. This list represents less than a sixth of the total number of about 400 questions/problems. So if you haven't studied calculus yet, this book can still be very useful. The chapter introductions also occasionally use calculus, but this won't prevent you from using the given results. We aren't counting Taylor series as a calculus topic, because as mentioned in Footnote 4 in Chapter 1, the application of Taylor series in this book involves only algebra.

1	M:	none
1	P:	none (3 appears to require calculus, but it doesn't actually)
2	M:	3
2	P:	none
3	M:	none
3	P:	2, 3, 8, 9, 10, 12, 22
4	M:	none
4	P:	none
5	M:	none
5	P:	1, 2, 3, 6, 14, 16, 19, 21, 23
6	M:	18, 20
6	P:	7, 13, 14, 22, 23, 24, 25 (also, 5 can be solved with sums or integrals)
7	M:	none
7	P:	6, 10, 11, 12, 14, 22, 33 (1, 2, 3, 4, 5 can be solved with sums or integrals)
8	M:	none
8	P:	2, 4, 11 (1 can be solved with sums or integrals)
9	M:	none
9	P:	none
10	M:	1, 2
10	P:	3, 4, 6, 17, 18, 19
11	M:	none
11	P:	1, 5, 7, 8, 11, 12, 13, 14, 18, 20
12	M:	none
12	P:	4, 6, 8

Made in the USA
Las Vegas, NV
05 September 2022